# ARCHIVES of heat transfer

# ARCHIVES of heat transfer

edited by
NAIM AFGAN

Volume 1

A publication of the International
Centra for Heat
and Mass Transfer

Belgrade, Yugoslavia

● Hemisphere Publishing Corporation
A member of the Taylor & Francis Group
New York   Washington   Philadelphia   London

**ARCHIVES OF HEAT TRANSFER: Volume 1**
Copyright © 1988 by Hemisphere Publishing Corporation. All rights reserved. Printed in the Yugoslavia by Beogradaski Izdavački grafički zavod, Beograd. Except as permitted under the Unated States Copyright Act of 1976, no part of this publication may be reproduced or distributed in any form or by any means, or stored in a data base retrieval system, without the prior written permission of the published.

1 2 3 4 5 6 7 8 9 0 X X X X 8 9 8

Library of Congress Cataloging in Publication Data

ISBN 0–89116–877–X    Hemisphere Publishing Corporation
ISSN–0899–5311

# ARCHIVES

## CONTENTS

Preface

Heat and Mass Transfer in Turbulent Boundary Layers .................................. 3
International Seminar 1968
*Naim Afgan*

On the Turbulent Boundary Layer with Vanishing Viscosity ........................... 7
*S. S. Kutateladze*

International Seminar 1969 ..................................................................... 47
*Zoran Zaric*

Gas Concentration Measurments in Boundary Layers .................................. 49
*E. D. Brun*

Heat and Mass Transfer in Rheologically Complex Fluids ............................. 69
International Seminar 1970
*W. R. Schowalter*

Suspension Rheology ............................................................................... 71
*H. Brener*

On Fast Sodium Cooled Braeder Related Thermophysical Investigations ........... 117
International Seminar 1971
*V. I. Subotin*

Film Condensation of Liquid Metals .......................................................... 121
*W. Rohsenow*

Recent Developments in Heat Exchangers .................................................. 137
International Seminar 1972
*E. Schluender*

The Optimal Design of Heat Exchanger Networks - a Review and Evalution of Current Procedures ................................................................................. 139
*T. W. Hoffman*

International Seminar 1973 ................................................... 165
*J. M. Beer*

First Estimates of Industrial Furnace Performance - the One-Gas-Zone Model
Reexamined ........................................................................ 167
*H. C. Hottel*

Transport Processes in Plant Environment ................................... 193
International Seminar 1974
*D. A. de Vries*

Heat and Mass Transfer within Plant Canopies ........................... 195
*B. Legg and J. Montheith*

Heat and Mass Transfer in Modern Energy Systems ..................... 215
International Seminar 1975
*M. A. Styrikovich*

Environment and Energy Production after the Year 2000 ............... 217
*R. Gibrat*

Heat Transfer and Turbulent Bouyant Convection ........................ 233
International Seminar 1976
*D. B. Spalding*

Turbulent Flow and Heat Transfer in Pipes under the Considerable Influence of
Thermogravitional Forces .................................................... 235
*B. S. Petukhow*

Energy Conservation in Heating, Cooling and Ventilation of Buildings ........ 251
International Seminar 1977
*C. J. Hoogendorn*

Thermal Processes of Solar Houses .......................................... 255
*Ken-Ichi Kimura*

Two Phase Momentum, Heat and Mass Transfer in Chemical Processes and Energy
Engineering Systems ........................................................... 273
International Seminar 1978
*F. Durst*

Heat and Mass Transfer at Liquid-Gas Interphase ........................ 275
*F. Mayinger*

Heat and Mass Transfer in Metallurgical Systems ........................ 293
International Seminar 1979
*D. B. Spalding*

Transport Processes in Agitated Ladles: Problems, Solutions and Experimental
Techniques ............................................................................................................ 295
*J. Szekely*

Heat Transfer in Nuclear Reactor Safety .................................................... 315
International Seminar 1980
*G. Bankoff*

The Analysis of Two-Phase Level in a PWR Core During Conditions of Severely
Reduced Liquid Mass Inventory ................................................................. 317
*R. T. Lahey*

Heat Exchangers - Theory and Practice ..................................................... 335
International Seminar 1981
*J. Taborek*

Fouling in Crude Oil Preheat Trains ........................................................... 339
*G. D. A. Lambourn, M. Durrieu*

Heat and Mass Transfer in Rotating Machinery ........................................ 351
International Symposium 1982
*D. E. Metzger*

Heat Transfer Problems in Aero-Engines .................................................. 353
*D. K. Hennecke*

Heat and Mass Transfer Measurment Techniques .................................... 377
International Symposium 1983
*O. G. Martinenko*

The Calculation of IR Radiation Transfer in Combustion Gases Containing $CO_2$
and CO for Temperature and Concentration Measurments ..................... 381
*S. I. Kryuchkov, N. N. Kudryavtsev, and S. S. Novikov*
*R.I.Solouhin*

Heat and Mass Transfer in Fixed and Fluidized Beds .............................. 395
International Symposium 1984
*W. P. M. van Swaaij*

The Effects of Pressure and Temperature on Heat Transfer to Gas-Fluidized Beds
of Solid Particles ......................................................................................... 397
*H. Martin*

High Temperature Heat Exchangers ......................................................... 411
International Symposium 1985
*Y. Mori*

High Temperature Heat Exchangers for Advanced Power Generating Facilities ..... 413
*A. E. Shiendlin*

Heat and Mass Transfer in Refrigeration and Cryogenic Engineering ................ 429
International Symposium 1986
*J. Bougard*

Heat Transfer in Low Temperature Insulation ............................................. 431
*C. L. Tien, A. P. Stratton*

Heat and Mass Transfer in Gasoline and Diesel Engines ............................... 443
International Syposium 1987
*D. B. Spalding*

The Importance of Heat Transfer to IC Engine Design and Operation ................ 445
*R. Pischinger*

# PREFACE

Archives are a collection of selected contributions to the scientific meetings organized by the International Centre for Heat and Mass Transfer marking the occasion of the Centre's 20th Anniversary. In nature, significant changes are difficult to observe over a time span of only twenty years. For modern science, the last two decades, were, however, a period of extremely rapid development. During our lifetime scientific achievements in a whole range of fields have changed our daily lives. Heat and mass transfer, is in the opinion of many experts, one of the most farreaching fields in this respect. Its development was the result of the need to understand physical phenomena underlying processes crucial to the development of a number of technologies. Thus, in the late sixties, a period favourable to the exchange of scientific information and noted for growing international cooperation in many scientific fields, people the world-over initiated activities aimed at promoting the further development of the field. For the last twenty years, first Herceg Novi and then Dubrovnik in Yugoslavia have been the settings for an international scientific forum for the free exchange of knowledge in the field of heat and mass transfer. The concept of scientific and technical information exchange grew and the forum became a platform for not only the exchange of ideas but also the broadening of international understanding and friendship, thereby overcoming political, religious and other bounderies constraining the modern world.

This collection of contributions to the International Symposia and Seminars organized by the International Centre for Heat and Mass Transfer was made by inviting all the Chairmen of the Symposium Committees to select one contributions representative of the state-of-the-art of the respective topic or that had the most impact on the field. In addition, the Chairmen were asked to write a two to three page overview giving their views on the meeting and its topic.

Archives are proof that genuine art and true science are not eroded by time. I would like to honor all our collegues whose contributions have stood the test of time and remain a lasting contribution to the field. Also, I would like to take this opportunity to extend our deepest appreciation to all the Chairmen for their contributions to the development of the International Centre for Heat and Mass Transfer as for their contunous interest in its activities.

ACKNOWLEDGMENTS

This volume could not have been completed without the enthusiastic and professional efforts of a whole team and therefore I would like to express a special thanks to: Computer Graphics Design Studio, Beograd, who patiently worked with us to typeset and compose the Archives; Ms. Nada Jevtić M. Sc., who worked on the compilation and english text editing, and Ms. Marina Deleon, who assisted me in preparation and editing of this volume.

The Editor

International Seminar 1968

# HEAT AND MASS TRANSFER TURBULENT BOUNDARY LAYERS

## N. Afgan

*The Boris Kidrich Institute for Nuclear Sciences, Belgrade, Yugoslavia*

Since 1904 when Prandtl published his first contribution to boundary layer theory, the incentive has been great to understand processes taking place in the fluid in the vicinity of a solid surface. With modern technology and its effort to intensify processes in the vicinity of any interface, problems relevant to boundary layer theory have become an attractive subject for a number of specialists in different fields, working to solve the respective problems encountered.

In the late sixties, it became obvious that progress of the research in boundary layer theory had reached such a level that further development would greatly benefit from an international gathering facilitating exchange of information between specialist the world over. In particular it was felt that an exchange of ideas between the Eastern and Western scientific communities would be a boon to further international cooperation.

A group of specialists in the field of heat and mass transfer involved in energy and space research related programs forwarded a proposal that an international meeting should be held on heat and mass transfer in turbulent boundary layers. The one hundred fifteen participants from eleven countries at the meeting held in 1968 with their sixty three contributions proved during very intensive discussion that the great interest existed in the topic and the meeting goal was indeed well served it beeing singularly helpful to professionals in the field.

At the meeting Prof A.I.Leontiev presented an approach to turbulent boundary layer model development that he fathered in 1968 (Fig. 1). It was then predicted that a sharp increase in the application of computers in turbulent boundary layer theory computations was to be expected after 1980. The currently large number of papers on the development of computational methods prove this assumption true.

Also, in addition to the above, a subdivision of the general theory has taken place with the attendant incorporation of particularities and details and the development of models applicable to specific boundary layer problems. Instead of reviewing all the new developments in turbulent boundary layer theory I would like to discuss two which are representative of the present state of the art. The first deals with turbulence anisotropy in the boundary layer and its recognition as a specific means of mass, momentum and energy transport in turbulent layers. Knowledge of flow patterns in the vicinity of solid surfaces and its experime-ntal verification make it a cornerstone for the evergrowing body of turbulent layer models and may lead to the deeper insight into turbulent layer theory. With the incor-poration of anisotropy into turbulent boundary layer theory a specific flow structure in the boundary layer was recognized [2]. Modern noninvasive techniques used in the experimental investigation of turbulent flow, together with powerful computers, enabled the acquisition of new knowledge, and have brought us closer to the real nature of these physical phenomena.

The second issue which characterizes the present status of turbulent boundary research is related to turbulence phenomena in two phase systems. The experimental evidence of a velocity fluctuation in the continuous phase and its specific pattern have lead to theoretical research of the related phenomena in an attempt to explain the evidence [3]. The recognition of characteristic patterns of fluid flow in the vicinity of the interphase in two phase flow, have made the development of specific models of turbulence in two phase systems. This approach has opened a new field and may lead to a general theory of turbulence [4].

In addition to the new information obtained on turbulent boundary layer phenomena, which improve our ability to model fluid flow and heat and mass transfer phenomena for specific applications, a need still exists to introduce new features in order to promote further intensification of heat and mass transfer in the boundary layer. Thus the engineering requirement for specific experimentally obtained correlations will spare the most important research needed. As a reflection of the state-of-the-art in the late sixties I have selected the paper by the late Academician S.S.Kutateladze "On the Turbulent Boundary Layer with Vanishing Viscosity", which is an integral approach to turbulent boundary layer theory and presents a powerful tool for the solving many different engineering problems. It is proof that a novel idea of high sophistication can result from a simple method, and sustain time erosion and the competition of modern numerical methods in conjuction with powerful computers.

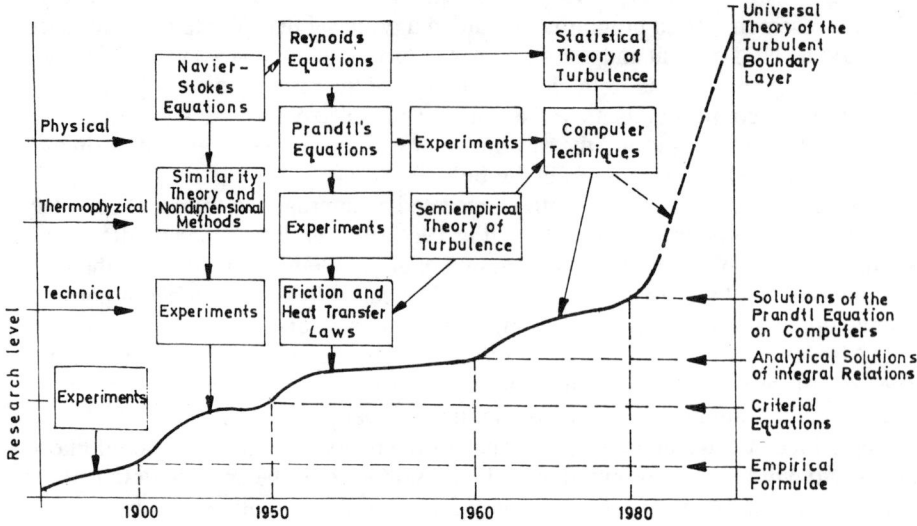

Fig. 1 Scheme of the development of calculation methods of the turbulent boundary layer

## REFERENCES:

1. S.J.KLINE, S.K. ROBINSON Turbulence Producing Coherent Structures in the Turbulent Boundary Layer, Proceedings of 1988 International Seminar on Near Wall Turbulence Structure, edited by S.J.Kline, Hemisphere Publishing Corporation, New York
2. Z.ZARIĆ, R.E. FALCO, C.P. GENDRICEH The Turbulence Burst Detection Algorithm of Ibid 1
3. R.T. LAHEY, JR. Turbulence and Phase Distribution Phenomena in Two Phase Flow Proceedings of 1987 International Seminar on Transient Phenomena in Multiphase Flow edited by N. AFGAN Hemisphere Publishing Corporation, New York
4. A. SERIZAWA, I. KATAOKA Phase Distribution in Transient Two Phase Flow. Ibid 3

# ON THE TURBULENT BOUNDARY LAYER WITH VANISHING VISCOSITY

## S. S. Kutateladze

*Thermal Physics Institute, Siberian Division of the USSR Academy of Sciences, Novosibirsk, USSR*

### 1. MODEL OF A FLUID WITH VANISHING VISCOSITY IN THE THEORY OF WALL TURBULENCE

#### 1.1. FLUID WITH VANISHING VISCOSITY

Investigation of the formation of separated flows in the laminar boundary layer indicates, in the general case, that the exact solution of the equations of hydrodynamics for viscous fluids when $\mu \to 0$ does not necessarily yield the movement of the body in an ideal fluid ($\mu \equiv 0$). This fact is a consequence of a difference in both the initial and boundary conditions.

The problem was first formulated in this form by Oseen [1].

However, the ideas of at work and its tools could not be applied to non-separated turbulent flow and to our knowledge no attempts were made in this direction. This seems rather strange, for the application of the model of a fluid with vanishing viscosity to the analysis of turbulent flow is, after all, quite natural.

Indeed in the turbulent flow along a solid body there is always a large region whose mean macroscopic motion does not depend on molecular viscosity. This so-called viscous sublayer thins rapidly when viscosity decreases.

The importance of the above fact to the theory of gas - liquid mixture flow structure stability was pointed out by the author in [2, 3, 4] and in a number of other publications, and to the theory of a single-phase boundary layer in [3] and further in [6, 7].

For a fluid with vanishing viscosity, and for finite body size and flow velocity the following condition is always satisfied:

$$Re \underset{\mu \to 0}{\longrightarrow} \infty \qquad (1.1)$$

i. e. the flow is turbulent.

## 1.2. DEGENERATON OF ISOTHERMAL BOUNDARY LAYER OF INCOMPRESSIBLE FLUID ON IMPERMEABLE PLATE

A number of important peculiarities in the behaviour of the boundary layer with vanishing viscosity can be clearly observed in the case of the simplest boundary layer appearing on the impermeable smooth plate longitudinally streamlined by an unlimited isothermal incompressible flow.

The friction law in this case is well described by von Karman's formula

$$C_f = 2 \, (2.5 \ln Re^{**} + 3.8)^{-2}, \qquad (1.2)$$

and for some intervals of the Reynolds number one can conveniently use power approximations in the form:

$$\left. \begin{array}{l} \omega \approx \xi^n \\[4pt] C_f \sim Re_\delta^{-\frac{2n}{1+n}} \\[4pt] \dfrac{v}{U} \approx \dfrac{1+2n}{2} C_f \xi^{1+n} \end{array} \right\} \qquad (1.3)$$

where $0 < n < 1$.

The region of the prevailing influence of molecular friction (viscous sublayer) is determined by the quantity

$$\xi_1 \approx \frac{16}{Re_\delta \sqrt{C_f}} \qquad (1.4)$$

In the turbulent core ($\xi_1 < \xi < 1$)

$$\tau \approx \rho \overline{u'v'} < \tfrac{1}{2} C_f \rho U^2 \qquad (1.5)$$

It may be seen from these correlations that when $Re \to \infty$ the following conditions hold:

$$\left. \begin{array}{l} n \to 0 \\ C_f \to 0 \\ \xi_1 \to 0 \\ \dfrac{v}{U} \to 0 \\ \dfrac{\overline{u'v'}}{U^2} \to 0 \end{array} \right\} \qquad (1.6)$$

## 1.3. DEGENERATON OF THE VISCOUS SUBLAYER

Let us consider the equation of motion of the two-dimensional boundary layer in the Reynolds - Prandtl form:

$$\rho_0 U \frac{\partial U}{\partial x} + \frac{\partial}{\partial y}\left(\mu \frac{\partial \bar{u}}{\partial y} - \overline{\rho u' v'}\right) = \overline{\rho u} \frac{\partial \bar{u}}{\partial x} + \overline{\rho v} \frac{\partial \bar{u}}{\partial y} \qquad (1.7)$$

$$\frac{\partial \overline{\rho u}}{\partial x} + \frac{\partial \overline{\rho v}}{\partial y} = 0 \tag{1.8}$$

Here

$$\tilde{\beta} = 1 - \frac{\overline{v}\,\overline{u'T'} + \overline{u'}\,\overline{v'T'}}{\overline{u'v'}\,\overline{T}} \tag{1.9}$$

is the coefficient which accounts for the influence of density pulsation on Reynolds stress.

When $y \to 0$ we have

$$\{\overline{u'v'} \to 0,\, u \to 0,\, \rho v \to j_{ct},\, \mu \to \mu_{ct}\} \tag{1.10}$$

$$\rho_0 U \frac{\partial U}{\partial x} + \mu_{ct} \frac{\partial^2 u}{\partial y^2} \to j_{ct} \frac{\partial u}{\partial y}; \tag{1.11}$$

$$\frac{2\omega}{Re_\delta} \xrightarrow[\xi \to 0]{} C_f \xi - f_\delta \xi^2 + \bar{j}_\alpha \int_0^\xi \omega d\xi, \tag{1.12}$$

where in this case $Re_\delta = \frac{\rho_0 U \delta}{\mu_{ct}}$. Let the region of the viscous sublayer be determined by a dimensionless coordinate $\xi_1$. By definition $\omega_1 < 1$ and the boundary of the viscous sublayer can be evaluated by inequality

$$(C_f + 2\,\bar{j}_{ct})\xi_1 - f_\delta \xi_1^2 < \frac{2}{Re_\delta} \tag{1.13}$$

Thus, the thickness of the viscous sublayer always decreases faster with increasing Reynolds number, then the thickness of the whole turbulent boundary layer near the wall.

We shall call this effect viscous sublayer degeneration.

## 1.4. DENSITY PULSATION DEGENERATION

The intensities of heat and momentum transfer in the turbulent core of the boundary layer are of the same order, i. e.

$$\overline{T'v'} \approx \Delta T \frac{\overline{u'v'}}{U} \tag{1.14}$$

Substituting this expression in (1.9) and taking into account the degeneration of parameters $\frac{v}{U}$ and $\frac{\overline{u'v'}}{U^2}$ with the increase of the Reynolds number we find that

$$\beta \xrightarrow[Re \to \infty]{} 1 \tag{1.15}$$

Thus, with the increase of the Reynolds number a degeneration of the influence of density pulsations upon turbulent friction also occurs.

## 1.5. RELATIVE LAW OF FRICTION

Let us introduce the sonic linear scale of turbulence according to Prandtl [9]:

$$l = \sqrt{\overline{u'v'}} \left| \frac{\partial \overline{u}}{\partial v} \right| \qquad (1.16)$$

retaining its traditional name: mixing–length. Then

$$\tau_t = \overline{\beta}\overline{\rho} \left( l \frac{\partial \overline{u}}{\partial y} \right)^2 \qquad (1.17)$$

so that in the domain $\xi_1 < \xi \ll 1$

$$l = \chi y \qquad (1.18)$$

According to the majority of experimental data the Prandtl-Karman constant $\chi = 0.4$ [10].

The analysis of extensive experimental data, carried out by Spalding and Escudier [11, 12] confirms the existence of a turbulent core in accordance with law (1.18) with a high degree of validity. Pikhal's [13] and Clauser's [14] experiments confirm that this region, although it becomes narrower, exists in the presence of strong external turbulence and a pressure gradient.

In the external part of the turbulent boundary layer the value of the mixing length is less than that according to formula (1.18) and may be regarded constant in the first order of approximation [11].

It is seen from (1.16) that the absolute values of all the parameters of the turbulent boundary layer tend to zero when viscosity vanishes. However, it seems that their relative changes due to different external factors tend to some limits other than zero.

Let us introduce the following somewhat artificial ratio:

$$\int_{\omega_1}^{1} \sqrt{\frac{\overline{\rho}\overline{\tau}_0}{\psi\overline{\tau}}}\, d\omega = \int_{\xi_1}^{1} \sqrt{\frac{C_p \overline{\tau}_0}{2\overline{\beta}}} \frac{d\xi}{\widetilde{l}} \qquad (1.19)$$

where:

$\widetilde{\rho} = \dfrac{\rho}{\rho_0}$ — relative medium density at a given point of the boundary layer

$\widetilde{\tau} = \tau/\tau_{ct}$ — relative value of the shear stress

$\widetilde{l} = l/\delta$ — dimensionless mixing length

$$\psi = \left(\frac{C_f}{C_{f0}}\right)_{Re^{**}} - \text{relative value of the friction coefficient}$$

The index 0 denotes standard conditions under which we are considering the boundary layer on the smooth impermeable plate streamlined by an isothermal flow of incompressible fluid with a zero pressure gradient and with no other external forces and sources.

With a high degree of accuracy:

$$\frac{\sqrt{\tilde{\tau}_0}}{\bar{l}} = \frac{1}{\chi\xi}(1 - \Sigma\, a_i \xi^i) \tag{1.20}$$

where $\Sigma a_i = 1$.

When $Re \to \infty$ we have:

$$\left.\begin{array}{c} \sqrt{\frac{C_{f0}}{2}} \to \chi/\ln Re^{**} \\ \xi_1 \to 0 \\ \omega_1 \to 0 \\ \beta \to 0 \\ \int\limits_{\xi_1 \to 0}^{1}\left(1 - \Sigma a_i \xi^i\right)\frac{d\xi}{\chi\xi} \to -\frac{1}{\chi}\ln\xi_1 \end{array}\right\} \tag{1.21}$$

Accordingly:

$$\int_0^1 \sqrt{\frac{\bar{\rho}\,\tilde{\tau}_0}{\psi\,\bar{\tau}}}\,d\omega \xrightarrow[Re\to\infty]{} -\frac{\ln \xi_1}{\ln Re^{**}} \tag{1.22}$$

In flows without pressure gradients, $\xi_1 \sim 1/Re^{**}$, and at the separation point of the turbulent boundary layer from an impermeable surface, $\xi_1 \sim 1/Re^{**2}$ thus in some cases exactly and approximately in others:

$$\int_0^1 \sqrt{\frac{\bar{\rho}\bar{\tau}_0}{\psi\bar{\tau}}}\,d\omega \xrightarrow[Re\to\infty]{} 1 \tag{1.23}$$

It is interesting to note that in this general form the limiting relative law of friction of the turbulent boundary layer does not contain any constants specific to turbulence.

## 1.6. STABILITY OF GAS - LIQUID SYSTEMS WITH VANISHING VISCOSITY

Two-phase flow is an even more appropriate object for the appllication of the model of a viscous fluid if compared to a single-phase boundary layer. In this system a surface tension force acts at the phase separating boundary. Thus not only do the relative hydrodynamic values remain finite, the absolute values do so as well. Reley's problem of the stability of a heavy liquid flow over a light inviscid flow was the first example of this kind described in the literature [15].

The stability criterion to which we can reduce Reley's formula has the following form:

$$k = \frac{U_{kp}'' \sqrt{\rho''}}{\sqrt[4]{\sigma g (\rho' - \rho'')}} \qquad (1.24)$$

where $U_{kp}''$ is the critical velocity at which a two - phase structure becomes unstable and

$\sigma$ is the coefficient of surface tension at the boundary of phase separation. In the general case, criterion (1.24) should be regarded as a measure of the of the dynamic pressure of the light phase with the surface tension and gravity in reconstructing the structure of the two-phase system [2, 3, 4]. This is rather evident in the case of gas barbotage through a net into a big volume of fluid.

When the rate of gas flow is sufficiently large there occurs a breakaway of the fluid from the grid to form a gas bell or layer.

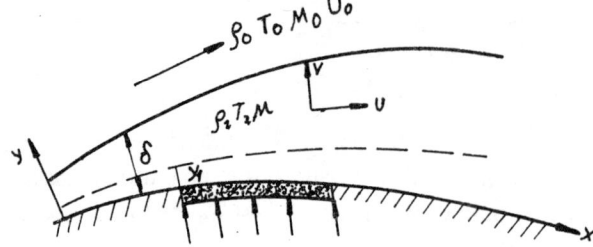

Fig. 1.1. A turbulent boundary layer of finite thickness.

When $\mu \to 0$ the thickness of the viscous sublayer can be arbitrarily smaller than that of freely appearing films and bubles, whose linear size in this case is conmeasurable with the Laplace constant:

$$L \sim \sqrt{\frac{\sigma}{g(\rho' - \rho'')}} \qquad (1.25)$$

The order of magnitude of the break-off work is:

$$g(\rho' - \rho'')L \approx \rho' \frac{U'^2}{2}, \qquad (1.26)$$

leading directly to criterion (1.24).

Turbulent friction does not change the interaction forces in the two–phase flow as it is proportional to the dynamic pressure. The appearance of the automodelity of a two–phase flow characteristic in reference to the Reynolds number is connected with the weak effect of the thin stagnated fluid layer near the wall upon the development of the two–phase boundary layer structure.

## 2. GAS BOUNDARY LAYER ON THE IMPERMEABLE PLATE

### 2.1. LIMITING RELATIVE LAW OF HEAT EXCHANGE

In the general case, the limiting law of heat exchange is significantly more complicated than the law of friction.

Let us specify the turbulent heat flux in the following form:

$$q_T = -\tilde{\beta}_T \rho l \frac{\partial u}{\partial y} l_T \frac{\partial i}{\partial y}, \qquad (2.1)$$

where

$l_T$ — heat mixing–length

$\tilde{\beta}_T$ — coefficient accounting for the influence of density pulsations turbulent heat transfer

In an ideal gas: $i = c_p T$.

Reducing (2.1) to the form of (1.19) we obtain:

$$\psi_s = \left( \frac{1}{Z_T} \int_{\xi_{1T}}^{1} \sqrt{\frac{\tilde{\rho} \tilde{q}_0}{\tilde{q}} \cdot \frac{\partial \omega}{\partial \xi} \cdot \frac{\partial v}{\partial \xi}} \, d\xi_T \right)^2, \qquad (2.2)$$

where:

$$Z_T = \int_{\xi_{T1}}^{1} \sqrt{\frac{St_0 \tilde{q}_0}{\tilde{\beta}_T \varepsilon_l}} \cdot \frac{d\xi_T}{\tilde{l}_T} \qquad (2.3)$$

Here $\tilde{q} = \frac{q}{q_{cT}}$ is the relative heat flux at the point $\xi_T = y/\delta_T$ where:

$\delta_T$ — thickness of the heat boundary layer

$\upsilon$ - dimensionless enthalpy difference

$\dfrac{q_{CT}}{\rho_0 U(i_{CT}-i_0)}$ - Stanton number

$St_0$ - Stanton number at given $Re_T^{**}$ under standard conditions

$\varepsilon_l = l/lT$ - degree of similarity of hydrodynamic and heat mixing–lengths.

The values $\tilde{\beta}_T$, $Z_T$, $\xi_{lT}$ have the same properties as their dynamic analogues and accordingly,

$$\int_0^1 \sqrt{\dfrac{\tilde{\rho}}{\psi_s} \dfrac{\tilde{q}_0}{\tilde{q}} \cdot \dfrac{\partial \omega}{\partial \xi} \cdot \dfrac{\partial v}{\partial \xi}}\, d\xi_T \xrightarrow[Re \to \infty]{} 1 \qquad (2.4)$$

When $\tilde{\beta} \to 1$ the following formulae are true:

$$\lambda_T = c_p \rho l_T l \dfrac{\partial u}{\partial y} \qquad (2.5)$$

$$\mu_T = \rho l^2 \dfrac{\partial u}{\partial y} \qquad (2.6)$$

i. e. quantity $\varepsilon_l$ acquires the meaning of the turbulent Prandtl number

$$Pr_T = \dfrac{\mu_T}{c_p \lambda_T} \qquad (2.7)$$

It is known that these values in the vicinity of the wall are close to unity [16, 17].

The consequence of the condition $Pr_T \approx 1$ and of the degeneration of the viscous sublayer is that when the boundary conditions are similar and the pressure gradient is zero, at any finite physical Prandtl number with $Re \to \infty$ a similarity of velocity and enthalpy fields occurs:

$$\left\{\dfrac{dp}{ax} = 0,\ \forall Pr,\ Re \to \infty,\ \dfrac{\partial \omega}{\partial \xi} \to \dfrac{\partial v}{\partial \xi}\right\} \qquad (2.8)$$

## 2.2. DEGENERATION OF FUNCTIONS $\tilde{\delta}*$, $\tilde{\delta}**$, H AND $\tilde{\tau}$

In an ideal gas flow when the velocity and enthalpy fields are similar we have

$$\dfrac{1}{\tilde{\rho}} = \psi - (\psi - \psi^*)\omega - (\psi^* - 1)\omega^2 \qquad (2.9)$$

where: $\psi = i_{cm}/i_0$ - enthalpy factor

$\psi^* = i^*/i_0$ - kinetic enthalpy factor.

If the velocity distribution may be approximated by power law (1.3)

$$\omega = \xi^n$$

so that $n<1$ when $Re \to \infty$. For $n \to 0$ we have the following expressions for the displacement thickness and energy deficit thickness:

$$\left. \begin{aligned} \tilde{\delta}^* &= \int_0^1 \left(1 - \frac{\xi^n}{\psi - (\psi - \psi^*)\xi^n - (\psi^* - 1)\xi^{2n}}\right) d\xi \\ \tilde{\delta}^{**} &= \int_0^1 \frac{\xi^n - \xi^{2n}}{\psi - (\psi - \psi^*)\xi^n - (\psi^* - 1)\xi^{2n}} d\xi \end{aligned} \right\} \quad (2.10)$$

The difference of the integrands in (2.10) when $\psi \equiv \psi^* \equiv 1$ and for nonisothermal conditions is proportional to the expression:

$$1 - \xi^{2n} + \psi(1 - \xi^n) + \psi^*(\xi^n - \xi^{2n}) \xrightarrow[n \to 0]{} 0 \qquad (2.11)$$

Thus, when $Re \to \infty$ the displacement thickness and energy–deficit thickness become automodel with reference to the nonisothermicity of the boundary layer in gas flow along the plate and they tend to the values corresponding to the isothermal boundary layer:

$$\left. \begin{aligned} Re &\to \infty \\ \forall \psi \end{aligned} \right\} \left\{ \begin{aligned} \tilde{\delta}^* &\to \frac{n}{1+n} \to 0 \\ \tilde{\delta}^{**} &\to \frac{n}{(1+n)(1+2n)} \to 0 \\ H &\to 1 + 2n \to 1 \end{aligned} \right. \qquad (2.12)$$

From the motion equation when $dp/dx = 0$ and when the surface is impermeable, for a power law for the velocity distribution it follows that:

$$\left. \begin{aligned} \tilde{\tau} &= -\frac{n}{(1+n)\tilde{\delta}^{**}} \int \bar{\rho}\, \xi^{2n} d\xi \\ \tilde{\tau} &\xrightarrow[\forall \psi, Re \to \infty]{} 1 - \xi^{1+2n} \to 1 - \xi \end{aligned} \right\} \qquad (2.13)$$

In the general case, the shear stress distribution across the boundary layer is not automodel either with respect to the Reynolds number or with respect to the non-isothermicity. Such an automodelity takes place only in the degenerated boundary layer.

It will be interesting to compare this result with the widely known approximation of Pohlhausen - Fedyavsky [18, 19] in which the coefficients of the approximating polynomial are calculated according to the conditions taking place at the internal and external boundaries of the boundary layer, i. e. they are, in principal, automodel with respect to the numbers $\psi$ and $Re$. For an impermeable plate the approximation by cubic parabola is [6]:

$$\bar{\tau} = 1 - 3\xi^2 + 3\xi^3 \qquad (2.14)$$

Figure 2.1 represents a comparison of the calculations in accordance with formulae (2.13) and (2.14); Mickley's experimental data [20] in Bernard's and Lindon's interpretation [21] are also show in this figure.

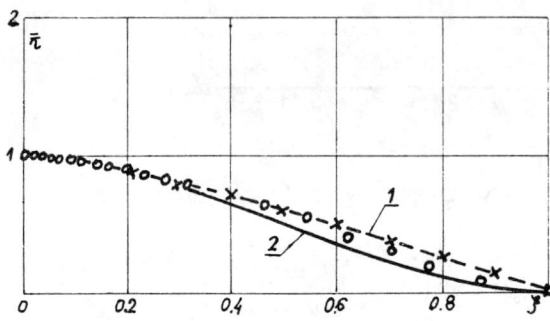

Fig. 2.1. Shear stress distribution in the isothermal turbulent boundary layer on the impermeable plate

1 - $\bar{\tau} = 1 - \zeta^{1+2n}$ ($n = 1/7$);
2 - $\tau = 1 - 3\xi^2 + 2\xi^3$;
o - Mickley's experimental data [20, 21]

It is evident that the change of the quantity $Re^{**}$ from $10^4$ ($n \approx 1/7$) to $\infty$ ($n = 0$), only slightly affects the function $\bar{\tau}(\xi)$.

## 2.3. LIMITING RELATIVE LAW OF FRICTION AND HEAT TRANSFER

We have: $\dfrac{dp}{dx} = 0$, $T_{cm} = $ const, $T_0 = $ const, $Pr \approx 1$, $Re \to \infty$, $\omega = v$, $\bar{\tau} = \bar{\tau}_0$, $\tilde{\rho}$ in accordance with formula (2.6),

$$\Psi = \Psi_s = \left( \int_1^2 \tilde{\rho}^{1/2} d\omega \right)^2 =$$

$$= \frac{1}{\psi^*-1}\left[ \arcsin \frac{\psi^*+\psi - 2}{\sqrt{(\psi+\psi^*)^2 - 4\psi}} - \arcsin \frac{\psi - \psi^*}{\sqrt{(\psi+\psi^*)^2 - 4\psi}} \right] \qquad (2.15)$$

when $\psi^* \to 1$, i.e. when the flow is markedly subsonic,

$$\Psi = \left( \frac{2}{1 + \sqrt{\psi}} \right) \qquad (2.16)$$

when $\psi = \psi^*$, i. e. for flow over an adiabatic surface ($q_{CT} = 0$),

$$\Psi = \frac{1}{\psi^* - 1}\left( \arcsin\sqrt{\frac{\psi^* - 1}{\psi^*}} \right)^2 \qquad (2.17)$$

or taking into account that $\psi^* - 1 = r\dfrac{k-1}{2}M^2$ \qquad (2.18)

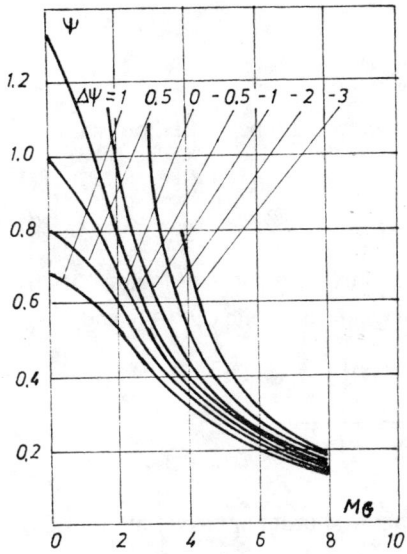

Fig. 2.2 Dependence of $\psi$ on $\psi^*$ and $\Delta\psi = \psi - \psi^*$ when $Re \to \infty$ on the impermeable plate streamlined by gas.

$$\Psi = \left( \frac{\arctg M \sqrt{r \frac{k-1}{2}}}{M \sqrt{r \frac{k-1}{2}}} \right) \quad (2.19)$$

Dependence (2.15) is shown in Fig. 2.2.

## 2.4. INFLUENCE OF THE FINITE REYNOLDS NUMBER (FIRST APPROXIMATION)

Let us analyze function (1.19) in the neighbourhood of $\psi = 1$. We have

$$\left. \begin{array}{l} \psi \to 1, \ \bar{\tau} \to \bar{\tau}_0, \ \omega_1 \to \omega_{10}; \\ \displaystyle\int_{\xi_1}^{1} \sqrt{\frac{C_p \bar{\tau}_0}{2 \bar{B}} \cdot \frac{d\xi}{\bar{l}}} \to 1 - \omega_{10} \end{array} \right\} \quad (2.20)$$

When density varies according to law (2.6)

$$\Psi \xrightarrow[\psi \to 1]{} \frac{1}{(\psi^* - 1)(1 - \omega_{10})^2}$$

$$\left[ \arcsin \frac{\psi^* + \psi - 2}{\sqrt{(\psi + \psi^*)^2 - 4\psi}} - \arcsin \frac{2(\psi^* - 1)\omega_{10} + \psi - \psi^*}{\sqrt{(\psi + \psi^*)^2 - 4\psi}} \right]^2 \quad (2.21)$$

where in the two-layer scheme of the boundary layer:

$$\omega_{10} = 11.6\sqrt{\frac{C_{f0}}{2}}$$

Naturally, the comparison of friction coefficients should be carried out with equalized values of the number:

$$Re^{**} = \frac{W_0 \delta^{**}}{\nu} \tag{2.22}$$

since when $\psi \to 1$ the values of $Re^{**}$ and $Re_\delta$ differ only by a constant factor. Strictly speaking when $\psi \to 1$ the limiting flow should always be subsonic ($\psi^* \to 1$) and:

$$\Psi \xrightarrow[\psi \to \psi^* \to 1]{} \left[\frac{2}{1 + \sqrt{\psi - (\psi - 1)\omega_{10}}}\right]^2 \tag{2.23}$$

Formula (2.21) may be considered as an exact first order approximation of the function:

$$\Psi(\psi, \psi^*, Re^{**}) = 0, \tag{2.24}$$

and formula (2.23) as an exact first-order approximation of the function:

$$\Psi(\psi, 1, Re^{**}) = 0, \tag{2.25}$$

Comparison with extensive experimental data shows that formula (2.21) qualitatively reflects the influence of the Reynolds number on the relative change of the friction and heat transfer coefficients as a function of the nonisothermicity of the gas flow in a correct form. The quantitative agreement is good if instead of the

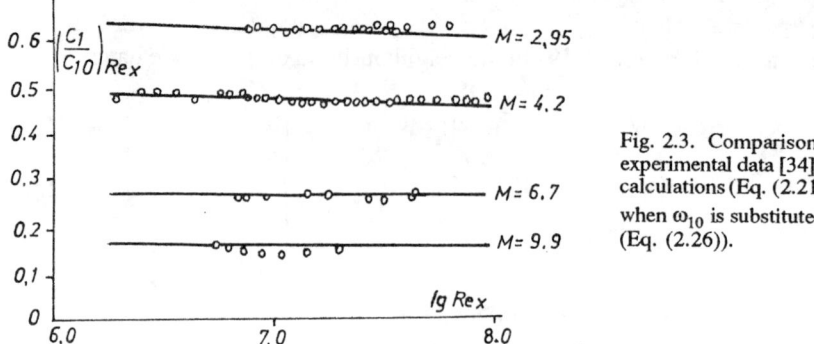

Fig. 2.3. Comparison of experimental data [34] with calculations (Eq. (2.21)) when $\omega_{10}$ is substituted by $\omega_1$ (Eq. (2.26)).

value $\omega_{10}$ in formulae (2.21) and (2.23) we introduce the quantity

$$\omega_1 = 11.6\sqrt{\frac{C_f}{2}} = 11.6\sqrt{\Psi \frac{C_{f0}}{2}} \tag{2.26}$$

Figure 2.3 is a comparison of such a calculation with experiments for supersonic gas flow over an impermeable plate.

# 3. GAS BOUNDARY LAYER ON A PERMEABLE PLATE

## 3.1. APPROXIMATION OF A SHEAR STRESS PROFILE

The use of a power law velocity distribution in the analysis of function $\tau(\xi)$ in the presence of transversal flow appears ineffective owing to the fact that in this case the function $\omega(\xi)$ behaves significantly correctly both for $\xi \to 0$ and $\xi \to 1$.

Nevertheless, it may be assumed that the degeneration properties observed in the previous chapter for $\bar{\tau}(\xi)$ also occur qualitatively for more complicated layers. This statement is confirmed by experience in the application of approximation $\bar{\tau}(\xi)$ for boundary conditions, accumulated in approximative boundary layer calculations. However, when studying the degenerated turbulent boundary layer the use of the exact value of $\frac{\partial \tau}{\partial y}$ at point $y = 0$ leads to some difficulties because the uncertainty of expression

$$\left(\frac{\partial \tau}{\partial y}\right)_{y=0} = \frac{\partial p}{\partial x} + j_{cm}\frac{\tau_{cm}}{\mu}$$

has to be determined when $dp = 0$ and $\mu \to 0$.

Therefore in the asymptotic theory of the turbulent boundary layer a very convenient approximation exists based on the following correlations:

$$\left.\begin{array}{l} \xi = 1, \tau = 0, \dfrac{\partial \tau}{\partial y} = 0; \xi = 0, \tau = \tau_{cm}: \\ \xi \to 0, \dfrac{\partial \tau}{\partial y} \to \dfrac{\partial p}{\partial x} + j_{cm}\dfrac{\partial u}{\partial y} \end{array}\right\} \qquad (3.1)$$

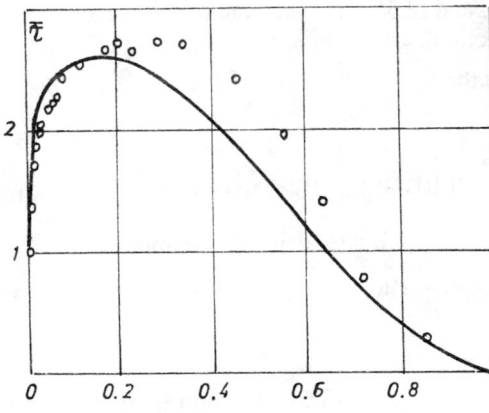

Fig. 3.1. Comparison of formula (3.2) with the experimental data [20, 21].
$j_{cm} = 0.003; b = 1.3; \Psi = 0.455$

Consequently when $dp = 0$

$$\bar{\tau} \approx \bar{\tau}_0 + b_1\omega(1-\xi)^2, \qquad (3.2)$$

where $\bar{\tau}_0$ is determined in accordance with (2.14).

A comparison of calculations Eq. (3.2) with Mickley's experiments [20, 21] is given in Fig. 3.1.

In the region adjacent to the wall, the agreement is not only qualitative, quite satisfactory quantitative agreement exists as well. However, according to experimental data the zone of the maximum is more extended than follows from (3.2). For the asymptotic theory, as shown, this is of no significance.

## 3.2. LIMITING LAW OF FRICTION

From (3.2) it follows that in the case analyzed:

$$\Psi \frac{\bar{\tau}}{\tau_0} \approx \Psi + \frac{b\omega}{1 + 2\xi} \tag{3.3}$$

when $Re \to \infty$, $\frac{\partial \omega}{\partial y} \to 0$ and, in accordance with (1.23):

$$\int_0^1 \frac{\bar{\rho}^{1/2} d\omega}{\left(\Psi + \frac{b\omega}{1 + 2\xi}\right)^{1/2}} \xrightarrow[Re \to \infty]{} \int_0^1 \frac{\bar{\rho}^{1/2} d\omega}{(\Psi + b\omega)^{1/2}} \to 1 \tag{3.4}$$

At the point of boundary layer separation from the wall, imposed by injection, the so-called effect of "sharp blowing" occurs and, according to the definition, $\Psi = 0$. The corresponding critical value of the blowing factor is:

$$b_{kp} \xrightarrow[Re \to \infty]{} \left[\int_0^1 \left(\frac{\bar{\rho}}{\omega}\right)^{1/2} d\omega\right]^2 \tag{3.5}$$

A remarkable peculiarity of formulae (3.4) and (3.5) is the fact that they can be solved without any knowledge of the velocity distribution in a boundary layer. It is sufficient to know the connection between the relative density of the flow $\bar{\rho}$ and dimensionless velocity $\omega$.

## 3.3. SUBSONIC FLOW WITH HYPERBOLIC DEPENDENCE OF $\bar{\rho}$ ON $\omega$

In many cases the following (exact or approximate) dependence applies:

$$\frac{\rho_0}{\rho} = \psi_1 - (\psi_1 - 1)\omega, \tag{3.6}$$

where $\psi_1 = \frac{\rho_0}{\rho_{cm}}$.

In the case of an ideal homogeneous gas $\psi_1 \equiv \psi$ (3.6), (3.4) and (3.5) yield: $\psi_1 < 1$

$$\Psi = \frac{4}{b_1(1 - \psi_1)}\left[\ln \frac{\sqrt{(1 - \psi_1)(1 + b_1)} + \sqrt{b_1}}{\sqrt{1 - \psi_1} + \sqrt{b_1 \psi_1}}\right]^2 \tag{3.7}$$

$$b_{kp} = \frac{1}{1 - \psi_1}\left(\ln \frac{1 + \sqrt{1 - \psi_1}}{1 - \sqrt{1 - \psi_1}}\right)^2 \tag{3.8}$$

$\psi_1 > 1$

$$\Psi = \frac{4}{b_1(\psi_1 - 1)} \left[ \arctg \sqrt{\frac{b_1}{(\psi_1 - 1)(1 + b_1)}} - \arctg \sqrt{\frac{b_1 \psi_1}{\psi_1 - 1}} \right]^2 \quad (3.9)$$

$$b_{kp} = \frac{1}{\psi_1 - 1} \left( \arccos \frac{2 - \psi_1}{\psi_1} \right)^2 \quad (3.10)$$

The value of $\psi_1$ for some processes ( $\tilde{R} = R_1/R_0$ - the ratio of the gas constants of the injected and main flow gas), is given in the following table:

| | |
|---|---|
| Gas homogeneous nonisothermal boundary layer | $\psi_1 = T_{cm}/T_o \equiv \psi$ |
| Gas heterogeneous isothermal boundary layer | $\psi_1 = 1 + \frac{b_1}{1 + b_1}(\tilde{R} - 1)$ |
| Heterogeneous nonisothermal boundary layer of gases of the same atomicity | $\psi_1 = \psi \left[ 1 + \frac{b_1}{1 + b_1}(\tilde{R} - 1) \right]$ |

Formulae (3.7) and (3.9) are well approximated by:

$$\Psi \approx 4 \left( \frac{1 - \tilde{b}}{\sqrt{\psi_1} + 1} \right)^2 \quad (3.11)$$

where $\tilde{b} = b/b_{kp}$ is the relative value of the blowing factor of the wall.
For the homogeneous isothermal boundary layer the limiting law of friction is ultimately simple:

$$\Psi = \left( 1 - \frac{b}{4} \right)^2; \quad (3.12)$$

$$b_{kp} = 4; \quad (3.13)$$

$$\tilde{j}_{cm,kp} = 2C_{f0} \quad (3.14)$$

If $Pr=1$ and the boundary conditions are given in similar form, the Reynolds analogy occurs and $St = C_f/2$.

We shall not consider more complicated cases here.

## 3. 4. VELOCITY DISTRIBUTION IN A TURBULENT CORE OF A GAS BOUNDARY LAYER AT HIGH REYNOLDS NUMBERS

Changing the limits of integration in (1.19):

$$\int_\omega^1 \sqrt{\frac{\tilde{\rho} \tilde{\tau}_0}{\psi \tau}} d\omega = \int_\omega^1 \sqrt{\frac{C_{f0} \tilde{\tau}_0}{2 \tilde{\beta}}} \frac{d\xi}{\tilde{i}} \quad (3.15)$$

Substituting expression (3.3) into the above expression and accounting for the well known asymptotic properties of the turbulent boundary layer:

$$\int_\omega^1 \sqrt{\frac{\tilde{\rho}}{\Psi + b\omega}} d\omega \xrightarrow[Re\to\infty]{} \int_\xi^1 \sqrt{\frac{C_{f0}}{2}\bar{\tau}} \frac{d\xi}{\bar{l}} \qquad (3.16)$$

The conservation of the function $\bar{l}(\xi)$:

$$\int_\xi^1 \sqrt{\frac{C_{f0}}{2}\bar{\tau}_0} \frac{d\xi}{\bar{l}} = 1 - \omega_0 \qquad (3.17)$$

yields the velocity distribution in the core of a turbulent boundary layer at large Reynolds numbers:

$$\int_\omega^1 \sqrt{\frac{\tilde{\rho}}{\Psi + b\omega}} d\omega \approx 1 - \omega_0 \qquad (3.18)$$

Here $\omega_0(\xi)$ is the velocity distribution in the isothermal boundary layer on an impermeable plate.

In the simplest case, when $\tilde{\rho} = 1$

$$\omega \approx 1 - \sqrt{\Psi + b}\,(1 - \omega_0) + \frac{b}{4}(1 - \omega_0)^2 \qquad (3.19)$$

When $Re \to \infty$ ($\tilde{\rho} = 1$, $\xi > \xi_1$)

$$\omega \to \left(1 - \frac{b}{4}\right)\omega_0 + \frac{b}{4}\omega_0^2 \qquad (3.20)$$

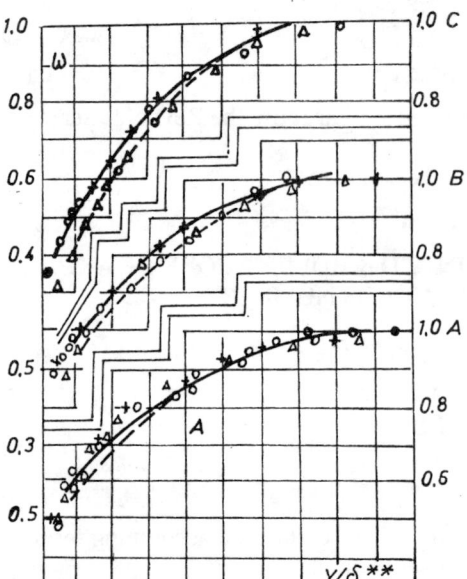

Fig. 3.2. Comparison of formula (3.19) with experimental data.
Calculation: ——— $C_{f0}$ by $Re^{**}$
— — — $C_{f0}$ by $Re^{**}_1$
(dots - the experiment, see the nomenclature).

|     |       | A    | B    | C    | author |
|-----|-------|------|------|------|--------|
| I   | b     | 1.31 | 2.5  | 4.89 | TPI    |
|     | Re**  | 5860 | 7200 | 8410 |        |
| II  | b     | 1.1  | 2.2  | 4.9  | [35]   |
|  +  | Re**  | 2630 | 2870 | 3410 |        |
| III | b     | 1.25 | 2.4  | 4.72 | [20]   |
|  O  | Re**  | 777  | 1120 | 1265 |        |

and at the point of separation,

$$\omega \to \omega_0^2 \qquad (3.21)$$

A comparison of formula (3.19) and experimental data [22] is given in Fig. 3.2.
It is natural that in the region $b \gg b_{kp}$ the velocity profile disagrees with (3.18).

## 3.5. THE INFLUENCE OF THE REYNOLDS NUMBER ON THE CRITICAL INJECTION PARAMETER

In approximation (3.1) the velocity distribution in the viscous sublayer ($\xi < \xi_1$) has the form [23, 22]:

$$\omega = \frac{\Psi}{b}\left[\exp\left(\frac{j_{cm} \cdot y}{\mu}\right) - 1\right] \qquad (3.22)$$

Consequently, in the two-layer scheme of the turbulent boundary layer with (3.22) velocity profiles, when $0 < \xi < \xi_1$, and according to formula (3.19) when $\xi_1 < \xi < 1$, at the point of separation reverse flow occurs in the direction of the $x$ axis:

$$\omega = 0 \ (b = b_{kp} \neq \infty, \ \xi < \xi_1) \qquad (3.23)$$

Assuming $\omega = 0$ in (3.21), we find the order of magnitude of the stagnated zone:

$$\xi_{kp} \approx \exp\left(-\chi\sqrt{\frac{2}{C_{f0}}}\right) \ll \xi_0 \qquad (3.24)$$

From (3.4), under conditions (3.23) and (3.24):

$$b_{kp} \geq \left(\int_0^1 \sqrt{\frac{\bar{\rho}}{\omega}}(1 + 2\xi)d\omega\right)^2 \qquad (3.25)$$

where $\xi = \xi(\omega_0^2)$.

The result of the calculations according to this formula are shown in Fig. 3.3. As seen, the value of $b_{kp}$ grows with the decrease of the Reynolds number. Experimental data obtained by the diffusion method [22] are also shown in the same figure.

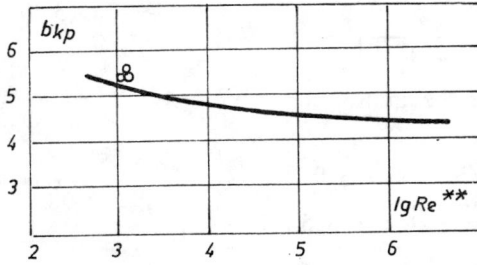

Fig. 3.3. $b_{kp}$:
I - formula (3.25).
o - the experiments of B. P. Mironov and P. P. Lugovskoy

## 3.6. FRICTION AND HEAT TRANSFER – EXPERIMENTAL DATA

Below formula (3.11) is compared with the experimental data, the value $\Psi$ being determined at $Re^{**}$ = idem and $b_{kp}$ according to formula (3.25).

The momentum equation at $dp = 0$ has the form:

$$\frac{dRe^{**}}{dRe_\infty} = (\Psi + b)\frac{C_{f0}}{2}, \qquad (3.26)$$

where:

$$Re_\infty = \frac{W_0 x}{\nu}$$

We shall denote the values related to the number $Re_x$ by an asterisk over the symbol. For example:

$$\overset{*}{\Psi} = \frac{C_f}{C_{f0}} \quad (Re_x = \text{idem}) \qquad (3.27)$$

The solution of (3.26) for the boundary layer, developing from the front edge of the plate ($x = 0$, $\delta = 0$) for $b$ = const. is of the following form:

$$\left.\begin{array}{l}\overset{*}{C}_{f0} = C_{f0}(\Psi + b)^{m_1} \\ \overset{*}{\Psi} = \dfrac{\Psi}{(\Psi + b)^{m_1}} \\ \overset{*}{b} = \dfrac{b}{(\Psi + b)^{m_1}}\end{array}\right\} \qquad (3.28)$$

Here $m_1 = \dfrac{m}{1+m}$ where $m$ is the exponent of the power law in the second formula of (1.3).

The injection parameter values at which the considered flow exists in the range of $Re$ with $m = 1/4$ lie within the limits

$$\left.\begin{array}{l} -5.2 + b < + 5.2 \\ -\infty < \overset{*}{b} < + 3.7 \end{array}\right\} \qquad (3.29)$$

If $j_{cm}$ = const., equation (3.26) may be solved with $b_{kp} = 4$ and $m = 1/4$. With an accuracy to the first five members of the series for the logarithm:

$$\left.\begin{array}{l}\overset{*}{\Psi} = \dfrac{(1 - 0.25b)^2}{(1 + 0.25b)^2} \\ \overset{*}{b} = \dfrac{b}{(1 + 0.25)^{0.2}}\end{array}\right\} \qquad (3.30)$$

The flow being analyzed exists within the limits:

$$\left.\begin{array}{l} -4 < b < +4 \\ (b = \text{const}) \; -\infty < b < + 3.0 \\ (j_{cm} = \text{const}) \; -\infty < \overset{*}{b} < + 3.5 \end{array}\right\} \qquad (3.31)$$

## 4. HEAT TRANSFER CRISIS AT BOILING WITH SUBCOOLING OF THE FLUID MASS TO THE TEMPERATURE OF SATURATION

### 4. 1. THE TRANSFER TO FILM BOILING UNDER FREE CONVECTION IN LARGE VOLUMES OF FLUID (POOL BOILING)

Substituting the rate of vaporisation, related to the unit area of a heating surface, in criterion (1.24):

$$U'' = \frac{q}{r\rho''} \quad (4.1)$$

we obtain an expression for the critical heat flux at boiling of the saturated fluid with vanishing viscosity under the conditions of free convection and automodelity of the process with respect to the geometrical dimensions of the system (1):

$$q_{kp} = kr\sqrt{\rho''} \sqrt[4]{\sigma g(\rho' - \rho'')} \quad (4.2)$$

Later, N. Zuber [24] calculated the value of $k \approx \frac{\pi}{24}$ by considering the equilibrium of a system consisting of spherical bubles, inscribed into the cubic cell. The vapour content of the layer near a wall $\varphi_* = \pi/6$ corresponds to this model. According to the experiments of G. G. Malenkov [25] the mean value is $\varphi_* = \pi/4$. However, the value of $k = 0.14$ is on the average close to the experimental data for many substances.

Formula (4.2) correctly models the influence of the steam phase density at boiling and the gas - phase at barbotage. It has been tested for the pressure interval:

$$5 \cdot 10^{-4} < \frac{P}{P_{kp}} < 1 .$$

This formula takes into account gravity effects. The validity of this formula was checked both for values which are much less than earth gravity and for those exceeding it by 2000 times [26].

The subcooling of the fluid to the temperature of saturation results in the increase of the critical heat flux. Formally, the effect may be interpreted by correlation:

$$q_{kp} = q_{kp,0} + \alpha(i'' - i_0), \quad (4.3)$$

where $\alpha$ is the heat transfer coefficient from the boundary of phase separations to the mass of supercooled fluid, and $q_{kp,0}$ is the critical heat flux in the saturated fluid.

According to Traibus and Zuber the second term in (4.3) is a consequence of non - stationary heat conduction from the surface of the vapor film to the bulk of fluid. Such a mechanism may be called a heat shock. The following relation holds:

$$(\bar{q}_{kp} - 1) \sim \left(\frac{a^2}{gL^3}\right)^{1/4} \left(\frac{\rho'}{\rho''}\right)^{3/4} \frac{\Delta i}{r} . \quad (4.4)$$

An expression differing from (4.4) by $Pr^{1/6}$ corresponds to the quasistationary laminar boundary layer at the boundary of phase separation on the fluid side. The precision of modern experiments is not high enough to elucidate this difference.

The values of the complex $(a^2/gL^3)^{1/4}$ for a number of nonmetallic fluids are rather close to each other and we can use a first - order approximation:

$$(\tilde{q}_{kp} - 1) \sim \left(\frac{\rho'}{\rho''}\right)^{3/4} \frac{\Delta i}{r}. \tag{4.5}$$

Figure 4.1 is the generalized diagram of the effect of the heater diameter and the parameter of fluid subcooling on the critical heat flow at boiling on horizontal cylinders [25].

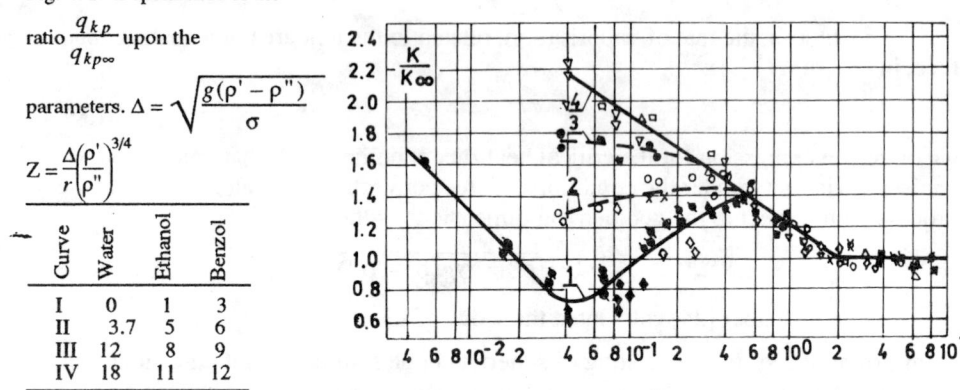

Fig. 4.1. Dependence of the ratio $\frac{q_{kp}}{q_{kp\infty}}$ upon the parameters. $\Delta = \sqrt{\frac{g(\rho' - \rho'')}{\sigma}}$

$Z = \frac{\Delta i}{r}\left(\frac{\rho'}{\rho''}\right)^{3/4}$

| Curve | Water | Ethanol | Benzol |
|---|---|---|---|
| I | 0 | 1 | 3 |
| II | 3.7 | 5 | 6 |
| III | 12 | 8 | 9 |
| IV | 18 | 11 | 12 |

The lower curve corresponds to boiling in an absolutely saturated fluid. Clearly seen are:

a) The region of automodelity with respect to heater size. Its existence is connected with a heating surface large enough in comparison with the steam bubble size ($\Delta > 2$) and because of this the formation of a critical steam layer is equally probable at any point on the wall.

b) The first region where the critical heat flux increases with the decrease of heater size ($0.5 < \Delta < 2$). Its existence is linked with the better evacuation of bubbles from the bottom of a horizontal cylinder.

c) The region of the decrease of the critical heat flux with decreasing heater diameter ($0.05 < \Delta < 0.5$). Its existence is connected with the fact that at such dimensions the appearing steam bubbles quickly envelop the whole circumference of the cylinder and cause its local destruction. This may be quite naturally considered as a dangerous crisis of heat transfer analogous to the occurence of a normal steam film.

d) The second region of the increase of the critical heat flow with decreasing heater size ($\Delta < 0.03$). Its existence is connected with the increase of free convection intensity, the critical regime practically coinciding with the beginning of boiling.

With intensive subcooling ($Z > 18$, where $Z$ is the r. h. of (4.5)) a certain degeneration takes place and the critical heat flux monotonously increases with the decrease of heater diameter in the range $\Delta < 2$.

## 4.2. BOILING - BARBOTAGE ANALOGY

The hydrodynamic of boiling–barbotage analogy is a natural consequence of the theory under consideration. This analogy was shown quantitatively for the first time in the experiments of the author and V. N. Moskvitcheva [26] and later in the contributions of the author and I. G. Malenkov [27, 28, 29] and Aktyurk [30].

In considering this analogy one should obviously take into account certain differences in the dynamics of boiling and barbotage.

It may be exemplified by multiple circulation of the fluid in the boiling layer and in the layer under barbotage. In the first case a part of the fluid approaching the wall evaporates and evacuates in the form of bubbles. In the second case all the mass of the circulating fluid does not change its aggregate state. However even the evaluation by the joined mass of the fluid in non - stationary movement of a bubble shows that this difference is to within ~ 10%.

The following table shows the value of the hydrodynamic stability factor of a layer near a wall under barbotage and for pool boiling on a horizontal surface oriented upwards.

As it is seen that there is rather satisfactory quantitative agreement with the data on the boiling crisis.

| Fluid | Methyl alcohol | Ethyl alcohol | Water | Acetone |
|---|---|---|---|---|
| The average value of the stability criterion k (light phase water) | 0.141 | 0.139 | 0.160 | 0.156 |

There is an important effect that was discovered during the study of the stability a layer adjacent to a wall under barbotage. It is the dependence of the criterion $K$ on the molecular weight of the injected gas. It should be stressed that the stability of this criterion with pressure change (i. e. gas density) for the given gas - fluid pair has been confirmed by experiments. Only the change of the pair leads to a different value of the hydrodynamic stability criterion for a two - phase boundary layer.

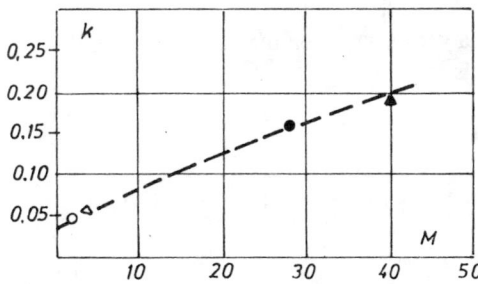

Fig. 4. 2. Dependence of the value $k$ on the molecular weight of the gas blown into the water.

Figure 4. 2 illustrates this fact with experimental data. It is likely that the mechanism is connected with the movement of steam bubbles at the moment of their growth on the wall immediately after departure. In any case further theoretical and experimental research of this effect is of doubtless interest.

## 4. 3. CRITICAL HEAT FLUX WITH LARGE STREAM VELOCITIES

Let us examine the flow along a flat plate by an unlimited fluid whose average temperature changes to equal that of saturation only near the downstream edge, (Fig. 4. 3).

A stable vapour layer can form near the outlet edge only, the main current is not encumbered by vapour bubbles.

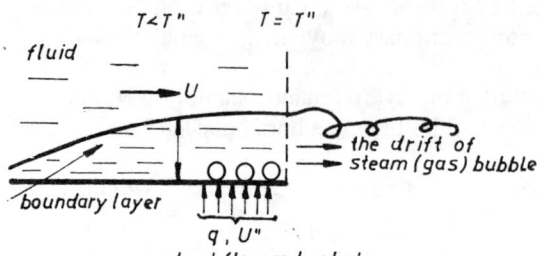

Fig. 4. 3. The appearance of film boiling at the outlet edge of the plate in the unlimited flow of the fluid.

A turbulent boundary layer develops on the plate whose thickness is inversely proportional to the stream velocity. The diameter of the bubbles departing from the heating surface at large stream velocities is inversely proportional to the square of the velocity. Thus, to within an order of magnitude:

$$\frac{D_0}{\delta} \sim \frac{1}{U} \qquad (4.6)$$

Consequently it is possible to theoretically envisage a flow in which the bubble size is significantly smaller than the thickness of the boundary layer of the liquid phase.

The heat transfer crisis begins at the ejection of the fluid between the vapour bubbles forming on the wall and the turn - out of the continuous vapour layer. The latter will be most stable, if at the moment of expulsion of the liquid layer adjacent to the wall, the longitudinal component of the fluid velocity near the wall is the smallest. Therefore it is possible to identify the situation with that which takes place when the boundary layer is blown off a permeable surface.

If the void fraction of a layer, close to a wall is $\varphi_*$, then the actual flow of the fluid ejected at the moment of the crisis equals to:

$$j_* = j_{kp}(1 - \varphi_*) \qquad (4.7)$$

In the flow with vanishing viscosity the value $j_{kp}$ is determined by formula (3.14) and consequently:

$$j_* = 2C_{f0}\rho'U(1 - \varphi_*) \qquad (4.8)$$

The energy required for the creation of flow $j_*$ originates in the kinetic energy of the vapour being generated (or kinetic energy of injected gas in the analogy with barbotage). On assuming:

$$\frac{j_*^2}{\rho'} \approx \left(\frac{q_{kp}}{\varphi_* r \rho''}\right)^2, \qquad (4.9)$$

we obtain formula:

$$q_{kp} \approx 2C_{f0}\varphi_*(1 - \varphi_*)\sqrt{\rho'\rho''rU} \qquad (4.10)$$

At large flow velocities:

$$\alpha \approx \rho'U\frac{C_{f0}}{2}Pr^n \qquad (4.11)$$

Then (4.3), (4.10) and (4.11) yield for high velocities of the supercooled fluid:

$$(\tilde{q}_{kp} - 1) = [4\varphi_*(1 - \varphi_*)K]^{-1}\left(\frac{\rho'}{\rho''}\right)^{1/2} \qquad (4.12)$$

Experimental studies carried out by L. S. Shtokolov, V. M. Adamovsky and S. I. Verkhman under the guidance of the author have confirmed that this dependence is in principal correct [28, 31, 32, 33].

## APPENDIX 1.

## 5. TURBULENT TRANSFER IN VICINITY OF THE WALL

### 5. 1. TURBULENT VISCOSITY WHEN $y \to 0$

It is known from the general convective heat transfer theory that the following correlation between the physical properties of the medium and relative thickesses of dynamic and thermal boundary layers applies:

| $C_p\mu < \lambda$ | $C_p\mu \approx \lambda$ | $C_p\mu > \lambda$ |
|---|---|---|
| $\delta < \delta_T$ | $\delta \approx \delta_T$ | $\delta > \delta_T$ |

Therefore, strictly speaking, the theory of the turbulent boundary layer analyzed above corresponds to the situation.

$$\mu \to 0; Pr \approx 1 \qquad (5.1)$$

The situation $Pr \to 0$ corresponds convective heat conductivity in an ideal liquid, as in this case $\delta \to 0$ and the ratio $\delta_\tau/\delta \to \infty$. Liquid metals are the real fluids in which turbulent transfer is significantly degenerated [36].

The situation $Pr \to \infty$ corresponds to turbulent heat conductivity (or diffusion) in the quasilaminar flow, as far as for $Pr \gg 1$, $\delta_\tau < y_1$, where $y_1$ is the thickness of the viscous sub-layer.

L. D. Landau and V. G. Levitch [37, 38] suggested that owing to the cessation of pulsation due to the strong influence of molecular viscosity the following conditions are fulfilled:

$$\{u' \sim \overline{u}; T = \text{Const}\}, \qquad (5.2)$$

where $T$ is the period of pulsation.

Figure 5.1 depicts the results of measurements made by V. V. Orlov and B. P. Perepelitsa at the Institute of Thermal Physics of the Academy of Sciences of the USSR. Their results were obtained by observing the microimpurities in a flow of

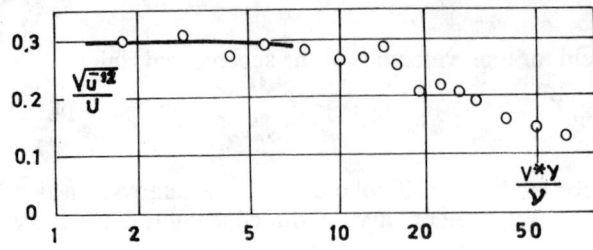

Fig. 5.1. Dependence of the value $\dfrac{\sqrt{u'^2}}{u}$ on dimensionless distance $\mu = \dfrac{v^* \upsilon y}{v}$.

fluid. It is evident that within the viscous sublayer the following condition is fulfilled with a great degree of certainty:

$$\frac{\sqrt{u'^2}}{u} = \text{Const} \tag{5.3}$$

Let us analyze a flow along a smooth impermeable wall.

$$\begin{cases} j_{CT} = 0, \ y \to 0; \ u' = l \dfrac{\partial \overline{u}}{\partial y} \sim \overline{u} \\ l = v'T \sim v'; \ v' = -\displaystyle\int_0^y \dfrac{\partial u'}{\partial x} dy \sim -\displaystyle\int_0^y \overline{u}\, dy \\ \overline{vu} \to \dfrac{C_f}{2} U^2 y - \dfrac{UU'}{2} y^2 \end{cases} \tag{5.4}$$

Here $U' = \dfrac{dU}{dx}$.

It follows that in the immediate vicinity of an impermeable wall:

$$\tau_T = \rho \overline{u'v'} \sim \rho v'^2 \dfrac{\partial \overline{u}}{\partial y} \tag{5.5}$$

$$\mu_T \sim \rho v'^2 \sim \rho \left( \int_0^y \overline{u}\, dy \right)^2 \tag{5.6}$$

After substituting the value $\overline{u}$ from (5.4), Eq. (5.4) is reduced to a dimensionless form by normalization over $\mu$, or as it appears to be the same formally over $v^* y_1$.

As a result:

$$\frac{\mu_T}{\mu}\underset{y\to 0}{\longrightarrow}\beta\left(\frac{Uy}{\nu}\right)^4\left(\frac{C_f}{2}-\frac{U'}{3U}y\right)^2 \qquad (5.7)$$

where $\beta$ is a constant.

When $U' = 0$ (the flow without longitudinal pressure gradients) formula (5.7) becomes the Landau and Levitch formula:

$$\frac{\mu_T}{\mu}\underset{U'=0,y\to 0}{\longrightarrow}\beta\eta^4 \qquad (5.8)$$

At the point of boundary layer separation ($C_f = 0$):

$$\frac{\mu_T}{\mu}\underset{C_f=0,y\to 0}{\longrightarrow}\frac{\beta y^6}{3\nu^4}(UU')^2 \qquad (5.9)$$

The ratio of the turbulent viscosity coefficient in the layer near the wall at the point of separation and in a flow with a zero pressure gradient, is equal to:

$$\frac{\mu_{T,C_f\ne 0}}{\mu_{T,\nu'=0}}\underset{y\to 0}{\longrightarrow}\frac{4}{3}\left(\frac{U'y}{UC_{f0}}\right)^2 \ll 1 \qquad (5.10)$$

## 5.2. HEAT TRANSFER (MASS TRANSFER) COEFFICIENT WHEN $Pr \to \infty$

Let us consider heat transfer under the following conditions:

$$\left\{Pr\to\infty,\ q_{CT}\gg\int u\frac{\partial i}{\partial x}dy;\ C_f\gg\left|\frac{U'}{U}y_1\right|\right\} \qquad (5.11)$$

The first two conditions denote that the thermal resistance of the system is actually concentrated in the viscous sub-layer and that the heat produced on the wall is absorbed in the turbulent core of the flow.

The third condition denotes that turbulent viscosity is determined by formula (5.8).

The coefficients of turbulent viscosity and heat conductivity are linked by the expression:

$$\frac{\lambda_T}{\lambda}=\varepsilon Pr\frac{\mu_T}{\mu}, \qquad (5.12)$$

where $\varepsilon$ is the turbulent Prandtl number.

For the conditions being analyzed:

$$\frac{q_{CT}}{T_{CT}-T_0}=\left(\int_0^{y_1}\frac{dy}{\lambda+\lambda_T}\right)^{-1} \qquad (5.13)$$

and after the necessary substitution and assuming law (5.8) to be valid over the whole of zone $\eta_1$, we obtain:

$$\frac{\alpha}{C_p\rho\upsilon^*} = \left(Pr\int_0^{\eta_1}\frac{d\eta}{1+\varepsilon\beta Pr\,\eta^4}\right)^{-1} =$$

$$= \left(\frac{4}{Pr}\right)^{3/4}\left(\ln\frac{1+x_1 2^{1/2}+x_1^2}{1-x_1 2^{1/2}+x_1^2} + 2\,\text{arc tg}\,\frac{x_1 2^{1/2}}{1-x_1^2}\right)^{-1}, \quad (5.14)$$

where $x_1 = (\varepsilon\beta Pr)^{1/4}\eta_1$.

There follows:

$$St \xrightarrow[Pt\to\infty,\,\frac{U'}{U}y_1\to 0]{} \frac{\beta_1 C_f^{1/2}}{Pr^{3/4}}, \quad (5.15)$$

where $\beta_1 = 4\pi(\varepsilon\beta)^{1/4}$.

Thus, at high Prandtl numbers:

$$\alpha \sim \frac{U}{\nu}C_f^{1/2}Pr^{1/4}, \quad (5.16)$$

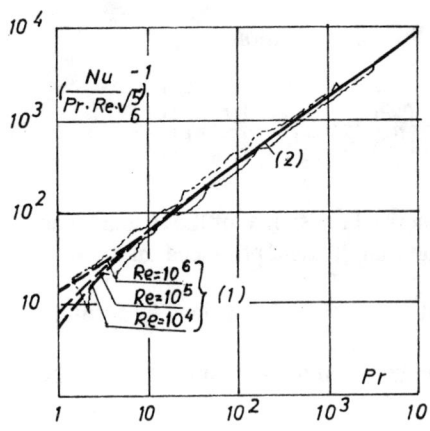

Fig. 5.2. Comparison of the theory with the experimental data systematized by Spalding and Jayatillaka.
1 - according to limiting formula (5.15).
2 - according to formula (5.18) for round tubes.
The dotted line limits the change of the experimental data.

while for Prandtl numbers close to unity:

$$\alpha \sim \frac{U}{\nu}C_f Pr^{2/5} \quad (5.17)$$

Figure 5.2 presents experimental data systematized by Spalding and Jayatillaka [39]. Line (1) corresponds to formula (5.15) when $\beta_1 = 0.07$, line (2) to the author's general formula [8] for flow in a cylindric tube when $Pr > 0.5$.

## 5.3. THE CONNECTION BETWEEN TURBULENT VISCOSITY AND THE REYNOLDS NUMBER

At $y \approx y_1$ the turbulent and molecular friction in the flow are commensurable, the mixing length is practically proportional to the distance from the wall. Therefore it is quite natural to take:

$$\frac{\tau_T}{\tau_l} \sim \left(y \frac{\partial u}{\partial y}\right)^2 / \nu \frac{\partial u}{\partial y} \tag{5.18}$$

as a measure of the interaction between these two mechanisms at a given point of the boundary layer. As shown by Loytsyansky this parameter is important in the theory of wall turbulence.

That is why we shall call

$$\dot{Re} = \frac{y^2}{\nu} \cdot \frac{\partial u}{\partial y} \tag{5.19}$$

a Reynolds number in the Loytsyansky sense.

As seen from (5.18) the value $\nu$ in $Re$ characterizes the state of the medium at a given point, i.e. it refers to the local thermodynamic parameters of the flow.

Prandtl's formula becomes:

$$\xi_1 < \xi \ll 1; \frac{\mu_T}{\mu} = \chi^2 \dot{Re}, \tag{5.20}$$

and formula (5.7):

$$\xi \to 0; \frac{\mu_T}{\mu} \to \beta \dot{Re}^2 \tag{5.21}$$

## APPENDIX 2.

## 6. INFLUENCE OF DENSITY PULSATIONS ON THE TURBULENT FRICTION AT FINITE REYNOLDS NUMBERS

### 6.1. CORRELATIONS

In the flat boundary layer turbulent shear stresses are equal to [8]:

$$\tau_T = -\overline{(\rho v)'u'}. \tag{6.1}$$

For a gas, the density pulsation is:

$$\rho' = \rho \frac{T'}{T}. \tag{6.2}$$

And for the gas boundary layer (6.1) yields:

$$-\tau_T \frac{T}{\rho} = \overline{u'v'}T + v\,\overline{u'T'}\,\overline{u'v'T} \tag{6.3}$$

Here, the local parameters of the flow as well as $\rho, v, u, T$ averaged in time are written without the averaging sign. The term $-\rho \overline{u'v'}$ is an analogue of the Reynolds stresses in the turbulent boundary layer of an incompressible fluid and:

$$\Delta \tau_T = -\frac{\rho}{T}(\nu \,\overline{uT'} + \overline{u'\nu T'}) \qquad (6.4)$$

is an additional term due to density pulsations. Introducing to the notion of the mixing length we can write:

$$\tau_T = \rho\left(\frac{\partial u}{\partial y}\right)^2\left(1 + \frac{\nu}{T}\frac{\partial T}{\partial u} + \frac{l}{T}\frac{\partial T}{\partial u}\right) \qquad (6.5)$$

Here in the gas layer near a wall:

$$l_T \approx l$$

## 6.2. GAS BOUNDARY LAYER ON A SMOOTH IMPERMEABLE PLATE

In a multiatomic gas $Pr = 1$ and under the conditions being considered the velocity field and the stagnation temperature field are similar:

$$\frac{T}{T_0} = \psi - (\psi - \psi^*)\omega - (\psi^* - 1)\omega^2 \qquad (6.6)$$

Hence,

$$-\frac{\partial T}{\partial y} = \frac{T_0}{U}\cdot\frac{\partial u}{\partial y}\left[\psi - \psi^* + \frac{2u}{U}(\psi^* - 1)\right] \qquad (6.7)$$

and,

$$\frac{\Delta \tau_T}{\rho \overline{u'v'}} = \left(\frac{\nu}{U} + \frac{l}{U}\frac{\partial u}{\partial y}\right)\frac{\psi - \psi^* + 2\omega(\psi^* - 1)}{\psi - (\psi - \psi^*)\omega - (\psi^* - 1)\omega^2} \qquad (6.8)$$

When $|\Delta \tau_T| \ll |\rho \overline{u'v'}|$, which is always fulfilled for large Reynolds numbers:

$$l\frac{\partial u}{\partial y} \approx \nu^*\sqrt{\frac{\rho_0}{\rho}\tilde{\tau}}, \qquad (6.9)$$

where $\nu^* = \sqrt{\frac{\tau_{CT}}{\rho}}$.

The quantity $\nu$ can be evaluated according to the formula:

$$\nu \approx \frac{1 + 2n}{2}C_f U \xi^{1+n} \qquad (6.10)$$

where $n$ is the exponent in the power law approximation of the velocity profile. With $Re \to \infty$, $n \to 0$

$$\frac{\nu}{U} \to \frac{C_f}{2}\xi \qquad (6.11)$$

Thus, we estimate that:

$$\frac{\Delta \tau_T}{\rho \overline{u'v'}} \approx \left\{\frac{C_f}{2}\xi + \sqrt{\frac{C_f}{2}\tilde{\tau}\left[\psi - (\psi - \psi^*)\omega - (\psi^* - 1)\omega^2\right]}\right\} \times$$

$$\times \frac{\psi - \psi^* + 2\omega(\psi^* - 1)}{\psi - (\psi - \psi^*)\omega - (\psi^* - 1)\omega^2}. \qquad (6.12)$$

When $\psi^* \to 1$, (subsonic flow):

$$\frac{\Delta \tau_T}{\rho \overline{u'v'}} \approx \left\{ \frac{C_f}{2} \xi + \sqrt{\frac{C_f}{2} \tilde{\tau} \left[ \psi - (\psi - 1)\omega \right]} \right\} \cdot \frac{\psi - 1}{\psi - (\psi - 1)\omega}, \quad (6.13)$$

when $\psi = \psi^*$ (flow along an insulated plate):

$$\frac{\Delta \tau_T}{\rho \overline{u'v'}} \approx \left\{ \frac{C_f}{2} \xi + \sqrt{\frac{C_f}{2} \tilde{\tau} \left[ \psi^* - (\psi^* - 1)\omega^2 \right]} \right\} \cdot \frac{2\omega(\psi^* - 1)}{\psi^* - (\psi^* - 1)\omega^2}, \quad (6.14)$$

It is possible to quantitavely judge the effect of density pulsations on the local turbulent shear stress in the boundary layer using the data given in the following tables. The calculations are carried out for the range of $Re^{**} \approx 10^4$, when $\omega \sim \xi^{1/7}$. With the increase of Reynolds number this effect becomes weaker owing to

$$C_f \underset{Re \to \infty}{\longrightarrow} 0.$$

Table 6.1.

| $\psi$ \ $\xi$ | $|\Delta\tau_T / \overline{\rho u'v'}|$ according to formula (6.13) | | | | |
|---|---|---|---|---|---|
| | 0.2 | 0.4 | 0.6 | 0.8 | I |
| 5 | 0.066 | 0.061 | 0.053 | 0.028 | 0.002 |
| 2 | 0.070 | 0.024 | 0.020 | 0.011 | 0.001 |
| I | 0 | 0 | 0 | 0 | 0 |
| 0.5 | 0.022 | 0.018 | 0.015 | 0.0078 | 0.0009 |
| | | | 0.0149 | | |
| 0.1 | 0.053 | 0.044 | 0.036 | 0.019 | 0.0029 |

Table 6.2.

| $M$ \ $\xi$ | $|\Delta\tau_T / \overline{\rho u'v'}|$ according to formula (6.14) | | | | |
|---|---|---|---|---|---|
| | 0.2 | 0.4 | 0.6 | 0.8 | I |
| 2 | 0.0312 | 0.03 | 0.027 | 0.014 | 0.0014 |
| 5 | 0.088 | 0.9 | 0.087 | 0.044 | 0.0036 |
| 10 | 0.12 | 0.13 | 0.14 | 0.095 | 0.005 |

APPENDIX 3.

## 7. CALCULATION OF CONSTANT $\chi$ OF WALL TURBULENCE

### M. A. Goldsthik, S. S. Kutateladze

Turbulent flows posses some conservative properties with respect to the mean flow characteristics, which make it possible to build a phenomenological theory of the mean flow without considering its detailed inner structure.

Three characteristic regions can be observed in the boundary layer near a wall: a viscous sub-layer, in which the molecular friction is substantially greater than the turbulent; the region of the quadratic law of friction in which the Prandtl - Taylor law is valid with a great degree of accuracy [40, 9]

$$\tau_T = \rho\left(\chi y \frac{\partial u}{\partial y}\right)^2, \tag{7.1}$$

where $\chi$ is Prandtl - Karman's constant [10] determined in semiempiric theories by means of experiments; and the external region of the boundary layer.

Of greatest importance for the formation of velocity profiles in the boundary layer near a wall is the region of the Prandtl - Taylor law, which is due to two important factors: First with the increase of Reynolds number the relative thickness of the viscous sub-layer tends to zero, and in the external region of the boundary layer the velocity of the flow tends to the velocity of the unperturbated flow, or to the velocity on the axis of the channel.

Second - the region of the Prandtl - Taylor law is rather conservative to external actions (longitudinal pressure gradient, the degree of turbulence of the external flow etc. ) [13, 14].

As a generalization of the above mentioned properties of the turbulent boundary layer it is possible to formulate the principle of maximum stability of the mean turbulent flow. With respect to the viscous sub - layer such considerations were forwarded by S. S. Kutateladze in [18] and for the whole flow by M. A. Goldsthik. In the Appendix 4 the mathematical formulation of the problem is given and a functional introduced:

$$\Pi = \sup_{\alpha} Y \tag{7.2}$$

as a measure of stability, where $Y$ is the logarithmic decrement of damping and $\alpha$ is the wave number.

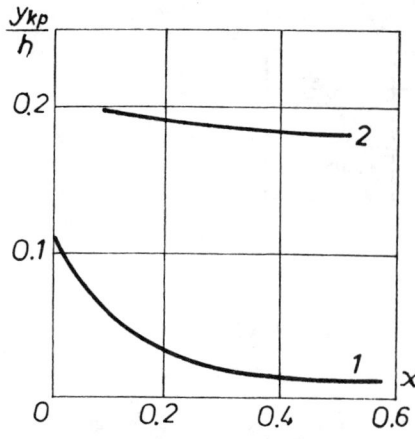

Fig. 7. 1. The position of the critical point for shortwave (1) and long wave (2) perturbations.

It was assumed that in the first approximation the turbulent stress tensor does not vary when small perturbations superimposed on the flow of are, and the value $Y$ may be calculated by solving the usual Orr - Sommerfeld equation. This is explained by the fact that for the most stable velocity profile, the function $Y(\alpha)$ has two equally pronounced maxima which are situated near the boundary of the spectrum. The presence of two "dangerous" perturbations corresponds to the two natural scales of the turbulent flow - the size of the boundary layer or channel and the thickness of the viscous sub - layer. It is evident that very long waves cannot interact significantly with turbulent pulsations, the maximum of their intensity laying in the short - wave part of the spectrum. As to the short - wave perturbations with the increase of the fullness of the velocity profile they tend towards the viscous sub - layer where the degree of turbulent friction is small. This is clear from Fig. 7. 1., illustrating the dependence of the position of the critical layer on the $\chi$ parameter, which is chosen here as the measure of the fullness of the velocity profile. As is known, in the critical layer the local velocity of the flow coincides with the perturbation propagation velocity and the perturbation position is the main factor which characterizes the stability of the flow. With the increase of turbulence the short - wave perturbation critical layer submerges into the viscous sub - layer. The long - wave perturbation critical layer remains in the region of intensive turbulence.

The dependence of functional (7.2) upon the parameters of the flow is not smooth accounting for the significant inconvenience of calculation. Therefore it seems quite reasonable to switch over to an integral functional. The latter can be constructed in the following way. Let the perturbation in the form of a $\delta$ function be superimposed at the starting moment in a certain section of the channel. The kinetic energy of the perturbation motion is:

$$E \sim \int_{-\infty}^{\infty} v^2 dx. \tag{7.3}$$

If the perturbation dampens down we may define:

$$J_1 = \int_0^\infty dl \int_0^\infty v^2 dx. \tag{7.4}$$

For the given perturbation the energy of a separate Furier harmonic is proportional to $\exp(2\alpha Yt)$ and the sought after functional may be given as:

$$J = -\int_0^\infty \frac{d\alpha}{\alpha Y} \tag{7.5}$$

Essentially, the main contributions to this integral are short and long wave perturbations. Due to this, as was shown by calculations, most stable profiles found according to (7.2) also exhibit the minimum of the functional (7.5).

For unstable profiles, when $Y(\alpha)$ changes sign, the functional $J$ does not have any meaning.

In flows along an impermeable wall with small pressure gradients, the shear stresses in the zone of the validity of law (7.1) are close to the friction at the wall, and, therefore, a linear law of the turbulent viscosity applies as follows:

$$\mu \ll \mu_T = \chi \rho v^* y \tag{7.6}$$

In the regions of the viscous sub-layer [3 - 7] when $y \to 0$:

$$\mu \gg \mu_T \sim y^4 \tag{7.7}$$

Thus the effective viscosity which forms the turbulent velocity profile:

$$\mu_* = \mu + \mu_T \tag{7.8}$$

is a complex function of the distance from the wall, its form being known only in the region of the validity of local laws (7.1) and (7.7).

However, because of the degeneration of the viscous sub-layer and the maximum fullness of the velocity profiles in the core of the flow with $Re \to \infty$, the profile calculated using the effective viscosity

$$\mu_* = \mu + \chi \rho v^* y \tag{7.9}$$

approaches the actual distribution of the velocity at the increase of the Reynolds number. At the same time such a profile has only one variable parameter, namely the Prandtl - Karman constant that at sufficiently large Reynolds numbers it is in fact, automodeled.

Consequently, the calculation of the value $\chi$ from the condition of the minimum of the functional $J$, when friction law (7.9) is used, will be formally correct with

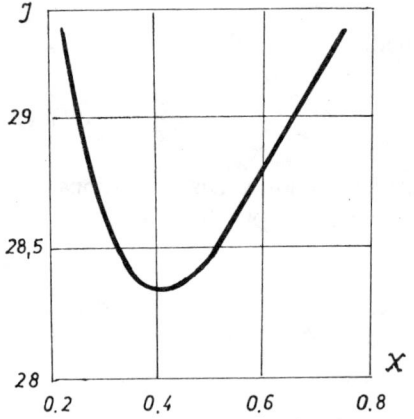

Fig. 7. 2. Dependence of the integral functional on the filling in of the profile.

$Re \to \infty$ practically giving an accurate result in the whole region of turbulent developed flows ($Re \gg Re_{kp}$).

Figure 7. 2 shows the $J(\chi)$ dependence for the effective viscosity (7.9) which is on purpose calculated for not very large values of $Re$. It is avident that functional $J$ is smooth and an obvious minimum occurs at $\chi = 0.4$ which is in good agreement with all the available experimental data [11].

The authors express their grattitude to V. A. Sapozhnikov and V. N. Shtern for their participation in this work.

## APPENDIX 4.

### 8. PRINCIPLE OF MAXIMUM STABILITY OF AVERAGED TURBULENT FLUXES

**M. A. Goldsthik**

As it is known there still exists no single concept which allows us to theoretically calculate turbulent flows. Moreover, in spite of certain progress turbulence statistical theory seems to be quite a distance from this objective. Therefore it seems rather logical to try to formulate a certain qualitative heuristic principle which would allow the calculation of the principal turbulent flow parameters without recourse to experiment.

The first attempt of this kind was made by Malkus [41] who assumed that the average turbulent flow is a neutrally stable flow for which the velocity of the energy dissipation is at maximum. We are inclined to consider his ideas inadequate, it being difficult to assume that there exists some particular perturbation of the mean profile of the turbulent flow which would not damp unlike the perturbation of the remaining wavelengths. As shown by experiments, any perturbations introduced into the averaged turbulent flow damp rather rapidly because of the intensive intermixing action of the turbulent pulsations. It is not surprising, therefore, that Malkus' theory did not sustain Lin's numerical check applied by Reynolds and Tiederman [42] in 1967. These authors have established that in the first the experimental turbulent profile is rather stable and second that the class of neutrally stable velocity profiles does not contain any elements whith the maximum energy dissipation velocity.

The purpose of the present note is to formulate a new extremal hypothesis as well as to account for the results of its numerical check.

The suggested hypothesis assumes that the mean turbulent flow possesses properties of maximum stability with respect to the most "dangerous" perturbation, i. e. to such a perturbation which damps slower than all the rest. It is possible to imagine that the mean turbulent flow is capable of accepting a number of stable states, however, taking the high dynamics of the turbulent system into account, having the greatest probability – the state with maximal stability, if such a state exists.

If we apply the small perturbation method and assume that the turbulent stressk tensor does not change in the first - order approximation with mean velocity perturbations (other possibilities are mentioned below), then for the one - dimensional case it is not difficult to obtain the usual Orr - Sommerfeld equation with the aid of standard transformations:

$$\varphi^{IV} - 2\alpha^2 \varphi''' + \alpha^4 \varphi = i\alpha Re[(u-c)(\varphi'' - \alpha^2\varphi) - u''\varphi] \tag{8.1}$$

Here: $\varphi(y)$ - the sought complex amplitude of the perturbated function of the flow; $\alpha^{-1}$ - a dimensionless wavelength of the longitudinally superimposed harmonic perturbation; $Re$ - Reynolds number, $u$ - a dimensionless velocity of the mean flow, the stability of which is under consideration.

If we consider the case of a flat channel and restrict ourselves to the study of symmetrical perturbations, only then it is necessary to set the following boundary conditions for the function $\varphi$:

$$\varphi'(0) = \varphi'''(0) = 0, \tag{8.2}$$

$$\varphi(1) = \varphi'(1) = 0. \tag{8.3}$$

For the fixed profile $(u)$ the problem of its stability consists in determining its complex eigenvalue $C = X + iY$, which has the maximum imaginary part. If $Y > 0$ the profile is not stable, when $Y < 0$ the profile is stable. The value $Y$ depends on the parameter $\alpha$. To the most dangerous perturbation corresponds such a value $\alpha$ which gives the maximum $Y(\alpha)$, denoted further by $\prod$:

$$\prod = \max_{\alpha} Y. \tag{8.4}$$

In the usual set up of the problem of stability, the function $u$ is the solution of the Navler - Stokes equation and is considered to be given.

In the suggested set - up the function $u$ can be an arbitrary element of the space $C_2$ of functions which are continuously twice differentiated and which satisfy the boundary conditions of adhesion, and the mass flow constancy condition in the tube:

$$\int_0^1 u \, dy = 0 \tag{8.5}$$

The formulated variation problem is reduced to the finding of the expression:

$$\text{Inf}_u \prod = \text{Inf}_u \max_{\alpha} Y. \tag{8.6}$$

under additional condition (8.5).

In this form the problem was formulated by the author in Novosibirsk at the All - Union Physics Seminar in 1963. Then an attempt was made to check the given hypothesis by means of a numeric experiment. However, calculating difficulties have not been overcome. By the end of 1967 the author in collaboration with V. A. Sapozhnikov developed a new algorithm of solving equation (8.1) which consists the following.

Let us choose two solutions from four linearly independent solutions of equation (8.1), which satisfy conditions (8.2). In order to satisfy condition (8.3) the fulfillment of the condition that the function:

$$\psi = \frac{\varphi'_2}{\varphi_1} - \frac{\varphi'_2}{\varphi_2} \text{ for } y = 1 \tag{8.7}$$

is equal to zero, is necessary.

Thus the task is reduced to solution of equation $\psi(c) = 0$. The function $\psi(c)$ is determined by solving transformed equation (8.1) by the Rungge and Kutta method. The zeros of $\psi(c)$ are obtained by the cross-cutting method, which is modified for complex functions.

The main advantage of this method consists in an arbitrary choice of the integration step and in the possibility of calculation controlled precision.

Since the aim of this work is only to check the suggested hypothesis, we shall restrict ourselves to consideration of the class of profiles used in [42].

$$\frac{du}{dy_1} = Re\, B\frac{1-y_1}{1+E}; \quad u(0) = 0 \qquad (8.8)$$

$$E = \frac{1}{2}\sqrt{1 + \frac{1}{9}\chi^2 Re^2 B\{y_1(2-y_1)(3-4y_1+2y_1^2)\}\left(1 - exp\left[-\frac{Re\sqrt{B}}{2A}y_1(2-y_1)\right]\right)^2 - \frac{1}{2}} \qquad (8.9)$$

where $y_1 = 1 - y$ - distance from the channel wall, $B$ - a parameter determined from (8.5), $A$ - a parameter ($A = 31$) fixed in the these calculations, $\chi$ - a variable parameter which characterizes the profile. When $\chi = 0$ the velocity profile is Poiseuille's parabola, when $\chi = \infty$, $u = 1$ and, at last when $\chi \approx 0.4$ correlations (8.8) and (8.9) approximate Laufer's experimental data for the turbulent flow in the flat channel well. The results of the calculations carried out on the BESM - 6 computer, for the value $Re = 10^4$, are shown in Fig. (8.1), where the parameter $\chi$ is plotted as a function of $\Pi(\chi)$. As seen, the function $\Pi(\chi)$ has two branches which correspond to the

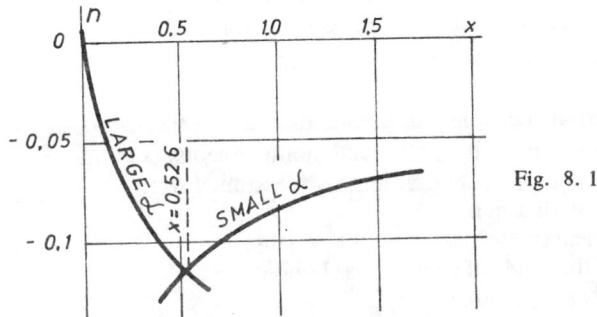

Fig. 8. 1.

presence of two maxima of the function for each fixed $\chi$, one of which lies in the region of short - wave perturbations (large $\alpha$) and the second corresponds to long - wave perturbations (small $\alpha$). When $\chi < 0.526$ the greatest danger is presented by the short - wave perturbations, when $\chi > 0.526$ the danger are long - wave perturbations. The very point $\chi = 0.526$ corresponds to the most stable profile with respect to the most dangerous perturbations which are both short - wave with $\alpha \cong 100$ and long - wave with $\alpha \cong 0.2$. It is not difficult to see that any other $\chi$ corresponds to less stable profiles.

The difference between the profile corresponding to $\chi = 0.526$ and the experimental one is comparatively small, therefore it is possible to conclude that in the first approximation the variation principle has passed the numerical test.

In the given calculations it was assumed that the role of Reynolds stresses is reduced only to forming the mean velocity field whose stability can be analyzed without accounting for the influence of these stresses the latter being considered a field of mass forces which does not depend on the state of motion. Therefore, strictly speaking, the obtained results should be interpreted as the solution of the problem of finding external forces which stabilize the laminar flow.

In order to make the next step (of the approximation) it is necessary to account for the immediate stabilizing action of turbulent stresses. In order to perform that introduce a variable "turbulent viscosity" and consider it an independent variable function which is to be determined from the condition of maximum stability of the corresponding mean profile of the mean velocity.

The extension of the Reynolds number range and the realization of the above mentioned step is likely to make physical sense of the suggested principle in more detail. Clearly, its final confirmation could open such interesting perspectives as the explanation of the phenomenon of alternating in tubes under transition conditions, the nonuniqueness of certain flows, the phenomenon of hydrodynamic histeresis. This principle might also appear useful in considering phenomena connected with plasma instabilities.

## NOMENCLATURE

| | |
|---|---|
| $\alpha$ | heat diffusivity coefficient |
| $b = 2j_{cm}/C_{f0}$ | wall permeability factor |
| $b_1 = 2j_{cm}/C_f$ | the same related to the real value of friction coefficient |
| $b_{kp}$ | critical value of the wall permeability factor |
| $\bar{b} = b/b_{kp}$ | relative value of wall permeability factor |
| $b^*$ | wall permeability factor related to value $C_{f0}$ |
| $C_f = 2\tau_{CT}/r_0 U^2$ | local friction coefficient related to $Re^{**}$, |
| $C_f$ | the same related to $Re_x$ |
| $C_{f0}$ | local friction coefficient in the streamlining of flat - smooth - impermeable plate by isothermal, homogeneous boundary layer without pressure gradient with assumed $Re^{**}$, |
| $C_{f*}$ | the same with assumed $Re_x$, |
| $C_p$ | specific heat capacity at constant pressure |
| $D_0$ | tear - off diameter of steam (gas) bubble |
| $D$ | diameter of the bubble |
| $f\delta = \dfrac{\delta}{U} \cdot \dfrac{dU}{dx}$ | form parameter |
| $g$ | acceleration due to gravity |
| $H = \delta^*/\delta^{**}$ | form parameter |
| $i$ | enthalpy |
| $i_0$ | flow enthalpy beyond the boundary layer |
| $i^*$ | stagnation enthalpy |
| $i_{CT}$ | medium enthalpy on the wall surface |
| $i''$ | enthalpy of boiling fluid saturation |
| $j_{CT}$ | mass velocity of substance through the permeable wall |
| $\bar{j}_{CT} = j_{CT}/\rho_0 U$ | dimensionless velocity of the substance through the wall |
| $k = C_p/C_v$ | Poisson adiabate index |
| $k = \dfrac{U_{kp}\sqrt{\rho''}}{\sqrt[4]{g\sigma(\rho' - \rho'')}}$ | the criterion of hydrodynamic stability of two - phase boundary layer (introduced first by this author in 1948) |

$K = \dfrac{r}{i'' - i_0}$    heat criterion of phase transformation (introduced first by this author in 1935)

$L = \sqrt{\dfrac{\sigma}{g(\rho' - \rho'')}}$    Laplace constant

- $l$    mixing length in Prandtl's formula
- $\tilde{l} = l/\delta$    relative mixing length
- $l_T$    heat mixing length
- $l_T/\delta_T$    relative heat mixing length
- $M = U/U_0$    Mach number where $U_0$ is sound velocity
- $p$    pressure
- $Pr$    physical Prandtl number
- $Pr_T$    turbulent Prandtl number
- $R$    gas constant
- $Re$    Reynolds number
- $Re^{**} = U\delta^{**}/\nu_0$    Reynolds number over the momentum thickness and kinematic viscosity outside the boundary layer
- $Re_\delta = \dfrac{U\delta}{\nu_0}$    the same over the thickness of the boundary layer (in the approximation of the theory of the finite depth)
- $Re_x = \dfrac{U\chi}{\nu_0}$    the same over the distance from the beginning of the turbulent flow in the boundary layer
- $r$    recovery factor
- $q$    latent heat of vaporization
- $q_T$    heat flux density conditioned by the pure turbulent transfer
- $q_{CT}$    heat flux density on the wall
- $\tilde{q} = q/q_{CT}$    relative heat flux
- $\tilde{q}_0$    relative heat flux on the standard plate
- $q_{kp}$    critical heat flux density under change over from bubble to film boiling
- $q_{kp,0}$    critical heat flux density in saturated fluid
- $\tilde{q}_{kp} = q_{kp}/q_{kp,0}$    relative critical heat flux
- $St$    Stanton number
- $St_0$    Stanton number for the standard plate
- $T$    absolute temperature (and the index which is identical with index over $i$)
- $U$    flux velocity at the external boundary of the boundary layer
- $U''$    rate of gas (steam) phase
- $U_{kp}''$    critical rate of barbotage or boiling
- $u$    velocity projection onto $x$ axis
- $u'$    longitudinal component of pulsation velocity
- $v$    velocity projection onto $y$ axis
- $v'$    transverse component of pulsation velocity
- $x$    coordinate directed down the flow along the wall surface
- $y$    coordinate directed along the normal to the wall surface into the depth of the flow

| | |
|---|---|
| $y_1$ | thickness of the viscous sub-layer |
| $\alpha$ | heat transfer coefficient |
| $\tilde{\beta}$ | coefficient of the influence of density pulsations on turbulent friction stresses |
| $\Delta$ | sign of difference |
| $\delta$ | thickness of dynamic boundary layer, steam film, etc. |
| $\delta_T$ | thickness of the thermal boundary layer |
| $\delta^*$ | displacement thickness |
| $\tilde{\delta} = \delta^*/\delta$ | relative thickness of displacement |
| $\delta^{**}$ | momentum thickness |
| $\tilde{\delta}^{**}$ | relative momentum thickness |
| $\delta_T^{**}$ | energy deficit thickness |
| $\tilde{\delta}_T^{**}$ | relative deficit thickness |
| $\varepsilon_l = l/l_T$ | degree of similarity of hydrodynamic and heat mixing lengths |
| $\upsilon$ | dimensionless enthalpy or temperature |
| $\lambda$ | heat conductivity coefficient |
| $\lambda_T$ | turbulent conductivity coefficient |
| $\mu$ | dynamic viscosity |
| $\mu_T$ | turbulent dynamic viscosity |
| $\nu = \mu/\rho$ | kinematic viscosity |
| $\mu_0, \gamma_0$ | viscosity values outside the boundary layer |
| $\mu_{CT}$ | viscosity value with parameters at the wall |
| $\tilde{\mu} = \mu/\mu_{CT}$ | relative value with parameters at the wall |
| $\xi = y/\delta$ | relative distance from the wall |
| $\xi_1$ | relative thickness of the viscous sub-layer |
| $\rho$ | medium density |
| $\rho_0$ | medium density outside the boundary layer |
| $\rho_{CT}$ | medium density with parameters at the wall |
| $\rho'$ | fluid phase density |
| $\rho''$ | density of steam (gas, in general, much lighter phase) phase |
| $\tilde{\rho} = \rho/\rho_0$ | relative density in the boundary layer |
| $\sigma$ | surface tension coefficient |
| $\tau$ | shear stress |
| $\tau_{CT}$ | shear stress at the wall |
| $\tilde{\tau} = \tau/\tau_{CT}$ | relative value of shear stress at a given point of the boundary layer |
| $\tilde{\tau}_0$ | the same on the standard plate |
| $\varphi$ | volume vapour content |
| $\varphi_*$ | volume vapour content of the boundary layer |
| $\chi$ | Prandtl-Karman constant |

$\Psi = C_f/C_{f0}$ relative friction coefficient with the assumed value of $Re^{**}$ ($Re^{**}$ = idem)

$\dot\Psi$ the same with $Re_x$ = idem,

$\Psi_s = \dfrac{St}{St_0}$ relative coefficient of heat transfer when $Re_T^{**}$ = idem,

$\psi = i_{CT}/i_0$ enthalpy factor (when $C_p$ = const., temperature factor),

$\psi_1 = \rho_0/\rho_{CT}$ density factor

$\psi^* = i^*/i_0$ kinematic enthalpy factor (when $C_p$ = const., kinematic temperature factor)

$\omega = u/U$ relative flux rate along $x$ axis

$\omega_1$ relative rate at the boundary of the viscous sublayer and turbulent core

$\omega_{10}$ the same under standard conditions

——— a line over the symbol is the sign of averaging

———→ limited transfer; asymptotic tendency

## REFERENCES

1. C. W. OSEEN, Hydrodynamik, Leipzig, 1927.
2. S. S. KUTATELADZE, Izv. Akad. Nauk SSSR, Otdel. Tekhn. Nauk, 10 (1950).
3. S. S. KUTATELADZE, *"Teploperedatcha pri kondensatsii i kipenii"*, Mashgis, Moskva - Leningrad, 1952.
4. S. S. KUTATELADZE, M. A. STYRIKOVITCH, *"Gidravlika gazo - zhidkostnikh sistem"*, Gosenergoizdat, Moskva - Leningrad, 1958.
5. S. S. KUTATELADZE, Zhurn. Prikl. Mekh. i Techn. Fiz., 1 (1960).
6. S. S. KUTATELADZE, A. I. LEONTJEV, *"Turbulent pogranitchnii sloi szhimaemogo gaza"*, Izdat. SO AN SSSR, Novosibirsk, 1962. E. Arnold Ltd., London, 1964; Acad. Press. Inc., New York, 1964.
7. S. S. KUTATELADZE, The concept of a fluid with disappearing viscosity and some problems of the phenomenological theory of turbulence near the wall; Invited Lecture, Third Int. Heat Transfer Conf., Chicago, August 11, 1966.
8. S. S. KUTATELADZE, *"Osnovi teorii teploobmena"*, Mashgis, Moskva - Leningrad, 1962. Academic Press. Inc., New York, 1963; E. Arnold Ltd., London, 1963.
9. L. PRANDTL, V. D. I. - Z. 77, 5 (1933).
10. TH. KARMAN, Some aspects of the theory of turbulent motion, Proc. Int. Congress for Appl. Mechanics, Cambridge, 1934.
11. M. P. ESCUDIER, The distribution of the mixing length in turbulent flows near the wall. Mech. Eng. Dept. Imperial College, London, March, 1965.
12. D. B. SPALDING, S. V. PATANKOR, *"Heat and mass transfer in boundary layers"*, Morgan - Grampian, London, 1967.
13. M. PIKHAL, *"Turbulentni mezni vratva na rovianne desko uproudani s vysokou intenzitou turbulence"*, Vydavatwlstvo Slovenskej Akademie Vied.
14. F. CLAUSER, J. AERONAUT. Sci., V. 21, 2 (1954).
15. RELEI, *"Theoriya zvuka"*, Gos. izdat. Tehniko - Teoretitcheskoi Literaturi, Glava XXI, Moskva, 1955.
16. L. A. VULIS, V. P. KASHKAROV, *"Teoriya strui vyazkoi zhidkosti"*, Izdat. "Nauka", Moskva, 1965.

17. ZH. BLOM, D. A. VRIZ, *"O velitchine turbulentnogo tchisla Prandtlya"*, Sbornik "Teplomassoperenos", tom 1, Izdat. "Energiya", Moskva, 1968.
18. K. POHLHAUSEN, ZAMM, I, (1921).
19. K. K. FEDYAVSKII, A. S. GINEVSKII, Zhurn. Tekhn. fiz., 27, 2 (1957).
20. H. S. MICKLEY, R. S. DAVIS, NACA TN N 4017, 1957.
21. E. R. BARTLE, B. M. LEADON, Convier scientific research laboratory, Research Report II, May, 1961.
22. S. S. KUTATELADZE, A. I. LEONT'EV, B. P. MIRONOV, Turbulent boundary layer with mass and longitudinal pressure gradient in the finite $Re$ number region, J. S. M. E. Semi - International Symposium, Tokyo, 1967.
23. W. D. RANNIE, Calif. aainst. Technol. Jet Propulsion Lab. Progress Rept., 4 - 30, Nov. 1947.
24. N. ZUBER, AECV - 4439, 1959.
25. S. S. KUTATELADZE, N. V. VALUKIN, I. I. GOGONIN, Teplofiz. vis. temp., 5, 5 (1967).
26. S. S. KUTATELADZE, V. N. MOCKVITCHYEVA, Zhurn. Tekhn. fiz., 9 (1959).
27. S. S. KUTATELADZE, I. G. MALENKOV, Zhurn. Prikl. Mekh. i Tekhn. fiz., 2 (1966).
28. S. S. KUTATELADZE, Proc. Symp. Two - Phase Flow, Vol. 1, Exeter (June 21 - 23), 1965.
29. S. S. KUTATELADZE, I. G. MALENKOV, V. N. NAKORYAKOV, B. G. POKUSAEV, Yu. L. SOROKIN, V. N. SHTERN, Process in boiling and barbotage. J. S. M. E.,Semi - International Symposium, Tokyo, 1967.
30. N. U. AKTURK, Proc. Symp. Two - Phase Flow, Vol. 2, Exeter (June 21 - 23), 1965.
31. S. S. KUTATELADZE, A. I. LEONT'EV, V. N. MOCKVITCHYEVA, G. I. BOROVITCH, I. I. GOGONIN, I. G. MALENKOV, L. S. SHTOKOLOV, A. G. KIRDYASHKIN, Nekotorie problemi gidrodinamitcheskoi teorii teploobmena pri kipenii, Tridi Ts. KTI, Kotloturbostroenie, vip. 58, Leningrad, 1965.
32. L. S. SHTOKOLOV, Zhurn. Prikl. Mekh. i Tekhn. fiz., 1 (1966).
33. V. I. ADAMOVSKII, I. N. SVORKOVA, L. S. SHTOKOLOV, Zhurn. Prikl. Mekh. i Tekhn. fiz., 2 (1967).
34. F. W. MATTING, D. R. CHAPMAN, J. R. NYHOIM, NASA, TR R - 82, 1961.
35. V. P. MUGALYEV, Eksper mentalnoe issledovanie dozvukovogo turbulentnogo pogranitchnogo sloa, Isvestiya Vuzov "Aviatsionnaya tekhnika", 1959, N 3.
36. S. S. KUTATELADZE, Zhurn. Tekhn. fiz., 4 (1958).
37. L. D. LANDAU, E. M. LIVSHITS, *"Mekhanika sploshnikh sred"*, Gostechizdat, 1944.
38. V. G. LEVITCH, *"Fiziko - khimitcheskaya gidrodinamika"*, Fizmatizdat, Moskva, 1959.
39. D. B. SPALDING, A. JAYATILLAKA, Survey of theoretical and experimental information on the resistance of the laminar sub - layer to heat and mass transfer, Imperial College, London, 1963.
40. G. J. TAYLOR, Proc. Roy. Soc., A, 135, 1932.
41. W. V. R. MALKUS, J. Fluid Mechanics, 1, 521 (1956).
42. W. C. REYNOLDS, W. C. TIEDERMAN, J. Fluid Mechanics, 27, 2 (1967).

International Seminar 1969

The editor deeply regrets that no introduction is included for the text selected from the 1969 Seminar. We find that this reflects the widely held opinion that Zoran Zarić's place is very hard to fill.

# GAS CONCENTRATION MEASUREMENTS IN BOUNDARY LAYERS

E. A. Brun

Faculté des Sciences, Université de Paris, France

## 1. DEFINITIONS

Let us recall some well known definitions.

Let a homogenous mixture of fluids occupy a domain $D$. In this domain in the volume $dv$ around the point $M$, at time $t$, the mass of the mixture is $dm$ and the density, at point $M$ and time $t$, is

$$\rho = \frac{dm}{dv}$$

Let $dm_i$ be the mass of species $i$ in the volume $dv$ at time $t$, the *mass concentration of species i, at point M and time t*, is:

$$\Gamma_i = \frac{dm_i}{dv} \qquad \text{with} \qquad \Sigma \Gamma_i = \rho \qquad (1)$$

*The mass fraction of species i, at point M and time t,* is:

$$\gamma_i = \frac{dm_i}{dm} = \frac{\Gamma_i}{\rho} \qquad \text{with} \qquad \Sigma \gamma_i = 1 \qquad (2)$$

*The molar concentration of species i, at point M and time t,* is:

$$C_i = \frac{dm_i}{M_i} \frac{1}{dv} = \frac{\Gamma_i}{M_i} \qquad (3)$$

Let

$$C = \Sigma C_i$$

be the molar concentration of the mixture, *the molar fraction of species i, at point M and time t*, is:

$$c_i = \frac{C_i}{C} \qquad \text{with} \qquad \Sigma c_i = 1 \qquad (4)$$

In the problem of interest, where the mixture is flowing, we also have to define quantities relative to mass transfer: velocities and mass flux. The component $u_i$

along the x axis of the *mass velocity of i species at* point $M = [x, y, z]$ *and time t*, is given by:

$$dm_i = \Gamma_i u_i dA dt$$

where $dm_i$ is the mass of species $i$ flowing, during the time $dt$, through the element of area $dA$, centered on $M$ and normal to the $x$ axis.

Therefore, *the mass flux density* $\vec{\varphi}_i$ *of species i, at point M and time $t_i$* is related to the velocity $\vec{U}_i = (u_i, v_i, w_i)$ of species $i$, at the same point and time, by

$$\vec{\varphi}_i = \Gamma_i \vec{U}_i \tag{5}$$

Here, the reference system is fixed in space.

*The mass average velocity of the mixture*, or barycentric velocity, at point $M$ and time $t$, is:

$$\vec{v} = \frac{\Sigma(\Gamma_i \vec{U}_i)}{\rho} = \Sigma(\gamma_i \vec{u}_i) \tag{6}$$

This is a fictitious velocity.

In order to characterize the motion of species $i$ in the flowing fluid, one introduces the relative velocity

$$\vec{V}_i = \vec{U}_i - U, \tag{7}$$

which is called *the mass diffusion velocity of species i in the mixture at point M and time t*.

Of course, we have the relation

$$\Sigma(\gamma_i \vec{V}_i) = 0$$

Here, the reference system is fixed on the velocity $\vec{U}$. *The mass diffusion flux density* of species $i$, at point $M$ and time $t$, is:

$$\vec{J}_i = \Gamma_i (\vec{U}_i - \vec{U}) = \Gamma_i \vec{V}_i \tag{8}$$

Similarly, one defines:
– the *molar flux density of species i*,

$$dn_i = C_i u_i dA dt \; ; \quad \vec{U}_i^* = C_i \vec{U}_i \tag{9}$$

– *the molar average velocity of the mixture*,

$$\vec{U}^* = \frac{\Sigma(C_i \vec{U}_i)}{C} = \Sigma(C_i \vec{U}_i) \tag{10}$$

– *the molar diffusion velocity of species i*,

$$\vec{V}_i^* = \vec{U}_i - \vec{U}^* \quad \text{with} \quad \Sigma(c_i \vec{V}_i^*) = 0 \tag{11}$$

Here, the reference system is fixed on the velocity $\vec{U}^*$.

– *the molar diffusion flux density of species i,*

$$\vec{J}_i^* = C_i(\vec{U}_i - \vec{U}^*) = C_i \vec{V}_i^* \tag{12}$$

## 2. MASS TRANSFER IN THE BOUNDARY LAYER

The aim of this paper is to describe the techniques of concentration measurement in the vicinity of boundaries. Such measurements rest on the principles of the thermodynamics of irreversible phenomena, such as that of the »local state«; to examine them would take me too far any they will be assumed known.

On the other hand, I find it useful to show, at the beginning, the interest of concentration measurements in the boundary layer by discussing a simple problem which will be dealt upon again later.

During missile reentry into the atmosphere, the components of the air undergo, when flowing through the shock wave and due to the strong rise in temperature, a number of transformations: dissociation, formation of nitric oxide, *NO* and ionization. In the vicinity of the wall the temperature, on the contrary, drops sharply, which should lead to transformations opposite of the above mentioned, but the gaseous mixture does not reach, at each point, the state of chemical equilibrium corresponding to the physical conditions, this, as the time required for a complete chemical reaction in an element of fluid is large compared to the time this element remains in the boundary layer.

However, generally, the wall acts catalytically and greatly increases the rate of reaction. Therefore, the species concentrations are not the same on the two frontiers of the boundary layer and diffusion of these species will occur, *mass transfer thus being superimposed onto heat transfer* through the boundary layer.

To investigate, for a simple case, this transfer mechanism, G. Lassau [1] let a nitrogen plasma ($N_2 + N + N_+$ + electrons) flow, at a velocity of the order of 250 m/s, on a steel flat plate parallel to the stream. The plasma, at a temperature on the order of 2000°C, is not in equilibrium and, practically, there is no chemical reaction in the free stream and in the boundary layer. However, in contact with the wall, maintained at a temperature on the order of 500°C, it is known by experience that a catalysis occurs, the initial reaction being monoatomic. The mechanism of the catalysis is probably that described by Rideal. An atom of gaseous nitrogen is adsorbed and another impinging gaseous atom, striking the adsorbed one, creates the molecular nitrogen, the adsorption sites on the surface being always saturated in adsorbed atoms [2]:

$$N + S \rightarrow N_{ads} \tag{a}$$

$$N + N_{ads} \rightarrow N_{2ads} \tag{b}$$

$$N_{2ads} \rightarrow N_2 + S \tag{c}$$

The initial monoatomic reaction occurs only forward; therefore the law of chemical kinetics with $C_N$ standing for the concentration in monoatomic nitrogen, at time *t* is:

$$-\frac{dC_n}{dt} = k\, C_n, \qquad (13)$$

where $k$ is the kinetic reaction constant. Integration leads us to consider $k$ the reciprocal of a time constant:

$$\tau = \frac{C_n}{dC_n/dt} \qquad (14)$$

On the plate, the time constant is very small due to catalysis; in the boundary layer, on the contrary, it is very high, the kinetic reaction constant being practically zero.

If instead of a steel plate, a pyrex plate is placed in the plasma jet, a catalysis does not occur.

More generally, let us deal with a *chemically reacting boundary layer* where, therefore, *mass diffusion phenomena occur*.

The diffusion of species $i$ can be characterized by a time $\tau_m$, which will be called simply »*mechanical time*«: it *represents* the time it takes species $i$ to go through the boundary layer:

$$\tau_m = \frac{\delta}{V_i} \qquad (15)$$

$\delta$ is the boundary layer thickness and $V_i$ the average diffusion velocity of species $i$ through the boundary layer.

Let $C_i$ be the concentration of species $i$, at a given point of the boundary layer, whereas the temperature at this point would lead to an equilibrium concentration $C_{ie}$. A «*chemical time*« can be defined by:

$$\tau_c = \frac{C_{ie} - C_i}{dC_i/dt}. \qquad (16)$$

It is the ratio of the shift from equilibrium over the reaction rate which effects the return to equilibrium. In the particular case of a first order forward reaction, $\tau_c$ is no more than the time constant defined in Eq. (14).

Consideration of these two characteristic times allows a classification of the different types of flows.

If $\tau_c \ll \tau_m$, the reaction is very fast; the *boundary layer is in equilibrium at every point*.

If $\tau_c \gg \tau_m$ the reaction does not have time to occur in the flow, the boundary layer is frozen. This is the case of fast catalysis at the wall, dealt with previously.

If $\tau_c$ and $\tau_m$ are of the same order of magnitude, *relaxation phenomena occur in the boundary layer*.

The importance of boundary layer profile concentration investigation is then understandable; it not only yields the constitution of the boundary layer, but furthermore it determines the type of flow and the reaction mechanisms which occur.

## 3. MEASUREMENTS BY SAMPLING

*The sampling probe* is, most often, a cylindrical tube, the axis of which is tangent to the stream. A drop in pressure in the tube, due generally to pumping, carries the gas sample to the *analyzer* placed outside of the stream.

Sampling methods have serious drawbacks: the insertion of the probe perturbs the stream and therefore the *analyzed sample is not necessarily that which would exist, at the center of the probe aperture is, without the probe.*

In order to minimize the perturbations due to the insertion of the probe, a number of measures must be taken.

a) Of course, *the probe must be as small as possible* [3]. Indeed, the amount of perturbation is related to the ratio of the tube diameter to the boundary layer thickness and this ratio can thus be kept small by increasing the boundary layer thickness: in particular, this is why the above mentioned experiments of Lassau were carried out at low pressures.

b) The *tube axis must be as parallel as possible to the flow direction* which existed before insertion of the probe, otherwise a distortion of the velocity field would be superimposed on the perturbation of the density streamlines due to this insertion.

c) The perturbation of the velocity along the outside wall of the sampling tube will be as small as possible if the sucking rate of the probe is equal to that which crossed the section of the tube before its insertion into the boundary layer (*isokinetic probe*). Thus we do not disturb the diffusion flux and chemical reaction rate too much in the vicinity of the tube and where the sampling is performed.

d) We must ensure that *the probe is not a heat sink or source* in the boundary layer and that thermal equilibrium must be nearly reached between the probe wall and surrounding fluid: a heat source would change, not only the temperature distribution, but also the concentration gradient due to a change in the chemical reaction rate.

e) *The probe wall must have no chemical effect on the gas* (catalysis for example in the case of a non-equilibrium gas); furthermore it must not undergo any change (fusion, sublimation, etc).

f) Finally *the sample, while being sucked, must not be submitted to any change*; if a chemical reaction occurs within the tube, the transfer time must be as small a fraction as possible of the chemical time; should absorption occur, it must be kept as small as possible by avoiding the use of metals (teflon walls, apertures drilled in sapphire plates) and by using a steady flow allowing the analysis to be carried out only after absorption equilibrium is reached [4].

We will deal more closely with the last condition, very specific to the problem of interest, which takes on great importance in the case of a sample not in chemical equilibrium.

Let us deal, for example, with boundary layer experiments in a nitrogen plasma out of equilibrium. The probe has a quartz tube: this material has a very low catalytic reaction constant for the transformation of atomic nitrogen; it can be assumed that it does not change the concentration gradient in the boundary layer much. However, the transfer of atomic nitrogen through the tube up to the analyzer lasts long enough for an appreciable catalytic action to occur and, upon arrival in the analyzer, we are dealing with a different sample. The solution is to transform the atomic nitrogen into a stable species before any transfer for analysis.

It is known that hydrocarbons react with activated nitrogen to give cyanhydric acid, traces of $CN$ radicals and cyanogen gas $C_2N_2$ [5,6]. With methane, which was the hydrocarbon used, the main reaction is:

$$N + CH_4 \rightarrow HCN + 3/2\, H_2 \tag{d}$$

The reaction rate is high enough for the reaction to be considered instantaneous [7]. The idea is to react the atomic nitrogen with the hydrocarbon, then to transport the stable products the analyzer.

The gases mix at the entrance of the probe and are sucked in by a vacuum pump, the analyzer is linked to this vacuum pump. The flow rate of methane compared to the rate of nitrogen must be such that chemical reaction (d) is complete, but there is evidently no point in diluting the sample in methane too much since the analyzer will yield more accurate measurements if the fractions of nitrogen and cyanhydric acid in the mixture are high. The methane flow rate is then increased

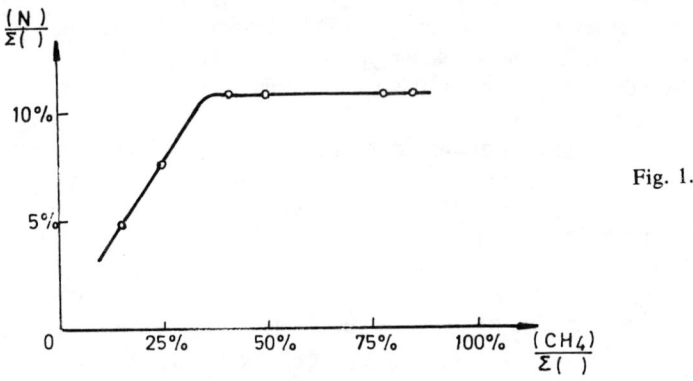

Fig. 1.

progressively: analysis shows that the fraction of atomic nitrogen detected increases proportionally to the rate of flow of methane until it becomes constant (Fig. 1). At this time the transformation of atomic nitrogen is complete; the results of the analysis corresponding to the minimum flow rate of methane allowing the complete transformation of $N$ into $HCN$ are retained.

This is given as an example but there are, as will be seen later, other cases *where a fast auxiliary chemical reaction can freeze the state of the mixture* when a reaction of the mixture components between the probe and the analyzer is feared.

## 4. DIFFERENT TYPES OF REMOTE ANALYZERS

The intention here isn't to give a lecture on chemical analysis. Different techniques will nevertheless be mentioned, because they are used in the problems of interest.

### 4.1. MASS SPECTROMETER

This is the apparatus most often used in our laboratory, the analysis being carried out at very low pressure, between $10^{-5}$ and $10^{-11}$ torr. It is of particular interest to our laboratory, where most of the analyses are carried out in rarefied gas flows.

The molecules of the mixture, introduced into a very low pressure container (Fig. 2), are ionized, for example by electron bombardment,

$$M + e \rightarrow 2e + M^+ \tag{e}$$

The ions thus created flow through a set of apertures in plates raised to electric potentials such that focusing of these ions occurs. The ion beam then enters an element which allows separation. These separators might involve either a combination of electric and magnetic fields as in the Omegatron, or grids raised to alternating electric potentials as in the Topatron, or a quadripolar tube, the electrodes

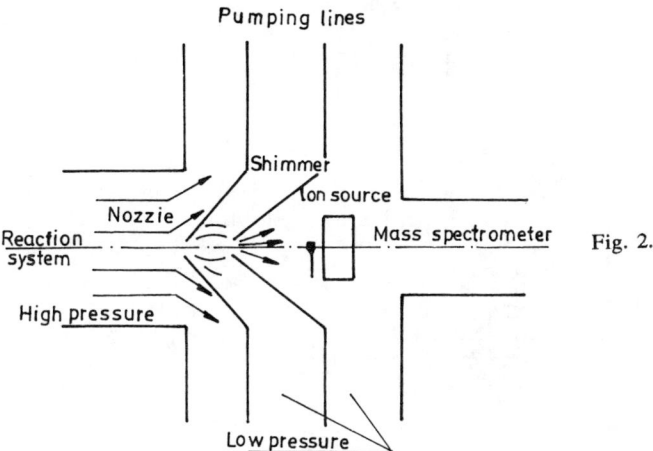

Fig. 2.

of which are subjected to a variable frequency electric potential, as in the Varian. In the latter case, which in our opinion, is the system best suited to the measurements of interest, for a given frequency, only those electrons corresponding to a mass/electric charge ($M/q$) ratio within a narrow band have a regular sinusoidal trajectory, whereas the others oscillate transversally with an increasing amplitude until they reach the walls of the quadripole, allowing their deionization and elimination. The ions, in resonance with the quadripole frequency (Fig. 3) reach an electron multiplier leading finally to a recording device. By changing the frequency, all values of $M/q$ are scanned; those corresponding to ions in the beam create peaks of a height proportional to the molar concentrations (Fig. 4).

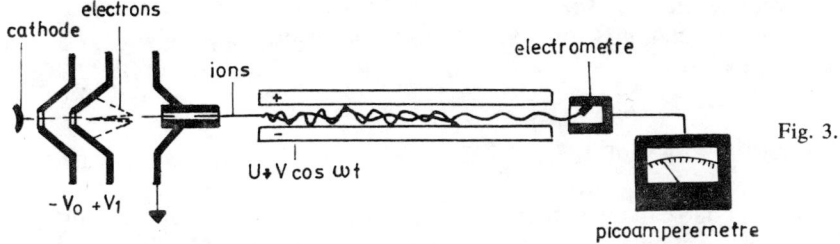

Fig. 3.

The resolution of a mass spectrometer is defined by the ratio $M/\Delta M$, where $M$ is the ion mass and $\Delta M$ the shift in mass corresponding to the peak width, measured at half height. In the range of molar mass between 1 and 50, the resolution is on the order of 50, whereas for higher molar masses it is on the order of 20.

The sensitivity is very high and a pressure of $10^{-14}$ torr of a component such as nitrogen can be detected.

The mass spectrometer is currently commonly used for the investigation of many gas reactions [8, 3]: its use for the determination of concentration profiles in a boundary layer, with or without chemical reactions, seems to be less frequent. Provided the necessary care is taken at sampling, the method yields, in most cases, good results which, in our laboratory, may be checked by other methods.

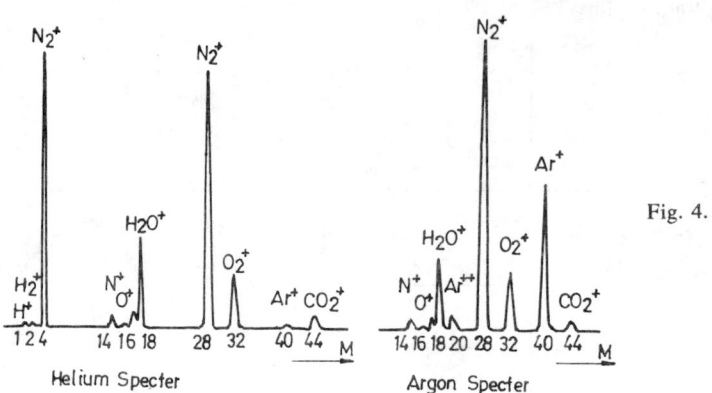

Fig. 4.

## 4.2. CHROMATOGRAPHY

The sample can of course be analyzed by chromatography. It is a method commonly used by chemists. However, in the present case, it is less interesting than the method based on the mass spectrometer and this for several reasons.

First, the pressure in the analyzer must be higher than 5 mm Hg and the mass flow rate in the chromatometer must be considerably higher than in the mass spectrometer; this may, in some cases, create large perturbations (case of free diffusion).

Second, the response time is long, considerably longer than in a mass spectrometer.

Finally, the chromatometer must be adjusted prior to work on mixture components, whereas the mass spectrometer determines all the components while scanning and, in particular, it is very useful in the analysis of gases containing atoms, free radicals, etc.

## 4.3. OTHER TYPES OF ANALYZERS

There are of course many types of analyzers based on different physical properties, but all have the above mentioned drawbacks of chromatography: high flow rates, lengthy experiments (response time on the order of a minute), specific components.

a) The *infrared radiation absorption analyzer* is based on the fact that many gases ($CO$, $CO_2$, hydrocarbons, $SO_2$, $NH_3$) absorb in the infrared: an absorbing component weakens, the infrared radiation of a given wave – length, which is characteristic of the gas and independent of its temperature and pressure, the weakening being proportional to its concentration.

The measurement is differential, as the energies involved are low. For example, two tubes of equal length contain one (1) dry pure air (reference tube), the other (2) the gaseous mixture with the component $i$ to be measured. This mixture flows through tube (2) (Fig. 5). The same infrared radiation passes through both tubes

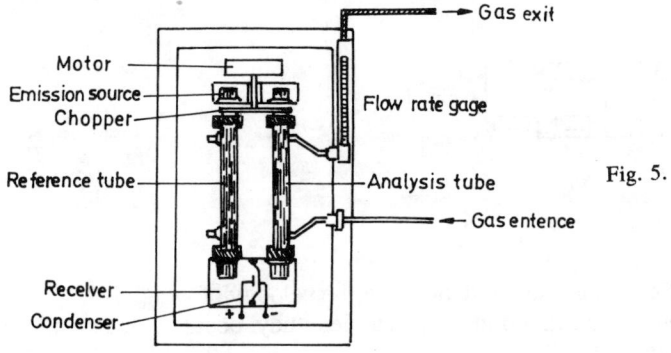

Fig. 5.

and it then enters two identical chambers, separated by a thin membrane, containing the component $i$ to be analyzed. The gas in the chambers will absorb the radiation characteristic of component $i$. As the emission is interrupted five times a second by a rotating chopper, the temperatures and, consequently, the pressures in two chambers are modulated at the frequency of the chopper. With the membrane, one measures the difference in instantaneous pressure which is nearly proportional to the selective weakening that the radiation undergoes in tube (2) where the mixture containing the component to be measured flows.

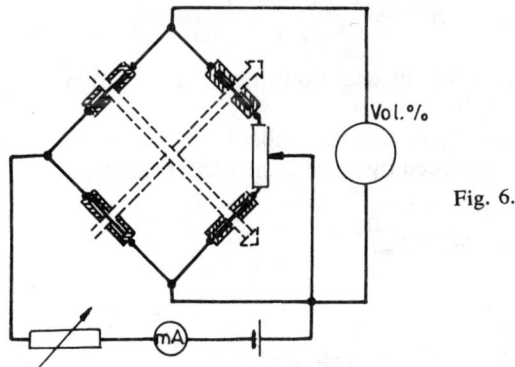

Fig. 6.

b) The *catharometer*, an analyzer based on the difference of the thermal conductivities of gases, is also a differential apparatus: four wires, whose resistance varies with temperature make up a Wheatstone bridge and are positioned in such a way that two opposite wires are swept by the gas mixture to be studied, the other two by the reference gas (Fig. 6). The difference in thermal conductivity between the gas mixture and the reference gas is enough to generate, in the bridge diagonal, a current proportional to the concentration to be measured. In order to protect against corrosion and catalysis, the measuring wires are inserted in perfectly sealed glass capillaries.

c) In the case of oxygen a *magnetic analyzer* can be employed, paramagnetism being a specific property of this element. In the apparatus in Fig. 7, due to the

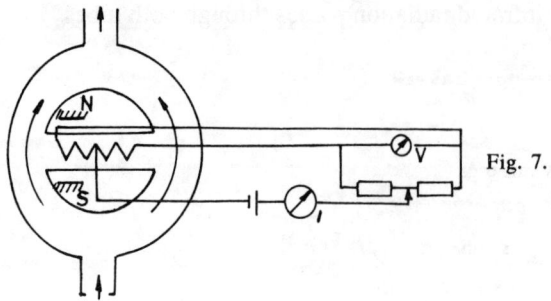

Fig. 7.

heterogeneous magnetic field a thermomagnetic flow, the velocity of which is proportional to the oxygen concentration in the gas under study, occurs in the small electrically heated transversal tube.

## 5. IN SITU MEASUREMENTS, WITHOUT PROBE IN THE BOUNDARY LAYER

These measurements are obviously the best, particularly if they do not require the introduction of gas into the boundary layer. Optical methods are usually used and thus they cannot, unfortunately, be applied in all cases.

### 5.1. INTENSITY OF EMISSION RAYS

This method is only applicable to two dimensional flow and for a fairly high temperature gas mixture (plasma for example).

The observation plane in the boundary layer is both parallel to the stream and normal to the wall (Fig. 8). The image is produced by lense L on a spectrometer slit

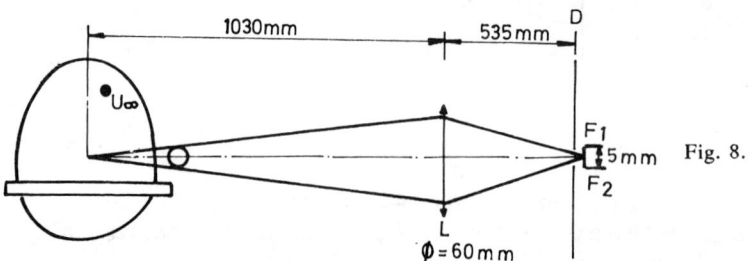

Fig. 8.

which is parallel to the observation plane and normal to the wall. The spectral rays of the component in the gas mixture appear on the plate. With a special microdensitometer, the intensity is determined at each point of each ray. As this intensity is proportional to the concentration of component $i$ of the mixture, the concentration is directly obtained as a function of distance from the plate.

The optics of the apparatus must be such that:

a) The entire thickness of the boundary layer is seen on the spectrometer slit.

b) Not only the observation plane, but practically every point of the boundary layer which illuminates the slit can be assumed practically at infinity and this requires that the width of the boundary layer on one the hand and the focal distance of the lense on the other hand be small compared to the distance of the lense to the middle of the boundary layer.

c) However, as the spectrometer integrates all the light coming out of the boundary layer, its width must be large enough compared to the part perturbed by the lateral edges of the plate.

This method is particularly suited to the study of plasmas [9]. It allows us to determine the electron density distribution up to the near vicinity of the wall in the boundary layer along flat walls and also atom and ion concentrations. However the interpretation of the spectra is accurate only for a boundary layer in equilibrium.

Instead of an optical spectrum, a Raman spectrum, or the electronic paramagnetic resonance, can also be used.

## 5.2 ABSORPTION OF A LIGHT BEAM

This method is again appropriate only for two dimensional flow.

A cylindrical light beam, parallel to the wall and normal to the stream, goes through a boundary layer; the light beam diameter is small compared to the boundary layer thickness. If the component $i$ of the gas mixture absorbs a light ray or an infrared band, the variation of the optical absorption, proportional to the concentration of component $i$, is investigated as a function of the distance to the wall.

The method is harder to apply than emission methods as the beam has to stay rigorously parallel while going through the boundary layer.

As in the case of emission, the length that the light beam goes through must be large enough for lateral edge effects to be negligible.

This method has been used in particular in the infrared for the investigation of $OH$ radical concentrations.

## 6. IN SITU MEASUREMENTS WITH PROBE IN THE BOUNDARY LAYER

### 6.1. DETERMINATION BY PHOTOCHEMISTRY

#### 6.1.1. $N$ Determination

We have already mentioned the complete reaction of *atomic nitrogen and methane*, with the formation of component $HCN$ the molar concentration of which is proportional to that of $N$. This reaction is luminous and gives out a blue light due to the $CN$ bands [10]. If the $CN$ concentration is high enough, the radiated energy, is proportional to the atomic nitrogen concentration [11].

The probe used is a quartz tube, its axis normal to the wall, where the methane comes out. In order to measure the luminous intensity of the reaction, the image of the area in the vicinity of the probe aperture is formed on a photomultiplier.

However, it must be noticed that the light emitted throughout the plasma located on the system optical axis is also integrated; the measurement will be accurate only if, in the radiation received by the photomultiplier, that part coming out of the chemical reaction dominates; in other words, the ratio of the intensity $I$ obtained with injection over the intensity $I_0$ without injection, must be large; it can almost reach 6 by a proper choice of the spectral region employed. The relative concentration of atomic nitrogen is then given by:

$$\frac{I - I_0}{I_\infty - I_{0\infty}}, \qquad (17)$$

where $I_\infty$ and $I_{0\infty}$ are the reference luminous intensities, for example, the intensities outside the boundary layer.

### 6.1.2. $O$ Determination

When a small quantity of nitric oxyde $NO$ is added to atomic oxygen, yellow–green light is emitted; it is due to the set of slow reactions.

$$O + NO + M \rightarrow NO_2^* + M \qquad (f)$$

$$\begin{cases} NO_2^* + M \rightarrow NO_2 + M & (g) \\ NO_2^* \rightarrow NO_2 + h\nu & (h) \end{cases}$$

$$\overline{O + NO \rightarrow NO_2 + h\nu}$$

but atomic oxygen reacts rapidly with $NO_2$

$$O + NO_2 \rightarrow NO + O_2 \qquad (i)$$

in such a way that nitric oxide is regenerated. Therefore, for small quantities of nitric oxyde, the light emitted during the above set of reactions is proportional to the atomic oxygen concentration [12]. If the mixture is sucked into a tube after $NO$ addition, a weakening of the luminous intensity $I$ is observed along the tube axis ($x$ axis), such that:

$$-\frac{dI}{dx} = -\frac{dC_0}{dx} \qquad (18)$$

In order to determine the absolute atomic oxygen concentration $C_0$ at a given point along the tube and, consequently, in any other point, it is sufficient to add nitric oxide in progressively increasing quantity until light disappears at the injection point; the molar concentration in $O$ is then, according to reaction (i) equal to that of $NO_2$, which is known if the flow rate of $NO_2$ is measured.

### 6.1.3. *H* determination

A similar method allows us to measure the molar concentration of atomic hydrogen with nitric oxyde, according to the set of slow reactions that produce red light.

$$H + NO + M \rightarrow HNO^* + M \tag{j}$$
$$\begin{cases} M + HNO^* \rightarrow HNO + M & \text{(k)} \\ HNO^* \rightarrow HNO + h\nu & \text{(l)} \end{cases}$$

Here again, if the quantity of nitric oxyde is low, the reaction rate is proportional to the molar concentration in atomic *H* and nitric oxyde is regenerated through the fast reaction.

$$H + HNO \rightarrow H_2 + NO. \tag{m}$$

### 6.2. DETERMINATION BY MICROCALORIMETRY

In principle, microcalorimeter inserted in the flow, may be used to measure the amount of heat produced in a complete chemical reaction which leads to the disappearance of one of the components of the mixture [13]. In this case, only equilibrium are dealt with. We will consider here only the case of catalytic reactions, for which the probes are small and consequently the suitable for use in boundary layers [14,15]. Even in this case, the size of the probes is too large for them to be used in the near vicinity of the wall, but we will see that, in the case of the thickened boundary layer investigated by Lassau, the microcalorimetric probe has given results in accordance with those of other concurrently concurrently used methods. The principle of this method is easy to understand.

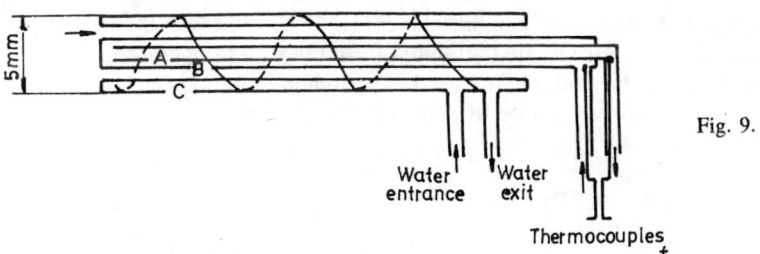

Fig. 9.

The heat of dissociation of nitrogen $N_2$ is produced when the nitrogen *N* recombines. Thus with a sample of the mixture taken at a point of the boundary layer, and the amount of heat due to this recombination by catalysis on the walls, yield the concentration of atomic nitrogen in the mixture.

An isokinetic probe is used for sampling (Fig.9): a given mass flow rate of gas is sucked through the annular tube of the probe, and the gas transmits heat to two water flows, one inside, the other external. By measuring the water and sucked gas flow rates, as well as the entrance and exit temperatures of the water in the two pipes, with or without pumping in the probe, the heat flux received by the walls of the annular tube can be obtained from the difference. Or course, the heat fluxes transmitted to the water through other mechanisms (convection, radiation) must be taken into account via various corrections.

### 6.3. COMPARISON OF THE METHODS

The three methods used by Lassau, mass spectrometer, sampling measurements, in situ photochemistry measurement and in situ measurement by microcalorimetry, gave, as shown on the curves of Fig. 10, very consistent results: the elevations represent the relative concentrations. The kinematic viscosities, proportional to the distance to the wall, if expressed in terms of the Reynolds number are computed

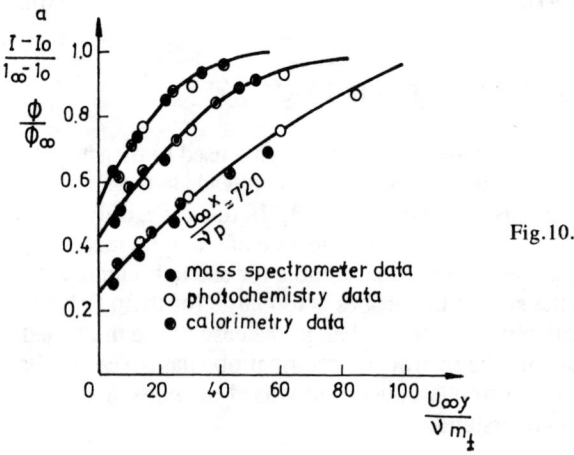

Fig.10.

from the average temperature of the boundary layer (1200°C), whereas those defining the Reynolds numbers of the curves are computed from the wall temperature.

In conclusion, it may be said that the used experimental methods are suitable, with good precision, to the investigation of concentration profiles.

## 7. RESULTS AND APPLICATIONS

I could not end this lecture without giving some results and applications of the measurements.

### 7.1. WALL CATALYSIS OF A FROZEN MIXTURE $N+N_2$

In Fig. 10, in the vicinity of the wall, the concentration profiles are practically straight; this quasilinearity is demonstrated by boundary layer theory [16]. It is then easy to obtain, the relative wall concentration by extrapolating these profiles as defined, for example, in Eq. (17). It is found then that the mass flux at the wall $(da/dy)_p$ is proportional to the wall concentration $a_p$ (Fig. 11), indicating that the reaction of atomic nitrogen disappearance is first order, as was said previously. In Fig. 12 the elevation represents the dimensionless number $da/dRy$, where $Ry = V_\infty y/\nu$.

The knowledge of the mass flux and the wall concentration allows us to determine the values of the catalysis rate constant which is about 45 m/s on a steel plate, under these experimental conditions.

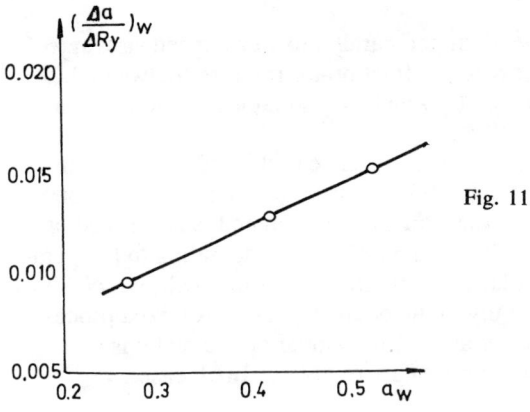

Fig. 11.

## 7.2. INVESTIGATION OF AN ARGON PLASMA

The work of Valentin [9] on argon plasmas allowed him to obtain, using in situ continuous background luminous intensity, measurement the electron concentration distribution curve $C_e/C_{emax}$ versus the distance normal to the plate at a distance

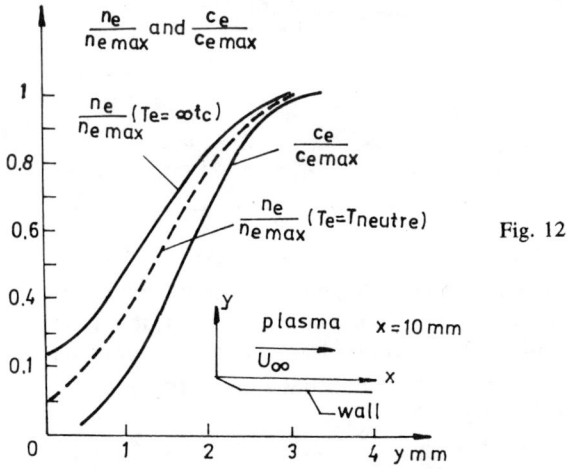

Fig. 12.

from the leading edge of 10 cm (Fig. 12). The value $C_{emax}$ is the free stream electron density of $2 \cdot 10^{15}$ $e/cm^3$. The full line is for a constant electron temperature; the dotted line corresponds to an electron temperature equal in every point to the temperature of the neutrals. The molar fraction distribution curve is also presented in the figure, assuming the electron temperature equal to the temperature of the neutrals at every point.

### 7.3. HYDROGEN COMBUSITON

A platinum plate is placed in a wind tunnel, parallel to a low speed stream (6.5 m/s for example), of a cold hydrogen low molar fraction-air mixture (between 0.02 and 0.06). The flow is laminar along the plate, which is maintained at a temperature between 300 and 1500°C [17] by Joule effect.

The curves of relative molar fraction versus distance to the wall are given in Fig. 13. If the wall temperature is lower than 600°C (curve 1), a laminar diffusion profile is observed: the hydrogen goes to the wall by diffusion and is consumed only on the wall by catalytic wall combustion. Curves 3 and 4, corresponding to temperatures higher than 900°C, have a nul molar fraction gradient at the wall, which shows that the hydrogen combustion occurs totally in the boundary layer. A mixed process (a fraction consumed in the boundary layer and the remainder on the plate) is displayed as curve 2, corresponding to a temperature of 880°C. In all cases, combustion is always complete.

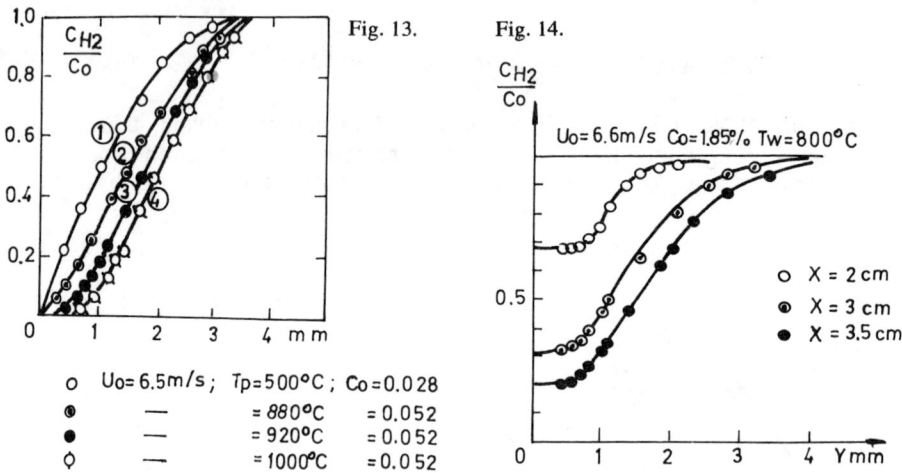

Fig. 13.　Fig. 14.

If the wall is non–catalytic (platinum plate coated with aluminium oxyde or gold), in the case where the reaction is incomplete in the boundary layer, it will not go to completion at the wall. Figure 14 shows the molar fraction profiles for different abscissae for a temperature of 800°C, a speed of 6.6 m/s and a molar fraction of 0.0185. The mass flux density is zero at the wall.

### 7.4. METHANE COMBUSTION

The results obtained for methane combustion in the boundary layer are not entirely analogous to those obtained for hydrogen combustion [18]. Added to the combustion in the homogeneous phase and catalytic wall combustion, there seems to be catalytic combustion in the homogeneous phase probably due to platinum evaporation products [19]. This is shown by the fact that the degree of homogeneous phase combustion depends on the nature of the plate, all other factors held constant (Fig. 15). The plate temperature is such (1320°C) that the reaction is complete in the boundary layer, whatever the nature of the plate.

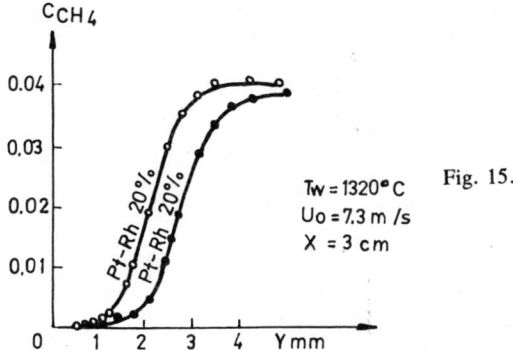

Fig. 15.

## 7.5. WALL INJECTION

Investigations of the concentration profile in boundary layers are of interest not only when chemical reactions occur close to the wall, but also when there is wall sublimation, evaporation or injection. As an example, I will mention Cornil's ex-

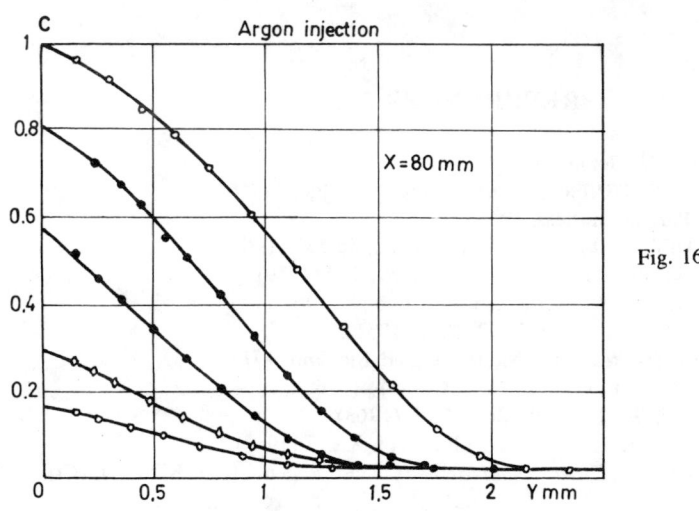

Fig. 16.

periments [20] on argon and helium injection through a flat plate immersed in a hypersonic (M = 5) air stream (Figs. 16, 17).

The dynamic or diffusion boundary layer thicknesses increase with injection:

for argon injection, the dynamic thickness increases form 1.5 to 2.5 when the injection rate $\rho_p v_p/\rho_e u_e$ increases up to $1.58 \cdot 10^{-3}$ and the diffusion thickness increases from 1.2 to 1.6 for argon and from 1.5 to 2.5 for helium as the injection rate changes from 0.11 to $0.495 \cdot 10^{-3}$.

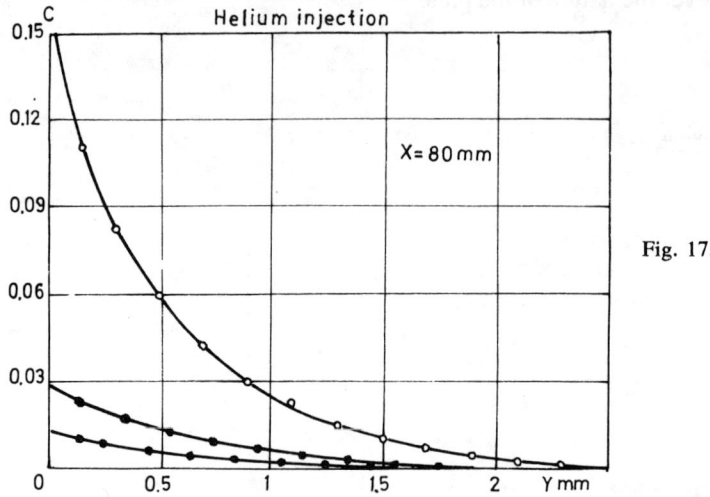

Fig. 17.

It can be seen that, at the same injection rate, the boundary layer diffusion thickness is larger for helium than for argon; on the contrary, the mass fractions at the wall are smaller in the case of helium.

## REFERENCES

1. LASSAU, Chimie et Industrie, Genie chimique, **100**, 7 (1968).
2. LE GOFF, CASSUTO and PENTENERO, Industrie chimique belge, 4 (9164).
3. TINE: Agardograph 47, Pergamon Press, 1961.
4. FRISTROM, GRUNFELDER and FAVIN, J. Phys. Chem., **65**, 587 (1961).
5. HERRON, FRANKLIN and BRADT, Canad. J. Chem., **37**, 579 (1959)
6. FONTIJN, ROSNER and KURZIUS, Aerochem. Res. Lab., **8** (1962).
7. BLADES and WINKLER, Canad. J. Chem., **29**, 1022 (1951).
8. STEVENSON, *Ion−Molecule−Reactions*, Mc Dowell, ed. Mc Graw−Hill, 1963.
9. VALENTIN, PIAR and LACASE, J. Phys., **29**, 4, C.3−44 (1968).
10. YOUNG and SHARPLESS, J. Chem. Physics, **39**, 4, (1963).
11. VAN TIGGELEN and FEUGIER, Revue de l'I.F.P., vol. XX, 7 (1965).
12. MELVILLE and GOWENLOCK, *Experimental Methods in Gas Reactions*, Mc Millan and Co, 248, 1964.
13. AU and SPRENGEL, Z. Flugwiss, **14**, 4 (1966).
14. HAENIG, A.R.S. J., **29**, 5 (1959).
15. ROSNER, AIAA J., **2**, 4, (1964).
16. LASSAU, C.R. Acad. Sci., **261**, 4617 (1965).

17. VALENTIN, Annales Phys. (1961).
18. CABANNES and VALENTIN, Bull. Société chimique de France, 166 (1962).
19. DEVORE, EYRAUD and PRETTRE: C.R. Acad. Sci., **248**, 1227 (1958), and **248**, 2345 (1959).
20. CORNIL, C.R. Acad. Sci., session. July 16, 1969.

International Seminar 1970

# HEAT AND MASS TRANSFER IN RHEOLOGICALLY COMPLEX FLUIDS

W. R. Schowalter

*Princeton University, Princeton, USA*

In the nearly two decades since the 1970 International Seminar in Herceg Novi on "Heat and Mass Transfer in Rheologically Complex Fluids", activity in the field has intensified and substantial progress has been made. Several of the Herceg Novi papers addressed issues which were foreseen as critical to our fundamental understanding of transport processes in rheologically complex fluids. Events of subsequent years have shown that the participants of the seminar exhibited remarkably clean insights of future needs. The rheology of suspensions is a case in point. This subject might be considered to have been the source from which modern activities in constitutive modeling have evolved. I am thinking of course of the work of Einstein in which he predicted the viscosity of a dilute suspension of rigid spheres. Fundamental studies of the rheology of suspensions after the pioneering work of Einstein, Jeffery, Frohlich and Sack, and others, were relatively infrequent until approximately 1970. The paper at the Herceg Novi seminar by Brenner (1) sets out some of the important questions which, since 1970, have been the stimulus of the role of rotational Brownian motion on bulk stress than was available at the time of Brenner's paper. We also know much more about geometric and physicochemical effects, both of which are addressed by Brenner.

Several of the seminar papers dealt with flow instabilities and/or secondary flows. The paper by Hayes and Hutton [2] continues to be one of the best sources for documentation of the analogue of Taylor instability in non-Newtonian fluids. That work is nicely balanced by Giesekus' [3] description of instabilities in other flow geometries, such as free jets issuing from axisymmetric or plane channels containing grooved walls. Vinogradov [4] documented the flowcurve discontinuity that can take place during conversion to a "high elastic state". This conversion is known to be related to the possibility of loss of adherence between a polymer and the wall of a conduit through which the polymer is flowing. The paper by Shulman, et al, [5] illustrates the special instabilities that complex rheology confers on mass transfer in rotating fluid systems.

During the past two decades there has been and increased reliance by scientists on scaling laws as a means for organization and explanation of physical phenomena. Such an activity is a branch of the general subject of dimensional analysis. An example of the interplay between dimensional analysis, mathematics, and physics is demonstrated in the paper by Pearson [6], who has shown how, through proper scaling, one has a hope of performing useful analyses of complicated problems in which close coupling exists between momentum and heat transfer.

There are areas of intense current rheological activity that were not clearly in evidence at Herceg Novi. One of these is computational fluid dynamics. To be sure, a paper by Spalding [7] contains algorithms for computation of transport coefficients in turbulent pipe flow of non-Newtonian fluids. However, the last decade has seen and enormous growth in interest, parallel to the development of computer capacity, in overcoming limits imposed by elastic behavior on the flow rate at which numerical schemes fail to converge. This "high Weissenberg-number problem" is alluded to in the context of heat transfer in boundary layers by Astarita and Nicodemo [8].

Because heat transfer was an important part of the seminar, several papers dealt with one or more aspects of thermodynamics, an example being the brief communication by Gorodtsov and Leonov [9]. Thermodynamics as it applies to dynamic processes of non–Newtonian fluids is a subject which has not progressed in the years since the Herceg Novi seminar as much as one might have hoped. The subject, requiring a foundation in general nonequilibrium thermodynamics, is difficult. Continuum thermodynamics has attracted the attention of several superbly qualified researchers. However, the connection with coupled transport problems of engineering interest has yet to be made in a way comparable to the connection of engineering problems with continuum mechanics. I hope that twenty years hence ans of server will be able to recount the blossoming of the thermodynamics as it applied to transport problems in rheologically complex fluids.

## REFERENCES

1. H. BRENNER, Suspension rheology. Progress in Heat and Mass Transfer, vol 5, pp. 89-129, W. R. SCHOWALTER, W. J. MINKOWYCZ, A. V. LUIKOV, AND N. H. AFGAN, eds., Pegamon Press, Oxford, 1972.
2. J. E. HAYES and E. F. HUTTON, The effect of very dilute polymer solutions on the formation of Taylor vortices. Comparison of theory with experiment. ibid., pp. 195-209.
3. H. GIESEKUS, On instabilities in Poiseuille and Couterre flows of viscoelastic fluids. ibid., pp. 187-193.
4. G. V. VINOGRADOV, Flow, high elasticity and relaxation characteristecs of polymer systems. ibid., pp. 51-71.
5. Z.P. SHULMAN, N. A. POKRYVAILO, V. I. KORDONSKII and E. B. KABERDINA, Rheodynamics and mass transfer in rotating flows of anomalous-viscous fluids. ibid., pp. 177-185.
6. J. R. A. PEARSON, Heat-transfer effects in flowing polymers. ibid., pp. 73-87.
7. D. B. SPALDING, A midel and calculation procedure for the friction and heat transfer behaviour of dilute polymer solutions in turbulent pipe flow. ibid., pp. 275-284.
8. G. ASTARITA and L. NICODEMO, Transport phenomena in turbulent flow of rheologically complex fluids. ibid.,pp. 37-50.
9. V. A. GORODTSOV and A. I. LEONOV, Non-equilibrium thermodynamics and rheology of viscoelestc fluids.ibid., pp. 173-176.

# SUSPENSION RHEOLOGY

H. Brenner

*Carnegie–Mellon University, Pittsburgh, U.S.A*

## 1. INTRODUCTION

The science of rheology is concerned with establishing constitutive relations connecting the dynamical and kinematical responses of materials. It deals, in essence, with general relations existing between the states of stress and strain in a substance undergoing deformation. Paradoxically, the great generality of these constitutive relationships represents both the principal strength and weakness of the science. The strength resides in the fact that the fundamental precepts underlying the theory of rheological constitutive equations are so broadly applicable as to be capable of embodying virtually any conceivable type of behavior, from the elementary to the bizarre. The weakness stems from the fact that the general theory by itself, being purely formal, furnishes no insight into which particular constitutive relation applies to a given substance. This poses a dilemma for the experimentalist who seeks, in an objective and unbiased fashion, to embed his data into a particular constitutive formulation. How is he to choose, especially in the initial phases of his search, among the myriad rheological formalisms that have been proposed?

In this sense, modern rheological theories tend to be sterile, at least in proportion to the heavy theoretical investment that has entered into their creation and subsequent development. A non–professional rheologist[*] like myself cannot help but feel stifled, indeed bewildered, by the endless array of contravariant and tensor affices that characterize the formal aspects of the discipline. The successful application of such theories to real substances depends, in large measure, upon the ability to supplement these formal rheological schemes with simple, but reasonably detailed, physical models of the phenomena, into which are incorporated the principal structural features of the substance under investigation. It is only through the use of such models that one can hope to understand and interpret the rich variety of rheological responses manifested by various classes of materials. It is only through the use of such models that one can hope to understand and interpret the rich variety of rheological responses manifested by various classes of materials. These models may merely constitute analogies as, for example, in the case of Maxwellian or

---

[*] I must also confess to being a professional non – rheologist.

Kelvin–Voigt spring–dashpot models of viscoelastic fluids or elasticoviscous solids. In other cases the model may represent an attempt at a more faithful portrayal of the detailed geometrical microstructure characterizing the system. Such is the case with the rheology of suspensions, the subject of this lecture.

Suspension rheology is a highly developed discipline in terms of the quantity and quality of knowledge currently available in the field. This is equally true of both the experimental and theoretical facets of the subject. Apart from the pioneering work of Einstein [1,2], concerned essentially with developing a kinetic theory of the liquid state [3], the impetus for the rapid development of the subject over the past thirty years came about largely from the needs of polymer science [4–8], where the "suspended" particles are macromolecular in size. Other applications of interest exist in the areas of emulsion rheology, slurry transport in pipelines, paint technology, ferrofluid rheology, blood flow, and a host of other areas involving non–Newtonian technology. In a survey lecture of the type demanded by the heterogeneous nature of the audience in attendance at this symposium, attention will be directed only to the fundamental, theoretical aspects of the subject. An equivalent version of essentially this same lecture will be published elsewhere [9].

## 2. GENERAL VIEW OF A SUSPENSION AS A CONTINUUM

### 2.1. CHARACTERISTIC LENGTH SCALES

Consider a suspension of identical, neutrally buoyant particles dispersed in an incompressible Newtonian fluid that is being sheared. The smallest linear dimension of a suspended particle is assumed to be large compared with intermolecular distances within the homogeneous fluid. Such a particle may therefore be regarded as a macroscopic object bathed in a fluid continuum. Its motion is thereby governed by the usual laws of viscous hydrodynamics. Three important length scales are associated with the suspension: (i) $a$, the maximum linear dimension of a particle; (ii) $l$, the average distance between the centers of adjacent particles; (iii) $L$, the length scale over which the bulk properties of the suspension change appreciably. Typically, $L$ is a linear dimension of the apparatus in which the rheological experiment is being performed, e.g. the distance between the plates of a Couette-flow apparatus.

Provided that $l/L \ll 1$, the fluid - particle mixture may be regarded from a sufficiently coarse viewpoint as a heterogeneous continuum.* This composite continuum will possess rheological properties that differ from those of either the homogeneous carrier fluid (the "solvent") or the suspended particles (the "solute"). Theories of suspension rheology aim at predicting these rheological properties from knowledge of the separate properties of the solute and solvent. This is basically a joint problem in hydrodynamics and statistical mechanics. Statistics enters by virtue of the fact that the detailed, instantaneous, geometric configuration of the suspension

---

* That is, an observer performing experimental measurements with instruments that record "local" phenomena on a length scale that is large compared with $l$ (but small compared with $L$) may be expected to observe essentially continuous changes in instrumental readings as the device is moved about to neighboring positions within the flowing suspension. A porous medium furnishes another example of a heterogenous continuum.

will not generally be reproduced in replicate experiments performed on the same system. Indeed, the "microscopic" behavior of the suspension in a single experiment is necessarily time dependent (due to changes in the relative positions of the suspended particles occasioned by the deformation of the suspension), even for a suspension undergoing a macroscopically steady mean flow.

Our main concern will be with dilute suspensions, for it is in this area that the greatest progress towards achieving the aims of this rheological program has been realized. By a dilute system is meant one in which hydrodynamic interactions among the suspended particles are negligible. This idealization permits attention to be focused on the motion of a single, isolated particle within the suspension. Thereby, one essentially by–passes the profound statistical problems implicit in the more general theory of concentrated suspensions. The criterion that the system be dilute is that the average distance between particles be large compared with their maximum linear dimension:

$$a/l \ll 1. \tag{1}$$

Equivalently, the volume fraction $\phi$ of suspended particles must be small compared with unity:

$$\phi \ll 1. \tag{2}$$

## 2.2. EINSTEIN'S RESULT FOR A DILUTE SUSPENSION OF SPHERICAL PARTICLES

The classical result in this field is Einstein's formula [1,2] for the viscosity $\mu$ of a dilute suspension of rigid spheres uniformly dispersed in an incompressible Newtonian fluid of viscosity $\mu_0$:

$$\mu = \mu_0(1 + \frac{5}{2}\phi) \tag{3}$$

That is, on a length scale which is large compared with the mean distance between suspended particles, the average (or bulk) deviatoric stress $\overline{t}_{ij}$ in the suspension is linearly related to the average rate-of-strain in the suspension, $\overline{d}_{ij} = \frac{1}{2}(\overline{v}_{i,j} + \overline{v}_{j,i})$, by the relation

$$\overline{t}_{ij} = 2\mu \, \overline{d}_{ij} \tag{4}$$

This is the rheological equation of state of an incompressible Newtonian fluid possessing a shear viscosity $\mu$.

Einstein's analysis of the problem is based upon a computation of the additional rate of mechanical energy dissipation engendered by the introduction of a single sphere into a homogeneous shear flow. Since energy is a scalar quantity, such energy-dissipation techniques are generally incapable of furnishing the complete tensorial properties of suspensions, except when it is known a priori that the suspension is Newtonian (and isotropic). Accordingly, more modern theories [10–19] approach the subject via dynamic, rather than energetic, methods.

## 2.3. RELATION BETWEEN BULK AND LOCAL FIELDS

The central statistical–mechanical problem in theoretical suspension rheology is that of establishing the general relationships existing between the ensemble–average (macroscopic) fields defined in the fluid interstices and in the interiors of the particles [20]. This fundamental problem has never been solved in a manner that is wholly convincing from the point of view of rigor. The usual approach, and the one that we shall follow here, is to choose a representative volume element of volume $V$ whose typical linear dimension, $V^{1/3}$, is small compared with the apparatus length scale $L$, but large compared with the interparticle length scale $l$. The latter insures that this volume element contains a statistically significant number of particles.

The bulk or average fields are than defined essentially to be volume averages of the corresponding local fields over this representative volume element. The domain over which this averaging is to be performed includes both the interstitial fluid and particle interiors. For example, the average velocity gradient $\overline{v}_{i,j}$ is defined as the volume average of the local velocity gradient $v_{i,j}$:

$$\overline{v}_{i,j} = \frac{1}{V} \int_V v_{i,j} \, dV \tag{5}$$

in which $dV$ is a microscopic element of volume.

The mean stress (or pressure) tensor $\overline{p}_{i,j}$ is, however, defined in a slightly different manner [11, 12, 16, 17] as the first surface moment of the local stress tensor* $p_{i,j}$ integrated over the area $A$ bounding the representative volume element externally:

$$\overline{p}_{i,j} = \frac{1}{V} \int_A x_j p_{ik} n_k \, dA \tag{6}$$

in which $n_k$ is a unit outer normal to the volume, $dA$ is an element of microscopic surface area, and $x_j$ is the local position vector of a point on the surface $A$ relative to the centroid of the volume element. In the absence of interfacial tensions this definition of the bulk stress can be shown to be identical to the volume average of the local stress [16,17]. In general, however, it is not, unless explicit recognition be given to the fact that the volume integral of the local stress over a region containing an interfacial surface is improper, and must therefore be appropriately interpreted [16]. The definition of stress adopted in Eqn. (6) is "energy-preserving" in the sense that the rate $\dot{E}$ at which mechanical energy is being dissipated, defined as

$$\dot{E} = \frac{1}{V} \int_V p_{ij} d_{ij} \, dV \tag{7}$$

may be shown to be [16,21]

$$\dot{E} = \overline{p}_{i,j} \, \overline{d}_{i,j}. \tag{8}$$

Thus, the average of the product is the product of the averages.

---

* This is related to the deviatoric stress $t_{ij}$ by the expression
$p_{ij} = -\delta_{ij} p + t_{ij}$
where $p$ is the pressure.

## 2.4. LOCAL FIELDS IN DILUTE SYSTEMS

In dilute systems we focus attention on the behavior of an isolated particle suspended in the undisturbed shear flow

$$\vec{v}^o = \vec{R} \cdot \vec{G}, \quad p^o = \text{const.} \tag{9}$$

where $\vec{R}$ is the position vector and $\vec{G}$ is the undisturbed velocity gradient dyadic. Moreover, in place of the representative volume over which integrations of the local fields are to be performed to find the corresponding bulk fields, we choose instead a large sphere of volume $V = v_p/\phi$ containing the particle at its center. Here, $v_p$ is the volume of a single particle. Physically, $V$ is the superficial volume associated with a single particle in the suspension; that is, it is the reciprocal of the number of particles per unit volume. As $\phi$ tends to zero, $V$ tends to infinity. It will be assumed henceforth that all the particles in the suspension are identical in size, shape, and other relevant physical properties.

A rigid suspended particle may undergo both translational and rotational motions. The translational velocity vector of an arbitrary point $O$ fixed in the particle will be denoted by $\vec{U}_o$ and its angular velocity vector by $\vec{\Omega}$. Upon neglecting the inertia of the fluid one has therefore to solve the quasistatic Stokes (creeping motion) equations,*

$$\nabla^2 \vec{v} = \frac{1}{\mu_0} \nabla p \tag{10}$$

and continuity equation

$$\nabla \cdot \vec{v} = 0 \tag{11}$$

in the essentially infinite fluid region external to the particle, so as to satisfy the boundary condition

$$\vec{v} = \vec{U}_o + \vec{\Omega} \times \vec{r} \tag{12}$$

on the particle surface, where $\vec{r}$ is a local position vector measured from $O$. Furthermore, the disturbance due to the presence of the particle must vanish at a sufficiently large distance from the particle. Thus, the local fields must asymptotically tend to the undisturbed shear flow at infinity:

$$\vec{v} \sim \vec{v}^o, \quad p \sim p^o \quad \text{as } r \to \infty \tag{13}$$

The undisturbed shear flow can be expressed as the sum of translational, rotational, and deformational contributions:

$$\vec{v}^o = \vec{v}^o_O + \vec{\omega}^o \times \vec{r} + \vec{d}^o \times \vec{r} \tag{14}$$

where $\vec{v}^o_O$ is the translational approach velocity to the center $O$ of the particle, $\vec{\omega}^o$ is half the vorticity of the undisturbed flow, and $\vec{d}^o$ is the undisturbed rate of strain dyadic.

---
*Inertial effects are discussed by Batchelor [16].

For a specified orientation $\vec{e}$, say, of the particle relative to the principal axes of strain of the undisturbed flow, Eqns. (10)–(14) lead to a unique solution for the perturbation fields $\vec{v} - \vec{v}^\circ$ and $p - p^\circ$ in terms of the three parameters $U_O - v_O^0$, $\Omega - \omega^\circ$, $\vec{d}^\circ$ and the particle orientation $\vec{e}$. In turn, this permits the local stress

$$p_{ij} = -\delta_{ij} p + \mu_o (v_{i,j} + v_{j,i}) \tag{15}$$

and, ultimately, the hydrodynamic force and torque exerted on the particle to be calculated in terms of these same three parameters. The condition that the particle experience no net force or couple then uniquely determines the unknown slip velocities $\vec{U}_o - \vec{v}_o^\circ$ and $\vec{\Omega} - \vec{\omega}^\circ$ of the particle relative to the surrounding fluid in terms of the remaining parameter $\vec{d}^\circ$, and the orientation $\vec{e}$.

## 2.5. ORIENTATION – SPECIFIC BULK FIELDS

Each specified orientation of the particle yields different values for the local fields $v_i$, $p$ and $p_{ij}$ at a given point in the fluid. In discussing the pertinence of particle orientation as an independent variable, we shall for simplicity restrict attention to bodies of revolution, as in Fig. 1.

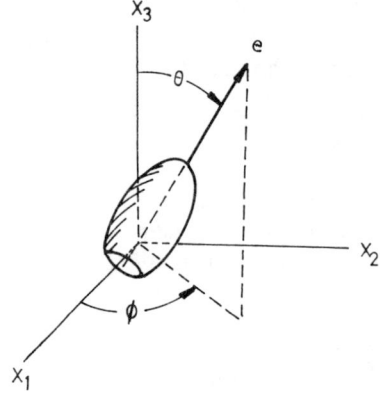

Fig. 1. Unit orientation vector for a body of revolution.

Consider a unit vector $\vec{e}$ locked into the axisymmetric particle, lying along its symmetry axis. This unit vector may be regarded as a unit radial vector in a system of spherical–polar coordinates, $(r, \theta, \phi)$. Knowledge of this vector $\vec{e}$ is therefore equivalent to knowledge of the angles $\theta$ and $\phi$, and vice versa, relative to some system of Cartesian axes fixed in space, e.g. the principal axes of strain of the undisturbed flow. Hence we may write $\vec{e} \equiv \vec{e}(\theta, \phi)$. The unit vector $\vec{e}$ may therefore be regarded as specifying the instantaneous orientation of the particle [22]. It will, for this reason, be termed the orientational vector of the particle.

As has been pointed out, the local fields $v_i$ and $p$ depend, *inter alia*, on particle orientation. They are, therefore, implicit functions of $\vec{e}$ :

$$p_{ij} \equiv p_{ij}(\vec{e}), \quad d_{ij} \equiv d_{ij}(\vec{e}). \tag{16}$$

Integration of these fields over the sphercial domain $V$ yields the bulk or average fields, $\overline{p}_{ij}(\vec{e})$ and $\overline{d}_{ij}(\vec{e})$, specific to particles of this orientation:

$$\overline{p}_{ij}(\vec{e}) = \frac{1}{V}\int_A x_j p_{ik}(\vec{e})n_k dA, \tag{17}$$

$$\overline{d}_{ij}(\vec{e}) = \frac{1}{V}\int_V d_{ij}(\vec{e})dV. \tag{18}$$

If all of the particles possessed precisely the same orientation $\vec{e}$, then these averages would be the true bulk values for the suspension as a whole. Inasmuch as the suspended particles do not, however, generally possess the same orientation, the distribution of particle orientations must be taken into account. This is done by introducing an orientational distribution function.

### 2.6. ORIENTATIONAL DISTRIBUTION FUNCTION

The (instantaneous) orientational distribution function $f(\vec{e})$ is defined such that if the scalar $d\vec{e}$ is a differential element of solid angle centered about the orientation $\vec{e}$ (Fig.2), then $f(\vec{e})d\vec{e}$ gives the fraction of the total number of

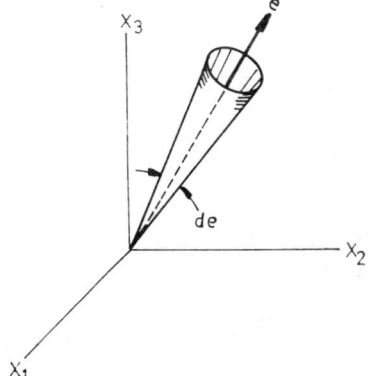

Fig. 2. Orientations lying in the range between $\vec{e}$ and $\vec{e} + d\vec{e}$.

particles in the suspension possessing orientations lying between $\vec{e}$ and $\vec{e} + d\vec{e}$ (at time $t$)– that is, the fraction of particles contained within the right circular cone of apex angle $d\vec{e}$ centered about the orientation $\vec{e}$. This distribution is normalized to unity by integrating over all orientations:

$$\int_{\vec{e}} f(\vec{e})d\vec{e} = 1 \tag{19}$$

and therefore represents a probability density with respect to orientation.

## 2.7. GENERAL RHEOLOGICAL CONSTITUTIVE EQUATION FOR A DILUTE SUSPENSION

The bulk value $\overline{\Psi}$ of any intensive property $\psi$ associated with the suspension may thereby be obtained by multiplying the corresponding orientation–specific bulk property $\overline{\Psi}(\vec{e})$ by the probability $f(\vec{e})d\vec{e}$ of a particle possessing that orientation, and subsequently integrating over all orientations. In this manner we have that

$$\overline{p}_{ij} = \int_{\vec{e}} \overline{p}_{ij}(\vec{e}) f(\vec{e}) d\vec{e} \tag{20}$$

and

$$\overline{d}_{ij} = \int_{\vec{e}} \overline{d}_{ij}(\vec{e}) f(\vec{e}) d\vec{e} \tag{21}$$

This procedure ultimately yields an expression for the symmetric* part, $\overline{t}_{ij}^{(s)}$, of the bulk deviatoric stress,** $\overline{t}_{ij}$, in the suspension in terms of the undisturbed rate of strain dyadic $\vec{d}^{\circ}$:

$$\overline{t}_{ij}^{(s)} = \text{function}(\vec{d}^{\circ}). \tag{22}$$

The bulk rate of strain may be similarly expressed in terms of this same auxiliary parameter:

$$\overline{d}_{ij} = \text{function}(\vec{d}^{\circ}). \tag{23}$$

By eliminating the parameter $\vec{d}^{\circ}$ between these two relations one eventually arrives at an explicit or implicit relation of the functional form [19]

---

* The antisymmetric part is expressible in terms of the angular velocity defference $\overline{\overline{\Omega}} - \vec{\omega}^{\circ}$, in which $\overline{\overline{\Omega}}$ is the mean angular velocity of the suspended particles [17]. The subject of antisymmetric stresses in suspensions, arising from the action of external couples, will be taken up later.

** The isotropic part of $\overline{p}_{ij}$, i.e. the pressure, is of no interrest in connection with incompressible suspensions. Only the deviatoric stress $\overline{t}_{ij}$, is relevant to rheology.

$$\overline{t}_{ij}^{(s)} = \text{function}(\overline{d}_{ij}).\qquad(24)$$

This relation between the (symmetrized) bulk deviatoric stress and the bulk rate of strain is the rheological constitutive equation for the suspension. Because of the assumption of non–interacting particles, the result is valid only to the first order in the volume fraction $\phi$ of suspended particles.

## 2.8. EFFECT OF ROTATIONAL BROWNIAN MOTION UPON THE ORIENTATIONAL DISTRIBUTION FUNCTION

Suspended particles of colloidal or macromolecular dimensions are sufficiently small to suffer appreciable rotary Brownian movements. This phenomenon refers to random fluctuations in the orientation of a particle due to thermal agitation. Rotary Brownian motion enters into the theory of suspension rheology via two distinct mechanisms [19].

Firstly, the rotational Brownian motion affects the orientational distribution function. The latter satisfies a conservation equation

$$\frac{\partial f}{\partial t} + \frac{\partial}{\partial \vec{e}} \cdot \vec{j} = 0 \qquad(25)$$

analogous to that for the concentration of a conserved chemical species in physical space transport phenomena. Here, $\vec{j}$ is the rotary flux vector and $\partial/\partial \vec{e}$ is the orientation space gradient operator [22]. In turn, the rotary flux vector for rigid particles may be written as the sum of diffusive and convective contributions:

$$\vec{j} = -D_r \frac{\partial f}{\partial \vec{e}} + f\,\dot{\vec{e}}. \qquad(26)$$

The leading term in Eqn. (26) is the Fickian rotary diffusion flux, stemming from the Brownian motion. The term $\partial f/\partial \vec{e}$ represents the orientational gradient, which constitutes the driving force for rotational diffusion. It vanishes only for a random distribution of orientations, $f = \text{constant}\ (\equiv \frac{1}{4}\pi)$. The phenomenological coefficient $D_r$ is the rotary diffusion coefficient for rotation about an axis perpendicular to the symmetry axis of the body. It is given in terms of the particle geometry and fluid properties by the Stokes–Einstein equation [23]. For example, in the case of a spherical particle of radius $a$,

$$D_r = \frac{kT}{8\pi\mu_0 a^3} \qquad(27)$$

in which $k$ is the Boltzman constant and $T$ the absolute temperature.

The second term in Eqn. (26) is the rotational convection flux, arising from hydrodynamic rotation $\vec{\Omega}$ of the suspended particles in response to the shear. The vector $\dot{\vec{e}} \equiv d\vec{e}/dt$ is the time rate of change of the orientation vector $\vec{e}$ due to particle rotation, as seen by an observer fixed in space. By elementary rigid-body kinematics it is related to the hydrodynamic angular velocity $\vec{\Omega}$ of a particle by the expression

$$\dot{\vec{e}} = \vec{\Omega} \times \vec{e} \qquad (28)$$

Since, for a specified orientation $\vec{e}$, the angular velocity $\vec{\Omega}$ is known from the condition of no net force and couple on the particle, Eqn. (28) furnishes the "the velocity" $\dot{\vec{e}}$ as a function of the instantaneous orientation $\vec{e}$ of the particle.

On substituting Eqn. (26) into (25) we obtain the following second-order, partial-differential, diffusion-convection equation governing the instantaneous distribution of orientations:

$$\frac{\partial f}{\partial t} + \nabla_e \cdot (f\, \vec{\Omega} \times \vec{e}) = D_r \nabla_e^2 f \qquad (29)$$

in which $\nabla_e = \partial/\partial \vec{e}$ and $\nabla_e^2 = (\partial/\partial \vec{e}) \cdot (\partial/\partial \vec{e})$. The only "boundary conditions" imposed upon $f$ are that it be positive, single-valued, finite, and that it satisfy the normalization condition (19). For a specified initial distribution of orientations at time $t=0$, Eqn. (29) may be solved to obtain

$$f \equiv f(\vec{e}, t;\ \vec{d}^\circ,\ \vec{\omega}^\circ)$$

For sufficiently long times, $t \to \infty$, one obtains a unique, steady-state distribution that is independent of the initial orientational distribution. If $G$ is a characteristic shear rate, then $\vec{\omega}^\circ$ and $\vec{d}^\circ$ are of order $G$. The general character of steady-state orientational distribution is thus governed in large measure by the magnitude of the rotary Péclet number [24].

$$Pé = G/D_r \qquad (30)$$

This dimensionless number is a global measure of the relative importance of convective transport compared with moleculars (diffusive) transport.

## 2.9. DIRECT CONTRIBUTION OF ROTARY BROWNIAN MOTION TO THE BULK STRESS

The second effect of the rotary Brownian motion upon the rheology of suspensions stems from the fact that it contributes directly to the orientation-specific bulk stress. This occurs by virtue of the fact that the Brownian motion induces an angular velocity given by [11, 19, 25]

$$\vec{\Omega}^{Br} = -D_r\, \vec{e} \times \frac{\partial \ln f}{\partial \vec{e}} \qquad (31)$$

This angular velocity contributes to the stress in exactly the same manner as the hydrodynamic angular velocity, $\vec{\Omega}$, of the particle; that is, in place of $\vec{\Omega}$ in the preceeding theory,* one now employs $\vec{\Omega} + \vec{\Omega}^{Br}$. This has its origin in the fact that

---

* However, in Eqn. [29] one still retains only the hydrodynamic angular velocity $\vec{\Omega}$. Alternatively, one may write Eqn. [29] in the form

(in the absence of external couples acting upon the suspended particles) the sum of the hydrodynamic couple $\vec{L}$ and Brownian couple $\vec{L}^{Br}$ acting upon a suspended particle is zero [8]:

$$\vec{L} + \vec{L}^{Br} = 0 \tag{32}$$

in which, for a particle of revolution [25],

$$\vec{L}^{Br} = -\vec{e} \times \frac{\partial V}{\partial \vec{e}} \tag{33}$$

where $V \equiv V(\vec{e})$ is the orientational chemical potential per particle:

$$V = kT \ln f + \text{const.} \tag{34}$$

Whereas in the absence of Brownian motion one has that $\vec{L} = \vec{0}$, Eqn. (32) applies to the more general case where Brownian motion is included. Failure to include the Brownian couple or, equivalently, to include the Brownian angular velocity given by Eqn. (31) has led to a number of errors in the literature [16,26,27] – especially for the case where the particles are randomly oriented; for it does not follow from Eqn. (31) that $\vec{\Omega}^{Br} = \vec{0}$ even when the particles are randomly oriented. This point is discussed by Scheraga [28] and Saito and Sugita [29]. When the Brownian motion is dominant, the solution of Eqn. (29) can be expressed as a power series in the Péclet number [cf. eqn. (30)]:

$$f(\vec{e}) = f_0 + \text{Pé} f_1(\vec{e}) + O(\text{Pé})^2 \tag{35}$$

in which $f_0 = \text{const.} \equiv \frac{1}{4}\pi$, corresponding to a random distribution of orientations, and $f_1 = O(1)$ with respect to Pé. Substitution into Eqn. (31) therefore yields, as $D_r \to \infty$,

$$\vec{\Omega}^{Br} \sim -G 4\pi \vec{e} \times \frac{\partial f_1}{\partial \vec{e}} = O(G). \tag{36}$$

in which

$$\frac{\partial f}{\partial t} + \nabla_e \cdot (f \vec{\Lambda} \times \vec{e}) = 0$$

in which

$$\vec{\Lambda} = \vec{\Omega} + \vec{\Omega}^{Br}$$

is the total angular velocity of a suspended particle. That this is identical to Eqn. [29] may be seen by writing

$$\vec{j} = f \vec{e} *$$

where

$$\vec{e} * \stackrel{\text{def}}{=} \vec{\Lambda} \times \vec{e}.$$

The equivalence of theis expression for $\vec{j}$ with the rotary flux vector defined in Eqn. [26] may be demonstrated by noting that $\vec{e} \cdot \vec{e} = 1$ and $\vec{e} \cdot \partial f/\partial \vec{e} = 0$, the latter relation being a consequence of the fact that the operator $\partial/\partial \vec{e}$ possesses no $\vec{e}$ component. Hence, by the vector triple product,

$$f \vec{\Omega}^{Br} \times \vec{e} = f\left(-D_r \vec{e} \times \frac{\partial \ln f}{\partial \vec{e}}\right) \times \vec{e} = -D_r \vec{e} \frac{\partial f}{\partial \vec{e}}.$$

This Brownian angular velocity does not vanish even for a random distribution of orientations, $(D_r = \infty)$. Moreover, being of order $G$, it is of exactly the same order as the hydrodynamic angular velocity $\Omega$, and hence cannot be neglected. The important lesson here is that this aspect of the rotary Brownian motion cannot be ignored. Moreover, computation of its contribution to the stress for a random orientational distribution requires explicit knowledge of the first-order effects of the departure from randomness due to the shear.

Though this point was repeatedly emphasized by Kirkwood and coworkers some twenty years ago [8], it has been overlooked in several recent analyses [16, 27].

## 3. RIGID PARTICLES

### 3.1. SUSPENSIONS OF ELLIPSOIDAL PARTICLES IN SIMPLE SHEAR FLOW

#### 3.1.1. No Brownian motion

The rheology of a suspension of neutrally buoyant ellipsoids of revolution dispersed in a Couette flow (simple shearing flow) was investigated by Jeffery [30] in 1922 for the case of negligible Brownian motion. Neglecting both fluid and particle inertia, each spheroid undergoes a periodic rotation in which the terminus of its $\vec{e}$ vector, locked into the particle, describes an elliptic trajectory relative to an observer translating with the particle [30, 31]. Each particle possesses the same period of rotation $T$ irrespective of its initial orientation. Accordingly, the orientational distribution function $f(\vec{e}, t)$ is a periodic function of time, the period being

$$T = \frac{2\pi}{G}\left(s + \frac{1}{s}\right) \tag{37}$$

in which $G$ is the shear rate (radians per unit time) and

$$s = a_{\parallel}/a_{\perp} \tag{38}$$

is the axis ratio of the ellipsoid; $a_{\parallel}$ and $a_{\perp}$ are, respectively, the lengths of the semiaxes of the symmetry axis and transverse axis.

Jeffery [30] calculated the time-average viscosity of the suspension by utilizing additional-energy dissipation arguments and integrating this instantaneous, orientation dependent quantity over one period. The "viscosity" computed in this manner is not a unique property of the suspension since the particular periodic trajectory traced out by a given particle, and hence its contribution to the energy dissipation, depends upon its initial orientation. Accordingly, this viscosity is a function of the initial distribution of particle orientations, and cannot be regarded as an intrinsic property of the fluid particle suspension.

The idealized concept of an infinitely dilute suspension, in which hydrodynamic particle-particle interactions are absent, is a state that can be only approximately realized experimentally. Small interactions must always exist in real systems. Mason et al. (cf. [31, pp. 193–208]) have performed a sequence of remarkable experiments on the distribution of orbital trajectory parameters for dilute suspensions of both circular rods and disks in Couette flow, wherein they measured by cinematography the instantaneous orientations of each of the many suspended particles. After a sufficient number of individual particle rotations they find that the distribution of orbital parameters approaches a unique, steady-state distribution, which is apparently independent of the initial orientational distribution. Moreover, in their experiments performed with rods at concentrations of $\phi = 3.8 \times 10^{-5}$ and $1.5 \times 10^{-4}$, the steady-state distribution of trajectory parameters was found to be identical in the two cases; however, it changed significantly at a higher concentration ($\phi = 6.1 \times 10^{-4}$). These observations suggest that the orbital-parameter distribution becomes independent of particle concentration as $\phi \to 0$, though the data are insufficiently extensive to render this conclusion unequivocal. Contrary to the isotropic distributional hypothesis of Eisenschitz (cf. [31]), which assumes that the particles are randomly oriented before the onset of motion, and that subsequently each particle moves in a fixed orbit dictated by Jeffery's [30] theory, highly anisotropic distributions were, in fact, observed.

The existence of a unique, concentration-independent distribution of orbital parameters cannot be reconciled with the isolated-particle treatment of Jeffery, nor can it be ascribed to small random effects such as rotational Brownian motion. For the motion of each particle in the suspension was observed to be "reversible"; that is, upon quasistatically reversing the direction of the shear, each particle in the suspension retraced its entire prior trajectory, backwards in time. The mechanism responsible for the existence of a steady-state, orbital distribution must therefore be of a purely mechanical nature. Mason et al. [31] attribute this to hydrodynamic interactions among the suspended particles, but no mechanism is advanced – even qualitatively – to explain why such interaction effects should lead to a unique distribution that is independent of particle concentration, i.e. independent of the average spacing between adjacent particles.

In any event, the apparent existence of a unique, steady-state, distribution of orbit trajectory parameters leads to a correspondingly unique set of rheological properties being assigned to a suspension of spheroidal particles in Couette flow. This contrasts markedly with the non-uniqueness appropriate to Jeffery's single–particle rheological analysis. Methods for computing the rheological properties from the experimentally observed distribution are partially discussed by Mason et al. (cf. [31, pp. 211–14]).

These remarks demonstrate possible theoretical shortcomings of the assumed one–to–one correspondence between infinitely dilute suspensions and isolated–particle behavior in suspensions of non-spherical particles; that is, it is possible for the rheological properties to approach a definite limit as $\phi \to 0$, even though the single–particle analysis does not furnish a unique prediction. Somewhat similar behavior arises in the case where rotary Brownian is considered (see the subsequent section dealing with this topic). There, the limiting result for the case where the rotary diffusion coefficient $D_r$ tends to zero yields a definite, unique result for the rheological properties of a dilute suspension of spheroidal particles in Couette flow. In contrast, the case where $D_r$ is identically zero, corresponding to Jeffery's purely mechanical analysis, fails to furnish a unique rheological result. These show, in circumstances where the single-particle theory fails to yield a definite rheological prediction, that one cannot wholly avoid statistical considerations – even in extremely dilute systems

### 3.1.2. Preferred orientations

It should not be assumed that similar conclusions with respect to the lack of uniqueness associated with isolated–particle analyses (in the absence of rotary Brownian motion) apply to particles of all shapes. Particles other than spheroids do not necessarily undergo a periodic rotation in a Couette flow. Bretherton [32] demonstrated theoretically in 1962 the existence of particles of revolution – other than ellipsoids of revolution – that do not perform a periodic rotation, but rather attain a definite, "preferred" orientation with their symmetry axes directed along the streamlines. That is, regardless of its initial orientation relative to the streamlines, such a particle ultimately attains a unique terminal orientation after sufficient length of time. The steady-state orientational distribution function for a dilute suspension of these particles is thus a Dirac delta function, in which the symmetry axis of every particle in the suspension is aligned along a streamline. This eventually leads to a definite result for the viscosity of a suspension of such bodies, in marked contrast to the comparable behavior of ellipsoidal particles.

Fluid inertial effects may also result in the existence of preferred modes of motion. In Stokes flow, where inertial effects are negligible, a dumbbell-shaped particle behaves exactly like an ellipsoid of revolution, in that its $\vec{e}$ vector undergoes a periodic rotation in a manner dependent upon the initial orientation of the dumbbell. However, when Reynolds number effects are included in the analysis this ceases to be true. In 1968 Harper and Chang [33] showed, both theoretically and experimentally, that, as a result of fluid inertia, a spherical dumbbell suspended in a Couette flow eventually achieved a unique periodic orbit corresponding to a maximum (time–average) energy dissipation rate; that is, irrespective of its initial inclination relative to the streamlines and vortex lines, the dumbbell eventually attains a preferred orbital trajectory. By averaging over one period, this ultimately leads to a definite value for the viscosity of a suspension of such bodies.

That small fluid inertial effects lead to the existence of preferred orbital modes, corresponding to maximum energy dissipation, has also been observed experimentally in the case of circular rods and disks by Karnis et al. [31, 34].

Inertial effects are present in all real systems, however small they may be. If one performs and experiment of sufficient duration, these cumulative inertial effects must ultimately manifest themselves. Theoretical results predicated solely on the hypothesis of a creeping flow must therefore be interpreted in the light of this remark. This suggests the possibility that the short- and long-time rheological properties of a given suspension may be appreciably different. Use of a Deborah number, which is essentially a measure of the duration of the experiment to the relaxation time of the system, may prove necessary in the correlation of data on such systems gleaned over long intervals of time.

### 3.1.3. Brownian motion

Inclusion of the effects of rotary Brownian motion in the analysis for ellipsoidal particles dispersed in a Couette flow completely removes the indeterminacy present in Jeffery's results, leading to unique values for the (steady-state) rheological properties of such suspenions. No matter what may be the initial distribution of particle orientations at zero time, if sufficient time be allowed, the suspension will ultimately attain a unique, steady-state distribution of orientations that is independent of the initial distribution. Such is the effect of the rotary diffusion, however small it may be. The viscosity of a suspension of spheroids in a simple shear flow, allowing for the Brownian motion was first correctly determined by Saito [35] in 1951 up to moderate values of the Péclet number, and for arbitrary values of the Péclet number by Scheraga [28] in 1955.

The steady-state orientational distribution function for this case, obtained by solving Eqn. (29) using Jeffery's [30] expression for the orientation-specific hydrodynamic angular velocity $\vec{\Omega}$ of a spheroid, is of the general form

$$f \equiv f(\vec{e}, s, \text{Pé}). \tag{39}$$

Since $s$ and $D_r$ are intrinsic properties of the suspension, this relationship shows that the distribution of orientations and, consequently, the rheological properties of the suspension are a function of the macroscopic shear rate $G$ characterizing the Couette flow. This gives rise to non–Newtonian behavior of the suspension, in particular to a shear dependent viscosity.

The suspension viscosity $\mu$ is defined for the simple shear flow

$$(\overline{v_1}, \overline{v_2}, \overline{v_3}) = (Gx_2, 0, 0) \tag{40}$$

as the proportionality coefficient in the relation

$$\overline{t}_{21} = \mu G. \tag{41}$$

This leads in the present case to the following expression for the intrinsic or specific viscosity:*

$$[\mu] \stackrel{\text{def.}}{=} \lim_{\phi \to 0} \frac{\mu - \mu_0}{\phi \mu_0} = \text{function}(s, \text{Pé}). \tag{42}$$

Values of this function are tabulated by Scheraga [28].

A plot of the intrinsic viscosity of the suspension vs. the Péclet number, for a fixed axis ratio $s$, as in Fig. 3, illustrates the fact that the viscosity $\mu$ decreases with increasing shear rate $G$. Such suspensions therefore display shear thinning. The slope of the curve is horizontal in the two asymptotic limits, Pé$\to 0$ and Pé$\to \infty$, corresponding to Newtonian behavior at both zero and infinite shear rates. Since Pé $= G/D_r$, the upper limit may also be regarded as arising from the case where the rotary Brownian motion is vanishingly small. That a unique result for the viscosity exists in this latter limit contrasts with the conclusions of Jeffery [30] for the case where $D_r = 0$. This confirms our previous remark; namely, that even an infinitesimal amount of Brownian motion, if allowed to act for a sufficiently long time, will produce a unique, steady-state viscosity. From a mathematical point of view, this peculiar limiting behavior may be traced to the fact that Eqn. (29) is singular in the limit where $D_r = 0$. In this limit the character of the partial differential equation changes from a second-order equation to one of the first-order. The asymptotic theory of such systems is well known in connection with boundary layers. Proper treatment requires use of matched asymptotic–expansion techniques [24].

---
\* The specific viscosity is defined as

$$\mu_{sp} = \frac{\mu - \mu_0}{\phi \mu_0}$$

This is identical to the intrinsic viscosity to the order in $\phi$ for which the analysis is valid.

The property of shear thinning in suspensions of rigid nonspherical particles can be understood physically as follows. The viscous stresses tend to align the particles in such a manner as to alleviate these stresses and, hence, to reduce the

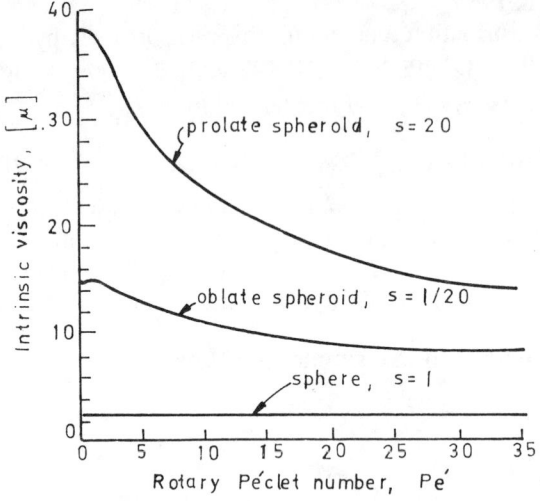

Fig. 3. Intrinsic viscosity vs. rotary Péclet number for prolate and oblate spheroids in simple shear flow. (After Scheraga [28].)

viscosity. In contrast, the rotary Brownian movement tends to randomize the particle orientations, thereby destroying this alignment and increasing the viscosity. It is for these reasons that the shear and the Brownian motion have opposite effects upon the suspension viscosity, as manifested by the fact that both effects can be correlated by the single dimensionless variable, $G/D_r$.

### 3.1.4. Normal stresses

In addition to a shear-dependent viscosity arising from the rotary Brownian motion, the latter also gives rise to normal stresses in a suspension of ellipsoidal particles. These were first calculated* by Giesekus [11] in 1962, at least for small Péclet numbers. Results covering the entire range of Pé appear to be nonexistent at the present time.

These results are most conveniently expressed in terms of the so-called "primary" and "secondary" normal stress differences:

$$\bar{\sigma}_1 \stackrel{\text{def.}}{=} \bar{t}_{11} - \bar{t}_{33} = \phi \frac{\mu_0 G^2}{D_r} F_1(s;|\text{Pé}|), \tag{43}$$

$$\bar{\sigma}_2 \stackrel{\text{def.}}{=} \bar{t}_{22} - \bar{t}_{33} = \phi \frac{\mu_0 G^2}{D_r} F_2(s;|\text{Pé}|), \tag{44}$$

in which $F_1$ and $F_2$ are dimensionless functions of the axis ratio and Péclet number, tabulated by Giesekus. Both are of order unity with respect to Pé for small Pé.

---

* Comparable results for normal stresses in suspensions of dumbbells were available earlier [36-38].

Accordingly, the normal stresses vanish in th limit as $D_r \to \infty$, corresponding to a random distribution of orientations. The normal stresses therefore stem from the nonrandom orientations, rather than from any elastic properties of the particles or carrier fluid.

Since $F_1 \neq F_2$ the normal stress differences are unequal, contrary to what is often assumed.

## 3.2. SUSPENSIONS OF ELLIPSOIDAL PARTICLES IN EXTENSIONAL AND CONTRACTILE FLOWS

### 3.2.1. No Brownian motion. Extensional flow

The previous calculations apply only to spheroidal particles suspended in a simple shear flow. Other rheologically interesting shear flows exist. One which has been rather extensively studied is elongational flow (extensional flow, uniaxial tensile flow, Trouton flow, radial hyperbolic flow). This flow is of special interest in connection with the fiber spinning of polymer solutions.

Elongational flow may be achieved experimentally by drawing out a cylindrical thread of viscous liquid via the application of tensile forces $F$ to the ends of the thread, as in Fig.4. The velocity field generated in this manner is the irrotational,

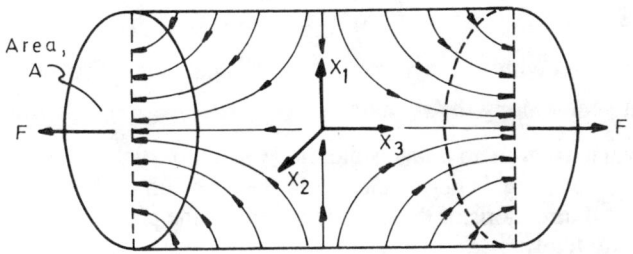

Fig. 4. Streamlines in a meridian plane for an extensional flow.

homogeneous, shear flow [31]

$$(\overline{v}_1, \overline{v}_2, \overline{v}_3) = \left(-\frac{1}{2}Gx_1, -\frac{1}{2}Gx_2, -Gx_3\right) \qquad (45)$$

in which $(x_1, x_2, x_3)$ are Cartesian coordinates measured from the center of the thread, with $x_3$ lying along the axis of tension. The macroscopic deformation- or shear- rate, $G$, is the fractional rate of elongation of the thread in the direction of tension. In the case of Newtonian liquids this is related to the tensile force per unit area, $T = F/A$, and the viscosity $\mu_0$ by Trouton's law:

$$T = 3\mu_0 G. \qquad (46)$$

The streamlines of the axisymmetric flow describe hyperbolas in meridian planes as indicated in Fig. 4.

In the absence of rotary Brownian motion, a prolate spheroid (s > 1), or rod-like particle, whose center is situated at the origin of the flow, $x_1 = 0$, $x_2 = 0$, $x_3 = 0$, ultimately aligns itself with its long axis (symmetry axis) parallel to the direction of tension [39], as shown in Fig. 5a. This stable orientation is achieved independently of the initial orientation of the spheroid. The existence of this preferred orientation is in marked contrast to the behavior of such particles in Couette flow, where they

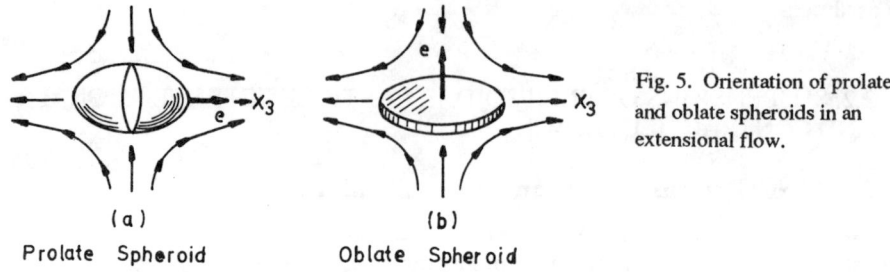

Fig. 5. Orientation of prolate and oblate spheroids in an extensional flow.

(a) Prolate Spheroid  (b) Oblate Spheroid

undergo a periodic rotation. The distinction stems form the fact that Couette flow is rotational, whereas elongation flow is irrotational. The steady–state orientational distribution function in a dilute suspension of prolate spheroids is thus the Dirac delta function

$$f(\vec{e}) = \delta(|\vec{e} \cdot \vec{x}_3| - 1)$$

in which all the particles are aligned with their symmetry axes parallel to the direction of tension. Here, $\vec{x}_3$ is a unit vector along the $x_3$-axis.

In the absence of Brownian motion an oblate spheroid ($s < 1$), or disk–like particle, aligns itself with its symmetry axis *perpendicular* to the direction of tension [39], as illustrated in Fig. 5b. All directions of the symmetry axis in the plane normal to the axis of tension are equally likely.

Because of the preferential alignment of either the prolate or oblate ellipsoids in elongational flow, it might be expected that suspensions composed of these particles would be non–Newtonian in consequence of the prevailing anisotropy extant in such flows. Surprisingly, such is not the case.* Rather, Newtonian behavior obtains in the sense that the bulk deviatoric stress $\overline{t}_{ij}$ is directly proportional to the bulk rate of strain $\overline{d}_{ij}$, the coefficient of proportionality, $2\mu$, being independent of the deformation rate $G$. That is,

$$\overline{t}_{ij} = 2\mu \overline{d}_{ij} \qquad (47)$$

in which

---

*Non-Newtonian behavior does, however, obtain for suspensions of these particles dispersed in the *two-dimensional* elongational flow, $(\overline{v}_1, \overline{v}_2, \overline{v}_3) = (\frac{1}{2}Gx_1, -\frac{1}{2}Gx_2, 0)$, generated within a "four-roller" apparatus [19].

$$[\mu] \equiv \frac{\mu - \mu_0}{\phi \, \mu_0} = \text{function } (s) = \text{const.} \tag{48}$$

The intrinsic viscosity is thus a function solely of the axis ratio, being independent of $G$. Th explicit form of the function in Eqn. (48) is given by Brenner [19]. The comparable result given by Takserman–Krozer and Ziabicke [27] is in error.

### 3.2.2. No Brownian motion. Contractile flow

Closely related to extensional shear flow is contractile shear flow [40], generated by compressing, rather than tensing, the cylindrical thread of viscous fluid, as in Fig. 6.

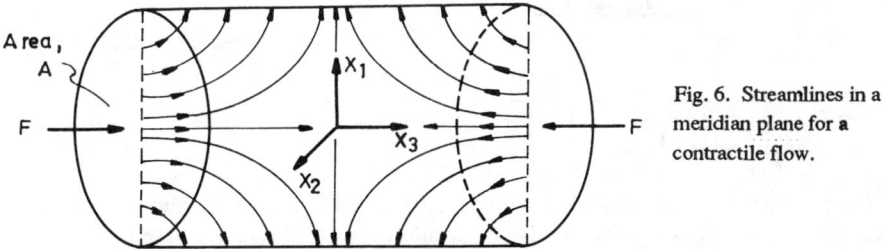

Fig. 6. Streamlines in a meridian plane for a contractile flow.

Contractile flow is kinematically identical to elongational flow, except for the fact that the algebraic sign of the macroscopic deformation rate $G$ is negative in the former case and positive in the latter. This change in algebraic sign has a profound effect upon the orientation of the suspended particles. For now, a prolate spheroid aligns itself with its symmetry axis normal to the direction of compression, while an oblate spheroid sets itself with its axis of revolution along the axis of compression [19] (see Figs. 7a and 7b). This represents a complete reversal of the behavior manifested by prolate and oblate spheroids in an elongational flow.

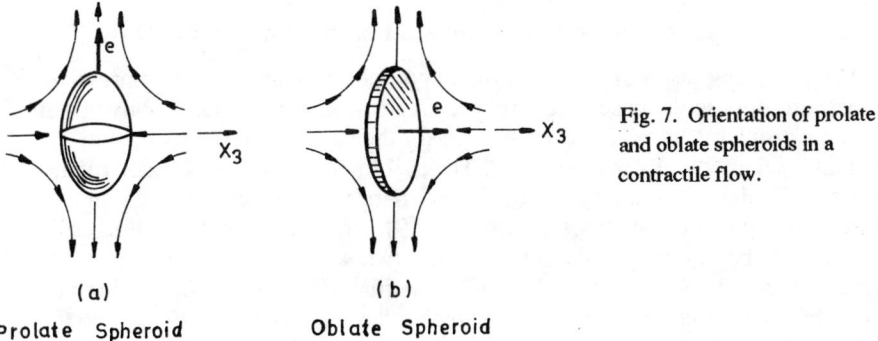

Fig. 7. Orientation of prolate and oblate spheroids in a contractile flow.

Dilute suspension of spheroidal particles in contractile flows are again Newtonian, in the same sense as for extensional flows. However, for a given axis ratio s, the intrinsic viscosity [μ] is different [19] from the comparable value for the elongational case. This is illustrated in Fig. 8, where the intrinsic viscosity is plotted against $G$ for a given particle (fixed $s$). The two distinct constant Newtonian viscosities for contractile flow ($G < 0$) and extensional flow ($G > 0$) are shown schematically.

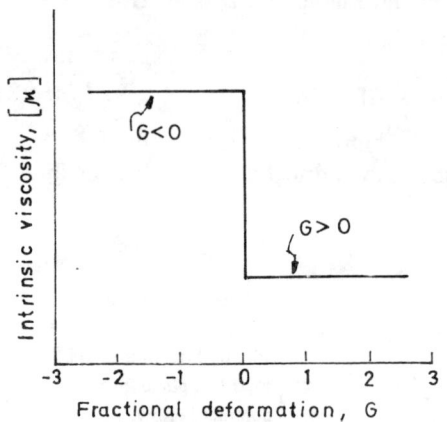

Fig. 8. Intrinsic viscosity vs. rate of deformation for a spheroid in extensional ($G > 0$) and contractile ($G < 0$) flows. No Brownian motion.

Despite the fact that the suspension is separately Newtonian for both positive and negative values of $G$, such a substance would have to be characterized as possessing a shear–dependent viscosity because of the abrupt change in viscosity that occurs in crossing the boundary at $G = 0$. A truly Newtonian fluid would not display such discontinuous behavior, but would, rather, possess the same viscosity for both elongational and contractile flows. The discontinuous change in viscosity at $G = 0$ is due, of course, to the abrupt change in particle orientation as one proceeds from positive to negative values of $G$.

The important lesson to be gleaned from this example is that the *viscosity of a suspension of nonspherical particles is not generally an intrinsic property of the suspension alone*, but depends upon the particular type of shear to which the suspension is being subjected. This comes about because of the dependence of the orientational distribution function upon the particular character of the shear flow.

### 3.2.3. Brownian motion. Extensional and contractile flow.

The preceding discussion of spheroidal particles in tensile and compressive flows pertains only to the case where rotational Brownian motion is wholly absent. These same problems have also been studied for the case where the Brownian movement is taken into account [19, 27]. The rheological analysis of Takserman – Krozer and Ziabicki [27] is based upon scalar energy dissipation methods, in contrast to the dynamical analysis of Brenner [19], which utilizes the techniques described at the beginning of this lecture. These two sets of results are in disagreement due, in part, to the failure of Takserman-Krozer and Ziabicki to take account of the contribution of the Brownian angular velocity [cf. Eqn. (31)] to the bulk stress. *

---

* An alternate dynamical theory of the rheology of a dilute suspension of ellipsoidal particles in elongational flows, including the effect of the rotary Brownian motion, is presented by Pokrovskii [41-

When the rotary diffusion is sensible, the steady-state distribution of orientations is no longer of the Dirac delta function type. Rather, it is given by the exponential distribution [19, 25]

$$f(\theta) = C \exp\left(\frac{3}{4}\frac{BG}{D_r}\cos^2\theta\right) \tag{49}$$

in which $\theta = \cos^{-1}(\vec{e}\cdot\vec{x}_3)$ $(0<\theta<\pi)$ is the polar angle between the symmetry axis of the spheroid and the axis of tension or compression of the flow; $C \equiv C(BG/D_r)$ is a normalization constant arising from Eqn. (19). The dimensionless constant $B$ is a function of the axis ratio:

$$B = \frac{s^2-1}{s^2+1} \tag{50}$$

and thus possesses the properties that

$$B > 0 \quad \text{for prolate spheroids} \tag{51}$$

and

$$B < 0 \quad \text{for oblate spheroids.} \tag{52}$$

For the case of a sphere ($s=1$), $B=0$ corresponding correctly to a random distribution of orientations. Since

$$G > 0 \quad \text{for tensile flow} \tag{53}$$

and

$$G < 0 \quad \text{for compressive flow} \tag{54}$$

all possible combinations of algebraic signs for $B$ and $G$ are thus contained within the distribution function, Eqn. (49). Note that in the absence of flow, $G=0$, Eqn. (49) correctly reduces to the random orintational distribution, $f=$ const.

In the limit where $D_r \to 0$, Eqn. (49) properly reduces to the delta function distributions discussed in previous sections. In contrast to the case of Couette flow, this asymptotic limit, corresponding to Pé $\to \infty$, is nonsingular; that is, the limiting form of the orientational distribution obtained by letting $D_r$ tend to zero in Eqn. (49) is exactly the same as that obtained by ignoring the rotary Brownian at the outset, and treating the orientational question as a deterministic problem involving only purely mechanical forces. This difference in limiting behavior may be traced to the irrotational character of elongational and contractile flows.

From a rheological point of view, the most important feature of Eqn. (49) is that $f$ depends upon the shear rate $G$. Upon performing the rheological calculation described at the beginning of this lecture, the suspension is found [19] to possess the properties of a quasi-Newtonian fluid,

$$\overline{t}_{ij} = 2\mu\,\overline{d}_{ij} \tag{55}$$

with a shear-dependent viscosity $\mu$:

$$[\mu] \equiv \frac{\mu-\mu_0}{\phi\mu_0} = \text{function}\left(\frac{BG}{D_r}\right) \tag{56}$$

---

43]. However, his theory also contains fundamental errors. In particular, when the suspended particles are spherical it yields [43] $\mu = \mu_0(1+\frac{3}{2}\phi)$, rather than the correct Einstein formula, Eqn. [3].

The explicit form of the function in this equation is given by Brenner [19]. Since $B$ and $D_r$ are intrinsic properties of the suspension, this shows that $\mu$ is a function of $G$. Figure 9 is a schematic plot of intrinsic viscosity vs. rotary Péclet number for a spheroid possessing the same axis ratio $s$ as in Fig. 8. The dashed curve corresponds

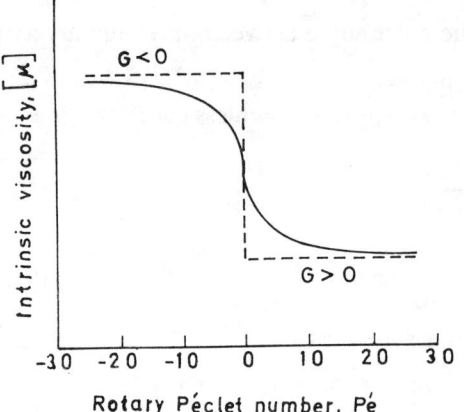

Fig. 9. Intrinsic viscosity vs. rotary Péclet number for a spheroid in extensional ($G > 0$) and contractile ($G < 0$) flows, including the effects of rotational Brownian motion.

to the latter figure, where the Brownian motion is absent. The effect of the rotary diffusion is to smooth out the discontinuity at $G=0$. Figure 9 also reveals more clearly than Fig. 8 the shear–dependence of the viscosity.

It is of interest to compare the case of Couette flow with extensional (or contractile) flow in the limit $|\text{Pé}| = |G|/D_r \to \infty$. For the same values of $\phi$ and $s$, one finds that the viscosity $\mu$ is different for these two different types of shear.* This conclusion once again emphasizes the fact that the viscosity of a suspension of nonspherical particles is not an intrinsic property of the suspension, but depends upon the particular type of shear to which the suspension is being subjected.

## 3.3. SUSPENSIONS OF DUMBBELLS AND LINEAR BEAD CHAINS

### 3.3.1. Burgers–Oseen tensor

In addition to ellipsoidal particles, several other simple rigid–particle shapes have been extensively studied in the literature, especially for the case where rotary Brownian motion is important. For the most part these other shapes have been confined to dumbbells and linear chains of beads [8], consisting of identical spherical beads of radii $a$, equally spaced a distance of $b$ units apart, as in Figs. 10a and 10b. The rigid connection links joining the beads are assumed to offer negligible hydrodynamic resistance compared with that of the beads themselves.

---

* Giesekus' [11] calculation of the normal stresses in a Couette flow are not sufficiently extensive in the limit as Pé$\to\infty$ to determine whether or not the normal stresses vanish in this limit.

The situation for which $b \gg a$ lends itself to especially simple analysis, for in this case hydrodynamic interactions among the beads are negligible. The force on the $i^{th}$ bead may then be regarded as given by Stokes law,

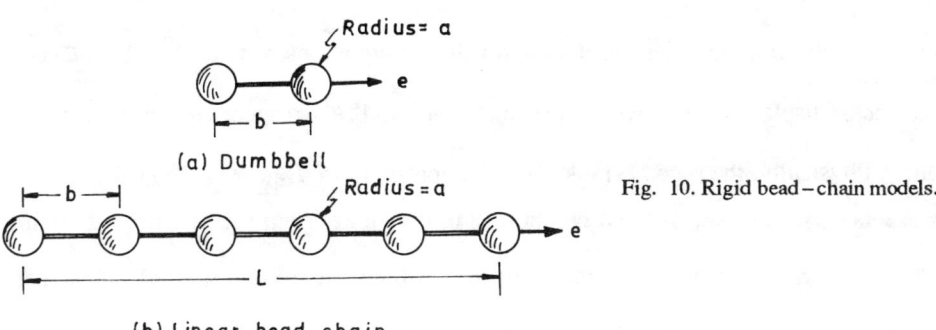

Fig. 10. Rigid bead–chain models.

$$\vec{F^i} = \zeta(\vec{v_o^i} - \vec{U^i}) \quad (i=1,2,...,n) \tag{57}$$

in which

$$\zeta = 6\pi\mu_o a \tag{58}$$

is the "friction constant" for a single bead; $\vec{v_o^i}$ is the undisturbed shear flow $\vec{v^o}$ evaluated at the center of the its bead, $\vec{U^i}$ is the translational velocity of the center of the $i^{th}$ bead, and $n$ is the number of beads in the chain. The case $n=2$ corresponds to a dumbbell.

When $n$ is large, convergence of the sums arising in the theory ** requires that at least first-order interactions among the beads in the array be taken into account [44]. This is normally done by means of the Burgers–Oseen tensor [8], wherein Eqn. (57) is replaced by the relation

$$\vec{F^i} + \zeta \sum_{\substack{j=1 \\ j \neq i}}^{n} \vec{T^{ij}} \cdot \vec{F^j} = \zeta(\vec{v_o^i} - \vec{U^i}) \tag{59}$$

in which

---

** Convergence problems arise from the fact that the Stokes translational velocity field is attenuated at a rate that varies inversely with distance from a particle. This slow arate of decay of the disturbance velocity leads to divergence of the appropriate sums in the limit as the number of beads in the chain becomes large. The first-order perturbation scheme described by Eqns. (59)-(60) is essentially a renormalization scheme designed to avoid this divergence. In the limit where $n \to \infty$ it leads to the appearance of logarithmic terms in the renormalized theory of the form $\ln(L/a)$, in which $L = (n-1)b$ is the total length of the linear chain.

$$\overrightarrow{T}^{ij} = \frac{1}{8\pi\mu_0 |\overrightarrow{R}^{ij}|} \left( \overrightarrow{I} + \frac{\overrightarrow{R}^{ij} \overrightarrow{R}^{ij}}{|\overrightarrow{R}^{ij}|^2} \right) \tag{60}$$

is the Green's function velocity dyadic for Stokes equations, where $\overrightarrow{R}^{ij} = \overrightarrow{R}^i - \overrightarrow{R}^j$ is the vector displacement between the centers, $\overrightarrow{R}^i$ and $\overrightarrow{R}^j$, respectively, of beads $i$ and $j$. Physically, the second-rank Cartesian tensor $T^{ij}_{\alpha\beta} = T_{\alpha\beta}(\overrightarrow{R}^i, \overrightarrow{R}^j)$ (the Burgers–Oseen tensor) is the $\alpha$ component of the Stokes fluid velocity field at point $\overrightarrow{R}^i$ due to the action of an isolated point-force singularity of unit strength situated at point $\overrightarrow{R}^j$ and acting in the $\beta$ direction ($\alpha; \beta = 1,2,3$).

If the superscript $O$ denotes and origin situated at the center of the chain, then

$$\overrightarrow{v}^k_o = \overrightarrow{V} + \overrightarrow{R}^{kO} \cdot \overrightarrow{G} \quad (k=1,2,\ldots,n) \tag{61}$$

and (in the absence of Brownian motion),

$$\overrightarrow{U}^k = \overrightarrow{V} + \overrightarrow{\Omega} \times \overrightarrow{R}^{kO} \quad (k=1,2,\ldots,n) \tag{62}$$

in which $\overrightarrow{V}$ is the translational velocity of the center of the chain; $\overrightarrow{G}$ is the velocity gradient dyadic of the undisturbed shear flow, and $\overrightarrow{\Omega}$ is the angular velocity of the rigid rod. The condition that the rigid chain experience no *net* total hydrodynamic force $\overrightarrow{F}$ or total couple $\overrightarrow{L}$ requires that

$$\overrightarrow{F} = \sum_{k=1}^{n} \overrightarrow{F}^k = \overrightarrow{0} \tag{63}$$

and

$$\overrightarrow{L} = \sum_{k=1}^{n} \overrightarrow{R}^{kO} \times \overrightarrow{F}^k = 0. \tag{64}$$

This system of equations may be solved so as to obtain the translational velocity $\overrightarrow{V}$ of the center of the chain and the instantaneous angular velocity $\overrightarrow{\Omega}$ of the chain as a function of the instantaneous orientation $\overrightarrow{e}$ of the rigid array. Brownian motion effects may be incorporated in essentially the same manner as described for ellipsoidal particles. In the limit where $n \to \infty$, the resulting behavior has been found by Kirkwood and Plock [8] to be essentially identical to that of a prolate spheroid of major semiaxis $L$ and minor semiaxis $a$. Logarithmic terms in the rationa $L/a$ arise as a matter of course in connection with ellipsoidal particles in the limit where $L/a \gg 1$.

Bead models are vastly simpler than corresponding ellipsoidal models. Calculating the force $\vec{F}^k$ on each discrete bead as a function of its relative position $\vec{R}^{kO}$ in the chain is, in effect, completely analogous to calculating the distribution of stresses on the surface of a continuous body, such as an ellipsoid. The advantages of bead models become apparent in cases where the rigid chain is no longer linear, so that its symmetry no longer corresponds to that of an ellipsoid of revolution. One can, in effect, model particles of more-or-less arbitrary shape by choosing appropriate bead configurations. Though analytical methods of solution are normally sought in the limit where $n \to \infty$, involving the replacement of Eqn. (59) by a corresponding integral equation, high–speed digital computers have much to offer in the discrete case.

Appart form the use of bead–chain arrays as models of continuous particles of specified geometry, the former are of interest in their own right in connection with the rheological properties of "stiff" polymer molecules, wherein each bead in the chain is essentially regarded as a single monomer unit.

Despite the success of the so-called Burgers–Oseen perturbation theory at the hands of Kirkwood and co-workers [8], it has recently been point out [45] that the theory may lead to physically impossible conclusions if it is applied to situations in which $b/a$ is too small. Though small values of this parameter are clearly outside the scope of a first-order perturbation theory, they are sometimes of interest in applications.

### 3.3.2. Time–periodic simple shear flow

Kirkwood and Plock* [8] have utilized the preceding theory (including the effects of rotary Brownian motion) to investigate the rheological properties of a dilute suspension of rigid, linear chains of beads for the case of a small amplitude, time–periodic shear flow,

$$\overline{v^o} = (Gx_2, 0, 0) \qquad (65)$$

in which

$$G = G_0 \text{ Real } \{\exp(i\omega t)\}, \qquad (66)$$

$i = \sqrt{-1}$; $G_0$ is the amplitude of the periodic shear wave and $\omega/2\pi$ is the frequency. This requires solution of the *unsteady* form of the orientational distribution equation, (29). For the case where $Pé = G/D_r \ll 1$, and wherein the number $N$ of chains per unit volume tends to zero, Kirkwood and Plock find that the shear viscosity $\mu$ of the suspension is given by and equation of the general form

$$\frac{\mu - \mu_0}{\mu_0} = N \frac{\pi L^3}{90 \ln(L/b)} [F_1(\omega \tau) + (G_0 \tau)^2 F_2(\omega \tau) + O(G_0 \tau)^4] \qquad (67)$$

in the limit where $n \to \infty$. The functions $F_1$ and $F_2$, tabulated by Kirkwood and Plock [8], are of order unity over the entire frequency range $0 \leq \omega < \infty$. The time constant $\tau$ appearing in this expression is a relaxation time, related inversely to the transverse rotary diffusion coefficient of the "rod" by the expression

$$\tau = \frac{1}{6} D_r \qquad (68)$$

---

* Corrections to the original work of these authors are discussed by Paul [46].

in which [8],

$$D_r = \frac{3kT\ln(L/b)}{\pi\mu_0 L^3}$$

The shear viscosity is a function of both the frequency $\omega$ of the oscillation and the time–average shear rate $G_0$.

The suspension also displays a rigidities,** given by the expression

$$\gamma = \frac{3}{5}NkT\,(\omega\tau)^2[F_3(\omega\tau) + (G_0\tau)^2 F_4(\omega\tau) + O(G_0\tau)^4]. \tag{69}$$

This properly vanishes in the limit where $\omega = 0$.

According to these relations the suspension possesses the characteristics of a viscoelastic, non–Newtonian fluid. The viscoelasticity is manifested by the frequency dependence of the response and by the existence of a rigidity. The non–Newtonian character is evidenced by the dependence of the shear viscosity upon the mean shear rate $G_0$. The latter persists even for the steady–state condition, $\omega = 0$. In addition to these properties, the suspension manifests normal stresses.

The existence of a relaxation time arises from the fact that, because the rotary diffusion coefficient is not infinite, the suspended particles cannot instantaneously change their orientations in response to the changing flow conditions. This relaxation phenomenon may therefore be regarded as a kind of "thermal inertia". Though neither fluid nor particle inertia are accounted for in the theory, such effects may be important at high frequencies, especially for relatively massive particles, e.g. for particles of colloidal dimensions. Relaxational phenomena associated with these types of inertia may, under certain circumstances, dominate those due to thermal inertia.

### 3.3.3. Complex bead–chain models

In addition to the spherical dumbbell and the rigid "pearl necklace", other related bead models have been studied in connection with suspension rheology. Indeed, the literature on this subject is voluminous, and no attempt will be made here to provide an exhaustive survey. Such studies take account of the Brownian motion, and include both steady and unsteady Couette and extensional flows. Most of these investigations have come about in connection with applications to solutions of high molecular-weight polymers.

Among these are the deformable bead-spring models of Rouse [47] and Zimm [48]. Here, the beads are joined together by Hookean springs. The separation between adjacent beads is not fixed, but can vary. Elastic restoring forces exist whenever the beads are not situated at their equilibrium separation distances. The analysis of the deformation of this model normally includes translational Brownian motion forces acting upon the beads; that is, each bead in the chain is subjected to a

---

** For small amplitude oscillataory motions the complex viscosity $\mu^*$ is defined by the relation $t_{12}^0 = 2\mu^* \, e_{12}^0$ in which the superscript zero here refers to the complex amplitudes of the corresponding real quantities, i.e. $t_{12} = \text{Real}\,\{t_{12}^0 \exp(i\omega t)\}$ and $e_{12} = \text{Real}\,\{e_{12}^0 \exp(i\omega t)\exp(\iota\omega\tau)\}$. The decomposition of the complex viscosity into real and imaginary parts, viz. $\mu^* = \mu - i\omega^{-1}\gamma$ gives the (dynamic) viscosity $\mu$ and the (dynamic) rigidity $\gamma$.

hydrodynamic force, an elastic force, and a *translational* Brownian motion force. A recent review article [49] provides the key to the modern literature in the field.

Another bead model is furnished by the "randomly - coiled" chain of Kirkwood and Riseman [8]. The junction between a bead and a link is treated essentially as a universal joint, thereby permitting free rotation of a link relative to the bead to which it is affixed. The distance between adjacent beads is fixed, but not the "bond angles".

Dumbbells in which the beads are ellipsoids of revolutions rather than spheres have been studied by Giesekus [11,12]. The essential difference between these and spherical dumbbells resides in the fact that the beads are now anisotropic, possessing different hydrodynamic resistances in different directions.

Giesekus has also investigated the cross dumbbell [12], consisting of two pairs of spherical dumbbells of unequal length, rigidly joined at their centers, at right angles to one another.

The variety of possible models of these general types is limited solely by one's imagination. In general, each different model leads to rheological characteristics that differ among one another in both their qualitative and quantitative behavior. Application of these models to real physical systems is, in large measure, a matter of curve fitting—but perhaps to a lesser extent than for purely phenomenological rheological "models".

## 3.4. ANTISYMMETRIC STRESSES IN SUSPENSIONS OF DIPOLAR PARTICLES IN AN EXTERNAL FIELD

### 3.4.1. General theory

Throughout the preceding analysis we have emphasized the importance of the external structure of the rigid suspended particles, as embodied in their geometrical shapes. The suspended particles may, however, possess internal structure too. This may occur in the form of a center of mass that is displaced from its center of buoyancy (gravitational dipole) due to inhomogeneities in the mass distribution, or in the form of permanent electric or magnetic dipoles locked into the particles. The presence of the suspension in an external gravitational, electric or magnetic field will then affect its rheological properties. Ferrofluids [50], for example, furnish a well-documented example of this phenomenon. The suspended particles are effectively small permanent magnets embedded in an approximately spherical polymeric matrix of colloidal dimensions that is dispersed in a conventional viscous fluid.

Consider the situation where the suspended particles are spherical in shape and possess internal structure in the form of a permanent dipole moment locked into them [51]. The properties of this dipole can be specified by giving the strength $D$ of the dipole moment and the direction $\vec{e}$ of the dipole. This $\vec{e}$ vector is a unit vector locked into the particle. It specifies the orientation of the dipole relative to, say, the direction of the uniform external field at infinity, or to the principal axes of shear. If the particle is present in an external field $\vec{E}$ it will experience an external couple whenever the dipole vector $\vec{e}$ does not lie parallel to the direction of the external field.

Two competing effects arise when the particle is suspended in a shearing flow. On the one hand, the external field tends to align the spherical particle such that its $\vec{e}$ vector is parallel to $\vec{E}$. On the other hand, the vorticity present in the shear field vector is parallel to $\vec{E}$. On the other hand, the vorticity present in the shear field tends generally to rotate the particle out of this alignment, thereby carrying the dipole with it. The instantaneous orientation $\vec{e}$ of the particle is thus governed by a balance between these two competing agencies. If rotary Brownian motion is sensible, this furnishes yet a third couple, which acts to randomize the distribution of dipole orientations. From a rheological point of view, the main feature engendered by the presence of these dipoles is that antisymmetric stresses will arise within the flowing suspension due to the action of external couples.

The external couple vector exerted upon each particle in the suspension (assumed identical) is

$$\vec{L}^{\text{ext}} = D\vec{e} \times \vec{E}. \tag{70}$$

In a suspension of particles whose orientations are characterized by the orientational distribution function $f(\vec{e})$, the average external couple per particle is obtained by multiplying the orientation–specific couple $\vec{L}^{\text{ext}} = \vec{L}^{\text{ext}}(\vec{e})$ exerted on a particle of orientation $\vec{e}$ by the probability $f(\vec{e})de$ that a particle possesses this orientation, and then integrating over all orientations:

$$\overline{\vec{L}}^{\text{ext}} = \int_{\vec{e}} \vec{L}^{\text{ext}}(\vec{e}) f(\vec{e}) d\vec{e} \tag{71}$$

The orientation–specific, hydrodynamic angular velocity $\vec{\Omega} \equiv \vec{\Omega}(\vec{e})$ required in Eqn. (29) for the distribution function may be obtained from the hydrodynamic couple balance,

$$\vec{L} + \vec{L}^{\text{ext}} = \vec{0} \tag{72}$$

in which the hydrodynamic couple exerted on a spherical particle of radius $a$ is

$$\vec{L} = 8\pi\mu_0 a^3 (\vec{\omega}^\circ - \vec{\Omega}) \tag{73}$$

where

$$\vec{\omega}^\circ = \frac{1}{2}(\nabla \times \vec{v}^\circ) \tag{74}$$

is half the vorticity of the undisturbed shear flow $\vec{v}^\circ$.

The bulk external couple density $\overline{\vec{N}}^{\text{ext}}$, i.e. the couple per unit volume of suspension, is gotten by multiplying the mean couple per particle by the number density $n$ of particles:

$$\overline{\vec{N}}^{\text{ext}} = n\overline{\vec{L}}^{\text{ext}} \tag{75}$$

in which $n$ is the number of particles per unit volume,

$$n = \phi/v_p \tag{76}$$

where $v_p = 4\pi a^3/3$ is the volume of a single spherical particle.

According to the angular momentum equation for the bulk continuum [52, 53], in the absence of rotary inertial effects and couple stresses the antisymmetric part, $\overline{t}_{ij}^{(a)}$, of the deviatoric stress is related to the external couple density by the expression

$$\overline{t}_{ij}^{(a)} = -\frac{1}{2}\varepsilon_{ijk}\overline{N}_k^{ext} \tag{77}$$

where $\varepsilon_{ijk}$ is the permutation symbol. On the other hand, for spherical particles, the symmetric part of the deviatoric stress continues to be given by Einstein' equation,

$$\overline{t}_{ij}^{(s)} = \mu_0(1 + \frac{5}{2}\phi)(\overline{v}_{i,j} + \overline{v}_{j,i}). \tag{78}$$

The latter is applicable even in the presence of rotary Brownian motion. The total deviatoric stress in the suspension is the sum of the symmetric and antisymmetric stresses:

$$\overline{t}_{ij} = \overline{t}_{ij}^{(s)} + \overline{t}_{ij}^{(a)}. \tag{79}$$

In the absence of linear inertial effects and external forces, the Cauchy linear momentum for the suspension is

$$\overline{t}_{ij,i} - \overline{p}_{,j} = 0 \tag{80}$$

in which $\overline{p}$ is the pressure. The equation that results from substituting Eqn. (79) into this momentum equation may be solved simultaneously with the continuity equation,

$$\overline{v}_{i,i} = 0 \tag{81}$$

subject to appropriate no-slip boundary conditions on the apparatus boundaries,[*] so as to obtain the velocity and pressure fields, $\overline{v}_i$ and $\overline{p}$. Knowledge of this velocity field permits calculation of the symmetric stress from Einstein's equation (78). The antisymmetric stress is already known from knowledge of the average external couple, $\vec{\overline{L}}_i^{ext}$. In this manner the complete state of stress in the suspension may be determined.

### 3.4.2. Simple shear flow

This theory has been applied [51] to various types of flows for the case of negligible Brownian motion. Of particular interest are the results for Couette flow between flat plates oriented at an arbitrary inclination relative to the external field, as

---

[*] That the no-slip boundary condition applies to a suspension is shown by Cox and Brenner [18] to be valid to at least the first order in $\phi$, which is the order of validity of the present theory.

in Fig. 11. This inclination may be specified by means of the angle $\gamma$ between the direction of the external field $\vec{E}$ and the undisturbed vorticity vector $\vec{\omega}^{\circ}$.

The principal result of this calculation is that the viscosity of the suspension is given by the expression

$$\mu = \mu_0 [1 + \frac{5}{2}\phi + \frac{3}{2}\phi\, F(\gamma;\lambda)]. \tag{82}$$

The last term represents the effect of the external field. The dimensionless function $F$ is a complicated function [51] of the inclination $\gamma$ and the nondimensional parameter

$$\lambda = \frac{ED}{4\pi\mu_0 a^3} \tag{83}$$

specifying the ratio of the strength of the dipole couple, $ED$, tending to prevent rotation of the sphere, to the strength of the hydrodynamic couple, $4\pi\mu_0 a^3 G$, tending

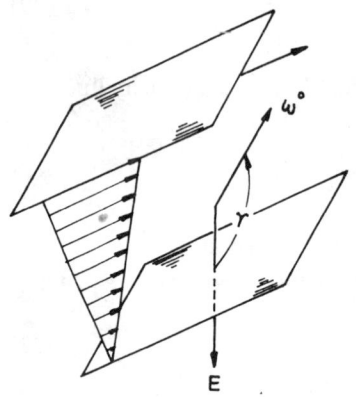

Fig. 11. Orientation of a Couette flow apparatus relative to a uniform external field.

to cause rotation. Here, $E = |\vec{E}|$ is the magnitude of the external field and $G$ is the macroscopic shear rate in the Couette apparatus.

Because of the dependence of $\lambda$ upon $G$, it follows that the suspension viscosity is shear-rate dependent. The quantative nature of this dependence is depicted in Fig. 12 for a specified apparatus orientation $\gamma$. The suspension displays shear thinning. At low rates of shear the external field is dominant, resulting in a strongly hindered rate of particle rotation. As the shear rate is increased indefinitely the viscosity asymptotically approaches the Einstein limit, corresponding to free or unhindered particle rotation.

In addition to a shear-dependent viscosity, there also exists a complex stress system. No normal stresses exist in the suspension, but the stress is antisymmetric in the sense that $\overline{t}_{ij} \neq \overline{t}_{ji}$. Such behavior is distinctly non-Newtonian. The treatment for spherical particles [51] has recently been extended to spheroidal particles by Leal [54].

Analyses of this type are of considerable interest in connection with ferrofluids. The rheological properties of colloidal suspensions of ferromagnetic particles can be varied over relatively wide limits by controlling the strength and orientation of the external magnetic field relative to the principal axes of shear.

### 3.4.3. Extensional and contractile flows

The behavior of a suspension of dipolar spherical particles in an extensional (or contractile) flow is markedly different than in a simple shear flow. The former is an irrotational flow, for which $\vec{\omega}^\circ = \vec{0}$. Accordingly, Eqn. (72) leads to the conclusion that spherical particles immersed in such a flow do not spin ($\vec{\Omega} = \vec{0}$), and that the equilibrium orientation of the dipole vector $\vec{e}$ lies parallel to the external field $\vec{E}$. No external couple arises in this case, so that the stress is symmetric. Indeed, the external field is wholly without influence. The rheological properties of the suspension are then those of a Newtonian fluid with a viscosity given by Einstein's

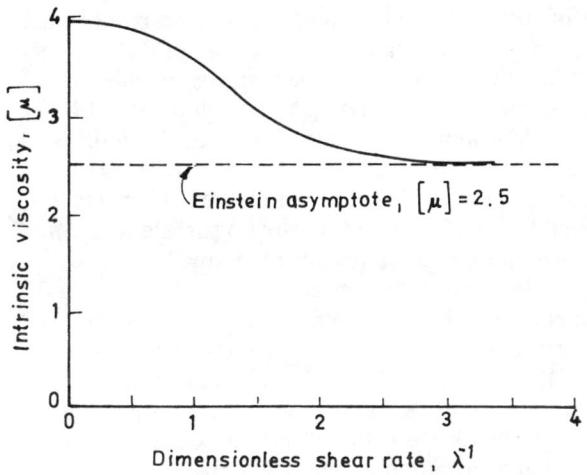

Fig.12. Intrinsic viscosity vs. dimensionless shear rate for an apparatus orientation angle of $\gamma = 75°$ or $105°$.

equation. This again illustrates, in an especially dramatic manner, the importance of the particular type of shear flow in governing rheological behavior.

## 3.5. MODERATELY CONCENTRATED SUSPENSIONS OF SPHERICAL PARTICLES

### 3.5.1. Radial distribution function

Our previous remarks have been confined exclusively to suspensions that are sufficiently dilute to permit neglect of hydrodynamic interactions among the suspended particles. Rheological results for such systems are valid only to the first order in the volume fraction $\phi$ of suspended particles. Considerable effort, both theoretical and experimental [55, 56], has been expended to extend the analysis to more concentrated systems – especially for suspensions of spherical particles in Couette flows.

The viscosity of moderately concentrated suspensions can be represented by the power series expansion [18]

$$\mu = \mu_0(1 + \frac{5}{2}\phi + K\phi^2 + \ldots). \tag{84}$$

Effects of particle–particle interactions are embodied in the dimensionless coefficient $K$. Just as the Einstein coefficient of 5/2 is governed by "one-body" particle interactions with the fluid medium, so the second-order coefficient $K$ is governed by "two-body" interactions. Thus, from a statistical view, its calculation requires knowledge of the two–body, radial distribution function [18, 57, 58]. This spatial distribution function describes the relative positions of pairs of particles in space. Accordingly, it depends upon the detailed hydrodynamics of two-body interactions. But the hydrodynamic interaction between twospheres depends markedly upon the particular type of shear flow being considered. The interaction in Couette flow will be very different than in, say, extensional flow. This leads immediately to the conclusion that $K$ is not an intrinsic property of the suspension alone, but also depends upon the particular type of shear flow to which the suspension is subjected. Thus, even for spherical particles, one cannot speak unambiguously of the viscosity of the suspension. The latter is not a unique property of the suspension without specification of the detailed macroscopic flow. Surprisingly, the significance of this observation does not appear to be widely appreciated in the literature [55, 56].

That the interaction in Couette flow is very different than in, say, elongational flow may be seen from the fact that a region of closed streamlines exists around a single sphere in Couette flow [59, 60]. That is, a dividing stream surface exists relative to an observer translating with the center of the sphere. A small* particle initially present within this closed surface will orbit permanently around the central sphere, behaving like a captive satellite. On the other hand, a small particle initially situated outside of this dividing surface will merely pass around the central sphere, ultimately receding to an infinite distance from the latter. The dividing surface between these inner and outer regions is axially symmetric about an axis passing through the sphere center that is perpendicular to both the undisturbed streamlines and vortex lines. The two-body radial distribution function will obviously be quite sensitive to the existence of these closed streamlines. No region of closed streamlines exists for an extensional flow – so that the doublet distribution can be expected to be quite different in character.

### 3.5.2. Translational Brownian motion as a source of non-Newtonian behavior

*Translational* Brownian motion effects become important when the suspended particles are of colloidal dimensions or smaller. This translational movement must obviously affect the radial distribution function and, hence, $K$. In turn, this effect affects the viscosity. The intensity of the translational Brownian motion is measured

---

* Particles of identical size appear to behave similarly [31]. Particles that are initially in close proximity remain permanently so, forming so-called "collision doublets". These behave essentially like a single, dumbbell-shaped body in that they undergo a rigid-body rotation about an axis passing through the midpoint of their line of centers.

by the translational diffusion coefficient, $D_t$. Consequently, the relative importance of the shear and the Brownian movement in governing the radial distribution function is measured by the translational Péclet number $a^2G/D_t$, where $a$ is the sphere radius. According to the Stokes–Einstein equation,

$$D_t = \frac{kT}{6\pi\mu_0 a}. \qquad (85)$$

It follows that

$$K = \text{function}\,(C) \qquad (86)$$

in which

$$C = \frac{\mu_0 a^3 G}{kT}. \qquad (87)$$

The dimensionless group has been named the "colloid number" by Krieger [61]. In essence it constitutes a ratio of viscous to thermal forces. Its appearance indicates that, even in a simple shearing flow, the viscosity of a concentrated suspension of spherical particles cannot be expressed in terms of $\phi$ alone, but depends also upon the particle size $a$ and the shear rate $G$. The latter gives rise to a form of non-Newtonian behavior involving at least a shear-rate dependent viscosity. Krieger and co-workers [62–64] have utilized this mechanism to explain the non-Newtonian behavior observed in concentrated suspensions ($\phi > 0.25$) of colloidal, spherical synthetic latex particles of uniform size.

## 4. DEFORMABLE PARTICLES

### 4.1. EMULSIONS

#### 4.1.1. Dependence of droplet shape upon shear rate

The rheological properties of a suspension are strongly influenced by whether the suspended particles are rigid or deformable. A simple example of the latter is furnished by a dilute emulsion composed of droplets of one Newtonian liquid dispersed in a second, immiscible Newtonian liquid [65].

Consider a single, neutrally-buoyant liquid droplet suspended in a steady-state Couette flow. Interfacial tension acting between the two phases tends to contract the droplet into a sphere of radius $a$. On the other hand, the tendency of the viscous shear acting across the droplet interface is to stretch the droplet out into a filamentary shape [66–71]. The equilibrium shape of the droplet surface, as well as its orientation relative to the principal axes of shear, is thus governed by a balance between these two competing forces.

The equation of the surface of the droplet,

$$f(x_1/a, x_2/a, x_3/a;\, \alpha;\, \beta) = 0 \qquad (88)$$

depends upon two dimensionless parameters:

$$\alpha = \frac{\mu_0}{\mu_i} = \frac{\text{continuous–phase viscosity}}{\text{droplet viscosity}}, \qquad (89)$$

$$\beta = \frac{\mu_0 G a}{\sigma} = \frac{\text{shearing force}}{\text{interfacial force}}, \qquad (90)$$

in which $a$ is the radius of the undeformed droplet, $G$ is the shear rate, and $\sigma$ is the interfacial tension between the phases.

It is assumed that the fluid motion both inside and outside the droplet satisfies the creeping motion and continuity equations. The vector velocity and tangential component of the stress vector are continuous across the interface, whereas the normal component of the stress vector satisfies the Laplace, interfacial tension boundary condition. Despite the fact that the differential equations governing the inertial and external flow fields are linear, the problem of determining the droplet shape is a nonlinear problem since the latter has to be determined simultaneously with the solution of the differential equations. Hence, the nonlinearity arises from the boundary conditions, rather than from the differential equations.

As the shear rate is increase from zero, the droplet departs more and more from the spherical shape it possesses at rest. Indeed, at sufficiently large rates of shear the droplet may burst [70,71] due to hydrodynamic instabilities. The nonlinearity of the single droplet problem ultimately reflects itself in a corresponding nonlinear rheological behavior of the emulsion of which it is a part [69, 72]. Emulsions, therefore, may be expected to manifest strongly non–Newtonian behavior, despite the Newtonian character of the separate liquids of which they are composed.

Calculation of the shape of the droplet normally proceeds by a perturbation technique involving expansion of the equation of the droplet surface into a series of spherical harmonics, the coefficients of which are functions of β and/or α, regarded as small perturbation parameters [67, 68]. This is tantamount to supposing that the shape departs but little from the spherical form. To terms of the first order in the deformation, the droplet contour is found to be ellipsoidal [66–68].

### 4.1.2. Newtonian rheology of a suspension of spherical liquid droplets

In the limiting case where the interfacial force $\sigma/a$ is effectively infinite compared with the viscous force $\mu_o G$ or $\mu_i G$, whichever is larger, the parameter ß is zero, whence the droplet is spherical. The rheology of a dilute suspension of such spherical liquid droplets was worked out by G. I. Taylor [65], who found that the emulsion behaves like a Newtonian fluid with a shear viscosity given by the expression

$$\mu = \mu_0 \left[ 1 + \frac{5}{2}\phi\left(\frac{1+\frac{1}{2\alpha}}{1+\alpha}\right) \right] \tag{91}$$

When the droplet is very viscous, corresponding to the case where the viscosity ratio $\alpha$ tends to zero, this reduces to

$$\mu = \mu_0 \left(1 + \frac{5}{2}\phi\right) \tag{92}$$

which is Einstein's formula for rigid spheres. In the opposite case, where the droplet is relatively inviscid, as for example in the case of a gas bubble, this yields

$$\mu = \mu_0 (1 + \phi). \tag{93}$$

Not surprisingly, this viscosity is considerably less than for a rigid sphere suspension.

### 4.1.3. Non–Newtonian rheology of a suspension of deformed droplets

Taylor's analysis does not take account of the deformation of the droplet. It is for this reason that the rheological behavior is linear, i.e. Newtonian.

Droplet deformation has been taken into account by Schowalter et al. [72] for the case in which the dimensionless shear rate $\beta$ is small compared with unity, but nonzero, and the viscosity ratio $\alpha$ is of order unity. Not restricting themselves to a simple shear flow, they consider the case of a general linear shear flow, characterized by an undisturbed rate of strain $\vec{d}^\circ$ and vorticity $2\vec{\omega}^\circ$, as in Eqn. (14). In contrast to the situation for rigid particles, the orientational distribution is not a factor for deformable particles. Rather, the orientation of the ellipsoidal droplet is uniquely fixed by the hydrodynamics of the problem,* each droplet possessing the same orientation. Accordingly, orientational questions need not be considered explicitly in determining the rheological properties of emulsions. Rather, one utilizes directly orientation–specific rheological formulas of the type given in Eqns. (17) and (18).

To the first order in the deformation (and the droplet concentration), the constitutive equation for the emulsion takes the form [72]

$$\vec{t} = 2\mu \vec{\vec{d}} + \phi\mu_0\tau\left[F_1(\alpha)\frac{\mathscr{D}\vec{\vec{d}}}{\mathscr{D}t} + F_2(\alpha)(\vec{\vec{d}}\cdot\vec{\vec{d}} - \frac{1}{3}I\vec{\vec{d}}:\vec{\vec{d}})\right] \tag{94}$$

where $\vec{t}$ is the bulk deviatoric stress and $\vec{\vec{d}}$ is the bulk rate of strain dyadic. The coefficient $\mu$ of the leading term is Taylor's viscosity, Eqn. (91), for the case where the droplets are spherical. $F_1$ and $F_2$ are complicated functions [72] of the viscosity rational, and

$$\tau = \frac{\mu_0 a}{\sigma} \tag{95}$$

is a "relaxation" time. The leading term of Eqn. (94) is the same as for spherical liquid droplets. The term in square brackets accounts for droplet deformation arising from the shear.

The time derivative $\mathscr{D}/\mathscr{D}t$ appearing in Eqn. (94) is the Jaumann derivative [73]. It is related to the conventational material derivative $\mathscr{D}/\mathscr{D}t = \partial/\partial t + \vec{v}\cdot\nabla$ and vorticity $2\vec{\omega}$ of the suspension by the expression

$$\frac{\mathscr{D}\vec{\vec{d}}}{\mathscr{D}t} = \frac{\mathscr{D}\vec{\vec{d}}}{\mathscr{D}t} + \vec{\vec{d}}\times\vec{\omega} - \vec{\omega}\times\vec{\vec{d}}. \tag{96}$$

---

*The principal axes of this ellipsoid lie along the principal axes of the undisturbed rate of strain dyadic $\vec{d}^\circ$.

This derivative accounts for the proper material–frame indifference when local rotation of the material is taken into account. This term arises naturally in the hydrodynamic rheological theory [69, 72].

The bracketed term in Eqn. (94) is nonlinear in the strain rate $\bar{\bar{d}}$ since it involves terms that are quadratic in this kinematic parameter. This deformational term renders the emulsion non–Newtonian.

The constitutive relation (94) applies to a general linear shear flow. For the particular case of a Couette flow [cf. Eqn. (40)], the nonlinear terms contribute only to the normal stresses, but not to the shearing stresses. Accordingly, for such a flow the shear viscosity continues to be given by Taylor's formula, Eqn. (91). * The normal stresses may be expressed in terms of the primary and secondary normal stress differences [cf. Eqns. (43) and (44)] :

$$\bar{\sigma}_1 = \phi\mu_0\tau G^2 f_1(\alpha), \tag{97}$$

$$\bar{\sigma}_2 = \phi\mu_0\tau G^2 f_2(\alpha), \tag{98}$$

in which $f_1$ and $f_2$ are functions [72] of the viscosity ratio $\alpha$. Since $f_1 \neq f_2$, these normal stress differences are unequal. The normal stress differences are quadratic in the shear rate $G$. Their existence confers upon the suspension properties comparable to those of a viscoelastic fluid. The elastic properties stem from the elastic–like character of the interface, originating from the interfacial tension.

In contrast with the viscoelastic–like behavior of the emulsion in Couette flow, the emulsion behaves in extensional flow {cf. Eqn. (45)] like a quasi–Newtonian fluid with a shear–dependent viscosity; that is, Eqn. (94) reduces to the form

$$\bar{t}_{ij} = 2\mu\bar{d}_{ij} \tag{99}$$

in which

$$\mu = \mu_0 \left[ 1 + \frac{5}{2}\phi\left(\frac{1+\frac{1}{2\alpha}}{1+\alpha}\right) + \phi\tau G F(\alpha) \right] \tag{100}$$

where $G$ is the macroscopic rate of deformation, and $F(\alpha)$ is a function of the viscosity ratio [72].

### 4.1.4. Rheological effects of droplet rupture

The preceding analysis applies only when the droplet deformation is small. For large deformations it is known experimentally [66, 70, 71] that the droplet may burst, at least in certain types of shear flow and for certain viscosity-ratio ranges. **

---

* That is, the only nonzero components of eqn. [94] are $\bar{t}_{11}, \bar{t}_{22}, \bar{t}_{33}$ and $\bar{t}_{12} = \bar{t}_{21}$. The latter is given by the relation $\bar{t}_{21} = \mu G$, in which $\mu$ is Taylor's viscosity.

** Experiments [71] show that droplet breakup in simple shear flow occurs only if the viscosity ratio l is greater than about 3. For $\alpha < 3$ no rupture occurs.

This necessarily affects the rheological properties of the emulsion in consequence of the altered value of the average droplet size.

As the shear rate $G$ in, say, a simple shear flow is increased from zero the droplet initially deforms into an ellipsoid. Further increase in the shear rate produces a filamentary shape [67], except for relatively viscous drops. Ultimately, as the shear rate is further increased, the droplet becomes unstable (in a hydrodynamic sense) and ruptures into approximately two equal fragments.*** The dimensionless shear rate $\beta$ at which this occurs depends upon the viscosity ratio $\alpha$. It may also depend to some extent upon the rate at which the shear is increased to the new value [71].

The phenomenon of droplet breakup thereby creates a "new" emulsion, composed of smaller droplets than existed in the "original" emulsion. This new emulsion possesses the same volume fraction $\phi$ of suspended droplets and viscosity ratio $\alpha$ as the original emulsion. However, the $\beta$ value is now less due to the fact that the radii $a$ of the new droplets are less than those of the original droplets. Consequently, the new emulsion possesses rheological properties that differ from those of the original emulsion.

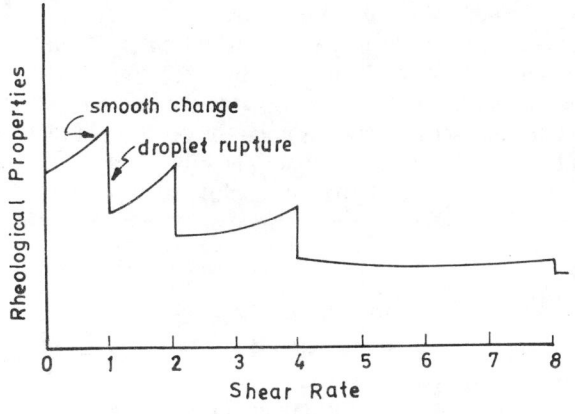

Fig. 13. Idealized change in rheological properties with shear rate in an emulsion containing droplets that undergo rupture.

On the supposition that coalescence of the suspended droplets does not occur, the rheological properties of the suspension as a function of shear rate may be expected to have the appearance depicted in Fig. 13. As the shear is increased quasistatically from zero the rheological properties change smoothly with increasing shear rate until the critical shear rate at which droplet rupture occurs is reached. At this point the rheological properties change, essentially discontinuously, to those appropriate to a new emulsion composed of smaller, less deformed, droplets. In turn, these smaller droplets deform further as the shear rate is increased. Ultimately, their deformation reaches a stage at which they too rupture, creating yer another "new" emulsion, composed off still smaller droplets.

---

*** Two equal fragments are produced for all $\alpha > 3$ [71], provided that the shear is increased slowly enough, i.e. quasistatically.

The cycle of droplet deformation and rupture presumably continues indefinitely, creating a never–ending sequence of emulsions. In the absence of coalescence the breakup process is irreversible. The emulsion therefore may be expected to manifest hysteresis in the sense that, when the shear rate is reduced to some prior value, the rheological properties are not restored to the values they originally possessed at that shear rate. This view of an emulsion is, of course, a considerable idealization, designed to illustrate the complex rheological behavior that formal rheological theories must be prepared to describe. One would expect such bizarre behavior only to be approximately realized in practice.

### 4.1.5. Non-uniform spatial droplet distribution due to lateral migration in Poiseuille flow

The preceding theory of emulsion rheology is based upon the assumption that the shear is homogeneous, or else that the spatial distribution of droplets is uniform throughout the fluid – in which case the rheological properties furnished by the preceding theory are those appropriate to the *local* state of affairs at a point in the inhomogeneous shear flow. Neutrally buoyant droplets suspended in a Poiseuille flow within a circular tube migrage across the streamlines until they reach the tube axis [74]. * This radial motion results in a nonuniform spatial distribution of droplets over the tube cross–section. Inasmuch as the *local* shear varies over the cross–section (being zero at the centerline), the contribution of a droplet to the overall rheological properties of the emulsion necessarilly depends upon the position of the droplet in the tube relative to the tube axis. This suggests that Poiseuille flow experiments cannot be utilized to investigate the fundamental rheological properties of emulsions.

### 4.1.6. Unsteady flow

Prior rheological results for the steady flow of emulsions have been extended to unsteady flows as well. Spherical liquid drops in unsteady flows are treated by Oldroyd [79, 80]. Deformed droplets in unsteady flows are treated by Frankel and Acrivos [69]. The novel feature of the latter analysis is that the shape of the droplet is now time dependent; that is, in a rapidly changing field of flow the droplet cannot instantaneously adjust its shape to conform to the changing flow conditions. This gives rise to relaxational phenomena dominated by a relaxation time $\tau$, which may be either $\mu_0 a/\sigma$ for $\alpha = O(1)$ or $\mu_i a/\sigma$ for $\alpha << 1$ [68, 69]. The latter corresponds to the case of a very viscous droplet. The principal new rheological characteristic of the emulsion is that, in addition to being viscoelastic and non–Newtonian, it displays finite "memory" effects.

---

* In contrast to the behavior of *rigid* spheres in a Poiseuille flow, which undergo lateral migration only as a consequence of inertial effects [75-77], liquid droplets undergo radial migration [74] even in creeping flow. This lateral motion occurs as a result of the particle deformation acting in concert with the non-uniform shear in the tube. Lateral motion of droplets also occurs when they are non-neutrally buoyant [78], but the mechanism is quite different thatn in the neutrally buoyant case. Motion of the former type would occur even for a uniform shear field.

## 4.2. ELASTIC PARTICLES

Another example of a deformable particle is furnished by a solid elastic body (obeying the classical stress–strain laws of linear elasticity theory), whose equilibrium shape is, say, spherical. The shape of such a particle in a shearing flow is governed by a balance between the viscous shear forces, tending to deform it, and the elastic forces, tending to restore it to its equilibrium spherical shape. The rheological behavior of dilute suspensions of such elastic particles has been studied for general shearing flows, including unsteady flows. Spherical elastic particles are treated by Fronlich and Sack [81], and deformed elastic particles by Roscoe [13] and Goddard and Miller [21].

The rheological equations of state deduced by these authors for elastic bodies are completely analogues in form to those already cited for liquid droplets [69, 72]. In particular, suspensions of such particles manifest viscoelastic behavior. Here, however, the elasticity derives from the elastic properties of the particle, rather than from the elastic-like properties of the interfacial film.

Whereas the relaxation time for the liquid droplets is either $\mu_0 a/\sigma$ or $\mu_i a/\sigma$, the corresponding relaxation time for the elastic particles is

$$\tau = \frac{\mu_0}{k} \tag{101}$$

in which $k$ is the shear modulus of the elastic particle. The principal difference between elastic particles and interfaces therefore lies in the fact that the relaxation time for the former is independent of particle size $a$. This difference is not surprising since in the former case the elasticity is a volumetric effect whereas in the latter is a surface effect.

## 4.3. COMPRESSIBLE GAS BUBBLES

Another aspect of particle deformability arises from the possibility that the suspended particles, if compressible, may undergo changes in volume. This gives rise to a dilatational or volume viscosity, in addition to the usual shear viscosity.

Consider a dilute suspension of spherical gas bubbles in pressure equilibrium with the incompressible liquid in which they are dispersed. When the external pressure exerted upon the suspension is changed, each bubble undergoes an appropriate alteration in its volume. The mixture of gas bubbles and liquid therefore behaves like a compressible fluid. Changes in the volume of a bubble are resisted by the ordinary viscosity of the liquid. The latter therefore governs the rate at which such changes can occur.

Analysis of the rheological behavior of a bubbly mixture of this type [82] reveals that, at least for sufficiently slow rates of change of volume, one can ascribe a volume viscosity to the dispersion, given by [82–84]

$$k = \frac{4\mu_0}{3\phi} \tag{102}$$

This formula possesses the unrealistic property of tending to infinity as the volume fraction $\phi$ of dispersed gas bubbles tends to zero. Davies [85] pointed out that this feature disappears if account is taken of the finite compressibility of the carrier liquid.

## 5. INERTIAL EFFECTS

### 5.1. SIMPLE SHEAR FLOW

Except for the work of Batchelor [16], general theories of suspension viscosity have ignored inertial effects. Such schemes are therefore applicable only in circumstances where the shear Reynolds number

$$\text{Re} = \frac{a^2 G}{v_0} \tag{103}$$

is effectively zero, $v_0$ being the kinematic viscosity of the continuous phase. Lin *et al.* [86] have utilized Batchelor's method in conjunction with a singular–perturbation analysis of the simple shear flow around an isolated sphere at small, nonzero values of Re to obtain the appropriate modifications of Einstein's formula arising from fluid inertia.

Their result for the viscosity of a dilute suspension of rigid spheres is

$$\mu = \mu_0(1 + \frac{5}{2}\phi + 1.34 \text{Re}^{3/2}) \tag{104}$$

valid for Re << 1. In view of the increase of $\mu$ with an increase in the shear rate $G$, inertial effects give rise to shear thickening. In addition to producing a shear–dependent viscosity, the effect of fluid inertia is also to bring into existence normal stresses. For the shear flow described by Eqn. (40), the primary and secondary normal stress differences [cf. Eqns. (43) and (44) have the forms

$$\overline{\sigma}_1 = -\phi\mu_0 G \text{ Re } (\frac{2}{3} - 0.035 \text{ Re}^{1/2}) \tag{105}$$

and

$$\overline{\sigma}_2 = \phi\mu_0 G \text{ Re } (\frac{2}{3} - 0.252 \text{ Re}^{1/2}) \tag{106}$$

Not only are these normal stress differences unequal, but in general they possess different algebraic signs.

### 5.2. INDIRECT EFFECTS ON FLUID INERTIA

In addition to explicit contributions to rheological formulae of the type displayed in Eqns. (104)–(106), fluid inertia may also make implicit contributions. The work of Harper and Chang [33] on maximum dissipation orbits in suspensions of rotating dumbbells, discussed earlier, furnishes an example of such indirect inertial effects. Implicit inertial effects may also arise in inhomogeneous shear flows, e.g. Poiseuille flow in a tube, due to nonuniform spatial distribution of the suspended particles stemming from radial migration. The rheological effects here are similar to those mentioned earlier in connection with the lateral migration of deformable particles, except that now the mechanism of radial motion resides in the inertia of the fluid [75–77]. In this context, Segre and Silberberg [87] observed a dependence of the apparent viscosity of a suspension of rigid spheres upon the tube Reynolds number during flow through a circular tube. This was attributed by them to inertial effects associated with variations in the spatial distribution of particles arising from lateral migration.

## ACKNOWLEDGMENT

This work was supported by the National Science Foundation under Grant NO. GK – 12583.

## REFERENCES

1. A. EINSTEIN, Eine neue Bestimmung der Moleküldimensionen. *Ann.Physik* **19**,289–306 (1906).
2. A. EINSTEIN, Berichtigung zu meiner Arbeit: Eine neue Bestimmung der Moleküldimensionen. *Ann. Physik* **34**, 591–2 (1911).
3. A. EINSTEIN, *Investigations on the Theory of The Brownian Movement* (R.Fürth, ed.) Dover: New York, 1956.
4. CH. SADRON, Dilute solutions of impenetrable rigid particles. Chapter in: *Flow Properties of Disperse Systems* (J.J. Hermans, ed.), pp. 131–98, Interscience: New York, 1953.
5. J.J. HERMANS, Dilute solutions of flexible chain molicules. Chapter in: *Flow Properties of Disperse Systems* (J.J. Hermans, ed.) pp. 199–265. Interscience: New York, 1953.
6. CH.SADRON, Methods of determining the form and dimensions of particles in solution. Chapter in: *Progress in Biophysics*, vol. 3 (J. A. V. Butler and J. T. Randall, eds.), pp. 237–304. Academic: New York, 1953.
7. H. L. FRISCH and R. SIMHA, The viscosity of colloidal suspensions and macromolecular solutions. Chapter in : *Rheology: Theory and Applications*, vol. 1 (F.R. Eirich, ed.), pp. 525–613. Academic: New York, 1956.
8. J. G. KIRKWOOD, *Macromolecules* (P. L. Auer, ed.) Gordon & Breach: New York, 1967.
9. H. BRENNER, Rheological properties of suspensions. *Proceedings of the "Fluid Dynamcis Symposium 1970"* (McMaster University, Hamilton, Ontario, Canada, August 25-27, 1970) (in press).
10. L. D. LANDAU and E. M. LIFSHITZ, *Fluid Mechanics*, pp. 76–79. Addison – Wesley: Reading, Massachusetts, 1959.
11. H. GIESEKUS, Elasto–viskose Flüssigkeiten, für die in stationären Schichtströmungen sämtliche Normalspannungskomponenten verschieden grosse sind. *Rheol. Acta* **2**, 50–62. (1962).
12. H. GIESEKUS, Statistical rheology of suspensions and solutions with special reference to normal stress effects. *Proc. Intern. Symp. Second–Order Effects Elasticity, Plasticity, Fluid Dynamics* (Haifa, Israel, 1962), pp. 553–084. Pergamon: New York, 1964.
13. R. ROSCOE, On the rheology of a suspension of viscoelastic spheres in a viscous liquid. *J. Fluid Mech.* **28**, 273–93. (1967).
14. H. S. LEW, Formulation of statistical equation of motion of blood. *Biophysical J.* **9**, 235–45 (1969),
15. L. SIROVICH and R. B. TURNER, *Flow of Suspension at Low Reynolds Numbers*. Techn, Rept. No Nonr 562(39)/10, 48 pp. Brown Univ., Providence, Rhode Island, 1966.

16. G. K. BATCHELOR, The stress system in a suspension of force–free particles. *J. Fluid Mech.* **41**, 545–70 (1970).
17. H. BRENNER, Rheology of two–phase systems. Chapter in: *Annual Review of Fluid Mechanics*, vol. 2 (M. van Dyke, W. G. Vincenti, and J. V. Wehausen, eds.), pp. 137–76. Annual Reviews: Palo Alto, California, 1970.
18. R. G. COX and H. BRENNER, The rheology of suspension of particles in a Newtonian fluid. *Chem. Eng. Sci.* **26**, 65–93 (1971).
19. H. BRENNER, Suspension rheology in the presence of rotary Brownian motion and external couples: Elongational flow of dilute suspensions. *Chem. Eng. Sci.* (in press).
20. M. J. BERAN, *Statistical Continuum Theories*, Interscience: New York, 1968.
21. J. D. GODDARD and C. MILLER, Nonlinear effects in the rheology of dilute suspension. *J. Fluid Mech* **28**, 657–73 (1967).
22. D. W. CONDIFF and H. BRENNER, Transport mechanics in systems of orientable particles. *Phys. Fluids* **12**, 539–51 (1969).
23. H. BRENNER, Coupling between the translational and rotational Brownian motions of rigid particles of arbitrary shape: II. General theory. *J. Colloid Interface Sci.* **23**, 407–36 (1967).
24. H. BRENNER, Orientation–space boundary layers in problems of rotational diffusion and convection at large rotary Peclet numbers. *J. Colloid Interface Sci.* **34**, 103–25 (1970).
25. H. BRENNER and D. W. CONDIFF, Transport mechanics in systems of orientable particles: III. Arbitrary particles. *J. Colloid Interface Sci.* (in press).
26. A. PETERLIN, Über die Viskosität von verdünnten Lösungen und Suspensionen in Abhängigkeit von der Teilchenform. *Z. Physik* **111**, 232–63 (1938).
27. R. TAKSERMAN–KROZER and A. ZIABICKI, Behavior of polymer solutions in a velocity field with parallel gradient: II. Viscosity of dilute solutions containing rigid ellipsoids. *J. Polymer Sci.* (Part A) **1**, 507–15 (1963).
28. H. A. SCHERAGA, Non–Newtonian viscosity of solutions of ellipsoidal particles. *J. Chem. Phys.* **23**, 1526–32 (1955).
29. N. SAITO and M. SUGITA, Energy dissipation and entropy production in irreversible processes of dilute systems. *J. Phys. Soc. Japan* **7**, 554–9 (1952).
30. G. B. JEFFERY, The motion of ellipsoidal particles immersed in a viscous fluid. *Proc. Roy. Soc. (London), A,* **102**, 161–79 (1922).
31. H. L.GOLDSMITH and S. G. MASON, The Microrheology of dispersions. Chapter in *Rheology: Theory and Applications*, vol. 4 (F.R. Eirich, ed.), pp. 85–250. Academic: New York, 1967.
32. F. P. BRETHERTON, The motion of rigid particles in a shear flow at low Reynolds numbers. *J. Fluid Mech.* **14**, 284–304 (1962).
33. E. Y. HARPER and I–DEE CHANG, Maximum dissipation resulting from lift in a slow viscous shear flow. *J. Fluid Mech.* **33**, 209–25 (1968).
34. A. KARNIS, H. L. GOLDSMITH and S. G. MASON, The flow of suspensions through tubes. V. Inertial effects. *Canad. J. Chem. Eng.* **44**, 181–93 (1966).
35. N. SAITO, The effect of the Brownian motion on the viscosity of solutions of macromolecules. I. Ellipsoids of revolution. *J. Phys. Soc. Japan* **6**, 297–301 (1951).
36. H. GIESEKUS, Das Reibungsgesetz der strukturviskosen Flüssigkeit. *Kolloid–Z.* **147**, 29–45 (1956); Erratum. See footnote 18 of *Rheol. Acta* **1**, 404–13 (1961).
37. S. PRAGER, Stress–strain relations in a suspension of dumbbells. *Trans. Soc. Rheol.***1**, 53–62 (1957).
38. T. KOTAKA, Note on the normal stress effect in the solution of rod–like macromolecules. *J. Chem. Phys.* **30**, 1566–7 (1959).
39. R. TAKSERMAN–KROZER and A. ZIABICKI, Behavior of polymer solutions in a velocity field with parallel gradient: I. Orientation of rigid ellipsoids in a dilute solution. *J. Polymer Sci.* (Part A) **1**, 491–506 (1963).

40. P. C. WANKAT, Flow during axial compression testing of materials. *Ind. Eng. Chem. Fund.*, **8**. 598 (1969).
41. V. N. POKROVSKII, Rheology of dispersed systems and polymers. Movement of rigid ellipsoids in the flow. *Kolloidnyi Zh.* **29**, 576–83 (1967). [English translation available in *Colloid Journal of the USSR* **29**, 428–33 (1967).]
42. V. N. POKROVSKII, Microrheology of dispersed systems and polymers. Stress tensor for a suspension of rigid ellipsoids. *Kolloidnyi Zh.* **30**, 881–8 (1968). [English translation available in *Colloid Journal of the USSR* **30**, 664–70 (1968).]
43. V. N. POKROVSKII, Refinement of the results of the theory of suspension viscosity. *Zh. Eksper. Teoret. Fiz.* **55**, 651–3 (1968).[English translation available in *Soviet Physics JETP* **28**, 339–40 (1969).]
44. R. M. SONSHINE and H. BRENNER, The Stokes translation of two or more particles along the axis of an infinitely long circular cylinder, *Appl. Sci, Res.* **16**, 425–54 (1966).
45. R. ZWANZIG, J. KIEFER and G. H. WEISS, On the validity of the Kirkwood – Riseman theory. *Proc. Natl. Acad. Sci. U.S.* **60**, 381–6 (1968).
46. E. PAUL, Non – Newtonian viscoelastic properties of rodlike molecules in solution: Comment on a paper by Kirkwood and Plock. *J. Chem. Phys.* **51**, 1271–2 (1969).
47. P. E. ROUSE, Jr., A theory of the linear viscoelastic properties of dilute solutions of coiling polymers. *J. Chem. Phys.* **21**, 1272–80 (1953).
48. B. H. ZIMM, Dynamics of polymer moleculs in dilute solution: Viscoelasticity, flow birefringence and dielectric loss. *J. Chem. Phys.* **24**, 269–78 (1956).
49. R. B. BIRD, H. R. WARNER, Jr. and D. C. EVANS, Kinetic therory and rheology of dumbbell suspensions with Brownian motion. Chapter in: *Advances in Polymer Science*, vol. 8, pp. 1–91. Springer Verlag: Berlin, 1971.
50. A. R. V. BERTRAND, Les ferrofluides. *Revue del' Institut Français de Pétrole et annales des combustibles liquides*, **25**, 16–35 (1970).
51. H. BRENNER, Rheology of dilute suspension of dipolar spherical particles in an external field. *J. Colloid Interface Sci.* **32**, 141–58 (1970).
52. J. S. DAHLER and L. E. SCRIVEN, Angular momentum of continua., *Nature* **192**, 36–37 (1961).
53. J. S. DAHLER and L. E. SCRIVEN, Theory of structured continua: I. General consideration of angular momentum and polarization. *Proc. Roy. Soc. (London)*, A,**275**, 504\d\fo23)–27 (1963).
54. L. G. LEAL, On the effect of particle couples on the motion of a dilute suspension of spheroids. *J. Fluid Mech.* **46**, 395–416 (1971).
55. R. RUTGERS, Relative viscosity of suspensions of rigid spheres in Newtonian liquids. *Rheol. Acta* **2**, 202–10 (1962); Errata, *ibid.* **3**, 118–22 (1963).
56. R. RUTGERS, Relative viscosity and concentration. *Rheol. Acta* **2**, 305–48 (1962).
57. J. M. PETERSON and M. FIXMAN, Viscosity of polymer solutions. *J. Chem. Phys.* **39**, 2516–23 (1963).
58. N. F. SATHER and K. J. LEE, Viscosity of concentrated suspensions of spheres. *Proceedings of the International Symposium on Two–Phase Systems, Technion, Haifa, Israel* (Aug. 29–Sept. 2, 1971).
59. R. G. COX, I. Y. Z. ZIA and S. G. MASON, Particle motions in sheared suspensions. XXV. Streamlines around cylinders and spheres. *J. Colloid Interface Sci.* **27**, 7–18 (1968).
60. N. A. FRANKEL and A. ACRIVOS, Heat and mass transfer from small spheres and cylinders freely suspended in shear flow. *Phys.Fluids* **11**, 1913–18 (1968).
61. I. M. KRIEGER, A dimensional approach to colloid rheology. *Trans. Soc. Rheol.* **7**, 101–9 (1963).
62. I. M. KRIEGER, A mechanism for non – Newtonian flow in suspensions of rigid spheres. *Trans. Soc. Rheol.* **3**, 137–52 (1959).

63. M. E. WOODS and I. M. KRIEGER, Rheological studies on dispersion of uniform colloidal spheres. I. Aqueous dispersions in steady shear flow. *J. Colloid Interface Sci.* **34**, 91–99 (1970).
64. Y. S. PAPIR and I. M. KRIEGER Rheological studies on disperson of uniform colloidal spheres. II.Dispersions in nonaqueous media. *J. Colloid Interface Sci.* **34**, 126–30 (1970).
65. G. I. TAYLOR, The viscosity of a fluid containing small drops of another fluid. *Proc. Roy. Soc.(London)*, A, **138**, 41–48 (1932).
66. G. I. TAYLOR, The formation of emulsion in definable fields of flow *Proc. Roy. Soc.(London)*, A, **146**, 501–23 (1934).
67. C. E. CHAFFEY and H. BRENNER, A second–order theory for shear deformation of drops. *J. Colloid Interface Sci.* **24**, 258–69 (1967).
68. R. G. COX, The deformation of a drop in a general time–dependent fluid flow. *J. Fluid Mech.* **37**, 601–20 (1969).
69. N. A. FRANKEL and A. ACRIVOS, The constitutive equation for a dilute emulsion. *J. Fluid Mech.* **44**, 65–78 (1970).
70. F. D. RUMSHEIDT and S. G. MASON, Particle motions in sheared suspensions. XII. Deformation and burts of fluid drops in shear and hyperbolic flow. *J. Colloid Sci.* **16**, 238–61 (1961).
71. S. TORZA, R. G. COX and S. G. MASON, Particle motions in sheared suspenisons. XXVII. Transient and steady deformation and burst of liquid drops. *J. Colloid Sci.***38**, 395–411 (1972).
72. W. R. SCHOWALTER, C. E.CHAFFEY and H. BRENNER, Rheological behavior of a dilute emulsion, *J. Colloid Interface Sci.* **26**, 152–60 (1968).
73. J. D. GODDARD and C. MILLER, An inverse for the Jaumann derivative and some applications to the rheology of viscoelastic fluids. *Rheol. Acta* **5**, 177–84 (1966).
74. H. L. GOLDSMITH and S. G. MASON, The flow of suspensions through tubes. I. Single spheres, rods and discs, *J. Colloid Sci.* **17**, 448–76 (1962).
75. G. SEGRÉ and A. SILBERBERG, Begavior of macroscopic rigid spheres in Poiseuille flow: Part II.Experimental results and interpretation. *J. Fluid Mech.* **14**, 136–57 (1962).
76. H. BRENNER, Hydrodynamic resistance of particles at small Reynolds numbers. Chapter in : *Advances in Chemical Engineering*, Vol. 6 (T.B.Drew, J.W.Hoopes, T. Vermeulen, eds.), pp. 287–438. Academic: New York, 1966.
77. R. G. COX and H. BRENNER, The lateral migration of solid particles in Poiseuille flow. Part I. Theory. *Chem. Eng. Sci.* **23**, 147–73 (1968).
78. S. HABER and G. HETSRONI, The dynamics of a deformable drop suspended in a long conduit. *J. Fluid Mech*, **49**, 257–78 (1971).
79. J. G. OLDROYD, The elastic and viscous properties of emulsions and suspensions. *Proc. Roy. Soc. (London)*, A, **218**, 122–32 (1953).
80. J. G. OLDROYD, The effect of interfacial stabilizing films on the elastic and viscous properties of emulsions. *Proc. Roy. Soc. (London)*, A, **232**, 567–77 (1955).
81. H. FRÖHLICH and R. SACK, Theory of the rheological properties of dispersions. *Proc.Roy, Soc.(London)*, A, **185**, 415–30 (1946).
82. G. I. TAYLOR, The two coefficients of viscosity for an incompressible fluid containing air bubbles. *Proc. Roy. Soc. (London)*, A, **226**, 34–39 (1954).
83. G. K. BATCHELOR, *An Introduction to Fluid Dynamics*, pp. 253–5, Cambridge: London, 1967.
84. G. K. BATCHELOR, Compression waves in a suspension of gas bubbles in liquid. *Fluid Dyn, Trans.* **4**, 425–45. Inst. Fund. Techn. Res., Polish Acad. Sci., Warsaw, Poland (1967).
85. R. O. DAVIES, A note on Sir Geoffrey Taylor's paper. *Proc. Roy. Soc. (London)*, A, **226**, 39 (1954).
86. C. J. LIN, J. H. PERRY and W. R. SCHOWALTER, Simple shear flow round a rigid sphere: inertial effects and suspension rheology.*J. Fluid Mech* **44**, 1–17 (1970).

87. G. SEGRÉ and A. SILBERBERG, Non – Newtonian behavior of dilute suspension of macroscopic spheres in a capillary viscometer. *J. Colloid Sci.* **18**, 312– 7 (1963).

## NOTE ADDED IN PROOF

Since completion of the draft of this chapter (in December 1970), several papers have appeared that pertain significantly to several of the topics covered in this review. A brief summary of the contents of these is offered in following paragraphs.

*Section* 2.4: For rigid particles of arbitrary shape, the general theory of suspension viscosity [19] requires knowledge of nine material tensors. These appear as proportionality coefficients in linear relations existing between the three "forces" (force, couple, and stokeslet [16] acting on the particle, and the three "fluxes" (translational slip velocity, rotational slip velocity, and undisturbed rate of strain). By means of a reciprocal relation, Hinch [88] demonstrates that only six of these are independet, and that of these, various internal kinetic symmetry relations exist within three of them.

*Section* 3.1.1. With regard to the last paragraph of this section of papers by Leal and Hinch [89 – 91] demonstrate how singular perturbation methods may be employed to calculate the rheological properties of dilute spheroid suspension in the singular limit where the rotary diffusion coefficient tends to zero.

*Section* 3.1.3.: Rheological results are furnished by Brenner [92] for the properties of a dilute suspension of dumbbells composed of equisized tangent spheres suspended in a simple shear flow, including the effects of rotary Brownian motion.

Brenner and Condiff [93] analyze the rheological properties of a dilute suspension of particles of revolution immersed in an *arbitrary* homogeneous linear shear flow, including the effects of rotary Brownian motion and induced dipoles in the presence of and external electric field.

Pokrovskii [94] summarizes work on the rheology of dilute suspensions of rigid or deformable Brownian spheroids, including the case where dipoles are induced by an electric field.

*Section* 3.2.1.: Prior results for infinitely dilute suspensions of spheroids in extensional flow without rotary Brownian motion have been extended by Batchelor [95] to *nondilute* suspensions of rodlike bodies.

*Section* 3.4.2.: Brenner and Weissman [96] incorporated rotary Brownian motion in their analysis of ferrofluid rheology.

*Section* 3.5.1.: By means of a comprehensive analysis of two–body interactions in a simple linear shear flow, Green [97] obtains a fairly rigorous estimate of $K= 10.7$ for the coefficient of the quadratic term in Eqn. (84). The contribution to this term arises from three distinct sources: far–field interactions, separating interactions, and permanent doublets. The coefficient applies to any viscometric flow.

## ADDITIONAL REFERENCES

88. E. J. HINCH, A remark on the symmetries of certain material tensors for a particle in Stokes flow. *J. Fluid Mech.* (in press).
89. L. G. LEAL and E. J. HINCH, The effect of weak Brownian rotations on particles in shear flow. *J. Fluid Mech.* **46**, 685– 703 (1971).
90. E. J. HINCH and L. G. LEAL, The effect of rotary Brownian motion on the rheological properties of a dilute suspension of rigid spheroids in steady shear flow. *J. Fluid Mech.* (in press).
91. L. G. LEAL and E. J. HINCH, A note on streaming double refraction in a dilute suspension of rigid spheroids subject to weak Brownian rotations. *Rheol. Acta* (in press).
92. H. BRENNER, Rheological properties of a dilute suspension of Brownian doublets in a simple shear flow. *Chem. Eng. Sci.* (in press).
93. H. BRENNER and D. W. CONDIFF, Transport mechanics in systems of orientable particles: IV. Convective transport. *J. Colloid Interface Sci.* (in press).
94. V. N. POKROVSKII, Stresses, viscosity, and optical anisotropy of a flowing suspension of rigid ellipsoids. *Usp. Fiz. Nauk* **105**, 625– 43 (1971).

95. G. K. BATCHELOR, The stress generated in a non-dilute suspension of elongated particles by pure straining motion. *J. Fluid Mech.* **46**, 813–29 (1971).
96. H. BRENNER and M. H. WEISSMAN, Rheology of a dilute suspension of dipolar spherical particles in an external field: II. Effects of rotary Brownian motion. *J. Colloid Interface Sci.* (in press).
97. J. T. GREEN, Properties of suspensions of rigid spheres. Ph.D. Dissertation, 195 pp.+ viii. University of Cambridge, Queen's College, July 1971.

# ON FAST SODIUM COOLED BREEDER RELATED THERMOPHYSICAL INVESTIGATIONS

V. I. Subotin

USSR Academy of Sciences, Moscow, USSR

The concept of economical supply has been considerably revised during the last ten years as a result of prevailing energy trends. Thus the fact that fast uranium-plutonium breeder reactors have a natural uranium efficiency of 60% as oppposed to the 1-2% for light water reactors has acquired particular significance. Despite the currently high unit energy cost, fast sodium cooled breeder reactors are indicated for all developed utility systems. Their only forseeable competition are future hybrid thermonuclear reactors.

The problems plagueing nuclear energy during the last decade, the Three Mile Island and Chernobil accidents, have created a climate adverse to the development of nuclear energy. If we refrain from speculation, the world has but one alternative, a global energy systems of fast sodium cooled breeder reactors and light water epithermal reactors. Such an energy system incorporating reprocessing complexes for spent fuel and plutonium – containing spent fuel, high-activity waste storage and fuel bundle transport has to be first and foremost safe against possible uncontrolled nuclear explosions and in respect to mechanical and thermal accidents. Total radiation safety must also be ensured.

According to modern concepts the probability of large-scale nuclear reactor accidents is small an the aftereffects of nuclear accidents are localizable if the appropriate safety systems are used and safety measures employed. That the effects of nuclear accidents can be localized has been convincingly demostrated at Three Mile Island. Up-dated current concepts are based on reactor inherent safety i.e. the reactor has to have a built-in capacity not only to not amplify accident-induced processes, it has to damp them before they reach the installation operational danger threshold. Experince world-wide indicates that such reactors can be designed and that the future is theirs It should be borne in mind that science answers the question: "Can it be done?" whereas engineers provide answers to : "How do we do it?" The econimics of any endeavor are dependent only on the answer to one question: "Should we build the installation today?". It is very wrong to consider only the economics of giant units and to disregard safety considerations. They deserve priority. Multi-unit high generating capacity installations are much safer.

The reliable operation of all reactor components, units and elements during normal operation and during accidents is ensured through complex and thorough research.

During the last 30 years, extensive, basically experimental research has been conducted the world-over towards the acquisition of sodium cooled breeder thermal physics knowledge as a basis for reactor design.

Unfortunately, many thermophysical papers end with recomendations on how to determine the average heat transfer coefficient. However, that is just the beggining. Local temperature values have to be known with accuracy as they determine nuclear reactor operation. Both the temperature map and temperature variation dynamics have to be known with the utmost precision. In nuclear reactors, in general, and in the active zone, in particular, no leeway exists nor will ever exist as far as the allowed temperatures are concerned.

Thanks to heavy nucleii fission, practically unlimited volume heat extraction is possible at arbitrarily high temperatures. Active zone designs depend on heat extraction constraints and on the maximum temperatures various structural materials can withstand during their design life.

The low heat capacity of sodium ensures an inlet-to-outlet temperature increase in the active zone at design flow rates of 150 to $200°C$. Due to the high hot wall-sodium heat transfer coefficients, that temperature drop is on the order of tens of degrees. This in turn, means that for sodium cooling, the decisive factor is not the heat transfer coefficient, it is the unit cell sodium flow rate. If, due to inadequate design or fuel, cladding or assembly wall expansion, the flow rate of sodium varies, fuel elements and assemblies may swell and rupture.

In order to increase the breeding factor, bundles may be made up of highly enriched and low-enrichment fuel elements as with their different heat generation capacity. Within such a bundle, the sodium in different cells will have different temperatures. Fuel elements in these compound flows will have non-uniform circumferential temperatures and may bend. Inter-cell flow mixing may result in flow temperature pulsations and thus surface temperature pulsations.

The majority of the investigations conducted deal with steady-state thermal dynamic and hydro-dynamic regimes. Special attention should be paid to transient start-up and shut-down, to variable coolant flow regimes, coolant loss, output variations etc.

If mechanical obstructions lower the sodium flow rate through an assembly, sodium boiling may result. In large active zones. the appearence of a void will lead to local reactivity increases. Local increases will accelerate the processes already taking place in that channel and may influence neighboring channels with stable flow, resulting in above-design heat generation. Part of the heat transfer from the boiling channel to neighboring channels may occur through walls as a result of temperature gradients. Boiling may occur in the surrounding assembly channels. In smaller active zones, such as those encountered in multi-unit systems, this is less probable. The locating local boiling in large active zones is one of the paramount problems.

The following are the most important accident related problems:

– The fuel element temperature variation at above-design loading.

– Sodium boiling probability and boiling onset speed at different sodium flow rates including free convection.

– Channel hydraulic resistnce dynamics in the above cases.

– The determination of the maximum power increase that will not cause damage to the fuel elements over the time interval needed for event detection and the decrease of output.

– The interaction of sodium and melted reactor fuel and cladding.
– The chemical reactions occuring and their energy.
– The possibility of heat shocks due to the rapid generation of large quantities of steam.
– The determination of fuel and structural material particle sizes in sodium vapour and liquid sodium and the study of their behaviour.
– The thermophysical properties of liquid and vapour fuel and cladding material towards an assesment of a maximum design accident.

It is important to note that since sodium is a conductor, its flow may be accompanied by self induction. It is also of note that sodium and stainless steel form a thermocouple. If during operation temparature gradients occur, a thermo-electromotive force coupled with the self induction will induce electromagnetic fields and wandering currents.

The significance of these phenomena should be investiquated as well as their effect on:

a) steady-state and transient sodium flow hydrodynamics,
b) reactor control sensors,
c) corrosion,
d) mass transfer of suspended particles in the sodium.

It should be bowne in mind that uranium dioxide is an insulator at room temperature, and at temperatures in the vicinity of its melting point (operating oxide fuel temperatures are 2000°), its conductivity increases by 5 orders of magnitude. Thus, wandering currents can effect fuel elements.

As far as active zone hydrodynamics are concerned, it is important to note that

a)Spacers are designed to prevent fuel rod contact and local temperature increases. Fuel element sodium flow induced vibrations have to be prevented in order to protect the cladding and this is effected by using spacers of the minimum size possible.

b) Undisturbed sodium flow should be inlet to the fuel element assembly to safeguard against flow and sodium heating pulsations.

c) Channel and active zone radial sodium mixing systems of low-mass are needed if we are to dispense with sealed channels and eliminated fuel rod local expansion and channel blocking in the active zone.

The author makes no claim to comprehensivness in this discussion of safe fast sodium cooled breeder reactor design problems.

# FILM CONDENSATION OF LIQUID METALS

W.M. Rohsenow

*Massachusetts Institute of Technology,
Cambridge, U.S.A.*

## 1. INTRODUCTION

Heat transfer associated with film condensation was analyzed by Nusselt [1] for a variety of geometries assuming the liquid condensate falls due to gravity in laminar flow, the liquid vapor interface temperature equals the saturation temperature of the vapor, momentum terms are negligible and the temperature distribution in the condensate film is essentially linear. His result for a vertical flat plate is

$$h(T_i - T_w) = q/A = 0.943 \sqrt[4]{\left(\frac{g\rho_l^2 k_l^3 h_{fg}}{L\mu_l}\right)} (T_i - T_w)^{0.75} \quad (1)$$

where $T_i$ = saturation temperature. Later Rohsenow[2] modified an analysis of Bromeley [3] to include the effect of a gravity pressure gradient and true nonlinear temperature distribution in the condesate layer and the subcooling effect on the enthalpy of the liquid layer with the following result:

$$h(T_i - T_w) = q/A = 0.943 \sqrt[4]{\left(\frac{g\rho_l(\rho_l - \rho_v)k_l^3[h_{fg} + 0.68(T_i - T_w)]}{L\mu_l}\right)} (T_i - T_w)^{0.75} \quad (2)$$

The analysis also provides the following relation for liquid film thickness

$$\delta = \sqrt[4]{\frac{4k_l(T_i - T_w)\mu_l Z}{g\rho_l(\rho_l - \rho_v)h_{fg}'}} \quad (3a)$$

or in terms of $q/A$ combine with eqn.(2) to obtain

$$\delta = \sqrt[3]{\frac{4\mu_l Z(q/A)}{g\rho_l(\rho_l - \rho_v)h_{fg}'}} \quad (3b)$$

Comparison of eqns. (1) and (2) shows the effect of the vertical gravity pressure gradient to be a change from a $\rho_l$ to a $\rho_l - \rho_v$, which is important only close to the critical pressure. The effect of the nonlinear distribution and the subcooling enthalpy in the liquid is to add $0.68 c_l(T_i - T_w)$ to $h_{fg}$. These are all very small effects for most applications. The effect of the momentum changes in the liquid film are seen in Fig. 1 to become important as $c_l \Delta T / h_{fg}$ increase for low Prandtl numbers, which represent liquid metals; however, for most applications the heat fluxes are low enough to produce $c_l \Delta T / h_{fg}$ magnitudes in the lower range where these momentum effects are not large and the simple Nusselt result is adequate.

Another effect which tends to increase the heat–transfer coefficient is ripples on the liquid interface; this is usually only a small increase. If the vapor passage is a confined channel, the vapor velocity may be high and produce a high shear stress at the liquid–vapor interface resulting in a thinner film; also if the surface is long enough in the flow direction the liquid layer may change to turbulent flow. In downward flow of vapor both of these effects significantly increase the heat transfer coefficient, Rohsenow et al. (7) and Duckler (8).

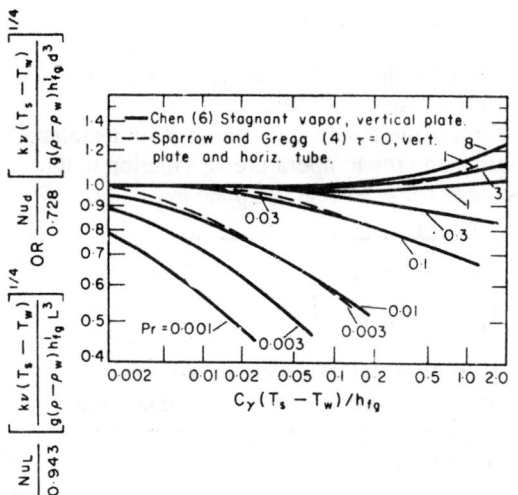

Fig. 1. Average Nusselt number for condensation on horizontal tubes and vertical plates

The heat–transfer coefficient will be decreased if the vapor flows upward along the liquid–vapor interface. A major cause of low heat–transfer coefficients is the presence of even small traces of non–condensable gas which get carried with the vapor to the cold surface and accumulate there providing a diffusion resistance to the vapor.

When proper account of these various effects is taken, the simple Nusselt type analysis with these modifications agree rather well with data obtained for condensation of non–metal vapors. However, data for condensation of metal vapors, even when all the above mentioned effects are considered, may fall well bellow the predicted magnitudes by as much as a factor of 50, the disagreement being larger at lower pressures. This, of course, assumes that $T_i$ is taken as $T_{sat}$ in defining $h$.

Eventually various people engaged in this research suggested with varying degrees of conviction that perhaps the interface temperature $T_i$, Figure 2, could be well below $T_{sat}$ and the mass exchange process of the liquid–vapor interface might account for the low apparent $h$ values when they are based on $(T_{sat}-T_w)$.

In the attempt to isolate one more effect, Sukhatme [9] attempted to measure the condensate layer thickness $\delta$ for condensing mercury by measuring the attenuation of gamma–ray radiation. At the measured heat fluxes the measured $\delta$ compared

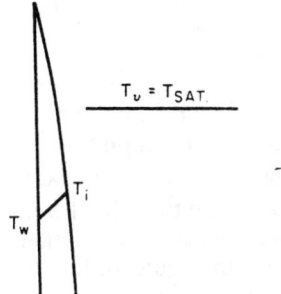

Fig. 2. Saturtated vapor.

favorably, within the precision of measurement (which was $+25\%$), with $\delta$ calculated at the measured heat flux from the Nusselt equation (3b). This suggests that the Nusselt result applies, eqns. (1) or (2), provided $T_i$ is the true liquid – vapor interface temperature which is less than saturation and that the measured $h$ is composed of two resistances in series

$$\frac{1}{h}=\frac{1}{h_{LF}}+\frac{1}{h_i} \qquad (4)$$

where $h_{LF}$ is the liquid film coefficient expressible by eqns. (1) and (2) and $h_i$ is the coefficient associated with the mass exchange process taking place at the interface.

## 2. TEMPERATURE DROP AT LIQUID–VAPOR INTERFACE

The equations developed previously all include the assumption that the liquid – vapor interface is at the saturation temperature. For nonmetal vapors this is a good assumption except at very low pressures, but for metal vapor the temperature difference $T_{sat}-T_i$, Fig.2, can be large compared with $(T_i-T_w)$. This difference becomes larger as system pressure is decreased. A condensation heat transfer coefficient based on $(T_{sat}-T_w)$ will not agree with the predictions of the previous equations where $h$ is based on $(T_i-T_w)$.

Schrage [10] used the results of elastic–collision, Maxwellian distribution kinetic theory to obtain the following relations for the mass exchange process at the liquid – vapor interface during condensation.

$$\frac{q}{A} = \frac{w}{A} h_{fg} \tag{5}$$

and

$$\frac{w}{A} = \frac{1}{\sqrt{(2\pi R)}} \left[ \sigma_c \frac{P_v}{\sqrt{T_v}} \Gamma - \sigma_e \frac{P_i}{\sqrt{T_i}} \right] \tag{6}$$

where

$$\Gamma = 1 + \frac{w/A}{P_v \sqrt{(2/\pi R T_v)}} \tag{7}$$

Here the first term in eqn. (6) represents the molecules from the vapor captured by the liquid surface and the second term represents those emitted from the liquid surface into the vapor. The term $\Gamma$ is present because the vapor has a "progress" velocity toward the surface, superimposed on the Maxwellian distribution; on the other hand, the vapor leaving the surface is considered to have a "half – Maxwellian" distribution in coming off the liquid surface. Here $p_v$ is the vapor pressure at the liquid surface which is essentially system pressure; $p_i$ is the saturation pressure corresponding to $T_i$ since this form of the second term would be the emission from the surface if the system were at a uniform $T_i$ and $p_i$ at equilibrium. It is assumed that the emission of molecules is not affected by the higher pressure in the vapor when condensation is occuring.

Combining eqns. (6) and (7)

$$\frac{w}{A} = \frac{1}{\sqrt{(2\pi R)}} \frac{2}{2-\sigma_c} \left[ \sigma_c \frac{P_v}{\sqrt{T_v}} - \sigma_e \frac{P_i}{\sqrt{T_i}} \right] \tag{8}$$

If $\sigma_c = \sigma_e = \sigma$

$$\frac{w}{A} = \frac{2\sigma}{2-\sigma} \frac{1}{\sqrt{(2\pi R)}} \left[ \frac{P_v}{\sqrt{T_v}} - \frac{P_i}{\sqrt{T_i}} \right] \tag{9}$$

These equations may be put into an alternative form if $(T_v - T_i) \ll T_v$:

$$\frac{w}{A} = \frac{P_v}{\sqrt{(2\pi R T_v)}} \frac{2\sigma_c}{2-\sigma_c} \left[ \frac{\sigma_e}{\sigma_c} \frac{\Delta p}{P_v} - \frac{1}{2} \frac{\sigma_e}{\sigma_c} \frac{\Delta T}{T_v} + \left(1 - \frac{\sigma_e}{\sigma_c}\right) \right] \tag{10}$$

or if $\sigma_c = \sigma_e = \sigma$

$$\frac{w}{A} = \frac{2\sigma}{2-\sigma} \frac{P_v}{\sqrt{(2\pi R T_v)}} \left( \frac{\Delta p}{P_v} - \frac{1}{2} \frac{\Delta T}{T_v} \right) \tag{11}$$

where $\Delta p \equiv (p_v - p_i)$ and $\Delta T \equiv (T_v - T_i)$.

For pressures as low as 0.001 atm for metal vapors and also water vapor the magnitude of $\Delta T/2T$ is around 3–4% of $\Delta p/p$ and is usually neglected. At higher pressures this relative magnitude is smaller.

There have been a number of attempts to improve the above result because of a general feeling of discomfort with the simple hard–sphere kinetic theory model. These attempts have not resulted in any major change in the equation.

Bornhorst and Hatsopoulos [11] and Adt [12] obtained results for this interface process through irreversible thermodynamics. The significant terms in their result reduced to the above Shrage equations.

Labuntsov [13] presented an analysis using the "double–flow" distribution function and the Lees [14] Maxwell moment method. Springer and Patton [15] extended this type of analysis to include the possibility that rarefied gas effects could be important. Their result is

$$\frac{w}{A} = \frac{p_v}{\sqrt{(2\pi RT_v)}} \left\{ \frac{\frac{\sigma_e}{\sigma_c}\frac{\Delta p}{p_v} - \frac{1}{2}\frac{\Delta T}{T}\left(\frac{1}{1+4/15Kn} + \frac{\sigma_e}{\sigma_c} - 1\right) - \frac{\sigma_e}{\sigma_c} + 1}{\frac{2-\sigma_c}{2\sigma_c} + \frac{1}{2+8/15Kn}} \right\} \quad (12)$$

Here $Kn$ is the Knudsen number defined as the ratio of the mean free path of the molecules to some system physical dimension. All available condensation data have been taken at pressures where $Kn$ is essentially zero. Then eqn. (12) reduces identically to the Schrage equation [10].

Another analysis by Huang [16] suggests that certain molecules at low grazing angles to the surface do not get captured. All other molecules are assumed to be captured. This analysis led to the following result:

$$\frac{w}{A} = \frac{\pi}{2} \frac{p_v}{\sqrt{(2\pi RT_v)}} \left(\frac{\Delta p}{p_v} - \frac{1}{2}\frac{\Delta T}{T_v}\right) \quad (13)$$

Comparison of eqn. (13) with eqn. (11) shows that $2\sigma/(2-\sigma)$ is replaced by $\pi/2$. Equating these gives $\sigma = 0.88$.

Before looking critically at data in the light of measurement precision [17] and the effect of non–condensable gas [18], existing data was analyzed [19] in terms of possible vapor subcooling near hte interface and a "temperature–jump" coefficient. We now believe that this does not occur.

Any of the preceding equations in $w/A$ may be written in terms of an "interfacial" heat–transfer coefficient with eqn. (5)

$$\frac{q/A}{T_{sat}-T_i} \equiv h_i = \frac{w}{A}\frac{h_{fg}}{(T_{sat}-T_i)} \quad (14)$$

which for eqn. (11) becomes

$$h_i = \frac{2\sigma}{2-\sigma} \frac{h_{fg}}{\sqrt{(2\pi RT_v)}} \left(\frac{P_{sat}-P_i}{T_{sat}-T_i}\right) \quad (15)$$

neglecting $p/2T$ compared with $\Delta p/\Delta T$.

Further from the Claussius–Clapeyron equation

$$\left(\frac{P_{sat}-P_i}{T_{sat}-T_i}\right) \cong \frac{h_{fg}}{v_{fg}T} \quad (16)$$

This approximation need not be employed.

## 3. EFFECTIVE OVERAL HEAT TRANSFER COEFFICIENT

An overal heat–transfer coefficient for the condensation process may be defined as

$$h \equiv \frac{q/A}{T_{sat}-T_w} \tag{17}$$

Considering the liquid film resistance to be in series with the interface resistance

$$\frac{1}{h} = \frac{1}{h_{LF}} + \frac{1}{h_i} \tag{18}$$

where $h_{LF}$ is given by eqn. (2) and $h_i$ by eqn. (15).

## 4. ACCOMMODATION COEFFICIENTS

Accommodation coefficients, $\sigma_c$ and $\sigma_e$, have been measured individually and together in a variety of experiments. These are summarized by Wilhelm [20] and Wylie and Brodkey [21]. Magnitudes from very low (0.001) to very high (1.0) have been reported. The many reviewers of this collection of data arrive at a variety of conclusions.

Looking selectively at this collection one finds that those experimenters who paid particular attention to eliminating noncondensable gas, and who took pains to vary the cleanliness of the system and the liquid found that as noncondensable gas was nearly eliminated and as cleanliness was increased the measured accommodation coefficients rose to magnitudes of close to unity.

These observations suggest that in a condensation process the accommodiaiton coefficients $\sigma_c$ and $\sigma_e$ ought to be unity since the liquid condensed at the liquid – vapor interface is surely quite clean in practically all systems. Of course, if Huang's suggestion [16] that molecules at small grazing angles are not captured is correct, perhaps this upper limit of $\sigma$ is around 0.88 or higher.

There is, quite naturally, an uneasy feeling in the minds of a number of people concerning the assumption that $\sigma_e = \sigma_c$. No generally accepted theory can prove or disprove this hypothesis.

In the next section it will be shown that condensation data taken in the range where measurement precision is adequate suggests that an appropriate conclusion to draw is that $\sigma_c = \sigma_e$ and their magnitudes are 1.0 or very close to it.

## 5. EFFECT OF NONCONDENSABLE GAS

The presence of traces of noncondensable gas in a condensation apparatus is perhaps the most perpleaxing problem an expreimenter faces. Gas may be present in the system from at least three sources: residual gas remaining after pumping down the system to as low as $10^{-7}$ to $10^{-8}$ torr, gas being brought into the system in the liquid, and gas given off by the inner walls of the system as the system is heated. This, of course, assumes the system is leak tight to any atmospheric in–leakage.

During the operation this gas is carried by the vapor to the cold surface and accumulates there. The condensing vapor at $p_v$ must diffuse through this gas reducing its pressure to $p_{vi}$, Fig. 3

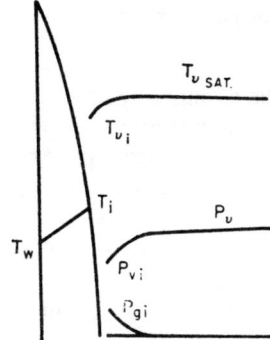

Fig. 3. Non–condensable gas.

The problem was studied by Kroger [22] who verified that the molecular diffusion equations predicted the effect of noncondensable gas. In most condensing systems the quantity of gas is so small that it practically all accumulates ina region about 0.01 in. thick; hence, it is necessary to identify this quantity of gas in terms of total mass of gas $M_g$ in the system rather than the percent of noncondensables in the vapor. This small amount of gas has a strong effect on $h$ and hence on the magnitude of the apparent $\sigma$, if it is ignored. Furthermore, the quantity of gas required to produce a serious effect on h and $\sigma$ is so small that it is virtually inpossible to detect its prsence its presence or to measure it.

Kroger [22] showed that the diffusion process between $p_v$ and $p_{vi}$, Fig.3, could be expressed as flollows:

$$\frac{w}{A} = \frac{(p_v - p_{vi}) D p_v}{(M_g/A) R_v R_g T_v^2} \tag{19}$$

where $M_g/A$ is the total mass of noncondensable gas in the system per unit area of condensing surface.

Since the driving force (potential) for both eqn. (13), neglecting $\Delta T/2T$, and eqn. (19) is $\Delta p$, it is appropriate to define the effect of the non–condensable gas in a heat transfer coefficient, $h_g$, in terms of $(T_{sat} - T_{vi})$ where $T_{sat}$ and $T_{vi}$ are saturation temperatures corresponding to $p_v$ and $p_{vi}$ respectively. Then with eqn. (5) and eqn. (19)

$$\frac{q/A}{T_{sat} - T_{vi}} = h_g = \frac{h_{fg} D p_v}{(M_g/A) R_v R_g T_v^2} \frac{(p_v - p_{vi})}{(T_{sat} - T_{vi})} \tag{20}$$

With the gas present, eqn. (15) must be written as follows:

$$\frac{q/A}{T_{vi} - T_i} = h_i = \frac{2\sigma}{2-\sigma} \frac{h_{fg}}{\sqrt{(2\pi R T_v)}} \frac{(p_{vi} - p_i)}{T_{vi} - T_i} \tag{21}$$

Then treating the three resistances, Fig. 3, in series the overall heat transfer coefficent, eqn. (15), may be written as

$$\frac{1}{h} = \frac{1}{h_{LF}} + \frac{1}{h_i} + \frac{1}{h_g} \tag{22}$$

where $h_{LF}$ is given by eqn. (2).

## 6. ACCELERATION OF VAPOR TO THE CONDENSING SURFACE

An effect neglected in treating existing condensation data, as suggested by Huang [16], is the static pressure drop in accelerating the vapor flow from some low velocity region to its velocity normal to the condensing surface. For an isentropic expansion

$$\frac{w}{A} = \rho_0 \sqrt{\frac{2g_0 \gamma R T_0}{\gamma - 1} \left[ \left( \frac{p}{p_0} \right)^{2/\gamma} - \left( \frac{p}{p_0} \right)^{(\gamma+1)/\gamma} \right]} \tag{23}$$

where $p_0$ is the stagnation pressure. The appropriate pressure to use in the preceding equations is the static pressure.

Associatied with this pressure drop is a temperature drop. However, the important quantity in eqns. such as (10) and (19) is $\Delta p/p$ not $\Delta T/2T$. Therefore, the true effect of the acceleration appears in the isentropic pressure drop. As discussed in connection with eqn. (20) the appropriate temperatures to consider here are the saturation temperatures corresponding to these two pressures.

While this acceleration effect is real, its actual magnitude in all available test data appears to have been very small. The correction to the $\Delta p/p$ quantity, and hence to the $\sigma$, has been small.

## 7. CONDENSATION DATA

Generally data is taken for measurements of $q/A$, $Tv_{sat}$ or $p_v$ and $T_w$. With this data and the fluid properties $\sigma$ is determined from eqn. (18) with eqn. (17), (2) and (15). The results of available determinations of $\sigma$ in this way are shown in Fig. 4 where s is plotted at viarious $p_v$ magnitudes. The magnitudes of $\sigma$ thus obtained may have errors due to lack of adequate precision of measurement and also due to the presence of traces of non-condensable gas. The magnitude of $\sigma$ in Fig. 4 is really an

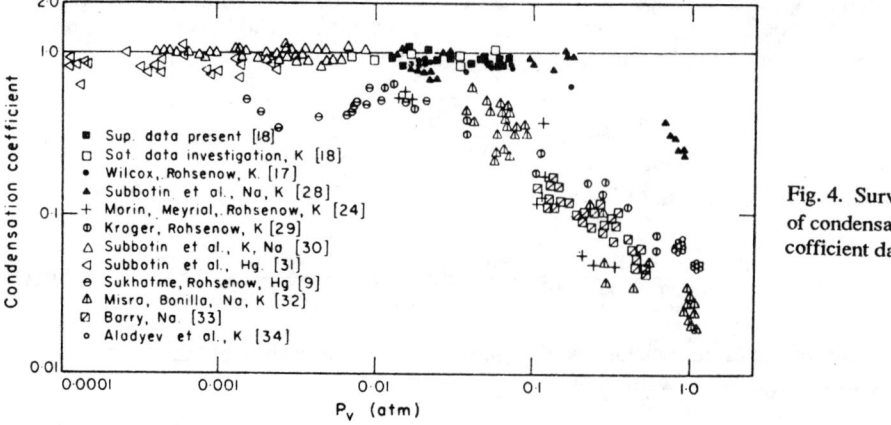

Fig. 4. Survey of condensation cofficient data.

"apparent" σ which may include these effects. Each of these effects will be considered separately. Careful consideration of each set of data in Fig. 4 in this light leads one to the conclusion that the true magnitude of σ is unity or perhaps 0.88 as suggested by Huang [16].

## 8. PRECISION OF MEASUREMENT

The major cause of error is in determination of $(T_v - T_i)$ or $(p_v - p_i)$. At a given $q/A$, this temperature or pressure difference decreases rapidly to very small magnitudes as pressure increases toward 1 atm. To illustrate this the following table shows magnitudes for $(T_v - T_i)$ for various metals if $\sigma \approx 1.0$ when $q/A = 50{,}000$ Btu/hr ft$^2$, which is the approximate level of heat flux in most experiments.

TABLE 1. $(T_v - T_i)$ (°F)

| $P_v$(atm) | Potassium | Sodium | Mercury |
|---|---|---|---|
| 0.001 | 21.5 | 16.0 | 31.0 |
| 0.01 | 3.2 | 2.3 | 4.7 |
| 0.1 | 0.5 | 0.4 | 0.7 |
| 1.0 | <0.1 | <0.1 | 0.1 |

In every system there is some error in the measurement of temperature or pressure making the determination of $(T_v - T_i)$ at the higher pressure very difficult or impossible to accomplish with the necessary precision.

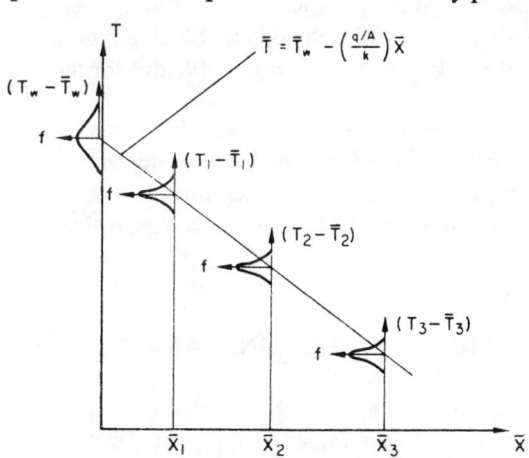

Fig. 5. Distribution of measurements errors.

While some experimenters have attempted to measure directly the temperature distribution in the liquid and vapor regions, most have measured temperatures in the cold block and extrapolated these readings to the surface to obtain a value for $T_w$. Then with eqn. (2) and a measured $T_v$ or $p_v$ the magnitude of $(T_v - T_i)$ or $(p_v - p_i)$ is deduced. In either case there is some measurement error to be expected casting doubt on the ability to determine $(T_v - T_i)$ magnitudes of 0.5 to less than 0.1 °F at pressures of 0.1 to 1.0 atm, Table 1.

Wilcox[17] studied precision of measurement associated with the measurements in the cold block as shown in Fig. 5. He postulated a Gaussian distribution in error which was related to thermocouple hole size at each location. He presents a relation for the standard deviation (error band) in the determination of the magnitude of $T_w$. This error band depends on hole size, number and spacing of the holes and very strongly on the thermal conductivity of the cold block. The error band is reduced by using larger conductivity cold blocks, small size and a greater number of thermocouple holes.

The following is the standard deviation error band in the determination of $T_w$ for three systems tested at M.I.T.

TABLE 2. STANDARD DEVIATION $(T_w - \overline{T}_w)$ FOR $q/A = 50,000$ Btu/hr $ft^2$

| Autor | Cold blok | No. holes | Diam. holes (in.) | $(T_w - \overline{T}_w)$ |
|---|---|---|---|---|
| Wilcox [17] | Copper | 6 | 0.046 | 0.24°F |
| Kroger [23] | Nickel | 3 | 0.062 | 2.9°F |
| Meyrial – Morin [24] | Stainless Steel | 3 | 0.062 | 8.0°F |

The measurement errors are of course independent of the presure level in the system. Certainly if the magnitude of this error band, standard deviation $(T_w - \overline{T}_w)$, is the same magnitude as $(T_v - T_i)$, we are expecting too much of our measurements to determine $(T_v - T_i)$ with any reasonable certainty. Inserting the numbers of Table 2 in Table 1 for potassium suggests that we should not have any faith in our determinations of $\sigma$ at pressures above about 0.3 atm for the copper block apparatus [17], above about 0.02 atm for the nickel block and above about 0.006 atm for the stainless–steel block.

In Fig. 4 the data with open points were taken on nickel or stainless–steel surfaces and the solid points on copper surfaces. Assuming the limit of measurement precision is similar in other systems, a close inspection of the magnitudes of $\sigma$, excluding data at pressures greater than the limits mentioned above , suggests that its magnitude is quite high close to unit.

## 9. EFFECT OF NON–CONDENSABLE GAS ON DATA

The effect of the presence of trace quantities of non–condensable gas on the determination of $\sigma$ from eqn. (18) is to result in lower values of apparent $\sigma$ even if the true value is unity.To illustrate the seriousness of this effect consider the data of Sukhatme [25] for condensing mercury. The magnitudes of apparent s calculated from data and eqn. (18) were in the range of $0.4 - 0.7$. If it is now assumed that the true magnitude of $\sigma = 1.0$ in $h_i$ of eqns. (21) and (22), then the magnitude of $h_g$ and $M_g/A$ in eqns. (20) and (22) can be calculated. The mass of gas $M_g$ in the system to account for Sukhatme's $\sigma$ being below 1.0 to $0.4 - 0.7$ was determinde to be equivalent to a gas pressure through the system volume of around $10^{-7}$ torr. It happens that this is the approximate level to which the system was evacuated prior to testing, suggesting that non-condensable gas could indeed have been the cause of the apparent σ being less than unity. Similar calculations can be made for each apparatus.

In later apparatus, Fig. 6, Wilcox [17] added a secondary condenser and Sakhuja [18] added a bleed just above the secondary condenser. They both used copper block condenses; hence their precision of measurement for potassium should have been adequate up to pressures of 0.3 atm (Tables 1 and 2). After evacuating the system to around $10^{-7}$ torr they first started the secondary condenser to gather any remaining non-condensables before operating the test condenser. Sakhuja also bled the gas off. Data obtained in this way at pressures below 0.3 atm yielded values of σ in the neighbourhood of unity, Fig. 4.

In another set of experiments Sakhuja [18], after operating the system as described above, shut off the secondary condenser with the bleed line closed and made

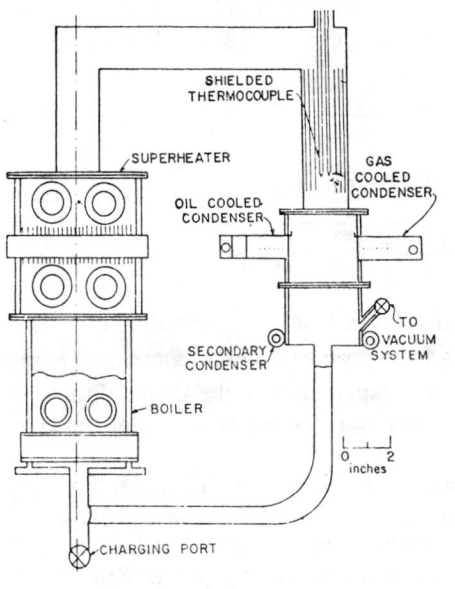

Fig. 6. Apparatus with secondary condensers and bleed [18].

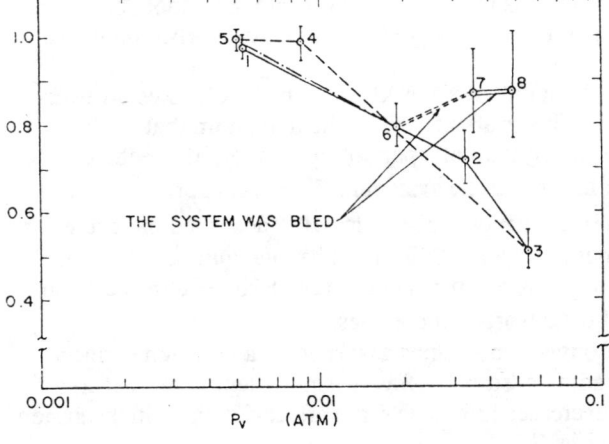

Fig. 7. σ with and without bleed [18].

a series of tests in the sequence shown in Fig. 7 – points 1 through 6. Here the apparent σ was determined from eqn. (18). Note that this data is in a pressure range less than 0.3 atm where measurement precision was adequate and where on previous runs with the secondary condenser in operation the magnitude of σ was found to be around unity. The curve in Fig. 7 looks like various data curves in Fig. 4.

At point 6, Fig. 7, Sakhuja placed the secondary condenser into operation. This would gather some of the noncondensable gas away from the test condenser. Then he took the data for point 7, where the value of σ rose toward unity. Subsequent operation by starting the secondary condenser before the test condenser yielded values close to unity, Fig. 4.

This series of tests strongly suggests that the decreasing σ with pressure in Fig. 7, and perhaps also in Fig. 4, is due to the presence of traces of non-condensable gas.

At a particular $q/A$, $P_v$, $T_v$ and $T_w$, the $h_i$ of eqn. (18) yields an apparent $\sigma_{app}$. The same data with the assumption that $\sigma = 1.0$ in $h_i$ of eqn. (22) yields a magnitude of $M_g/A$. Then equate the following:

$$\frac{1}{h_{i_{\sigma,app}}} = \frac{1}{h_{i_{\sigma}=1}} + \frac{1}{h_g} \tag{24}$$

Then with eqns. (15), (20) and (21) and (16) this vecomes

$$\frac{2\sigma_{app}}{2-\sigma_{app}} \cong 1 + \sqrt{\left(\frac{\pi}{2R_v R_g^2}\right)\left(\frac{DP_v}{T^{3/2}}\right)\frac{1}{M_g/A}} \tag{25}$$

where $(DP_v/T^{3/2})$ is essentially constant. This shows that if $\sigma_{app}$ less than unity is due to noncondensable gas then $\sigma_{app}$ decreasing from unity requires an increasing amount of gas $M_g/A$ at the cold surface. Therefore, this explanation of the shape of the σ vs. P curves in Fig. 7 or Fig. 4 requires the presence of more gas in the vapor space at the higher pressures.

Since the solubility of air in potassium liquid increases with tempreature the source of gas is probably not the liquid. On the onther hand, the area of dry metal surface inside the system is large. It is quite probable that the metal surface has gas adsorbed on it or absorbed in it. As pressure is increased the temperature of all surfaces of the system increases; hence gas may be driven off into the vapor space. Then when the system pressure (temperature) is lowered the gas is readsorbed on the surface.

A reasonable estimate of the number of weakly adsorbed molecules on a surface is $10^{14}$ molecules/cm$^2$ [35]. This is about 10% of the maximum that could be adsorbed. Noncondensable gas in a system at a partial pressure of $10^{-10}$ atm can cause the decrease in σ shown in Fig. 7. The system in Fig.6 has approximately 1500 cm$^2$ of solid surface area in the vapor space or $15 \times 10^{16}$ adsorbed molecules. It has a vapor space volume of approximately 4000 cm$^3$. The total number of molecules of gas in this volume required to produce a partial pressure of $10^{-10}$ atm is around $5 \times 10^{13}$ or only about 0.1% of the adsorbed molecules.

Recent measurements of oxygen molecules adsorbed on a tungsten surface show a reversable change in amount adsorbed – decreasing as temperature increases and increasing as temperature decreases [36]. The same effect if present in the system of Fig. 6 can explain the dat in Fig. 7.

## 10. EFFECT OF SUPERHEAT

The effect of superheat in the vapor was also measured by Sakhuja [18].

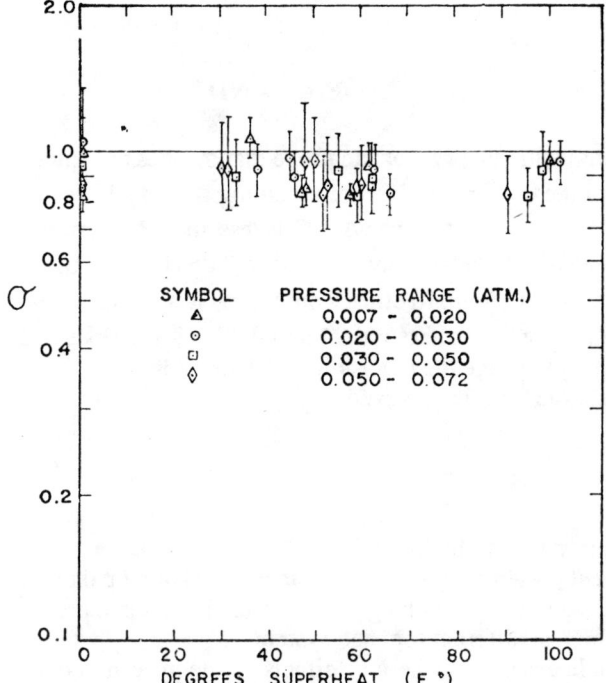

Fig. 8. Effect of superheat [18].

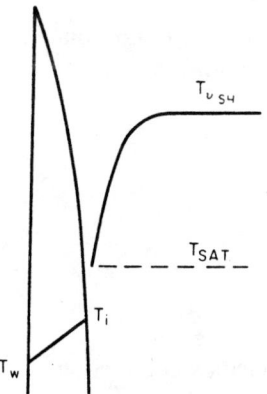

Fig. 9. Temperature for superheated vapor.

The presence of noncondensable gas had a greater effect on the data in the superheat vapor than it did in the saturated vapor case. With the secondary condenser in operation the data yielded magnitudes of $\sigma =$ unity if eqn. (18) was used and $T_v$ in eqn. (15) was taken as $T_{sat}$, Fig. 8. In other words, condensation of superheat liquid metal vapor is predicted by the same equations as saturated vapor – using system pressure and $T_{sat}$. The impliction of this is that just prior to reaching the liquid – vapor interface the vapor acts as though its pressure is system pressure and its temperature is $T_{sat}$, Fig. 9.

## 11. ON THE POSSIBLE INEQUALITY OF $\sigma_c$ AND $\sigma_e$

There appears to be no convincing theory that can prove that $\sigma_c$ does not equal $\sigma_e$. Let us consider for a moment the disastrous effect of assuming $\sigma_c \neq \sigma_e$. In eqn. (10) the primary terms are $\Delta p/p$ and $(1 - \sigma_e/\sigma_c)$ since $\Delta T/2T$ is less and 3% of $\Delta p/p$. For practically all of the avialable data, $\Delta p/p$ lies between 0.01 to 0.06. Then $\sigma_e/\sigma_c$ between 1.01 and 1.06 would cancel the $\Delta p/p$. In other words, if $\sigma_e$ is assumed equal to 1.0 then all data could be made to lie between $\sigma_c = 0.94$ and 0.99. This looks good on a graph [26,27] but does not at all change the precision of determining the condensation rate. It tends to hide the trouble observed in Fig. 4.

## 12. CONCLUSIONS

A careful review of the data suggests the following:
1. Precision of measurement presents a greater problem when obtaining data at higher pressure levels and is inadequate for obtaining precise data above some pressure levels, the magnitude of which is different for each system.
2. Traces fo noncondensable gas produce serious effects on the measured data. Eliminating these gases is difficult. They and their effects are present to a greater or lesser degree in all condensing test apparatus.
3. The true value of the condensing coefficient $\sigma$ in the absence of noncondensable gas is close to unity, within the range of precision of our measurement.

## NOMENCLATURE

$c_l$    liquid specific heat
$D$    diffusivity – vapor and gas
$g$    acceleration due to gravity
$h$    condensing side heat – transfer coefficient (eqns.(4),(22))
$h_g$    condensing coefficient through a noncondensable gas region
$h_i$    condensing side heat – transfer coefficients (predicted by considering liquid – vapor resistance)
$h_{LF}$    heat – transfer coefficient predicted from theories which take account only of the condensate film's thermal resistance
$h_{fg}$    latent heat of vaporization
$h'_{fg}$    equivalent latent heat
$k_l$    liquid thermal conductivity
$Kn$    Knudsen number

$L$     length of condenser surface
$M_g$     mass of noncondensable gas at condensing surface
$p$     pressure
$p_i$     saturation pressure corresponding to temperature $T_i$
$p_v$     pressure of pure saturated vapor in bulk space
$Pr$     Prandtl number
$q\backslash A$     average heat flux (units of heat/unit area)
$R$     gas constant
$Re_L$     film Reynolds number at $z=L$
$T$     temperature
$T_i$     temperature of liquid at liquid–vapor interface
$T_{SH}$     temperature of superheated vapor
$T_v$     saturation temperature corresponding to pressure $p_v$
$T_w$     condenser wall temperature at outer surface
$w/A$     mass flux (mass/unit time, unit are)
$z$     distance along condenser
$\delta$     condensate film thickness
$\gamma$     ratio of specific heats
$\Gamma$     nondimensional correction factor, eqn. (7)
$\mu_i$     liquid viscosity
$\rho_1$     liquid density
$\rho_v$     density of saturated vapor at conditions $\rho_v$ and $T_v$
$\sigma_c$     condensation coefficient
$\sigma_e$     evaporation coefficient
$\sigma_{app}$     $\sigma$ including effect of noncondensable gas

## REFERENCES

1. NUSSELT, W., *Z. ver. deut. Ing* **60**, 541, 569 (1916).
2. ROHSENOW, W.M., *Trans. ASME* **78**, 1645–8 (1956).
3. BROMLEY, L.A., *Ind. Engng. Chem.* **44**, 2966 (1952).
4. SPARROW, E.M. and GREGG, J.L., *Trans. ASME, J. Heat Transfer*, ser. C, 1959, pp. 13–18, including discussion by R.A. Seban.
5. KOH, J.C.Y., SPARROW, E.M. and HARTNETT, J.P., *Int.J. Heat Mass Transfer*, **2**, 69–82 (Mar.1961).
6. CHEN, M.M., *Trans ASME, J. Heat Transfer*, ser. C, **83**, 48–60 (Feb. 1961).
7. ROHSENOW, W.M., WEBBER, J.H. and LING, A.T., *Trans. ASME*, **78**, 1637–44 (1956).
8. DUKLER, A.E., Fluid mechancs and heat transfer in vertical falling films. 3rd National Heat Trans. Conf. ASME/AIChE, Storrs, Conn., Aug. 1959, Preprint 101.
9. SUKHATME, S.P. and ROHSENOW, W.M., Heat transfer during condensation of liquid metal vapors, *ASME J. Heat Trans.*, **C88**, 19–28 (1966).
10. SCHRAGE, R.W., *A Theoretical Study of Interphase Mass Transfer*, Columbia University Press, New York, 1953.
11. BORNHORST, W.J. and HATSOPOULOS, G.N., Analysis of a liquid vapor phase change by the methods of irreversible thermodynamics. *ASME J. Appl. Mech.*, **34E**, 840 (Dec. 1967).
12. ADT, R.R., A study of the liquid–vapor phase change of mercury base on irreversible thermodynamics. Ph.D.Thesis, MIT Cambridge, Massachusetts, 1967.

13. LABUNTSOV, D.A., An anaslysis of evaporation and condensation processes, *Teplofizika Vysokikh Temperatur* **5** (4), 647–54 (1967).
14. LEES, L.,J. *Soc. Ind. Appl. Mat.* **13**, 278 (1965).
15. SPRINGER, G.S. and PATTON, A.J., A kinetic theory description of liquid–vapor phase change. Sixth Rarefied Gas Dynamics Conference, 1968.
16. HUANG, Y.S., Heat transfer by condensation of low pressure metal vapors. Ph.D. Thesis, Case Western Reserve Univ., 1971. (Professors F.A. Lyman and W.J. Lick).
17. WILCOX, S.J. and ROHSENOW, W.M., Film condensation of potassium using copper condensing block for precise wall temperature measurement *ASME J. Heat Transfer* **92C**, 359–71 (Aug. 1970).
18. SAKHUJA, R.K., Effect of superheat on film condensation of potassium. Sc.D. Thesis, Mech. Engng., MIT, Sept. 1970.
19. FEDOROVICH, E.D. and ROHSENOW, W.M., The effect of vapor subcooling on film condensation of metals. *Int. J. Heat Mass Transfer* **12**, 1525–9 (Nov. 1969)
20. WILHELM, D.J., *Condensation of Metal Vapors: Mercury and the Kinetic Theory of Condensation*. USAEEC Rept. No. ANL 6948,Oct. 1964; also Ph.D. thesis, Chem. Engng. Ohio State Univ., 1964.
21. WYLIE, K.F. and BRODKEY, R.S., Transport phenomena at the liquid–vapor interface of mercury using radioactive tracers. *Int. Symp. on Two–phase Flow, Haifa*, Aug. 1971.
22. KROGER, D.G. and ROHSENOW, W.M., Condensation in the presence of non–condensable gas. *Int. J. Heat and Mass Transfer* **11**, 15–26 (1968).
23. KROGER, D.G. and ROHSENOW, W.M., Film condensation of saturated potassium vapor. *Int. J. Heat Mass Transfer*, **10** (Dec. 1967).
24. MEYRIAL, P.M., MORIN, M.L. and ROHSENOW, W.M., *Heat Transfer During Film Condensation of Potassium Vapor on a Horizontal Plate*, Report No. 70008–52, Engineering Projects Laboratory, MIT, Cambridge, Mass., 1968.
25. SUKHATME, S. and ROHSENOW, W.M., Film condensation of liquid metal, *ASME J. Heat Transfer*, **88c**, 19–29 (Feb. 1966).
26. BARRY, R.E. and SPRINGER, G.S., Vapor phase resistance filmwise condensation.ASME Paper 69–WA/HT–26, Los Angeles, Nov. 1969.
27. SARTOR, A.M., BALYHISER, R.E. and BARRY, R.E., Condensing heat transfer considerations relevant to rubidium and other alkali metals. 11th Natl. Heat Transfer Conf., Minneapolis, AIChE Paper no. 5, Aug. 1969.
28. SUBBOTIN, V.I., IVANOVSKII. M.N., SAROKIN, V.P. and CHULKOV, V.A. *Teplophizika Vysokih Temperatur*. no. 4, p. 616 (1964).
29. KROGER, D.G. and ROHSENOW, W.M., Film condensation of saturated potassium vapor. *Int.J. Heat Mass Transfer* **10** (Dec. 1967).
30. SUBBOTIN, V.I., BAKULIN, N.V., IVANOVSKII, M.N.and SOROKIN, V.P., *Teplofizika Vysokih Temperature*, vol. 5 (1967).
31. SUBBOTIN, V.I., IVANOVSKII, M.N. and MILOVANOV, A. I., Condensation coefficient for mercury *Atomn. Energ.* **24**, no. 2 (1968).
32. MISRA, R.and BONILLA, C.F., Heat transfer in the condensation of metal vapors: mercury and sodium up to atmospheric pressure. *Chem. Engng.Prog.Symp.*, ser. 18, **52** (7) (1965).
33. BARRY, R.E. and BALZHISER, R.E., Condensation of sodium at high heat fluxes, in the *Proc. 3rd Int. Heat Transfer Conference*, vol. 2, p. 318, Chicago, Illinois, 1966.
34. ALADYEV, I.T., KONDRATYEV, N.S., MUKHIN, V.A., MUKIN, M.E., KIPSHIDZE, M.E., PARFENTYEV, I. and KISSELEV, J.V., Film condensation of sodium and potassium vapor, *3rd Int. Heat Transfer Conference*, vol. 2, p. 313, Chicago, Illinois, 1966.
35. STICKNEY, R.E., MIT, Personal communication.
36. ARAMATI, V.S., Auger electron spectrometer for high temperrature adsorption studies. S.M. Thesis, Mechanical Engineering Department, MIT, June 1971.

International Seminar 1972

# RECENT DEVELOPMENTS IN HEAT EXCHANGERS

E. Schlüender

*Universität Karlsruhe, Karlsruhe, FR Germany*

The 1972 International Seminar on "Recent Developments in Heat Exchangers", held at Trogir, Yugoslavia, covered almost all the espects of heat exchanger design and operation. This may be elucidated by summarising the topics of the various sessions:

A   Heat Exchangers - Research and Development
B   Heat Exchanger Design
C   Heat Exchanger Elements
D   Heat Exchanger Systems
E   Transient Behaviour
F   Regenerators
G   Heat Exchangers Under Extreme Conditions
H   Heat Exchangers with Two-Phase Systems
I   Heat Exchangers with Two-Phase Systems and General Topics
J   Packed Beds
K   Heat Exchangers with Direct Contact
L   New Concepts in Heat Exchangers

More than 50 papers were presented, reflecting the status of knowledge and technology at the time with comments on solved and unsolved probles. Thus future trends in research and development were stimulated, especially, in the areas of fluid flow and heat transfer within heat exchangers, condensers, reboilers, packed and fluid bed equipment and regenerators. At the end of the seminar it was felt, that there was an urgent need for something like a 'Heat Exchanger Design Handbook" offering a screened collection of recommended equations and design methods to the practising engineer in the industry. An editorial board was formed and 12 years later a seven volume set was brought on the market. The editorial board is still working on supplements to be added to the core edition year by year thus keeping this standard source book up to date.

# THE OPTIMAL DESIGN OF HEAT EXCHANGER NETWORKS - A REVIEW AND EVALUATION OF CURRENT PROCEDURES

T. W. Hoffman*

*MacMaster University, Hamilton, Canada*

In most chemical and metallurgical processes, it is essential on economic grounds to recover and recycle energy. The most important part of this energy recovery is by way of heat transfer between outgoing hot and incoming cold streams. Moreover, temperature specifications on the hot and cold streams must be met and at the design stage a decision must be made regarding the use of a process stream or/and external service (e.g., cooling water or steam) to accomplish the required heat duty. In even relatively simple situations, the problem of pairing and sequencing exchanger streams becomes a large one and the use of the computer or short-cut techniques is necessary.

This paper reviews and evaluates the essential details and evaluates some techniques which have appeared in the recent literature to solve this combinatorial problem.

## 1. THE GENERAL PROBLEM

In the design problem considered here there are $s$ fluid streams, $n$ of which are to be heated and $m$ to be cooled to specified temperatures. The following specifications and/or assumptions pertain:

(*i*)   All information concerning flowrates, initial temperatures, heat capacities of the fluid streams is known.

(*ii*)   The final temperature requirements of the product, feed and intermediate process streams have been identified beforehand by process considerations. It is recognized that the optimal design of the overall process, of which this heat exchanger network is only a small part, may be a strong function of these temperature specifications. *

---

\* NOTE: SMALL FIGURES QUOTED IN TEXT ARE AT THE END OF THE CHAPTER.
\* Note that decomposition procedures for optimizing processes as suggested by Lasdon [1] still require the solution of the network optimization problem.

(*iii*) The availability of and costs for service heating or cooling (e.g., steam cooling water ) are known. It is assumed that the energy generated within the process under study cannot be "sold" outside it, so that recovery is only possible within it.

(*iv*) Heat exchange is effected in a shell – and – tube heat exchanger and the overall heat transfer coefficient is known. This poses a problem since the overall coefficient depends on exchanger geometry and fluid properties (and hence temperature level) of the exchanging streams. since the pairing of streams and temperature level are not known until the network has been established and further, the geometry of the heat exchanger is not known until its size is known, initial estimates of overall coefficients may have to be revised and the network reestablished.

This allows the economic design of any individual heat exchanger to be considered after the network has been established. Also, the interaction of the pressure drop with the overall process economics can be established later as well.

(*v*) Heuristics, such as the minimum allowable approach temperature and maximum allowable temperature differences between hot and cold streams may be included.

(*vi*) Operating cost data such as cooling water, steam, furnace fuel costs and capital cost data such as heat exchangers, coolers, heaters, furnaces are known. Capital costs of heat exchangers, furnaces etc. are correlated by a non-linear function of some characteristic, Z, by equations of the form:

$$C_{c_i} = a_i Z_i^{b_i} + k_i \qquad (1)$$

where $a_i$, $b_i$ and $k_i$ are constants depending upon the device $i$, $Z_i$ is exchanger area for heat exchangers and heat load for furnaces, etc.

A constant amortization rate on capital equipment is specified.

(*vii*) The controllability and cost of control and start-up which are related to process complexity are not considered in the economic objective function.

## 2. LITERATURE REVIEW

A brief "State-of-the-Art" review of previous work is presented below.

The early attempts at optimizing heat exchanger arrangements concentrated on the optimization of the exchanger arrangement and intermediate temperatures between heat exchangers, coolers and heaters for a single cold and/or hot stream. Bosnjakovic et al. [2], Mickley and Korchak [3], and Happel [4] consider the problem of optimizing the outlet temperature of a waste heat exchanger or exchangers trains. Ten Broeck [5] considered the optimization of a battery of waste heat exchangers. Westerbrook [6] applied dynamic programming to optimize a train of heat exchangers and a furnace. Bragin [7] employed the Discrete Maximum Principle to optimize the heat allocation and sequencing of hot streams to heat a cold stream. In each case however, the basic network was assumed known. Moreover, the methods are not applicable to interlocking networks where there are multiple hot and cold streams.

Hwa [8] was the first to consider tthe problem of optimal structure of multiple hot and cold streams. His formulation relies on piecewise linearization of the objective function and then solution by separable programmng. A serious shortcoming of his method is that it requires the synthesis of possible network structures* beforehand.

Kesler and Parker [9] break up hot and cold streams into small finite elements of unit heat flow (e.g., $1 \times 10^6$ kcal./hr.) which they call exchangelets. They then consider the transfer of heat between exchangelets or groups of exchangelets. This allows a set of linear equations to be formulated along with the costs for each heat exchange combination and hence overcomes the problems associated with a nonlinear, nonconvex objective function. The optimal network is found by searching via alternate use of an assignment algorithm and a modified Linear Programming algorithm. The fact that finite elements are used may mean that global optimality has not been attained.

A similar linear programming approach has been reported by Kobayashi et al. [10]. They divide the process into an internal and external system. The internal system is comprised of process stream - to - process stream exchangers; the external system contains external sources and sinks. Similarly to Kesler and Parker, they break each stream up into finite heat elements and then apply an optimal assignment algorithm in linear programming to establish the optimum network structure of the internal system. Box's complex-method optimizing technique [11] is used to optimize the economic objective function of the combined internal and external system. Some iteration within this two-level approach may be necessary since some of the assumptions involved with the formulation of the interrnal structure may be invalid after the second optimization.

Nishida and coworkers [12] working like Kobayashi et al. with internal and external systems have proposed a graphical method for obtaining the best network structure. They then recommend that the optimum sizes and splitting of streams within this structure be found by use of Box's Complex method for constrained optimization problems. The rules for the graphical technique have been developed from the mathematical analysis of the special circumstance where all overall coefficients are equal and the heat capacity-mass flowrate product is the same for all streams. The method when applied to other cases does not guarantee optimality but does serve as a useful guide in structuring at least a near optimum network.

Masso and Rudd [13] use a heuristic method for building networks. It has a built-in learning capability which enables the system to move towards the optimal structure. The heuristics permit incorporation of very practical considerations but destroy any guarantee of optimality.

King, Gantz and Barnes [14] have applied heuristic methods to an evolutionary approach to the design problem. Their method is quite general in that it can be applied to any process design problem; however, it does require a basic flowsheet at the outset.

Lee, Masso and Rudd [15] employ the branch and bound mathematical technique to reduce the amount of calculation required in this combinatorial problem. Their method guarantees optimality within the limitations of the heuristics that are included. This technique is ideal for heat exchanger networks where the same type of the heuristics that are included. This technique is ideal for heat exchanger networks

---

* The term network structure means the matching and sequencing of streams to achieve their desired temperatures.

where the same type of equipment is used in a relatively large number of places in the process.

Menzies and Johnson [16] have automated the branch and bound thechnique and applied it to optimize energy exchange networks where pressure and temperature effects are included. They also coupled this with a modular simulation program and Lagrangian decomposition techniques to break large systems down into subsystems of more manageable size. They applied it successfully to high and low pressure ethylene plants.

McGalliard and Westerberg [17] and Takamatou, Hashimoto and Ohno [18] apply Lagrangian-based sensitivity analysis to the problem of optimizing a given exchanger network. This method thus serves as an alternative to the Complex method employed by the Japanese workers mentioned above. As will be shown later, except for the branch and bound approach, these methods should find use in optimizing the apparent optimum structure found by the other methods.

The purpose of this paper is to review the essential details of the methods of Kesler and Parker, Lee, et al., Nishida, et al. and Kobayashi, et al. This will be done by applying these methods to a simple system. The predictions and limitations of the methods and some of the difficulties encountered in using the methods will be delineated by applying these techniques to obtain the optimal network of a real design problem. The techinques have been used as presented in the papers with little or no modification and hence results are presented from the standpoint of the user.

## 3. REVIEW OF METHODS

### 3.1. THE BASIC PROBLEM

The design problem to be used as a vehicle to demonstrate the methods is the one presented by Lee et al. and refferred to in their paper as 4SPI. The details of the two hot and two cold streams are indicated in Table I; the basic economic and operating data are indicated in Table II.

### 3.2. THE GRAPHICAL METHOD OF NISHIDA, KOBAYASHI AND ICHIKAWA.

Nishida et al. considered the heat exchange system to be broken into two subsystems (Figure 1): the interior subsystem where the process stream - to - process stream heat transfer occurs and the exterior sybsystem where the heating and cooling by auxiliary services allows the temperature specifications to be achieved. This decomposition allows the interior and exterior subsystems to be synthesized separately. The key to the method, and indeed the link between the two subsystems, is the assignment of a total heat duty to the interior subsystem. This total interior heat duty determines not only the overall cost of the total heat exchange network but also the structure of the internal system. The initial step in applying this method is to determine the internal arrangement of exchangers and streams, given its total heat duty. The optimization of the whole system given this internal structure follows.

The basic problem in determining the optimum interior structure is defined as: For a given total heat duty in the interior subsystem and given process streams, synthesize a feasible exchanger network which will minimize the total area in this subsystem and given process streams, synthesize a feasible exchanger network which will minimize the total area in this subsystem. This is stated as their defined optimum system.

## TABLE 1 – DESIGN PROBLEM 4SPI

| Stream No. | Flow Rate | Inplut Temp. | Output Temp. | Heat Cap. |
|---|---|---|---|---|
| 1($S_{C_1}$) | 20,643 | 140 | 320 | 0.70 |
| 2($S_{H_2}$) | 27,778 | 320 | 200 | 0.60 |
| 3($S_{C_2}$) | 23,060 | 240 | 500 | 0.50 |
| 4($S_{H_1}$) | 25,000 | 480 | 280 | 0.80 |

## TABLE 2 – DESIGN DATA

| | | |
|---|---|---|
| Stream (saturated) pressure | | 962.5 p.s.i.a. for problem 4SPI |
| | | 450 p.s.i.a. for problems 5SPI and 6SPI |
| Cooling water temperature | $t_w^i$ | 100°F |
| Maximum water output temp. | $t_w$ | 180°F |
| Minimum allowable approach | $\Delta T$'s | |
|    Heat exchanger | $\tau_{HE}$ | 20°F |
|    Steam heater | $\tau_H$ | 25°F |
|    Water cooler | $\tau_C$ | 20°F |
| Over-all transfer coefficients | | |
|    Heat exchanger | $U_{HE}$ | 150 Btu/(hr)(sq ft)(°F) |
|    Steam heater | $U_H$ | 200Btu/(hr)(sq ft)(°F) |
|    Water cooler | $U_C$ | 150 Btu/(hr)(sq ft)(°F) |
| Equipment down time | $\alpha$ | 380 hr/yr |
| Heat exchanger cost parameters | a,b | 350,0.6 |
| Cooling water cost | $C_w$ | $5 \times 10^{-5}$ \$/lb |
| Steam cost | $C_s$ | $1 \times 10^{-3}$ \$/lb |

In establishing this definition they have had to make the following additional limiting assumptions:

(*1*) The cost of and exchanger is a linear function of heat transfer area; this means that the cost index, $b=1$, and $k=0$ in Eq. (1).

(*ii*) Only sensible heat is transferred.

(*iii*) Each exchanger with its area calculated from the well-known expression.

$$A = \frac{q}{UF\Delta T_{l,m}} \quad (2)$$

has the same effective overall heat transfer coefficient, $UF$.

In order to syntesize this optimum internal subsystem, they develop a number of rules based on analytically derived necessary conditions. An example of one of their necessary conditions with it's inherent assumptions is: Given a system of equal number of hot and cold streams, all with equal hourly thermal capacity (weight flowrate-heat capacity product), assuming each stream is heat exchanged at most once and equal heat duty occurs in each exchanger, then it is shown that the streams should be matched.

$$(S_{h_1}, S_{C_1})\ (S_{h_2}, S_{C_2}) \text{----}\ (S_{h_m}, S_{C_m}) \tag{3}$$

where
$$T_{h_1} \geq T_{h_2} \geq \text{---} T_{h_m} \tag{4}$$

and
$$T_{C_1} \geq T_{C_2} \geq \text{---} T_{C_m} \tag{5}$$

If the number of hot and cold streams is unequal, the matches are made in the same way with the coldest set matched with the hot streams if the number of hot streams is less than the number of cold streams and vice versa if the number of cold streams is the smaller.

Additional necessary conditions are summarized as follows:

• If multiple exchange is to oxxur between a hot and cold stream then it should be done countercurrently.

• If splitting of streams into substreams is to occur and each of these substreams if exchanged only once, then for optimal assignment of heat duty to each exchanger, the same fraction of heat should be assigned to each exchanger. Under these conditions the total area is independent of the number of exchangers.

These conditions become the rules for a graphical synthesis of the network on what they refer to as a heat content diagram (Figure 2). This is a plot of input and output temperpatures of all streams in the system against their respectively hourly thermal capacity. Hot streams are shown above the horizontal axis, cold streams below. The origin of the horizontal axis is separate for each stream. The area of the block so formed represents the amount of heat to be removed from or added to each stream in order to satisfy its temperature specifications. This representation of the problem gives a very clear picture of any network design problem and is recommended to network designers.

Heat exchange is represented on the diagram by matching a hot block (or part thereof) with and equal area block in the cold section; this ensures heat balancing in an exchanger. Note that each horizontal division of a block corresponds to another heat exchanger for this stream; each vertical division corresponds to splitting a stream.

The three necessary conditions indicated above are used to establish the following rules for constructing the heat exchange network on this diagram:

(*i*) Hot and cold blocks are matched (and numbered) consecutively in decreasing order of stream temperature (Theorem 1).

(*ii*) For the hot blocks, the outlet temperature of exchangers in the *i*-th block is never lower than the inlet temperature of the $(i + 1)$-th block and similarly for the cold blocks the inlet temperature of the *j*-th block is never lower than the outlet temperature of the $(j+1)$-the block.

(*iii*) If the total heat duty of the internal subsystem is smaller than both the total heat removed from hot streams or total heat added to cold streams, the highest temperature portion of hot blocks and coldest temperature portion of cold blocks are exchanged in the internal subsystem. The remainder is assigned to the heating and cooling external subsystem.

The application of these rules is demonstrated via the 4SPI problem; Figures 3. and 4. demonstrate the steps in solving this problem.

(i) Figure 3 shows the heat content diagram for the streams listed in Table 1.

(ii) An arbitrary total heat duty is assumed. This is chosen to allow a 20°F approach to be achieved between the inlet to hot stream $S_{h_1}$ and cold stream $S_{C_1}$; hence a horizontal line is drawn at 460°F and the total heat duty on the hot streams provides an outlet temperature of 251.5°F on $S_{h_2}$.

(iii) From rule (ii), horizontal lines are drawn at the boundary temperatures as shown by the dotted lines in Figure 3.

(iv) Boundaries between streams are ignored for the moment and the blocks of heat are separated as shown by the dotted lines in Figure 3.

(v) Figure 4 shows the construction. By rule (1) the highest temperature hot stream is combined for exchange with the highest temperature cold stream (since the number of streams are equal). By the heat balance, the outlet temperature of the hot stream is 399.2°F.

The remainder of the cooling of $S_{h_1}$ to 320 °F is provided by part of combined blocks 2 and 3, their outlet temperature by heat balance being 259°F. Since there are two streams, $S_{h_1}$ must be split, the split (vertical line) or unknown weight flowrate in each stream is determined by the heat balance on blocks 2 and 3. The remainder of heat requirements on $S_{C_1}$ and $S_{C_2}$ between 240 and 259 are provided by the combined thermal capacity of $S_{h_1}$ and $S_{h_2}$ between 320 and 306.5. This leads to exchangers 4, 5 and 6 with an additional split on stream 1. The remaining heat in all of $S_{h_1}$ is exchanged with part of $S_{C_2}$ as shown. Exchanger 8 completes the network.

Figure 5 is the process flowsheet that arises out of the direct application of these rules. Obviously this network is much too complicated for a practical design since the control problems and costs may offset any expected savings.

A much simpler network results if one step in the rules is relaxed, namely rule (ii) concerning the mimimum inlet and outlet temperatures. If rule (ii) is applied to the hot streams, but the heat requirement of cold stream $S_{C_1}$ is allowed to be met entirely by $S_{h_1}$ and if the rules are applied rigorously for the remainder of the calculation, the heat content diagram as shown in Figure 6 results. This gives use to the much simpler network, Figure 7.

This procedure may seem simewhat arbitrary but actually arises by considering ways to reduce the complexity of the rigorous network, as Nishida et al. have suggested. Note that the lower portion of the process flowsheet (Figure 5) shows almost a true countercurrent exchange between $S_{h_1}$ and $S_{C_1}$ except for the flow to exchangers 3 and 5.

The next step in the procedure is to determine the economic objective function for the entire process. Nishida et al. do not present details of their method but the GEMCS modular simulation routine has been found to be ideal for this purpose. The details of this technique are readily available* [19].

Basically, this simulation technique involves translating the process flow sheet into an information flow diagram in which information concerning the flow material and thermodynamic state of any stream is transferred from one unit computation or module to another. All incoming information to the module is assumed known and this incoming information is modified in the unit computation; the outgoing infor-

---

* A listing of the Fortran IV program and a user's handbook are available from the author.

mation reflects these changes and becomes the incoming information for the next module in a specified calculation order. Here the unit computation describes heat exchangers and computes the heat and material balance when two or more streams are joined or split. Table III indicates a brief description of the modules. The procedure used is to size the exchangers in the network using the temperature specifications an each exchanger generated by the graphical procedure; this is done by the HEX2 modules (Figure 8a). The process is then simulated using the HEX1 modules in those situations where the outlet temperatures must not meet a temperature specification (Figure 8b). These exchanger areas are adjusted to ensure that all temperature specifications are met and the objective function evaluated. The optimization routine optimizes the entire network through manipulation of the fraction splits. If the heat duty of the internal subsystem is very much different from what was assumed in constructing the original network, the graphical procedure should be repeated to ensure that the internal network structure does not change. If it does, the entire procedure is repeated. In the example under consideration here the objective function was calculated to be $15,340 although the structure generated by the procedure was not optimized.

Global optimality is not ensured by this procedure but the simplicity of the procedure is its major advantage. Experience to date suggests that although the network structure is defferent from that generated by other techniques, the objective function is very close to that obtained by the other procedures to be discussed.

## TABLE III - DESCRIPTION OF MODULES USED IN NETWORK DESIGN. SIMULATION AND OPTIMIZATION

| | |
|---|---|
| HEX1 | Calculates the outlet temperature geven the heat transfer area and overall transfer coefficient by the effectiveness factor method (a simulator). |
| HEX2 | Calculates the area given incoming information and either the outlet temperature specification on one stream or the minimum approach temperature desired (a disign model). |
| HEX3 | Calculates the area and stream requirements in order to meet a required temperature on the process stream. |
| FURNI | Calculates fues requirements in a furnace to meet a required temperature on the process stream given the heating value of the fuel and furnace efficiency. |
| COOLI | Calculates the area and water requirements (given inlet and outlet temperatures) to meet a required temperature. |
| JUNCOI | Is a stream splitting module which requires specification of the fraction of the incoming flow going to each of the outgoing streams. |
| JUNCO2 | Is a convergence tester. Since the networks usually involve recycle of stream information and all modules assume that incoming information is known, an iterative calculation is required on these streams. This module tests if the assumed and calculated information is within a given tolerance. |
| CVRGY | Is a convergence promotion module. Speed of convergence is increased over direct iterative calculations on the recycle streams. Model uses the method of Orbach and Crowe [20]. |
| TESTI | Tests if the temperature constraints have been met with the conditions of the simulation. If not ensures appropriate increases/decreases in heat exchanger areas. |
| OBJI | Evaluates the objective function (cost/year) for any feasible network. |
| OPTI | Is a continuous optimizing routine to change the fraction split or other decision variable in the optimization. |

## 3.3. BRANCH AND BOUND PROCEDURE

Lee, Masso and Rudd [15] were the first to suggest and demonstrate successfully the application of the branch and bound technique to the design of heat exchanger networks. Their application is reviewed below.

The mathematical foundation for this method may be simply stated as follows [15] [16]. If a given optimazation problem $A$ (maximization of an objective function, $O_A(D_A)$) is excessively difficult to solve, replace it by branching to a problem or set of problems, $B$, which are similar but much more easily solved than $A$. Problem $B$, however, must be selected to bound the original problem $A$. This means that if the optimal solution for problem $A$ were available and inserted in the problem $B$, it must be a solution for problem $B$ (satisfy all technical feasibility constraints) but not necessarily be optimal for $B$. This solution of $B$ via the design $A$ must indicate an equal or greater objective function $O_B(D_A)$ than the objective function for solution $A$, $O_A(D_A)$, in order to be a valid upper bound for $A$. This is expressed mathematically as

$$O_B(D_A) \geq O_A(D_A) \tag{6}$$

Furthermore, if the optimal solution of problem $B$ is found and is feasible for problem $A$ and gives equal values of the objective function when applied to both $A$ and $B$, then it is also the required optimal solution to the original problem $A$.

The strategy in applying this technique can be summarized in the following steps, with each step and the mathematical foundations of the method demonstrated by considering the 4SPI problem:

(*i*) Formulate all possible stream matches for each primary stream without regard to the feasibility of any total network that may be formed from a combination of these matches. Table IV shows a number of examples of this stream matching. In matching the streams (for example, match 2) a specified heuristic has been introduced, namely the minimum approach temperature of 20°F. Note that only residuals that are technically feasible (hot stream hotter than cold stream) are matched: however, technically feasible residuals which result from processing the same stream previously are considered stream infeasible and hence are not matched. This procedure is repeated until all temperature specifications are met for each primary stream. In this case there are forty - one such stream matches. The operating and capital costs for each exchanger are recorded.

(*i*) Two approaches may be followed at this point:

(*ii*) Combine the stream matches into stream processing paths for each primary stream, i.e., sequence the exchangers that have resulted from the stream matches of a particular primary stream. The cost of each processing path is easily determined; half the exchanger cost for a process stream/process stream match is assigned to any one path. For the 4SPI problem $34 = (10+5+7+12)$ processing paths are found. Note that although each path is feasible because streams are only used once, combination of paths to form the final network may not be possible because of multiple stream use.

Unfortunately this method leads to a large number of possible networks: if there are $m$ primary streams and $n_i$ ($i = 1,m$) possible paths the number of possible networks is

$$N = \prod_{i=1}^{m} n_i \tag{7}$$

Although most of these are infeasible, the computation task may be much too formidable even if automated.

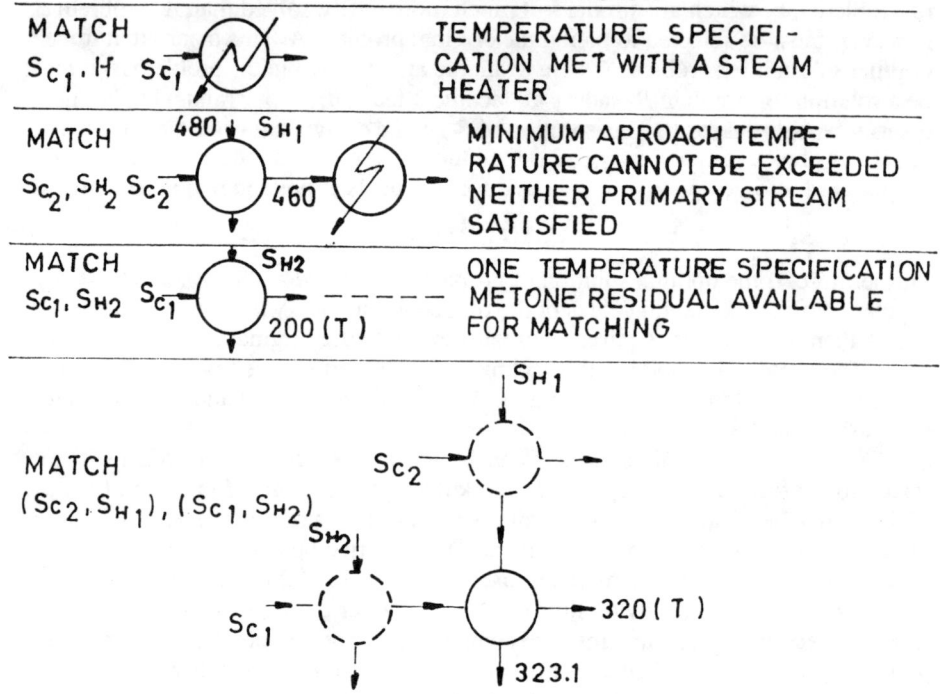

TABLE IV EXAMPLES OF STREAM MATCHE
(NOTE: MAXIMUM HEAT EXCHANGE OCCURS IN EACH HEAT EXCHANGER

| MATCH $S_{C_1}, H$ $S_{C_1}$ | TEMPERATURE SPECIFICATION MET WITH A STEAM HEATER |
| MATCH $S_{C_2}, S_{H_2}$ $S_{C_2}$ | MINIMUM APROACH TEMPERATURE CANNOT BE EXCEEDED NEITHER PRIMARY STREAM SATISFIED |
| MATCH $S_{C_1}, S_{H_2}$ $S_{C_1}$ | ONE TEMPERATURE SPECIFICATION MET ONE RESIDUAL AVAILABLE FOR MATCHING |
| MATCH $(S_{C_2}, S_{H_1}), (S_{C_1}, S_{H_2})$ | |

RESIDUAL MATCHED WITH RESIDUAL, NO FURTHER MATCH POSSIBLE WITH PROCESS STREAMS

This problem of evaluating all the feasible networks becomes the difficult design problem and the branching and bounding procedure is utilized again. In this case the difficult problem can be replaced by a set of problems that bound the one indicated above. Each set may be solved for the optimum network within it and then the optimum network is easily found from among the set.

In this example, a set of five problems is formulated as shown in Figure 9. This procedure leads to a drastic reduction in the number of primary stream processing paths in each problem. Similar branching can be initiated from any or all of these problems in the same way, although similar reductions with some eliminations will not result. The reader is referred to the original paper for further details.

This stream matching, path costing, branching procedure has been fully automated by Menzies and Johnson [16]. The 4SPI problem as outlined in Table I was solved using this program. The network is shown in Figure 10; its cost was found to be $ 15,454.*

## 4. SOLUTION BY LINEAR PROGRAMMING

Two independent solutions of this network optimization problem which utilize linear programming procedures have been reported. Kobayashi, Umeda, and Ichikawa [10] used the standard solution of the assignment problem in linear programming to determine the best internal subsystem network and then optimized the combined internal and external systems by Box's Complex optimization procedure. Kesler and Parker [9] formulated a special assignment algorithm procedure to determine the optimum total network. No further optimization was suggested.

### 4.1. KESLER AND PARKER'S SOLUTION

To cast the problem in the context of linear programing, Kesler defined the network problem in the following way.

Consider $m$ hot streams and $n$ cold streams as defined earlier. For the $m$ hot streams, assuming each cold stream exchanges with each hot stream:

$$\sum_{j=1}^{n} Q_{ij} = \Delta H_i \qquad i = 1,2 \ldots m \qquad (8)$$

where $Q_{ij}$ is the heat duty in the $ij^{th}$ exchanger and $\Delta H_i$ is the total heat removed in the $i^{th}$ hot stream.

Similarly for the $n$ cold streams,

$$\sum_{i=1}^{m} Q_{ij} = \Delta h_j \qquad j = 1,2 \ldots n \qquad (9)$$

These linear expressions satisfy the heat balance on each stream and the overall balance is indicated by

$$\sum_{i=1}^{m} \Delta H_i = \sum_{n}^{j=1} \Delta h_j \qquad (10)$$

The non-linearities are contained in the cost function, which is to be minimized:

$$\sum \sum \rho_{ij}(Q)(Q_{ij}) = C \qquad (11)$$

where the cost of removing heat, $\rho_{ij}$, is some non-linear function of the amount of heat removed, the physical properties of the streams and the temperatures of the streams.

The key to this method is the linearization of Equation (11) by fracturing each stream into heat elements or exchangelets, $q$, sufficiently small** that that objective function becomes a linear function of the elements.

---
* This figure is different from Lee's [15] since the amortization rate was assumed to be 25%; Lee did not report his value in his paper.
** For example, $1\times10^6$ B. t. u. /hr. or $3\times10^6$ kcal. /hr. etc.

Mathematically this means,

$$\sum_{k=1}^{K_i} q_{ik} = \Delta H_i \qquad (12)$$

for each of the $i = 1, 2 ---m$ streams each of which has $K_i$ elements and for the $n$ cold streams, each with $L_j$ elements

$$\sum_{l=1}^{L_j} q_{jl} = \Delta h_j \qquad (13)$$

As before

$$\sum_i \Delta H_i = \sum_j \Delta h_j \qquad (14)$$

Each exchangelet can be defined uniquely by the parameter $X_{ikjln}$ where $i$ is the hot stream number, $k$ is the number of the element (hence defining the origin of the exchangelet), $j$ is the cold stream number with which it is exchanging heat and $l$ is the number of elements of stream $j$; $n$ is the number of elements involved. By this definition;

$$\sum_{j,\,l,\,n} X_{ikjln} = q_{ik} \qquad (15)$$

$$\sum_{i,\,k,\,n} X_{ikjln} = q_{jl} \qquad (16)$$

$$\sum_{i,\,j,\,k,\,l,\,n} \rho_{ikjln} X_{ikjln} = C \text{ (minimum)} \qquad (17)$$

and

$$X_{ikjln} = 0 \text{ or } 1 \qquad (18)$$

The set of linear equations (12) through (18) in variables $X$ fully defines the problem. The number of equations in the set to be optimized depends upon the number of streams in the problem and the number of elements into which each stream is fractured; this does not represent a large number (34 in the 4SPI problem to be discussed). The number of possible exchanges of single or multiple exchangelets, i.e., the number of $X$'s, can be very large. By introducing certain limitations such as the minimum approach temperature or maximum difference in temperature between a hot and cold stream and the maximum size of a heat exchanger, this number can be significantly reduced. In the 4SPI problem (with a 10° approach temperature and a 300° maximum temperature difference between heat source and heat sink) the number was 444.

Basically Kesler's procedure is to evaluate the cost of exchanging heat between each possible cold and hot exchangelet or group of exchangelets to provide a main matrix. This main matrix is then used to form an assignment matrix with each row representing one of the cold elements and each column representing one of the hot elements. This procedure ensures the integer character of the problem as expressed by Equation (18). Entries in this matrix other than zero indicate an exchange between the respective cold and hot elements. This procedure formulates the problem as a classical assignment problem in linear programming. Its solution is solved by standard methods although some modification was required to assure feasible solutions.

## 4.2. KESLER'S SOLUTION TO THE 4 SPI PROBLEM

The initial step in Kesler's method is to choose the size of a heat element. Kesler provides no guidance in this selection although its size cannot be greater than the smallest hourly heat capacity of any one stream. Since it is highly unlikely that each stream will form an intergral number of exchangelets, some heat imbalance will arise because of the integer requirements. Decreasing the size of the elements will ensure greater accuracy but will increase the dimensions of the main and assignment matrices significantly. In the 4SPI problem the element size was $4\times10^5$ B.t.u./hr. The fracturing of the streams is best represented on the heat content diagram, Figure 11. The horizontal lines indicate the temperature limits for each exchangelet. The area of each block should be the same. Note, however, that the cold streams really do not have an integer number of elements.

In this solution additional heating and cooling elements must be included to satisfy heating and cooling demands that cannot be met with the process streams. Furthermore, this method requires an equal number of hot and cold elements since each element must exchange heat with another element in order to meet the temperature specifications of all streams. Since an equal number of hot and cold elements arose with our choice of element size, there is a temptation to add a steam heater element (and hence a cooling water element to balance the element) as Kesler suggests and to search for a solution. No feasible solution could be found, however, since upon inspection it will be noted that the upper two elements of $S_{C_1}$ cannot be exchanged with process fluids. The actual demands are clearly indicated by superimposing the cold part of the heat content diagram on the hot. Hence, it became necessary to modify Kesler's original program to allow for an unequal number of cold and hot exchangelets and to allow more than one external heat unit to be supplied. Unfortunately increasing the number of external hot units meant that the number of external cold units must be increased accordingly. This then leads to different networks that are found by the other methods and represents a serious limitation of the method.

The optimal solution of the 4SPI network is shown as Figure 12. It's cost is appreciably greater than that obtained by the previous methods. This network was optimized using the GEMCS procedure indicated earlier and the 20° minimum approach temperature. The resulting cost was $15,977/year.

## 4.3. THE METHOD OF KOBAYSHI UMEDA AND ICHIKAWA

As indicated earlier, like Nishida, Kobayashi et al. devide the system into the internal and external subsystem and then the internal network is determined through solution of the classical optimal assignment problem in linear programming.

Their linear progammming formulation for the internal subsystem is almost identical to that of Kesler and Parker except that they require each exchanger to transfer only one unit of heat (one exchangelet); hence all exchangers have the same heat duty. Moreover, in this formulation they require an equal number of hot and cold exchangelets, although they have overcome this problem by introducing hypothetical streams which carry a high processing cost so, that they will not enter into the final structure. Similarly their solution allows for fewer exchangers than the number of exchangelets. A major difference arises in this method since it allows for each stream to be processed in more than one heat exchanger. This is achieved by splitting hot and cold streams into equal hourly heat capacities and hence doubling the number of streams to be considered. No guidance is given as to when a stream should be divided nor how large each exchangelet should be.

Box's Complex method [11] is recommended for the final optimization. This method coupled with linearized modelling of the heat exchange network seems to be much more complex and restrictive than the simulation-optimization scheme suggested earlier in the paper.

No direct evaluation of this method has been made except to note that if the streams are split in the way suggested in this paper, Kesler's method can be utilized to find a solution. This was demonstrated with the refinery problem presented in their paper and utilized as a case study in the next section.

## 5. A CASE STUDY

An evaluation of these procedures would be icomplete without an application to a real problem. The case presented by the Japanese workers [10] [12] * involves the heat exchange system around a topping tower in a 80,000 bbl./day oil refinery. A cold crude oil stream is heated to 135 °C at which temperature it enters a desalting unit. After leaving the desalter at 130 °C it must be preheated to 355°C before it undergoes further processing. Seven hot streams of rather small hourly thermal capacity must be cooled to specified temperatures. A pipe still furnace is availabe for auxiliary heating and water coolers are used for auxiliary cooling. The inlet and specified temperatures along with their flows and heat capacities are indicated in Table V.
Table VI presents the cost and operating data for the system. (The reported conventional system is shown in Figure 13** ); the syntesized networks along with their costs are presented in Figures 14 and 15. The following point summarizes this experience.

### 5.1. SOLUTION VIA KESLER AND PARKER'S METHOD

An exchangelet size of $3 \times 10^6$ kcal./hr. was used in this calculation which corresponded to the hourly heat capacity of the smallest hot stream, $S_{h_2}$. Considerable difficulty was experienced in expanding the original program to accept this problem. Approximately 100 K of memory on a CDC 6400 computer was required; solution was reasonably fast performing about 5 iterations/sec. If a smaller heat element were used it is doubtful whether the available 140 K of memory would be sufficient to solve the problem with the present program.

---

* Both authors use the same example except that there are inconsistencies in the data both between and within the papers. Nishida's description is used here.
** The costs are not exactly as presented in their paper. This is partly because of inconsistent and incomplete data and also because the solution reported by Nishida et al. seemed to be in error.

## TABLE 5 – STREAM PROPERTIES OF THE APPLICATION PROBLEM

| (Hot streams) | $T_h$ (°C) | $T^*_{hi}$ (°C) | $w_{hi}$ (ton/hr) | $C_p(T_{hi})$ (kcal/kg°C) | $C_p(T^*_{hi})$ |
|---|---|---|---|---|---|
| $S_{h_1}$ | 340 | 90 | 200 | 0.69 | 0.46 |
| $S_{h_2}$ | 330 | 195 | 30 | 0.74 | 0.63 |
| $S_{h_3}$ | 270 | 55 | 60 | 0.72 | 0.53 |
| $S_{h_4}$ | 265 | 215 | 110 | 0.71 | 0.67 |
| $S_{h_5}$ | 235 | 145 | 95 | 0.75 | 0.66 |
| $S_{h_6}$ | 205 | 150 | 55 | 0.72 | 0.67 |
| $S_{h_7}$ | 150 | 50 | 315 | 0.72 | 0.62 |

| (Cold streams) | $T_{cj}$ (°C) | $T^*_{cj}$ (°C) | $w_{cj}$ (ton/hr) | $C_p(T_{cj})$ (kcal/kg C) | $C_p(T^*_{cj})$ |
|---|---|---|---|---|---|
| $S_c^1$ | 25 | 135* | 425 | 0.50 | 0.60 |
| $S_c^2$ | 130* | 355 | 425 | 0.60 | 0.79 |
| $S_w$ | 30 | 55 | | 1.0 | 1.0 |

## TABLE 6 – NUMERICAL DATA OF THE APPLICATION PROBLEM

*Operating conditions*
|   |   |   |
|---|---|---|
| Thermal efficiency of furnace | 0.80 | |
| Heating value of fuel | 9500 | (kcal/kg) |
| Annual operating hours | 8000 | (hr/year) |

*Economic data*
|   |   |   |
|---|---|---|
| Annual return of investment cost, σ | 0.25 | |
| (Operating cost) | | |
| Unit cost of cooling water | 0.83 | (cent/ton) |
| Unit cost of fuel | 1.94 | (Cent/1) |

*nvestment cost*
|   |   |   |
|---|---|---|
| Shell and tube type heat exchangers | 55.56×A | ($/unit) |
| Furnace | 2.50×$Q^{0.7}$ | ($/unit) |
| Annual return of the investment cost | 0.25 | |

*Operating cost*
|   |   |   |
|---|---|---|
| Unit cost cooling water | 0.83 | (cent/ton) |
| Unit cost fuel | 1.94 | (cent/1) |

Only the solution indicated in Figure 14 was found. The program repeated the same calculations each time it found this solution, so that more iterations would have not effected another solution. This network could be optimized considerably however but the simulations would have not effected another solution. This network could be

---

* Input temperature to the desalting unit.

optimized considerably however by the simulation method presented earlier, so that the cost/year could be reduced to $ 8.74 $\times 10^5$/year.

Many networks could be generated using Kobayashi's stream splitting technique; the best of these was not optimized so it is not reported.

### 5.2. SOLUTION BY THE METHOD OF NISHIDA ET AL.

The ease with which a network can be synthesized and simulated (and hence optimized) even with a relatively difficult system makes this technique an extremely powerful one. Moreover, the computer requirements are only nominal, so that good networks can be synthesized even on small computers. The cost for control and implementation of the rather complicated networks that result should be evaluated and included in the objective function before a direct comparison with other networks can be made.

### 5.3. THE METHOD OF KOBAYSHI ET AL.

The complicated structure that results from this procedure is a serious drawback. Furthermore other than arbitrary rules for deciding stream splitting must be established before the method will prove useful.

### 5.4. THE BRANCH AND BOUND METHOD

No solution could be obtained with the automated branch and bound procedure because of the high demands the method places on computer storage. Furthermore, although Lee et al. suggest that their method is amenable to hand calculation, this would not be possible in this case. The basic problem is in determining the stream matches. On the order of 650 separate exchangers were required (representing the feasible primary stream and residual matches). This meant that all the cost information of the heat exchangers and the stream history information associated with each of these matches had to be stored and required in excess of the 140 K storage available on a CDC 6400 computer. Moreover since most of the storage was required for storage of information, overlaying of programs did not alleviate the problem. Considerable economy of storage could be effected by packing more than one number in each memory location. This modification is planned for the future since the branch and bound technique seems to be the most powerful optimization method available at present. Computer time to solve this problem would be less than 100 sec. on the CDC 6400 computer.

## 6. SUMMARY AND CONCLUSIONS

A review and and evaluation of the current methods of syntesizing heat exchanger networks has been presented. A number of shortcomings of these methods has been uncovered when they were applied to a real problem. In particular the high demand on computer memory of these methods mitigates against their use for designing large systems.

The method proposed by Nishida coupled with the optimization program presented here seems to be worthy of more evaluation mainly because of its nominal computer requirements. Since global optimality is not assured by this method only experience will dictate its real usefulness.

## 7. ACKNOWLEDGMENTS

This work was supported through a grant from the National Research Council of Canada. The Kesler program was programmed with some modification by Mr. J. Li on the CDC 6400 computer through a contract from the Atomic Energy of Canada. Dr. M.A. Menzies supplied his version of the branch-and-bound program and helped in modifying it to accomodate general heat exchanger networks. His help in receiving the current literature is also gratefully acknowledged.

## REFERENCES

1. LASDON, L.S. *"Optimization Theory for Large Systems"*, MacMillan (1970).
2. BOSNJAKOVIC, F., VILICEC M. and SLIPCEVIC, B. VDI. Forschungsheft 432, Ausgabe B, Band 17 (1951).
3. MICKLEY, H.S. and KORCHAK, E.I., Chem. Eng., 69, No 23, 239, Nov. 1962.
4. HAPPEL, J. *"Chemical Process Economics"*, John Wiley (1969).
5. TEN BROCK, H., Ind. Eng. Chem. 36 64 (1944).
6. WESTERBROOK, G.T., Hyd. Proc. Pet. Ref. 40 20 (1961).
7. BRAGIN, M.S. *"Optimization Multistage Heat Exchanger Systems"*, Ph.D. Thesis, New York University (1966).
8. HWA, C.S., A.I.Ch.E. - I.Ch.E. Symp. Ser. No 4, 111 (1965).
9. KESSLER, M.G. and PARKER, R.O., CEP Symp. Series, 65 No 92, 111 (1969).
10. KOBAYSHI, S, UMEDA, T. and ICHEKAWA, A., Chem. Eng. Sc. 26, 1367 (1971).
11. BOX, M.J., Computer Journ. 8 42 (1965).
12. NISHIDA, N., KOBAYASHI. S. and ICHIKAWA, A., Chem. Eng. Sci. 26 1841 (1971).
13. MASSO, A.H. and RUDD, D.F., A.I.Ch.E.J., 15, No.1,11 (1969).
14. KING, C.J., GANTZ, D.W. and BARNES, F.J., Ind. Eng. Chem. Proc. Des. Dev. 11 272 (1972).
15. LEE, K.F., MASSO, A.H. and RUDD, D.F., I& EC Fundamentals, 9, No. 1, 48 (1970)
16. MENZIES, M., and JOHNSON A.I., Paper IC Presented at the Dallas Meeting of A.I.Ch.E. February 1972.
17. MC GALLIARD, R.L. and WESTERBERG, A.W., Paper 2A Presented at the Dallas Meeting of A.I.Ch.E. February 1972.
18. TAKAMATSU, T., HASIMOTO, I. and OHNO, H., Ind. Eng. Chem. Proc. Des. Dev. 9,368 (1970).
19. CROWE, C. M., HAMIELEC A.E., HOFFMAN, T.W., JOHNSON, A.I., SHANNON, P.T., and WOODS, D.R., *"Chemical Plant Simulation"*, Prentice Hall (1971).
20. ORBACH O. and CROWE, C.M., Can. Journ. Chem. Eng. 49 509-13 (1971).

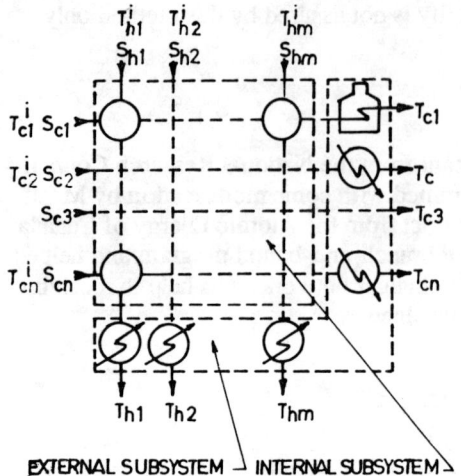

Fig. 1. Schematic representation of the combinatorial heat exchanger network design problem.

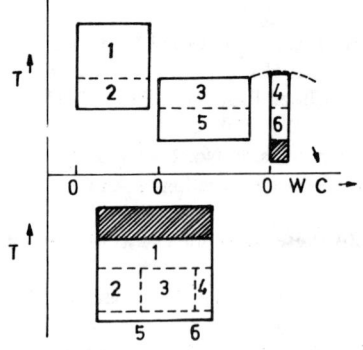

Fig. 2. Heat content diagram as suggested by Nishida et al. [12].

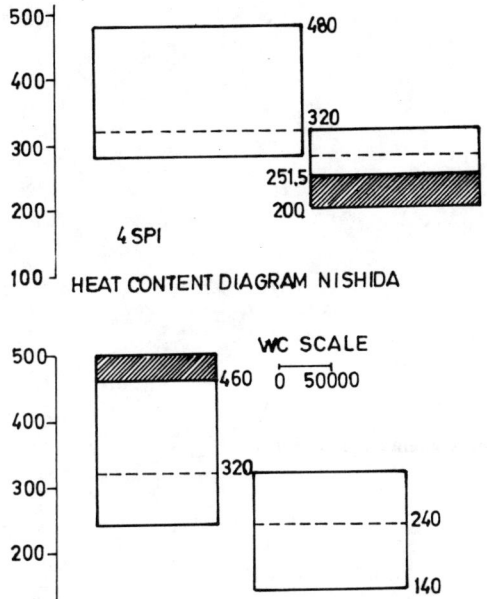

Fig. 3. The 4SPI Problem represented on a heat content diagram with preliminary construction.

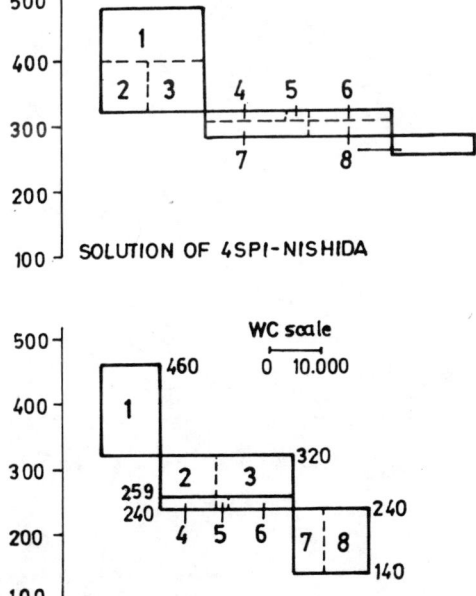

Fig. 4. Solution of the 4SPI problem by the method of Nishida et al. [12].

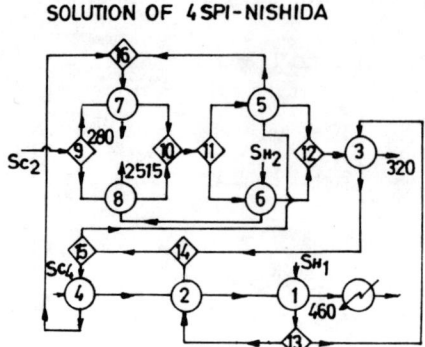

Fig. 5. Process flowsheet for 4SPI problem from heat content diagram (Fig.4).

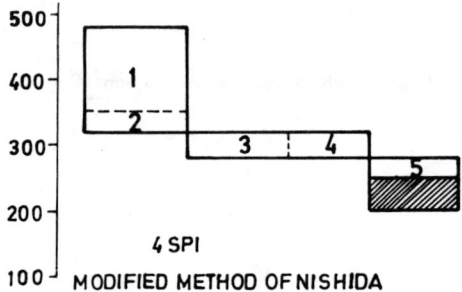

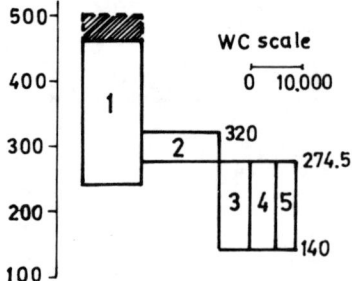

Fig. 6. Heat content diagram for modified procedure of Nishida et al. [12].

KESLER ET AL.

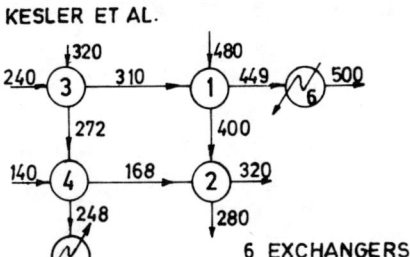

6 EXCHANGERS

Fig. 7. Process flowsheet arising out of modified method.

DESIGN MODE

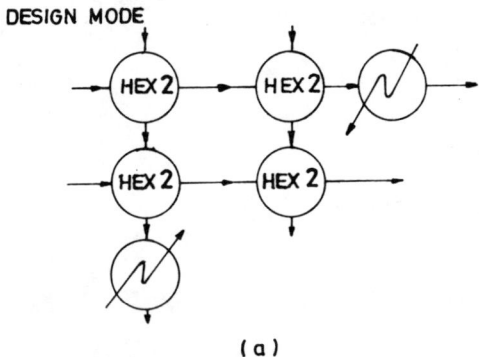

(a)

SIMULATION MODE

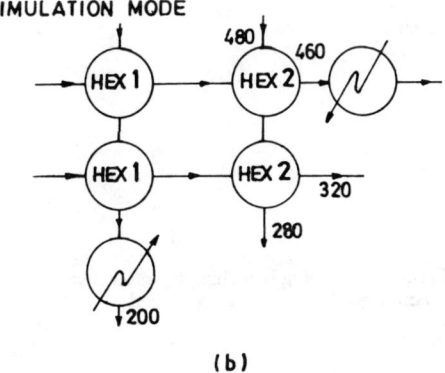

(b)

Fig. 8 (a) Using modular simulation executive computer program in the design mode.
(b) Using the executive program in the simulation (optimization) mode.

## CONSIDER THOSE NETWORKS ONLY WHERE:

1. NO STREAM MATCHING-STEAM OR WATER TO SATISFY $T_{SPEC}$

2. ALL NETWORKS COMPATIBLE WITH A MATCH OF $S_{h_1}$ AND $S_{c_1}$

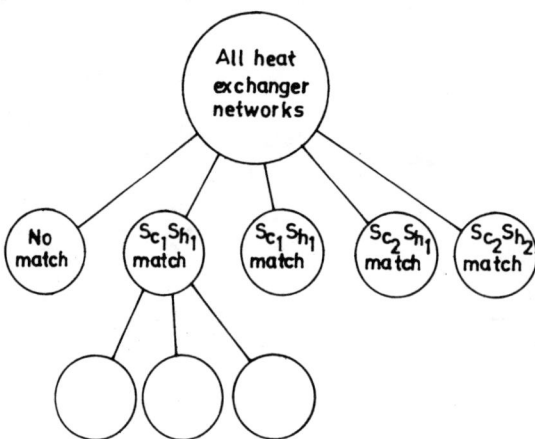

3. COMPATIBLE WITH A $S_{c_1}$-$S_{h_2}$ MATCH

4. COMPATIBLE WITH A $S_{c_2}$-$S_{h_1}$ MATCH

5. COMPATIBLE WITH A $S_{c_2}$-$S_{h_2}$ MATCH

Fig. 9. Branch and bound representation of the 4SPI problem showing branching to smaller size problems where networks are considered which are only compatible with certain specified matches.

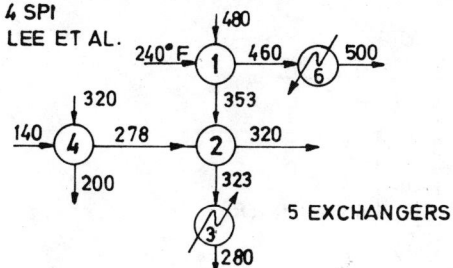

Fig. 10. 4SPI Process flowsheet from branch and bound method.

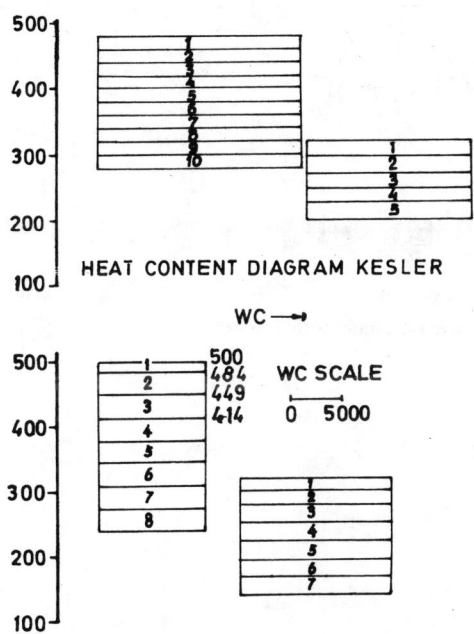

Fig. 11. The solution of the 4SPI problem by the method of Kesler and Parker [9]. Exchangelets used in the solution represented on a heat content diagram.

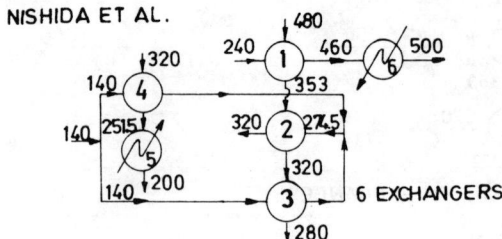

Fig. 12. 4SPI Process flowsheet by the linear programming method of Kesler and Parker [9] after further optimization of flowsheet.

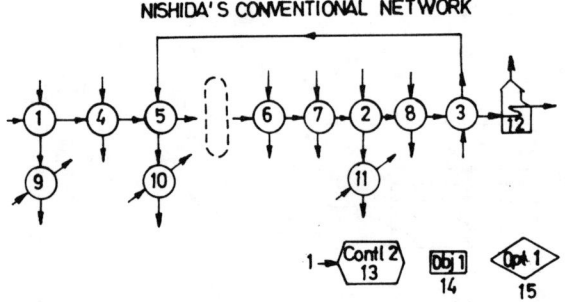

Fig. 13. The conventional process flowsheet for the refinery heating/cooling system (Cost = $11.02 \times 10^5$/year).

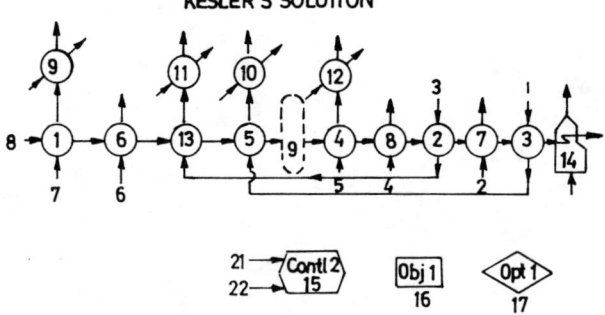

Fig. 14. Information flow diagram for design/simulation program showing solution of the refinery Problem by the method of Kesler and Parker (Cost by Kesler Method = $13.87 \times 10^5$; Optimized Network Cost = $8.74 \times 10^5$/year).

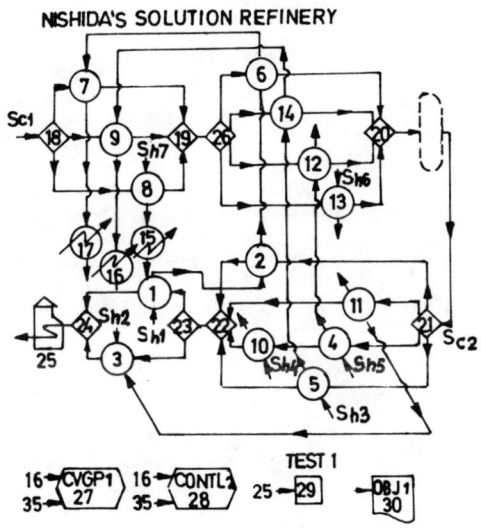

Fig. 15. Solution of the refinery problem by Nishida's method. (Information flow diagram for Desin/Simulation Program). (Cost = $7.33 \times 10^5$/year).

International Seminar 1973

# HEAT TRANSFER FROM FLAMES

**J. M. Beer**

*Sheffield University, Sheffied, U. K.*

Asked for his comments on the 1973 seminar, J. M. Beer sent H. C. Hottel's current comments on his paper dating from 1973. In the Editor's opinion, this in itself represents a unique commentary.

# FIRST ESTIMATES OF INDUSTRIAL FURNACE PERFORMANCE - THE ONE-GAS-ZONE MODEL REEXAMINED

H. C. Hottel

*Massachusetts Institute of Technology,
Cambridge, Mass., U.S.A.*

## 1. INTRODUCTION

The design of a furnace, more specifically the prediction of the performance of a chosen design, can be carried out at several levels of sophistication. Although a determination of the distribution of heat-flux density over the surface of the stock is desirable - sometimes, in high flux-density systems, necessary - the attainment of the simpler objective of determining the total heat transfer rate as a function of firing rate and excess air is a proper orienting first step; and often it suffices. Even if that is the sole objective, knowledge of the detailed interaction of radiation and convection with mass transfer and combustion is in principle necessary. But integral formulations are tolerant of casual treatment of detail, especially in the presence of the leveling effect of radiation, responsive to a high power of temperature; and a surprisingly accurate overall performance is predictable from a relatively simple model. Even though the knowledge of flux distribution over the stock may be the ultimate objective, it is still good engineering practice to start with an almost-quantitative understanding of the overall process. In fact, it may be asserted that prospects of success with the zone method are poor if the simpler and less ambitious approach, which is after all a one-gas-zone example of the zone method, is not thoroughly understood.

It is the object here to set up a simple overall furnace performance model, in form as general as is consistent with the assumption of a single-gas-radiating temperature, a single equilibrium refractory temperature and, a single equilibrium refractory temperature and a single term characterizing the exchange area between the combustion gases and the sink, which allows for:

1. The effect of the adiabatic flame temperature $T_F$, dependent on the entering fuel and air enthalpies and the gas heat capacity
2. The effect of stock or sink temprature $T_1$, measured by the ratio of its mean value to $T_F$.
3. The effect of stock temperature variation, measured by the ratio r of stock temperatue rise to its arithmetic mean temperature.

4. The value of the characteristic or effective sink area $A_S$, that area which, multiplied by the difference of black-body emissive powers of the gas and stock temperatures, gives the flux from gas to stock. The major problem of making the model describe realistically the effects of gas composition, furnace shape, and disposition of heat sinks in the furnace clies in the evaluation of $A_S$, the total gas-sink exchange area.

5. The use of a single gas-radiating temperature $T_g$.
6. The use of a difference $\Delta$ between the gas-radiating and leaving-gas-enthalpy temperature, which varies with firing rate.
7. The loss of heat through the refractory walls.
8. The loss of heat by radiation through openings.
9. Other factors, contributing to the evaluaiton of $A_S$.

## 2. THE ONE-GAS-ZONE FURNACE MODEL

Although parts of the following development have appeared before [1,2a], parts are new; and for completeness the full derivation will be presented. The well stirred furnace gases are at temperature $T_g$ in consequence of loss of heat (a) by radiative exchange with the stock or sink at $T_1$, (b) by convection to that part of the sink $A_{1,e}$ which does not include any curtain tubes across the gas exit from chamber, (c) by convection to and through the refractory walls, and (d) by radiation through furnace openings of area $A_0$.

a) The net radiative flux to the sink - direct as well as with the aid of refractory surfaces which reflect diffusely or absorb and reradiate - must, if the refractory is radiatively adiabatic and the gas is gray, be proportional to the difference in black-body emissive power of the gas and sink. The proportionality constant, having the dimensions of area, is called the gas-surface total-exchange area $(\overline{GS})_R$ the formulation of which will be discussed later. (The subscript indicates that allowance has been made for the aid given by refractory surfaces). The flux is then $((\overline{GS})_R \sigma(T_g - T_1))$.

b) Convective flux to those surfaces $A_{1,e}$ which affect the stirred-gas enthalpy is $hA_{1,e}(T_g - T_1)$. Because this term is quite small compared to (a), it is convenient to combine the two by forcing the convection into a fourth-power form:

$$[ hA_{1,e}(T_g - T_1) \sim hA_{1,e}\sigma(T_g^4 - T_1^4)/4\sigma T_{g1}^3 ]$$

where $T_{g1}$ is the arithmetic mean of $T_g$ and $T_1$. Then

$$(\overline{GS})_R \sigma(T_g^4 - T_1^4) + hA_{1,e}(T_g - T_1) = [(\overline{GS})_R + hA_{1,e}/4\sigma T_{g1}] \sigma(T_g^4 - T_1^4)$$

The bracket has in other contributions been called $(\overline{GS})_{R,c}$ to indicate that convection has been included. The simpler term $A_s$. the effective area of the sink, will be used here.

c) Convection to and through refractory walls. If the walls are in rediative equilibrium, convection - gas to wall - equals conduction through the wall. The flux is $U_r A_r (T_g - T_0)$, where $T_0$ is the ambient tempreature and $U_r$ is given, conventionally, by $U_r = \dfrac{1}{\dfrac{1}{h_i} + \dfrac{W}{K} + \dfrac{1}{h_{c+r,0}}}$

The inside flux is not in fact equal to the flux through the wall, but the difference is so small compared to the radiative flux as hardly to negate the assumption of radiative equilibrium. Without that assumption one would need to introduce an additional unknown and an additional equation, and the slight improvement in final accuracy does not justify the complication.

d) Radiation through peep holes or other openings, of area $A_0$. Rigorous allowance for this usually small effect would introduce such comlexities as to prevent obtaining a solution capable of easy engineering use. Altough the view from the outside through furnace openings is a view of sink and refractory surfaces seen dimly through partly diathermanous gas, the assumption will be made that the effective furnace tempreature (the inside plane of the openings) is $T_g$. With $\overline{F}$ representing the exchange factor to allow for wall thickness [3], the loss through the openings becomes $A_0 \overline{F} \sigma (T_g^4 - T_0^4)$. Furnaces with openings large enough to make this casual treatment inadequate are rare.

The equation of transfer from the gas is, from the above,

$$\dot{Q}_G = A_S \sigma (T_g^4 - T_1^4) + U_r A_r (T_g - T_0) + \overline{F} A_0 \sigma (T_g^4 - T_0^4) \tag{1}$$

An energy balance on the gas is needed. Although a single gas radiating temperature has been postulated, it can be a space-mean value rather than the uniform gas temperature of a perfectly stirred chamber; and the gas temperature measuring the gas enthalpy leaving the chamber is usually lower. Let $T_g - \Delta$ represent the leaving-gas temperature, between which and the base tempreature $T_0$ the mean heat capacity is $\overline{C}_{p,g}$. Then the energy balance is

$$\dot{Q}_G = \dot{H}_F - (T_g - \Delta - T_0) \dot{m}_g \overline{C}_{p,g} \tag{2}$$

where $\dot{H}_F$ is the hourly entering enthalpy, chemical plus sensible, in the fuel and air and recirculated flue gas, if any, To is the enthalpy-base temperature, and $\dot{m}_g$ is the mass flow rate of gas/hour. Let the same mean heat capacity be used to define an adiabatic pseudo-flame tempreature $T_F$

$$(T_F - T_0) = \dot{H}_F / \dot{m}_g \overline{C}_{p,g}$$

($T_F$ will in general be much higher than the true adiabatic flame temperature which allows for a temperature-varying $C_p$ and for dissociation). The energy balance may then be written in the form

$$\dfrac{\dot{H}_F - \dot{Q}_G}{\dot{H}_F} = \dfrac{T_g - \Delta - T_0}{T_F - T_0} \tag{2a}$$

The additional relation needed is that giving furnace efficiency $\eta$.

$$\eta = \frac{\dot{Q}_G - \text{wall losses}}{\dot{H}_F} = \frac{\dot{Q}_G - [U_r A_r (T_g - T_0) + F A_0 \sigma (T_g^4 - T_0^4)]}{\dot{H}_F} \qquad (3)$$

Equations (1) and (2a) contain as unknowns $T_g$ an $\dot{Q}_G$ (assuming rules available for finding $(\overline{GS})_R$ and choosing $\Delta$). Solution for these and insertion into (3) gives the furnace efficiency. But a much deeper understanding of the nature of furnaces can be obtained by further manipulation. Let (1) be made dimensionless by division through by $A_S \sigma T_F^4$ and let the dimensionless ratios of various tempreatures to $T_F$ be denoted by their primes.

$$\frac{\dot{Q}_G}{A_S \sigma T_F^4} = T'^4_g - T'^4_1 + \frac{U_r A_r}{\sigma A_S T_F^3}(T'_g - T'_0) + \frac{\overline{FA_0}}{A_S}(T'^4_g - T'^4_0) \qquad (4)$$

Equation (2a) may be written

$$\frac{\dot{Q}_G}{\dot{H}_F}(1 - T'_0) = 1 - T'_g + \Delta' \qquad (5)$$

where $\Delta' = \Delta/T_F$. To complete the normalization, let

$$\frac{\dot{Q}_G}{\dot{H}_F}(1 - T'_0) = Q', \text{ the reduced gas efficiency}$$

$$\frac{\dot{H}_F}{\sigma A_S T_F^4 (1 - T'_0)} = Q', \text{ the reduced firing density}$$

$$\frac{U_r A_r}{\sigma A_S T_F^3} = L'_r, \text{ the refractory loss factor}$$

$$\frac{\overline{FA_0}}{A_S} = L'_0, \text{ the furnace-opening loss factor}$$

Equations (4) and (5) then become

$$Q'D' = T'^4_g - T'^4_1 + L'_r(T'_g - T'_0) + L'_0(T'^4_g - T'^4_0) \qquad (6)$$

and

$$Q' = 1 - T'_g + \Delta' \qquad (7)$$

When loss factors $L'_r$ and $L'_0$ are small enough to be neglected and $\Delta'$ is assumed zero, $Q'$ becomes the reduced furnace efficiency - the furnace efficiency times $(1-T'_0)$- given by the solution of the equation

$$Q'D' = (1-Q')^4 - T'^4_1 \qquad (8)$$

This extraordinarily simple relation, giving furnace efficiency as a function of two dimensionless parameters, $D'$, proportional to the firing rate per unit of effective sink area, and $T'_1$, the ratio of sink temperature to adiabatic flame temperature, is the basic relationship governing the efficiency of furnaces of almost any class. Performance data on furnaces of a wide variety of types, from gas turbine combustors to openhearth furnaces, can be put on a diagram of efficiency versus firing rate, with sink temperature as a parameter [1,2b] (efficiency here refers to transfer of heat from the combustion gases rather than to the stock).

Better agreement between prediction and experiment, however, may be expected if allowance is made for some difference between mean-radiating temperature and leaving enthalpy temperature. Experience with furnaces of various types as well as with computations based on the multi-zone model indicate that $\Delta$ varies inversely with $H_F$, and the assumption that $\Delta'$ is proportional to $Q'$ is much more realistic than treating $\Delta$ as constant. For a reason which will emerge later let the proportionality constant be $(1-1/d)$.(There is some evidence that $d$ is about 4/3, putting $\Delta$ in the range of 150 to 240C for heavily fired cracking coils or marine boilers). The energy balance, Equation (7) then becomes

$$T'_g = 1 - \frac{Q'}{d} \qquad (9)$$

An additional modification is desirable. In many modern high-output furnaces - e.g., reformers and ethylene furnaces - the stock is often heated through a significant temperature interval: and the stock temperature leaving the combustion chamber may come to within a few hundred degrees of the leaving-gas temperature. It may readily be shown that, if the stock has a constant specific heat and $T_1$ rises from $T_{1,i}$ to $T_{1,o}$ within the combustion chamber, the term $(T'^4_g - T'^4_1)$ in Eq. (6) should be replaced by

$$\frac{8T'^3_g T'_1 r}{\ln\frac{T'_g - T'_1(1-r)}{T'_g + T'_1(1-r)} \cdot \frac{T'_g + T'_1(1+r)}{T'_g - T'_1(1+r)} + 2\left(\tan^{-1}\frac{T'_1(1+r)}{T'_g} - \tan^{-1}\frac{T'_1(1-r)}{T'_g}\right)} \qquad (10)$$

where $T_1$ is now $(T_{1,i}+T_{1,o})/2$ and $r = (T_{1,o}-T_{1,i})/(T_{1,o}+T_{1,i})$. As before, $T'_1 = T_1/T_F$. In the limit as $r \to 0$, (10) approaches $(T'^4_g - T'^4_1)$.*

---

* Although chemical or phase change within the stock invalidates the assumption of constant specific heat, it is always possible to find an equivalent r to make (10) numerically correct. With $H(T_1)$ representing the stock enthalpy as a function of $T_1$ varying from $H_{1,i}$ at entry to $H_{1,o}$ at exit, the true value of $Q_{stock}/\sigma A_S$ is

$$(H_{1,o} - H_{1,i}) / \int(H_{1,i}, H_{1,o}, \mathsf{f}(dH(T_1), T_g^4 - T_1^4)) ,$$

readily evaluated graphically. The result must equal $T_F^4$ times (10) or (10) with the primes missing. For the $T_g$ and $T_1$ of interest, r can be found by trial and error. Since the ratio of r defined as $(T_o - T_i)/(T_o + T_i)$ to r from above equality depends on the $H-T_1$ relation and on $T_G/T_1$ and the latter is substantially constant for a particluar furnace type, the ratio of the two r's requires but infrequent determination. It is to be remembered that if the stock is a fluid in a tube, $T_1$ in the $H-T_1$ relation is the outer tube surface

With (10) replacing $(T'^4_g - T'^4_1)$ in (6) and with $T'_g$ replaced by its value from (8), Equation (6) becomes

$$\left(\frac{Q'}{d}\right)(dD\,') = \frac{8\left(1-\frac{Q'}{d}\right)^3 T'_1 r}{\ln\frac{1-\frac{Q'}{d}-T'_1(1-r)}{\frac{Q'}{d}+T'_1(1-r)} \cdot \frac{1-\frac{Q'}{d}+T'_1(1+r)}{1-\frac{Q'}{d}-T'_1(1+r)} + 2\left(\tan^{-1}\frac{T'_1(1+r)}{1-\frac{Q'}{d}} - \tan^{-1}\frac{T'_1(1-r)}{1-\frac{Q'}{d}}\right)}$$

$$+L'_r\left(1-\frac{Q'}{d}-T'_0\right)+L'_0\left[(1-\frac{Q'}{d})^4-T'^4_0\right] \tag{11a}$$

with $r=0$, this reduces to

$$\left(\frac{Q'}{d}\right)(dD\,') = (1-\frac{Q'}{d})^4 - T'^4_1 + L'_r\left(1-\frac{Q'}{d}-T'_0\right)+L'_0\left[(1-\frac{Q'}{d})^4-T'^4_0\right] \tag{11b}$$

The error in the use of (11b) rather than (11a) is less then one percent when $T_1/T_g = 0.8$ and $r < 0.05$ or $T_1/T_g = 0.9$ and $r < 0.02$. $r$ is often many times these values and the error mounts rapidly.

If loss coefficients $L'_r$ and $L'_0$ are both zero, (11b) has the same structure as (8), with $(Q'/d)$ replacing $Q'$ and $(D'd)$ replacing $D'$. Thus the form of relation chosen for $\Delta$ has permitted inclusion of allowance for it without increasing the number of dimensionless groups.

The furnace efficiency $\eta$, when there are wall losses, may be shown from (3) to take the form

$$\frac{\eta}{d} = \frac{Q'/d}{1-T'_0} - \frac{L'_r}{(D'd)}\left(1-\frac{Q'/d}{1-T'_0}\right) - \frac{L'_0}{(D'd)(1-T'_0)}\left[(1-\frac{Q'}{d})^4-T'^4_0\right] \tag{12}$$

The desired relation between efficiency $\eta$ and reduced firing density $D'$ - or rather between $\eta/d$ and $D'd$ - is obtianable from (11a or b) and (12) considered as parametric equations in $Q'/d$. They express the relation

$$\eta/d = f(dD', T'_1 \text{ (or } T'_1 \text{ and } r\text{)}, L'_r \text{ and } L'_0) \tag{13}$$

which is in a form involving no commitment as to what value will be assigned to $d$, the measure of the difference between radiating and enthalpy temperature of the combustion gases.

## 3. OVERALL FURNACE PERFORMANCE - GRAPHICAL PRESENTATION

To indicate the character and use of relation (13), three graphs have been presented. Figure 1 shows, on logarithmic coordinates, the consequences of allowing for wall losses, for the case of $L'_0$ and $r$ both zero, and for $T'_0 = 1/8$, $T'_1 = 0.5$, and $L'_r = 0.02$ ( a realistic wall-loss coefficient). The lower curve is the furnace efficiency,

---

temperature, not the bulk temperature of the fluid. Enough is generally known in advance to construct an $H$–$T_1$ diagram allowing for temperature drop through the fluid film and tube wall.

efficiency, going to 0 when the normalized firing density drops to 0.015, * passing through a

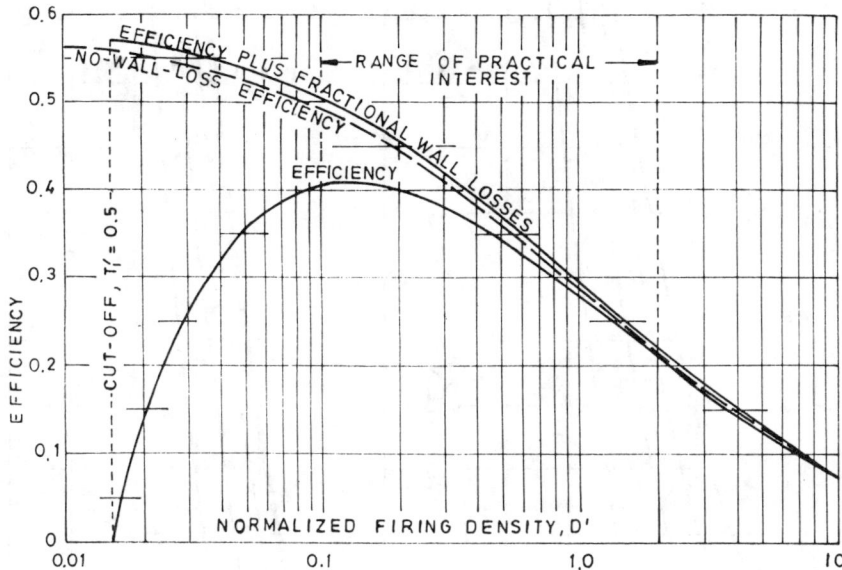

Fig. 1. Effect of wall-loss factor $L'_r$ on combustion chamber performance. $L'_r = 0.02$; $T'_1 = 0.5$; $T'_0 = 1/8$

maximum, and dropping at high firing rate. The top curve is the sum of efficiency and fractional wall loss. The middle curve is the no-wall-loss efficiency, which increases continuously as the firing rate drops, and at $D' \to 0$ is asymptotic to $(1-T'_1)/(1-T'_0)$.

Figure 2 shows, on cartesion coordinates, the $(\eta/d)$ versus $D'd$ relation, and gives the effect of sink temperature ($T'_1 = 0.2$ to $0.7$) and $L'_r$ (0.02 and 0.04).

The same picture on logarithmic coordinates is much more useful, Figure 3. Its importance is that in this form the relationship may be used to correlate and extrapolate furnace data without concern for the sometimes difficult technique of accurately evaluating the quantity $A_S$. Let us assume that furnace performance data are available in the form of effeciency $\eta$ versus firing rate $H_F$. The quantitaties $T_0, T_1, T_F, U_r A_r, A_0, \overline{F}$ are known or readily calculable, and a vlaue of $A_S$ sufficiently accurate for the determining the not too important quantity $L'_r$ and $L'_0$ should also be calculable. Let the efficiency $\eta$ be plotted vs $H_F/\sigma T'^4_F(1-T'_0)$ on transparent logarithmic paper which matches Fig. 3, and let the plot be superimposed on an equivalent of Fig. 3 which has been constructed for the values of $r$ and $L'_0$ which characterize the furnace. Let the plot be displaced vertically and horizontally until tha data fit the proper $T'_1$ and $L'_r$ curves. The relative vertical displacement of the two plots yields the value of $d$; their relative horizontal displacement yields the value $D'd/[H_F/\sigma T'^4_F(1-T'_0)]$ which is $d/A_S$. The two quantities $d$ and $A_S$ are characteristic of the furnace; one is an

---

* $(dD')_{\eta=0} = [L'_r(T'_1-T'_0)+L'_0(T'^4_1-T'^4_0)]/(1-T'_1)$; $(Q'/d)_{\eta=0} = 1-T'_1$

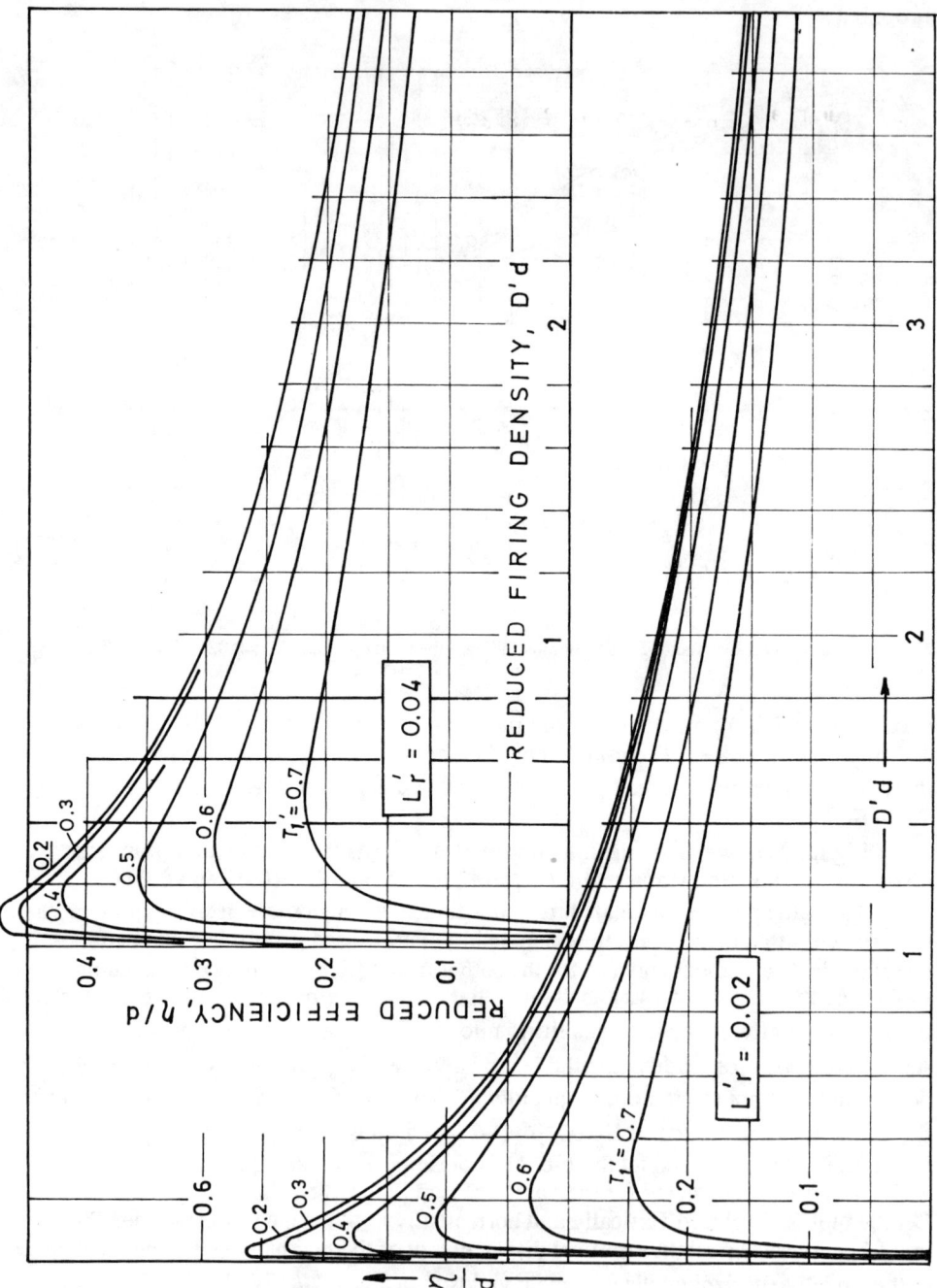

Figure 2. Efficiency vs firing density for 2 values of the wall-loss factor $L'_r$. $T'_1$ = reduced sink temperature, $T_1/T_F$. $L'_0 = 0$, $r = 0$.

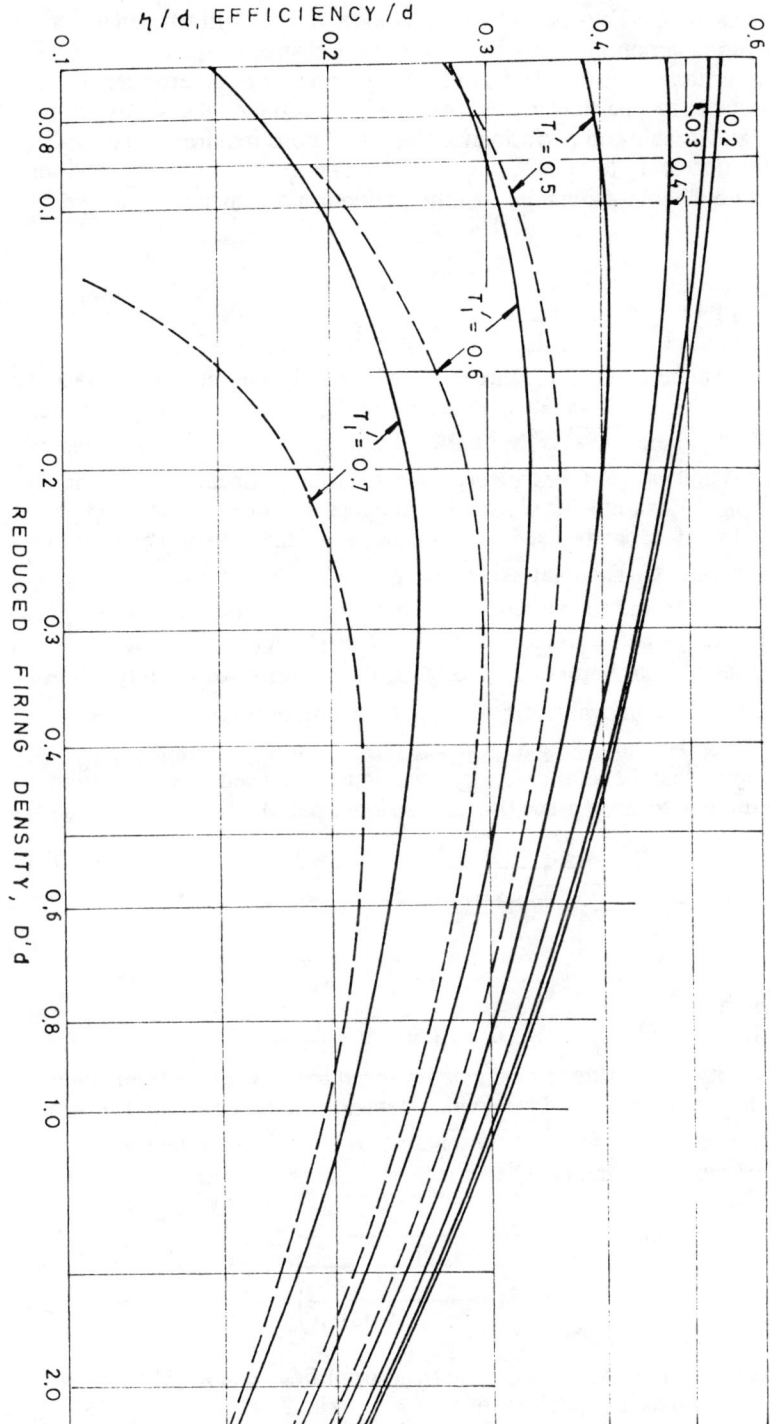

Figure 3. $(\eta/d)$ vs $(D'd)$ for furnace chambers. For comparison with data on efficiency vs. firing rate, to determine furnace performance characteristics $d$ and $A_e$ by curve fitting. Conditions: $T'_0 = 1/8$, $L'_0 = 0$, $r = 0$, $L'_r = 0.02$ (solid lines) and 0.04 (dotted lines).

empirical constant taking account of $\Delta$, the other a measure of the many factors affecting radiative exchange.They permit a computation of what will happen under other furnace operating conditions (provided these don't change the general character of the flow pattern or the system's radiation geometry); and when determined for each of several furnaces in a particular class they lead to an empirical determination of the effects of design variables on performance.For predictions requiring no prior data, however, it is necessary to be able to calculate $A_S$ from first principles, and this depends primarily on the total-exchange area the evaluation of which will now be considered.

## 4. THE TOTAL-EXCHANGE AREA

Given: a gray isothermal gas volume enclosed by an isothermal sink surface $A_1$ and by a refractory surface $A_r$ in radiative equilibrium. The net flux from gas to sink is $\overline{GS_1}\,\sigma(T'^4_g - T'^4_1)$ where $\overline{GS_1}$ is the total-exchange area. The restriction must be imposed that the gas and two surfaces are each treated as a single zone, implying not that all of surface $A_1$, for example, is segregated into a single area but that a single mean view that $A_1$ has of $A_r$ can be used in evaluating all of the radiation emitted or reflected from $A_1$ toward $A_r$. The total-exchange area $\overline{GS}$ allows for multiple reflections off all surfaces and for assistance given by the refractory in absorbing gas radiation and reradiating a part of it to the sink. In the model under discussion it is clear that $\overline{GS}$ carries a major burden of making the model agree with reality. Many degrees of complexity exist in evaluating $\overline{GS}$, and the engineer has the choice of advancing as far along the path as the importance of his problem warrants.

It may be shown that the total-exchange area depends on sink emissivity and on the inter-zone direct-exchange areas (lower case letter pairs)

$$\overline{GS}_1 = \cfrac{1}{\cfrac{1-\varepsilon_1}{A_1 \varepsilon_1} + \cfrac{1}{\overline{gs_1} + \cfrac{1}{\overline{gs_r} + \cfrac{1}{\overline{s_r s_1}}}}} \tag{14}$$

with $\overline{gs_1} \equiv A_1[\varepsilon_g(L_{m,1})]$ and $\overline{gs_r} \equiv A_r[\varepsilon_g(L_{m,r})]$ and $\overline{g_r s_1} \equiv A_r F_{r1}\{1-[\varepsilon_g(L_{m,m,r1})]\}$. Here the gas emissivity $\varepsilon_g$ is written to indicate its dependence on mean beam length $L_m$, and the latter in turn to indicate that its value depends on the source and sink of radiation. If the three $\varepsilon_g$'s are assumed representable by a single gas emissivity applicable to the whole enclosure, (14) becomes

$$\overline{GS_1} = \cfrac{A_1}{\cfrac{1}{\varepsilon_1} - \cfrac{1}{\varepsilon_g\left(1 + \cfrac{A_r/A_1}{1 + \varepsilon_g/(1-\varepsilon_g)F_{r1}}\right)}} \tag{15}$$

This ancient relation* was replaced early by a simplified version [5] based on what later became known as the speckled-furnace approximation. If $A_1$ and $A_r$ are

---

*Its black – sink equivalent appeared in 1928 [4], perhaps earlier.

intimately mixed over the surface of the enclosure rather than more or less segregated the view factor either surface has of $A_1$ is $A_1/(A_1+A_r)$. Subsituttion of this value for $F_{r1}$ in (15) together with the replacement of $A_1$ by $CA_T$ and $A_r$ by $(1-C)A_T$, where $C$ is the fraction of the total envelope area $A_T$ which is "cold" (i.e., which is $A_1$), yields the very simple relation

$$\overline{GS}_1 = \frac{A_T}{\frac{1}{C\varepsilon_1}+\left(\frac{1}{\varepsilon_1}-1\right)} \tag{16}$$

If instead of $A_1$ and $A_r$ being intimately mixed $A_1$ lies in a single plane, $F_{1r}$ becomes 1 and, since $A_1F_{1r}=A_r/F_{r1}$, $F_{r1}$ is $A_1/A_r$ or $C(1-C)$. Substitution of this into (15) yields a result which may be shown to have the structure of (16) except that the parenthesis in the denominator is now multiplied by

$$\left(\frac{1-\varepsilon_g}{1-C\varepsilon_g}\right) \tag{16a}$$

The object of converting (15) to (16) or to (16) modified has of course been to dodge the often tedious formulation of $F_{r1}$ (or $F_{1r}$). From the two results which appear to be limiting cases it is tempting to seek an interpolation procedure, i.e., to find a function by which to multiply the parenthesis in (16) which varies from 1 for speckled furnaces to (16a) for the case of $A_1$ in a single plane.

It may be shown that Eq. (15), with $A_1$ and $A_r$ replaced by $CA_T$ and $(1-C)A_T$, becomes

$$\frac{GS_1}{A_1} \equiv \frac{GS_1}{CA_T} = \frac{1}{\frac{1}{\varepsilon_1}+\left(\frac{1}{\varepsilon_1}-1\right)C\left[\frac{\varepsilon_g+F_{r1}(1-\varepsilon_g/C)}{C\varepsilon_g+F_{r1}(1-\varepsilon_g)}\right]} \tag{17}$$

The bracketed term can be written in a form more nearly symmetrical with respect to $C$ and $(1-C)$ by substitution of the exchange area $\overline{\imath r}$ for $A_1F_{1r}(\equiv A_rF_{r1})$ to give

$$\left[\frac{\varepsilon_g/C+(1-\varepsilon_g/C)\,\overline{\imath r}/A_TC(1-C)}{\varepsilon_g+(1-\varepsilon_g)\,\overline{\imath r}/A_TC(1-C)}\right] \tag{18}$$

The term takes the three limiting (?) forms

(a)

If $A_1$ lies in a single plane,$(C\leq 1/2)$, substitution of $F_{1r}=1$ yields

$$[\,] = \frac{1-\varepsilon_g}{1-C\varepsilon_g}$$

(b)

If the walls are speckled, $(0<C<1)$ substitution of $F_{r1}=C$ or $F_{1r}=1-C$ yields

$$[\,] = 1$$

(c)

If $A_r$ lies in a single plane, $(C>1/2)$ substitution of $F_{r1}=1$ yields

$$[\,] = \frac{1-\varepsilon_g(1-C)/C}{1-\varepsilon_g(1-C)}$$

The first two of these three results have already been given. Since $F_{1r}$ changes from $(1-C)$ for case (b) above - the speckled system - to 1 for $A_1$ segregated in a plane, let us make the heuristic assumption that a term $S$ - called the speckledness - measures the shift from complete speckledness to $A_1$ in a plane, according to the relation

or
$$\left. \begin{array}{r} F_{1r} = 1 - SC \\ \overline{1r}/A_T = C(1-SC) \end{array} \right\} \quad C \leq 1/2 \qquad (19)$$

with $S = 1$ for a speckled enclosure and 0 for $A_1$-segregation. Substitution of this into the bracket of (17) gives

$$[\,] = \left[ \frac{1 - \varepsilon_g - S(C - \varepsilon_g)}{1 - C\varepsilon_g - SC(1 - \varepsilon_g)} \right] \qquad (20)$$

Values of $S = 1$ and 0 substituted into (20) yield the results of cases (b) and (a) above respectively.

Similarly, since $F_{r1}$ changes from $C$ for the speckled system to 1 for $A_r$ in a single plane, we make the parallel assumption that $S'$ measures the shift from speckledness to $A_r$-segregation according to the relation

or
$$\left. \begin{array}{r} F_{r1} = 1 - S'(1-C) \\ \overline{1r}/A_T = (1-C)(1-S'(1-C)) \end{array} \right\} \quad C \geq 1/2 \qquad (21)$$

with $S' = 1$ for a speckled enclosure and 0 for $A_r$ in a plane. Substitution of this into the bracket of (17) give

$$[\,] = \left[ \frac{\varepsilon_g + (1-\varepsilon_g/C)(1-S'(1-C))}{\varepsilon_g/C + (1-\varepsilon_g)(1-S'(1-C))} \right] \qquad (22)$$

Values of $S' = 1$ and 0 substituted into (22) yield the results of cases (b) and (c) above, respectively.

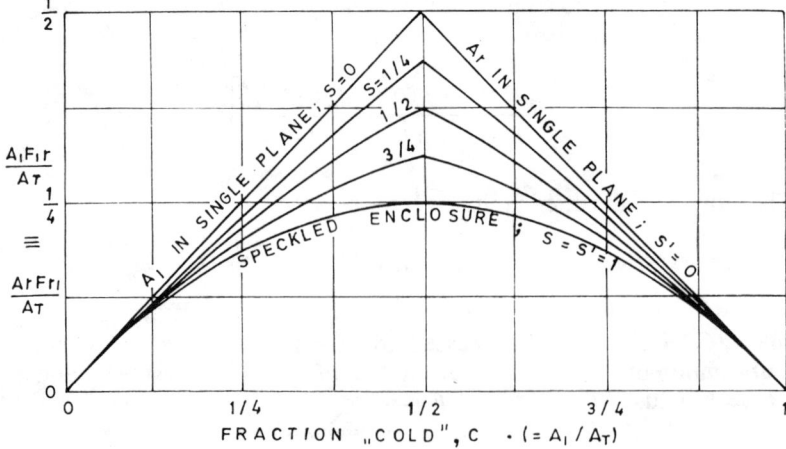

Figure 4. Evaluation of ratio of Sink-refractory direct-exchange area to total furnace envelope area, in terms of fraction cold $C$, and speckledness $S$.

In Fig. 4 the value $\overline{Ir}/A_T$ appears as a function of $C$ for the three limiting values:

speckled-wall enclosure, lower parabola ($S = S' = 1$)
$A_1$ in a plane, upper left straight line ($S = 0$)
$A_r$ in a plane, upper right straight line ($S' = 0$)

The values of $\overline{Ir}/A_T$ acoording to (19) and (21) are given in Fig. 4 at $S$ and $S'$ of 1/4, 1/2, 3/4. The question arises as to the meaning of these curves intermediate between $S$(or $S'$) = 1 and 0 where the meanings are quantitatively identifiable. Or if the questions is rephrased, does the $S$-concept have utility in quick identification of $\overline{Ir}$ for use in (17) to evaluate $\overline{GS_1}$ ? How significant is an error in $\overline{Ir}/A_T$ ?

An examination of a few geometrical shapes is illuminating:

1. Spheres. The view that $A_1$ has of $A_r$ depends on areas only, not on location. ($A_1 F_{1r} = A_1 A_r/A_T$). No matter what the disposition of surfaces or their degree of segregation, $S = 1 = S'$, and all spherical enclosures are in the "speckled" category.

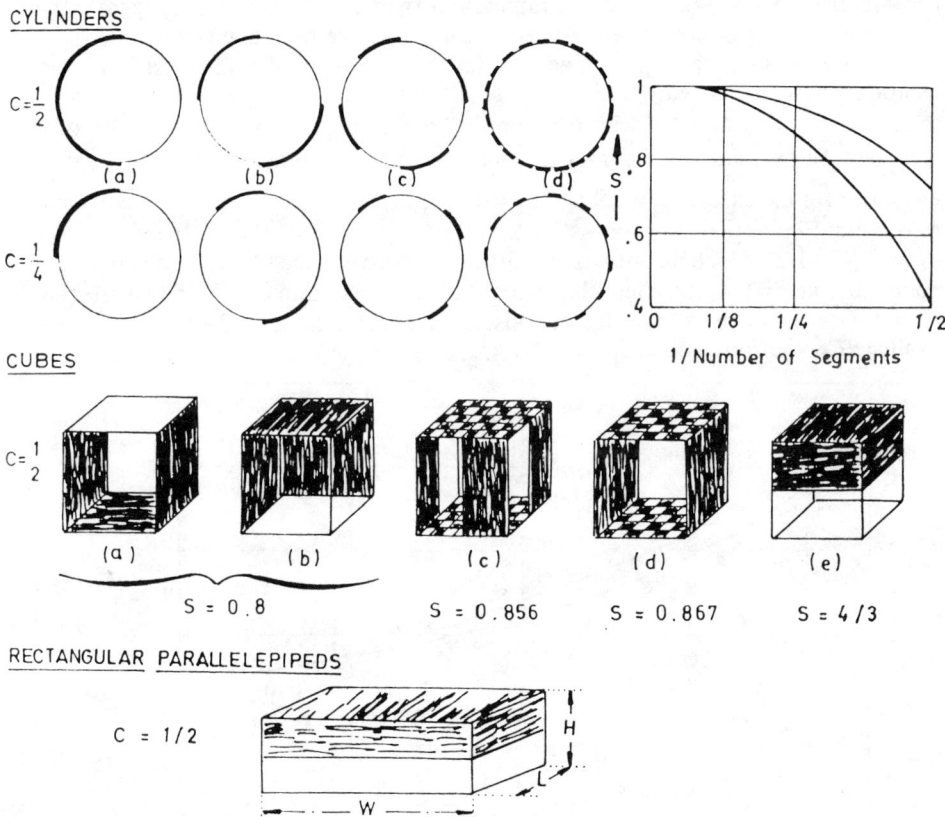

Figure 5. Effect of division of the surface of various enclosures into $A_1$ (sink) and $A_r$ (refractory) on the speckledness necessary to predict the exchange area $A_1 F_{1r} (\equiv A_r F_{r1})$.

2. Long cylinders. Figure 5 shows two sets of cylinders, $C = 1/2$ and $1/4$, with $A_1$ successively divided into smaller segments symmetrically disposed. The value of $S$ needed for the known value of $\overline{Ir}/A_T$ is plotted versus the reciprocal of the number of segments into which the surface is divided. Note that division of $A_1$ and $A_r$ into only two segments each is almost sufficient to put $S$ above 0.9.

3. Cubes, $C = 1/2$. When three of the six faces are $A_1$, the only two possible arrangements will yield an $S$ of 0.8, cases (a) and (b). Case (c), the four sidewalls symmetrically banded, with $A_1$ occupying one-half the area, and the roof and floor speckled. The overall $S$ is 0.856. When this case is modified to (d), with $A_1$ occupying all opposite sides and a speckled half of the roof and floor, $S = 0.867$. But when a value of $C = 1/2$ is reached (e) by a plane parallel to one of the faces, with all $A_1$ on one side of it, $S = 4/3$. It becomes obvious that $S$ values of 0 and 1 are not bounding values (see next case).

4. Rectangular parallelepipeds, $C = 1/2$, $A_1$ and $A_r$ on either side of a plane parallel to one face. It may be shown that in this case $S = 2-2/[1+H(1/W+1/L)]$. The larger $S$ the poorer the performance. When $H/(W+L)$ is small, $S$ is small, zero in the limits as $H \to 0$ (the case of infinite parallel planes). Under these conditions $A_r$ makes a maximum contribution to the transfer of heat to $A_1$. As $H$ increases to $1/2W$, with $L = W$, $S$ becomes 1. With $H$ great compared to $W$ or $L$, $S$ is increasingly greater than 1. Actual furnaces of such geometry as to make $S > 1$ are rare, and should be.

A study of the cases presented justifies the conclusion that for most furnace chambers $S$, which is near 1 unless $C$ is small (when $A_1$ may be in a single plane), may generally be estimated within 0.2 on the basis of a qualitative examination of the chamber. To examine the accuracy to which $S$ need be established, $\overline{GS_1}/A_1$ was evaluated for the case of $\varepsilon_1 = 0.9$ and $\varepsilon_g = 0.3$, realistic values for cracking coils and reformers. Figure 6 shows curves for $S = S' = 1$ and for $S = 0$ ($C < 1/2$) and $S' = 0$ ($C > 1/2$). Since the maximum difference between the two curves is only about 10% at $C = 0.4$, it is clear that estimation of $S$ to within 0.2 should be adequate for most design purpose, and that in consequence it should rarely be necessary to evaluate $F_{r1}$ rigorously.

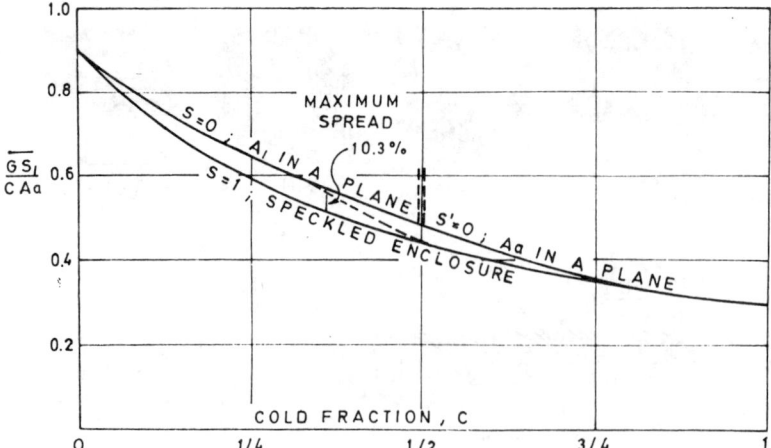

Figure 6. Example of the effect of the variation of speckledness from 0 to 1 on the relation between total gas-sink exchange area and "cold" fraction of a furnace chamber. ($\varepsilon_1 = 0.9$; $\varepsilon_g = 0.3$).

It is interesting to note that Lobo and Evans [6], in an early application of the equivalent of Eq. (15) to cracking-coil furnaces, recommended a procedure which can be shown to be the equivalent, in Fig. 6, of assuming that $\overline{GS_1}/CA_T$ is uniquely determined by $C$. The relation recommended was the equivalent of assuming that $S = 0$ for $C$ up to 1/3, that $S$ undergoes transition from 0 to 1 as $C$ varies from 1/3 to 1/2, and that $S = 1$ from $C = 1/2$ to 1. The transition is shown as a dotted line in Fig. 6. In Fig. 4, the Lobo and Evans recommendation is equivalent to following the top line from the left at $C = 0$ to $C = 1/3$, dropping from there to the $S = 1$ curve at $C = 1/2$.

## 5. EFFECT OF NON-GRAY GAS ON $\overline{GS}$

Although refractory surfaces are in overall radiative equlibrium, they are capable of absorbing radiaton from a gas and then reradiating it with a different spectral distribution, some of the radiation passing through the spectral windows of the gas directly to the sink $A_1$. The mixed-gray-gas assumption [7,2c] is consequently more realistic than the gray-gas assumption on which Eqs. (14) and (15) are based.

The simple mixed-gray gas is a gray-plus-clear system, with $a_g$ and $(1-a_g)$ representing the energy-fractions of the black-body spectrum in which the gas is gray (or the absorption coefficient is constant) and clear, respectively. The quantity $a_g$ is obtained from a determination of gas emissivity $\varepsilon_{g,L_m}$ at the mean length $L_m$ which characterizes the enclosure, and of $\varepsilon_{g,2L_m}$ at twice that length.

$$a_g = \frac{(\varepsilon_{g,L_m})^2}{2\varepsilon_{g,L_m} - \varepsilon_{g,2L_m}} \qquad (23)$$

The rather involved derivation [2d] leads to a reasonably simple expression for $\overline{GS_1}$.

$$\overline{GS}_1 = \frac{A_T}{\dfrac{1}{C\varepsilon_1} + \dfrac{1}{\varepsilon_{g,e}} - \dfrac{1}{a_g} + \left(\dfrac{1}{a_g} - 1\right)\dfrac{1}{C\varepsilon_1 + (1-C)\varepsilon_r}} \qquad (24)$$

Note that for a gray gas $a_g = 1$, and (24) reduces to (16).

The new subscript $e$ on gas emissivity requires explanation. For a gray gas, emissivity and absorptivity are the same; and the net gas-sink flux is the black-body emissive-power difference $(E_g - E_1)$ multiplied by a $\overline{GS_1}$ dependent on $\varepsilon_g$. For a non-gray gas $E_g$ and $E_1$ are separately multiplied by different values of $\overline{GS_1}$, based on gas emissivity and gas absorptivity, respectively. This complication may be avoided, however, by using a modified gas emissivity in (24), equal to the value at the arithmetic mean of $T_g$ and $T_1$, then multiplied by the factor $[1 + (a' + b - c)/4]$ which allows primarily for the way emissivity varies with absorption strength and tempreature [6.2c].

$$a' = \partial \ln \varepsilon_g / \partial \ln pL$$

$$b = \partial \ln \varepsilon_g / \partial \ln T_g$$

$c = 0.65$ for $CO_2$, 0.45 for $H_2O$, 0.5 for average flue gas.

When $T_1 < T_g/2$, the simpler $\varepsilon_g$ evaluated at gas temperature replaces $\varepsilon_{g,e}$ in (24).(Note that $\overline{GS_1}$ now depends on refractory emissivity, about 0.5).

## 6. SPECIAL PROBLEMS ASSOCIATED WITH TUBULAR HEATERS

The heat sinks of many furnaces are in the form of a row or rows of tubes, often mounted in a plane parallel to and near a refractory backing wall. When so mounted each tube row - backwall combination acts, with respect to radiative interchange with the remainder of the chamber - gas and walls - like a plane gray surface of area equal to the continuous tube plane $A_p$ and of effectove emissivity given [7,2f] by

$$\varepsilon_1 = \frac{1}{\frac{1}{F+(1-F)F} + \frac{B}{\pi}\left(\frac{1}{\varepsilon'_1} - 1\right)} \qquad (25)$$

Here $\varepsilon_1'$ is the true emissivity of the tube metal (often taken as 0.9, lower for high-quality alloys), $B$ is the ratio of tube center-to-center distance to diameter, and $F$ is the fraction of radiation incident on the plane through $2\pi$ steradians which is intercepted by the tubes. The fraction $(1-F)$ impinges on the backwall and is reradiated or reflected.

The fraction $F$, the view-factor for radiation incident on a row of tubes, is conventionally evaluated for incident radiation which is isotropic, of which black-body radiation is an example. In that case

$$F_{iso} = 1 - \frac{1}{B}[(B^2-1)^{1/2} - \cos^{-1}\frac{1}{B}] \qquad (26)$$

(The numerical equivalent of (26), clumsily obtained, first appeared in 1930[8]). The question unavoidably arises as to how much error is involved in using $F$ from (26) when the incident radiation on a tube row is non-isotropic, as from a gray gas (the result for a real gas is then readily evaluated by use of the mixed gray gas concept).

Consider a two-dimensionally infinite tube row, Fig. 7, irradiated by a slab of

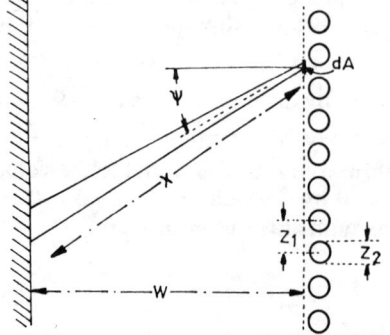

Figure 7. Section through a gas slab adjacent to a row of tubes $z_1/z_2 \equiv B$

gas of width $W$ and absorption coefficient $K$. The interchange-area $\overline{ss}$ between surface elements bounding the ends of the wedge-shaped space [2g] is

$$\frac{d^2\overline{ss}}{dA\cos\psi\,d\psi}=\frac{2}{\pi}\int_0^{\pi/2}e^{-Kx/\cos\theta}\cos^2\theta\,d\theta \tag{27}$$

Call the definite integral $f_3(Kx)$; its numerical value is given in [2g]. The value of $\overline{ss}$ if the gas is non-absorbing is, from (27) with $K=0$, $(2/\pi)f_3(0)=1/2$. Since

$$\overline{ss}_{clear}-\overline{ss}_{absorbing\ gas\ present}=\overline{gs}$$

one may express $\overline{gs}$, with $x$ replaced by $W/\cos\psi$, as

$$\frac{d^2\overline{gs}}{dA\cos\psi\,d\psi}=\frac{1}{2}-\frac{2}{\pi}f_3(KW/\cos\psi) \tag{28}$$

As $dA$ is moved along the plane of the tube row without changing $\psi$, the interception of the beam varies intermittently from 0 to 1. Let the mean fractional interception for $\psi$-oriented radiation be $F_\psi$, given by

$$F_\psi=\begin{cases}\dfrac{\sec\psi}{B} & \text{when }\sec\psi\leq B \\ 1 & \text{when }\sec\psi\geq B\end{cases}$$

The interchange area ratio $\overline{gs}/A$ - the flux from gas to tubes per unit area of tube plane and per unit black emissive power difference of gas and tube surface -- is then given by multiplying the r.h.s. of (28) by $F_\psi$ and integrating.

$$\frac{\overline{gs}}{A}=2\int_0^1\left[\frac{1}{2}-\frac{2}{\pi}f_3(KW/\cos\psi)\right]F_\psi\,d\sin\psi \tag{29}$$

The same interchange-area ratio for a continuous plane receiver would be (29) evaluated with $F_\psi=1$; this may be shown to be $1-2\varepsilon_3(KW)$, where $\varepsilon_3$ is the third exponential integral. The ratio of these two values of $\overline{GS}$ is the desired mean fractional interception $F_{gas}$ of radiation from a gas slab to a bounding tube row.

$$F_{gas}=\frac{2\int_0^1\left[\frac{1}{2}-\frac{2}{\pi}f_3(KW/\cos\psi)\right][F\psi(B)]d\sin\psi}{0.5-\varepsilon_3(KW)} \tag{30}$$

The ratio $F_{gas}/F_{isotropic}$, from (26)/(30), is the correction factor to the conventionally used graphs giving $F_{iso}$ as a function of $B$. This ratio appears in Fig. 8. Since, for the gray-plus-clear gas model, tube furnaces have a $KW$ of the order of 1, the correction for non-isotropic incidence on the tube row is generally small.

If the tube-row is mounted on a backwall, $F$ is required in (25) in order to obtain the effective plane emissivity $\varepsilon_1$. The term $[F+(1-F)F]$, representing interception of the incoming beam plus interception of the beam returning from the refractory, is called $\overline{F}$. Because the returning beam is either almost-isotropic emission or almost-isotropic diffuse reflection,

$$\overline{F}_{gas} = F_{gas} + (1-F_{gas})F_{iso} \tag{31}$$

Clearly, the ratio $\overline{F}_{gas}/\overline{F}_{iso}$ is even nearer 1 then the direct-radiation ratio $F_{gas}/F_{iso}$; and the correction can almost always be ignored.

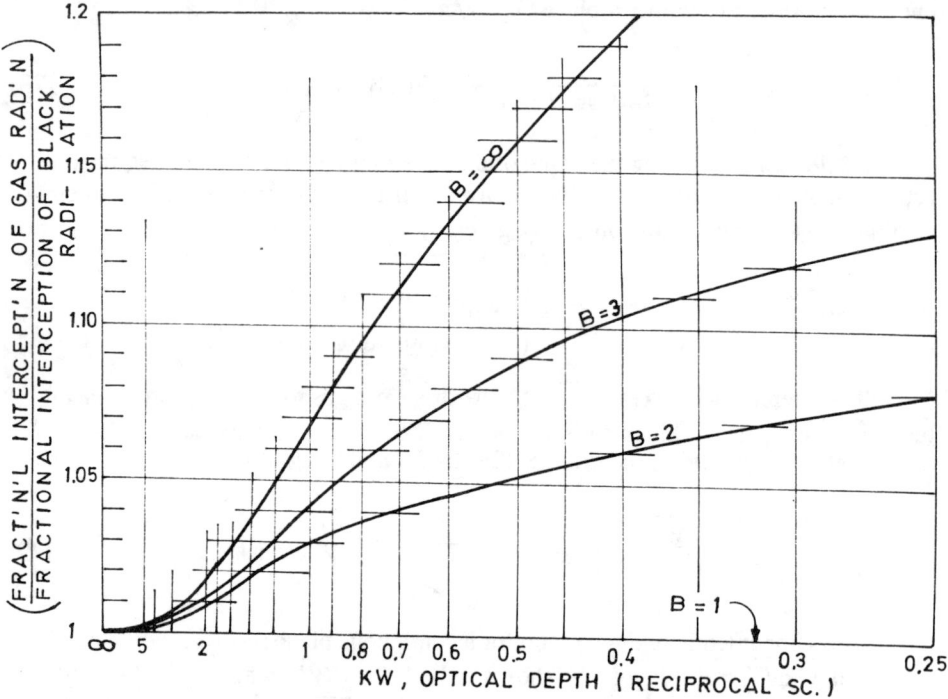

Figure 8. Ratio $F_{gas}/F_{iso}$, (fractional interception of a gas radiation)/ (fractional interception of isotropic radiation) by a tube raw. Tube center-to-center distance / diameter $=B$; optical thickness of gas a slab $=KW$.

From the above discussion it is clear that in using Equation (15)-(22) and (24) to predict the performance of a furnace with tubes mounted on walls, $A_r$ refers only to bare refractory, $A_1$ is the total area of the tube planes which, with their refractory backing act like a surface at $T_1$ and of emissivity $\varepsilon_1$ given by Eq. (25). There remains the evaluation of gas emissivity, which depends on the mean beam length. for rectangular parallepipeds varying from cubes to the space between infinite parallel planes an average mean beam length of 0.83x4 times the system mean hydraulic radius is an excellent approximation for absorption strengths in the range of industrial furnaces $(KL_m \sim 1\text{-}3)$.

The use of the one-gas-zone model for furnace chambers in which the tubes are enveloped in combustion gases rather than mounted in planes near refractory backwalls raises some difficult questions, such as how to define the sink area $A_1$ and what mean beam length $L_m$ to use in evaluating $\varepsilon_g$. No longer is the system representable as an equivalent box, with the sink $A_1$ represented by continuous plane surfaces. A rigorous treatment of this problem has not to my knowledge been made. The multizone method could be used to determined the best recommentations for a one-zone model. In the interim an approximation will be suggested.

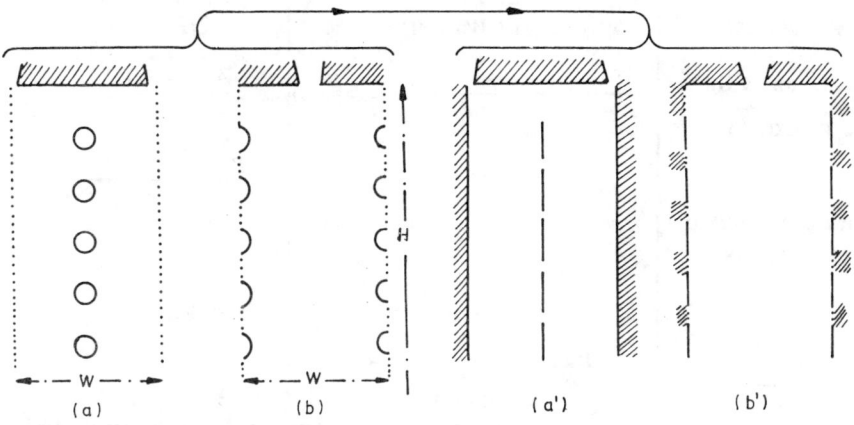

Fig. 9. Section through one cell of a furnace chamber filled wiht spaced parallel vertical tube screens, and its two alternative visualizaitons

Consider a furnace in which equally spaced parallel tube screens are mounted, with firing between each pair, sufficient in number to justify the assumption that a repeating pattern is typical. Figures. 9(a) and (b) depict the two choices avaiable for picturing the repeating pattern. If there are several zones on either side of the one pictured, the dotted lines are planes of no net radiative flux. Specular mirrors replacing the dotted lines would make both sketches truly representative, and therefore completely equivalent. It is tempting to visualize perfectly diffuse mirrors instead, because for a gray-gas system with refractory surfaces in radiative equilibrium such surfaces are the equivalent of perfectly diffuse mirrors. A little consideration shows, however, that diffuse and specular mirrors produce different distributions of flux,particlularly when the top and bottom refractory surfaces of the enclosure are significant. Because the scale of variation in image detail is small compared to $W$ and particluarly because $H/W$ is usually large, the assumption that the dotted lines are replaceable by refractory surfaces, thereby defining a chamber independent of the rest of the system, is probably a good one. But it provides two alternative proceedures between which to choose.

Step one is to replace the tubes by plane interrupted surfaces - vertical strips - which intercept the same radiation that the tubes would. Since most of the radiation is from the gas, the height of a replacing strip should be the tube pitch $P$ (the center-to-center spacing) multiplied by the $F_{gas}$ of Eq. (30) - the $F_{iso}$ of Eq. (26), multiplied if warranted, by a correction factor from Fig. 8. (This will hereafter be called F, without subscript). Figures. 9(a) and (b) then become (a') and (b'). With the number of tubes in a screen equal to $n$, the total screen height is $nP$, generally a little less than $H$; and the area $A_1$ is $2nPF$ (all areas are per unit dimension normal to the plane of the illustration). The tabulation below gives, for each of the two cases, values of $A_1, A_r, A_T, \overline{Ir}/A_T(\equiv F_{r1}A_r/A_T)$ for use in Eq. (15) or (18), and the transverse dimension involved in determining mean beam length $L_m$.

|  | Case (a') | Case (b') |
|---|---|---|
| $A_1$ | $2nPF(\cong 2HF)$ | $2nPF(\cong 2HF)$ |
| $A_r$ | $2(H+W)$ | $2[H-nPF+W]$ |
| $A_T=A_1+A_r$ | $2(H+W)+nPF$ | $2(H+W)$ |
| $\dfrac{\overline{Ir}}{A_T}$, clear-gas value for use in Eq. (15) | $\dfrac{A_1F_{1r}}{A_T} = \dfrac{nPF}{H+W+nPF} \equiv C\ (=0=S=0)$ | $\dfrac{A_1(1-F_{11})}{2(H+W)} =$ $\dfrac{F\{nP-[\sqrt{(nP)^2+W^2}-W]F\}}{H+W}$ |
| Basis for $L_m$, related to transmittances associated with; | | |
| $\overline{s_r s_1}$ | Based on $W/2$ | Based on $W$ |
| $\overline{s_1 s_1}$ | Inapplicable. $\overline{s_1 s_1} = 0$ | Based on $W$ |
| $\overline{s_r s_r}$ | Based on $W$ | Based on $W$ |

Some of the terms need explanation. For case (b'), the term $[\sqrt{(nP)^2+W^2}-W]F$ in the expression for $\overline{Ir}/A_T$ comes from setting $\overline{Ir}(\equiv A_1F_{1r})$ equal to $A_1(1-F_{11})$ and from visualizing $A_1F_{11}$ to consist of the exchange-area between two continuous parallel planes of height $nP$ and separated by $W(=\sqrt{(nP)^2+W^2}-W)$, then multiplying once by $F$ to allow for emitter area and once more to allow for receiver area.

The gas emissivity affects $\overline{GS}_1$ through its appearance in the the direct-exchange areas in Eq.(14). These may be written in terms of gas transmittance $\tau$ as follows, with primes indicating that gas absorption in present.

$$\overline{gs}_1 = A_1 - \overline{ll}\,' - \overline{lr}\,' = A_1 - \overline{s_1 s_1}\,\tau_{11} - \overline{s_1 s_r}\,\tau_{1r}$$
$$\overline{gs}_r = A_r - \overline{rl}\,' - \overline{rr}\,' = A_r - \overline{s_r s_1}\,\tau_{1r} - \overline{s_r s_r}\,\tau_{rr} \qquad (32)$$
$$\overline{s_r s_1}\,' = \overline{rl}\,' = \overline{s_r s_1}\,\tau_{1r}$$

All exchange areas on the r.h.s. are clear-gas values; those in Eq. (14) include gas emission or absorption. It is seen that $\overline{GS}_1$ involves three different values of gas transmittance $\tau(\equiv 1-\alpha_g, = 1-\varepsilon_g$ if the gas is gray), differentiated by the subscripts identifying the surfaces at the two ends of a beam.

Case (b'). In this case the three $\tau$'s are identical, and Eq.(15) may be used directly, with $\varepsilon_g$ based on a rectangular parallelepiped $H$ high and $W$ wide. Substitution of values from the table into (15) yields

$$\frac{1}{\overline{GS}_1} = \frac{1}{2nPF}\left[\left(\frac{1}{\varepsilon_1}-1\right) + \frac{1}{\varepsilon_g\left(1 + \dfrac{nPF}{H+W-nPF} + \dfrac{\varepsilon_g}{1-\varepsilon_g}\dfrac{nP}{nP-F(\sqrt{(nP)^2+W^2}-W)}\right)}\right]$$

(33)

As a numerical example, if $nP = H$, $W/H = 2/3$, $\varepsilon_g = 0.3$, $B = 2$, and $F = 0.6576 \times 1.02 = 0.67$ (use of Fig. 8), the above yields

$$\frac{1}{\overline{GS}_1} = \frac{1}{2nPF}\left[\left(\frac{1}{\varepsilon_1}-1\right)+1.91\right]$$

If $\varepsilon_1 = 0.9$, $\overline{GS}_1 = 2nPF/2.021$.

Case (a'). Although two different τ's (and $\varepsilon_g$'s) are involved, assume temporarily that a single value of $\varepsilon_g$ may be used. Substitution from the table into Eq. (15) yields

$$\frac{1}{\overline{GS}_1} = \frac{1}{2nPF}\left[\left(\frac{1}{\varepsilon_1}-1\right) + \frac{1}{\varepsilon_g\left(1+\dfrac{nPF}{H+W}+\dfrac{\varepsilon_g}{1-\varepsilon_g}\right)}\right] \qquad (34)$$

The same numerical example gives

$$\frac{1}{\overline{GS}_1} = \frac{1}{2nPF}\left[\left(\frac{1}{\varepsilon_1}-1\right)+1.511\right]$$

If $\varepsilon_1 = 0.9$, $\overline{GS}_1 = 2nPF/1.622$, which is 24.5% higher than case (b').

Consider the other $\varepsilon_g$ involved. $\tau_{r1}$ is based on the slab width $W/2$, $\tau_{rr}$ on $W$ (the screen effect is allowed for in the $F$'s). If the gas is gray and $A_r$ is dominantly on the walls,

$$\tau_{rr} = \tau_{r1}\tau_{1r}, \text{ or } [1 - \varepsilon_g(W)] = [1-\varepsilon_g(W/2)]^2.$$

Then

$$\varepsilon_g(W/2) = 1-\sqrt{1-\varepsilon_g(W)} \qquad (35)$$

Introducing $\varepsilon_g(W/2)$ into (33) instead of $\varepsilon_g(W)$

$$\frac{1}{\overline{GS}_1} = \frac{1}{2nPF}\left[\left(\frac{1}{\varepsilon_1}-1\right)+2.286\right]$$

If $\varepsilon_1 = 0.9$, $\overline{GS}_1 = 2nPF/2.397$, which is 15.7 percent lower than case (b').

Clearly, the two bounds set by use of the limiting values of $\varepsilon_g$ are too far apart to be useful, and it is necessary to go back to more basic Eq. (14). Either the substitution of exchange areas from (32) into (14) or the use of a different approach [7] yields

$$\frac{1}{\overline{GS}_1} = \frac{1}{A_1}\left[\left(\frac{1}{\epsilon_1}-1\right)+\frac{1}{1-\dfrac{\overline{ll\,'}}{A_1}-\dfrac{(\overline{lr\,'})^2/A_1}{A_r-\overline{rr\,'}}}\right] \quad (36)$$

From first principles and definitions

$$\overline{ll}\,' = A_1(1-F_{lr})\tau_{ll}$$
$$\overline{lr\,}\,'^2/A_1 = A_1(F_{lr}\tau_{lr})^2$$
$$A_r - \overline{rr}\,' = A_r[1-(1-F_{rl})\tau_{rr}] = A_r(1-\tau_{rr}) + A_l F_{lr}\tau_{rr}$$

Insertion of these into (36) yields

$$\frac{1}{\overline{GS}_1} = \frac{1}{A_1}\left[\left(\frac{1}{\epsilon_1}-1\right)+\frac{1}{\cdot 1-(1-F_{lr})\tau_{ll}-\dfrac{(F_{lr}\tau_{lr})^2}{(A_r/A_1)+F_{lr}\tau_{rr}}}\right] \quad (37)$$

This is still general. For case (a'), from the relation before (35), $\tau_{lr}^2 = \tau_{rr} = 1 - \epsilon_g(W)$. In addition, $F_{1r} = 1$ and $A_r/A_1 = (H+W)/nPF$; and substitution into (37)

$$\frac{1}{\overline{GS}_1} = \frac{1}{A_1}\left[\left(\frac{1}{\epsilon_1}-1\right)+\frac{1}{1-\dfrac{1}{1+\dfrac{H+W}{nPF}\dfrac{\epsilon_g}{1-\epsilon_g}}}\right] = \frac{1}{A}\left[\frac{1}{\epsilon_1}+\frac{1-\epsilon_g}{\epsilon_g}\frac{npF}{H+W}\right] \quad (38)$$

Numerical substitution gives

$$\frac{1}{\overline{GS}_1} = \frac{1}{2nPF}\left(\frac{1}{\epsilon_1}-1\right)+1.938$$

If $\epsilon_1 = 0.9$, $\overline{GS}_1 = 2nPF/2.049$, which is but 1.3% lower than case (b'). The excellent agrement between Eqs. (33) and (38) is no measure of the error introduced by the diffuse-mirror assumption (As $W/H \to 0$, both (a') and (b') reduce to: $A_1/\overline{GS}_1 = (1/\epsilon_1) + (FnP/H)(1-\epsilon_g)/\epsilon_g$. Because case (a'), Eq. (38), is simpler, its use is recommended. It is to be remembered that both derivations apply rigorously to two-dimensional systems; many reformer furnaces approximate that condition. Approximate allowance for the third both derivations apply rigorously to two-dimensional systems; many reformer furnaces approximate that condition. Approximate allowance for the third dimension comes from adding the end enclosure areas to $A_r$ and adjusting $\overline{ll}$ in the conventioanal manner of allowing for opposed rectangles rather than opposed strips.

There remains the term $\epsilon_1$, the effective emissivity of the strips which have replaced the tubes. The recesses formed by the tubes will give them an effective emissivity greater than that of plane metal. In the absence of gas, the interchange area between a tube row of true area $A_2$ and true emissivity of $\epsilon'$ and two parallel black plates on either side of it, of total area $A_B$ and coextensive with the plane of the tubes, is given by

$$\frac{1}{\dfrac{1}{A_2}\left(\dfrac{1}{\epsilon'}-1\right)+\dfrac{1}{A_B F}}$$

For black plates surrounding gray strips of emissivity $\varepsilon_1$ and area $A_BF$ replacing the tubes, the interchange area is $A_BF\varepsilon_1$. The two interchange areas are then equated to find $\varepsilon_1$. Replacement of $A_B/A_2$ by $2B/\pi$, where $B$ is the ratio of center-to-center distance to diameter, gives

$$\varepsilon_1 = \frac{1}{1 + \frac{2BF}{\pi}\left(\frac{1}{\varepsilon_l'} - 1\right)} \qquad (39)$$

This is the effective emissivity of $A_1$. When $B = 2$ and $\varepsilon'_1 = 0.8$, $F = 0.66$ and $\varepsilon_1 = 0.084$.

<u>Other tube arrangements.</u> Since furnaces with one central vertical screen of horizontal tubes between refractory walls are the complete equivalent of case (a'), they do not present a new problem. If the central screen consists of two rows of tubes the only change is the increase in the value of the interception factor $F$ defining the fraction of the central plane occupied by equivalent strips. If one cell consists of a fired section bounded by a refractory wall on one side and a tube screen on the other, with another cell on the other side of the tube screen, there is a lack of symmetry which makes the replacement of the tube row by mixed strips of refractory and sink not quite right; but such a model, if modified, is recommended. If, when calculations of the impinging performance of that cell and the one next to it have been completed, the flux densities onto the two sides of the tube row are different, there is then justification to for repeatining the calculations allowing for net flux between the two cells.

The statement has been made that the gray-plus-clear gas model is more realistic and by implication better than the gray-gas model; but the only solution given for the former was restricted to the speckled box-type furnace, Eq (24). It may be shown [2,h] that, if the constraint is put on the gray-plus-clear gas model that its refractory surfaces diffusely reflect all radiation incident on them (thereby failing to take advantage of shifting the incident gas radiation to the spectral windows of the gas on re-emission through gas to sink), the flux is obtained quite simply from the relation.

$\overline{GS}$ (gray-plus-clear-gas model, with refractory $= a_g \times (\overline{GS}$ based on a gray gas of surface perfect diffuse reflectors) emissivity $\varepsilon_g/a_g$) (39)

Furthermore, the result for the real-gas real-refractory model-when it is available -, with $\varepsilon_r = 0.5$, always lies roughly half-way between (39) and that obtained for the simple gray-gas mode. The arithmetic mean of (39) and the gray-gas model is therefore suggested for those cases, such as the ones last discussed above, which are too complex to formulate rigorously. How important it is to make such a correction can only be established by correlation of a considerable body of furnace data.

Models (a') and (b') presented here have not been tested with furnace data. It will be especially necessary to allow for the temperature difference $\Delta$ in tube-screen furnaces with fuel fired between screens because of the absence of strong back-mixing with a reach comparable to the total gas-flow path. The fitting by Lobo and Evans [6] of a model similar to that based on Eq. (15) was done without the use of a $\Delta$. Whether the striking success of that model-when fitted to and applied to box furnaces-indicated that $\Delta$ was in fact zero in such furnaces or whether there were compensating errors, such as the gray-gas assumption, was never established.It is reported that that model has in a number of cases been less than satisfactory in application to modern furnaces with interior tubes. The reason may lie in the incorrect evaluation of $\overline{GS}_1$.

A correlation of performance data on modern furnaces, using a model such as that herein described, is highly desirable.

## NOMENCLATURE

| | |
|---|---|
| $A_0$ | area of furnace openings losing radiation |
| $A_r(A_p)$ | area of refrectory (of tube plane) |
| $A_S$ | effective sink area (old $(\overline{GS})_{R,c}$) |
| $A_1$ | area of stock or sink in furnace chamber. If tubes on wall, plane of tubes |
| $A_{1,e}$ | that part of $A_1$ exclusive of any tube curtain across the gas-exit passage. |
| $a_g$ | energy fraction of black-body spectrum occupied by gray gas in a gray plus clear mixture |
| $B$ | ratio of center-to-center tube spacing to diameter |
| $C$ | "cold" fraction of furnace enclosure are, $A_1/(A_1+A_r)$ |
| $\overline{c}_{p,g}$ | specific heat of combustion gases, mean value from gas-chamber exit to base temperature |
| $D'$ | reduced firing density, $\dot{H}_F/\sigma A_S T_F^4 (1-T_0')$ |
| $d$ | dimensionless constant in relation $\Delta' = (1-1/d)Q'$ |
| $\varepsilon 3$ | third exponential integral, $\int_1^\infty (e^{-xt}/t^3)dt$ |
| $B$ | black emissive power, $\sigma T^4$ |
| $\overline{F}$ | factor for radiation loss through walls, based on inside and outside temperatures and area of opening |
| $\overline{F}_{iso(gas)}$ | fraction of isotropic (gas) radiation intercepted by tube row, directly plus by interception of returning beam from background |
| $F_{xy}$ | fraction of radiation from surface $x$ which is intercepted by surface $y$ |
| $\overline{g_x s_y}$ | direct-interchange area between gas $x$ and surface $y$ |
| $\overline{G_x S_y}$ | total-interchange area, ratio of net radiative flux between gas zone $x$ and surface zone $y$, allowing for reflections products at all surfaces, to the difference in black emissive powers ($\sigma T^4$) of $x$ and $y$. Sometimes, allowance made for refractory aid without appending sub-R. |
| $(\overline{GS})_R$ | total-interchange area between gas and surface zones, with aid given by equilibrium refractory surfaces included |
| $H(\overline{T_1})$ | enthalpy of stream 1 (sink stream) dependent on temperature |
| $H_{1,i(o)}$ | enthalpy of stream 1 at inlet (outlet) |

| | |
|---|---|
| $H_F$ | enthalpy of any entering streams affecting firing rate, including fuel, air, and recirculated flue gas if any, above dead state of completely burned gaseous products at $T_0$. |
| $h_i$ | convection heat-transfer coefficient on inside surface of refractory walls |
| $h_{c+r,0}$ | heat transfer coefficient by convection plus radiation, on outside walls of refractory surfaces |
| $K$ | absorption coefficient, $l^{-1}$ |
| $k$ | thermal conductivity of refractory walls |
| $L$ | dimension of parallelepipes |
| $L_m$ | mean beam length for gas radiation |
| $L'_O$ | dimensionless loss coefficient for radiative flux through furnace openings, $\overline{F}A_0/A_s$ |
| $L'_r$ | dimensionless loss coefficient for heat loss through refractory walls, $U_r A_r / \sigma A_S T_F^3$ |
| $m_g$ | mass flow rate of combustion gases |
| $n$ | number of tubes in a tube row |
| $P$ | pitch of tubes in row, center-to-center distance |
| $p$ | partial pressure of gas-radiating components, atm. |
| $Q_G$ | rate of heat transfer from combustion gases |
| $Q'$ | reduced rate of heat transfer from combustion gases, $(Q_G / \dot{H}_F)(1-T_0')$ |
| $r$ | ratio of temperature rise of stock surface to the sum of inlet and outlet temperatures, $(T_{1,o}-T_{1,i})/(T_{1,o}+T_{1,i})$. |
| $S$ | speckledness; 1 for surface with $A_1$ and $A_r$ intimately mixed |
| $\overline{s_x s_y}$ | direct-interchange area for radiative exchange between surfaces $x$ and $y$ no gas absorption included |
| $\overline{s_x s_y}'$ | same as above, except that gas absorption is included |
| $\overline{xy}\,(\overline{xy})'$ | shorthand for $\overline{s_x s_y}\,(\overline{s_x s_y}')$ |
| $T_F$ | adiabatic pseudo–flame temperature, based on Eq. following [2] |
| $T_g$ | mean radiating temperature of combustion gases |
| $T_1$ | mean temperature of stock or sink surface |
| $T_0$ | base temperature (also ambient) |
| $U_r$ | overall coefficient of heat transfer from combustion gases through refractory wall to ambient |
| $W$ | refractory wall thickness. Also thickness of gas slab, Figs. 7-9 |
| $\overline{I_r}$ | shorthand for $\overline{s_1 s_r} \equiv A_1 F_1 \equiv A_r F_{r1}$ |
| $\alpha$ | gas absorptivity |
| $\Delta$ | gas-radiating temperature minus gas temperature leaving combustion chamber |
| $\varepsilon_g(\varepsilon_1)$ | gas emisivity (effective emissivity (emittance, absorptance) of surface $A_1$) |
| $\varepsilon_1'$ | true emissivity of tube surface |
| $\eta$ | furnace efficiency, (flux to stock or sink)/$H_F$ |
| $\sigma$ | Stefan-Boltzman constant |
| $\tau$ | gas transmissivity ($\equiv 1$-absorptivity, $= 1-\varepsilon_g$ if gas gray) |
| $\psi$ | angle. See Fig. 7 |

Subscripts

| | |
|---|---|
| $i,o$ | inlet, outlet |
| $r$ | refractory |
| $l$ | sink surface, or stock surface |
| $T$ | total, applied to area |

Primes

On $T_g$, $T_1$, $T_o$ designate the ratio those temperatures to $T_F$.

## REFERENCES

1. HOTTEL, H.C., Melchett Lecture for 1960. Jl. of Inst of Fuel, 220-234, June, 1961.
2. HOTTEL, H.C., and SAROFIM, A.F., Radiative Transfer, McGraw-Hill, New York, 1967. Page references as follows: (a), 311, 459; (b), 462; (c), 247; (d), 316; (e), 300; (f), 113; (g), 271
3. HOTTEL, H.C., and KELLER, J.D., Tr.Am.S.Mech.Eng., IS 55, 39-49 (1933)
4. HASLAM, R.T., and HOTTEL, H.C., Tr.A.S.M.E., FSP 50 (1928)
5. HOTTEL, H.C., M.I.T. Notes on Heat Transfer in the Combustion Chamber of a Furnace, Cambridge, Mass., April, 1940
6. LOBO, W.E., and EVANS, J.E., Tr.A.I.Ch.E., 35, 743 (1939)
7. MCADAMS, W.H., Heat Transmission, 3rd Ed., Chapter 4, McGraw-Hill, New York, 1954
8. HOTTEL, H.C., Mech.Eng., 52, 699-704 (1930); Tr. A.S.M.E., FSP 53, 265-73,

*Author's comments, after 13 years, on the paper "First Estimates of Industrial Furnace Performance – the One – Gas – Zone Model Reexamined"*

The dominant theme, combining an energy balance based on the gas – exit temperature $T_E$ with a rigorously formulated heat transfer relation based on a mean gas – radiating temperature $T_G$ through an empirical relation between $T_E$ and $T_G$ is, by hindsight, clouded by much chaff. A few desired changes or omissions:

1. Since the energy balance and heat transfer relations force the solution of a limited quartic in $T_G$, there is no particular merit in forcing the convection term $hA_1(T_G - T_1)$ into a fourth–power temperature expression for combining with $(GS_1)_R$.

2. The recommended relation between $T_E$ and $T_G$ could have been stated much more clearly: $T_G = AT_E + (1-a)T_F$, with "$a$" in the neighborhood of 3/4 but increasing toward 1 as the firing rate goes down.

3. The attempted quantitative definition of "spleckledness" – material following Eq. 18 and running to two paragraphs before Eq. (23) – was an abberation, better forgotten. Arrangement of the sink and refractory surfaces of a furnace chamber in "speckled" vs. sink–in–a–plane form are not bounding arrangements determining the view factor $F_{1R}$. Speckledness is near the center of possible arrangements, and Eq. (24) is recommended for determining $(GS_1)_R$, except for metallurgical or glass furnaces where the sink is generally planar.

International Seminar 1974

# TRANSPORT PROCESSES IN PLANT ENVIRONMENT

J.R. Philip and D.A. de Vries*

Eindhoven University, Eindhoven, The Netherlans

Since the 1974 ICHMT Seminar the activity in this field has been continued and even increased. The present interest is exemplified by an international Symposium held in Canberra, August 31 - September 4, 1987 [1]. We return to this below, but first remark on combined transfer of heat and moisture in porous media, a topic was not discussed at the Canberra meeting, but has received considerable attention.

The theory published by the authors in 1957/8 is generally applicable, mainly in connexion with soils, but also in drying. A review of its present status was presented by de Vries at the Euromech 194 Colloqium on "Simultaneous Heat and Mass Transfer in Porous Media", held at Nancy, 1985 and was recently published [2]. The main unsolved problem is the theoretical incorporation of the influence of hysteresis. Certain criticisms, based on experimental observations interpreted without taking account of latent heat, have been refuted [3].

A special application is connected with the current rating of buried cables for the transmission of electricity. An international workshop on this subject was held in The Netherlands, 1984 [4]. A relatively new subject of interest is the storage of solar heat in a body of soil.

A subject outside the theory of Philip and de Vries is that of free convection in porous media. It has received attention from soil physicists (e.g. [5]) and engineers. Heat transfer in porous media is now a separate subject in the bibliographic reviews published annually in the Int. J. of Heat and Mass Transfer [6].

The major themes of the Canberra Symposium were: Soil Physics; Physical Ecology; Micrometeorology; and Applications (to physically similar industrial processes). They embrace most subjects treated in the 1974 ICHMT Seminar, and we refer to papers in [1] in discussing developments since 1974. Conveniently, the Canberra Symposium expressly addressed progress over the last 20 years.

W.A. Jury reviewed "Solute transport and dispersion" in soil- and groundwater on scales from, say, $10^{-1}$ to $10^5$ m. The present-day recognition of the profound difficulties associated with heterogeneity on many scales in real-world problems contrasts strongly with the simplistic perceptions of 1974. J.R. Philip treated

---
* retired professor

"Quasianalytic and analytic approaches to unsaturated flow" in soils, indicating many advances, so that the article brings up-to-date an earlier extended review [7]. The 1974 Seminar reported recent work on flow and volume change in swelling soils. The promising beginnings of this work have not been followed up adequately, though a useful summary is given in [8]. The article "Filtration and sedimentation" by D.E. Smiles treated the application in chemical engineering of concepts pioneered in the context of swelling soils and soil suspensions. I. White's "Measurement of soil physical properties in the field" reports a most promising development of recent years: for the first time, measured parameters relate directly to the physics of flow, and measuring techniques are optimized through physical principles.

In 1974 it was hoped that "Canopy transport processes" might be adequately describable by appropriate eddy diffusivities. As M.R. Raupach's review of this title reveals, instances of "counter-gradient transport" are now well-known and essentially understood. It remains for understanding to lead to cost-effective prediction. R.D. Jackson's "Surface temperature and the surface energy balance (on a variety of scales)" addressed the difficult task of reconciling interpretations of satellite "observations" of surface conditions with ground truth and the awkward facts of the diurnal cycle. The 1974 Symposium was silent on this problem, which has since burgeoned. "Stomatal physiology and gas exchange in the field" by I.R. Cowan reported on studies of the linked transfer of $CO_2$ and $H_2O$ between plants and the lower atmosphere, aimed at assessing efficiency of water use in photosynthesis.

The 1974 Seminar paid some attention to the micrometeorological consequences of horizontal differences at the earth's surface, but most attention mainly centered on ideal one-dimensional micrometeorology with uniform conditions at a horizontal plane surface. There has been surprisingly little progress on small-scale advection, but see [9]. The problem of "Air flow over complex terrain", has, however, excited much interes in the last decade. J.J. Finnigan's thoughtful and critical review with this title should be read by all interested in this question. "Convective processes in the lower atmosphere"by J.C. Wyngaard gave an authoratitive account of progress on this very important topic in the last decade or so. By 1987 standards, the techniques, observations, and concepts aired at the 1974 Seminar look rather naive. We have, however, yet to discover how to use this recent work predictively and economically. P.A. Taylor's article, "Turbulent wakes in the atmospheric boundary layer", indicated that the aerodynamics and transport properties of shelter belts, discussed briefly in 1974, are still far from well understood.

## REFERENCES

1. W.L. STEFFEN, O.T. DENMEAD, AND I. WHITE (Eds.) "Flow and Transport in the Natural Environment: Advances and Applications." Proc. Canberra International Symposium, 1987. In press, Springer, Berlin. (To appear mid-1988.)
2. D.A. DE VRIES, The theory of heat and moisture transfer in porous media revisited, Int. J. Heat Mass Transfer 30, 1343-1350, 1987.

# HEAT AND MASS TRANSFER WITHIN PLANT CANOPIES

B. Legg* and J. Monteith†

*Rothamsted Experimental Station, Harpenden, Herts, U.K.,
†University of Nottingham School of Agriculture,
Sutton Boninton, Loughborough, Leics. U.K.

## 1. OBJECTIVES

Investigations of heat and mass transfer in vegetation have two broad, complementary objectives, one physical and the other biological. The physical objective is to describe how parts of the earth's surface which are covered with grass, arable crops, swamps, or forests partition the radiant energy which they absorb, and in particular, to specify the input of heat and water vapour to the lower atmosphere. Realistic boundary conditions are an essential component of models of atmospheric behaviour and may eventually contribute to more accurate predictions of weather, both daily and seasonally. The biological objective is to understand how the growth, survival and reproduction of higher plants depends on the environment of foliage. Encouraged by a stimulus from ecology, micrometeorologists have attempted to measure, to analyse and to model processes of transfer which occur within plant communities and which are intimately related to physiological behaviour. This review is concerned with progress and with problems which inhibit progress. We shall try to determine whether micrometeorological methods of estimating heat and mass transfer in plant stands are worth pursuing, either in preference to other techniques or to provide complementary information.

## 2. SOURCES AND SINKS

### 2.1. DETERMINANTS

When sunlight strikes the surface of a leaf, part of the absorbed energy, usually a major part, is dissipated by the evaporation of water, and a much smaller part is stored in the products of photosynthesis. The leaf may therefore be regarded as a source of water vapour (or latent heat) and a sink of carbon dioxide. The mean temperature of the leaf adjusts to a value at which the net gain of heat is equal to the heat loss, i.e. the leaf may be a source or a sink of sensible heat depending on whether it is warmer or cooler than the surrounding air. At night, a leaf exposed to the sky loses heat by radiation and usually cools a few degrees (Celsius) below air temperature. Small amounts of heat are produced by respiration and by condensation when the temperature of the leaf is less than the dew–point of the ambient air.

Proceeding from the simple heat balance of a single leaf to the much more complex system of a leaf canopy, it is necessary to take account of the following factors:

1) The spatial distribution and arrangement of foliage specified by a leaf area index L and by the elevation and azimuth angles of leaf surfaces (1). The leaf area index per unit depth of canopy $a(z)$, often referred to as the foliage area density, usually increases from zero at the top of the stand, height $h$, to a maximum value at approximately $2h/3$ and then decreases to a very small value near the ground (Fig. 1). Stephens (3) showed that the foliage in Pine forests follows a normal frequency distribution with respect to height and the foliage of many agricultural crops is arranged in a similar way. In consequence the layer of foliage between $2h/3$ and $h$ is the 'active' layer for most types of mature vegetation, containing all the significant sources and sinks of heat, mass and momentum.

In some types of vegetation it is necessary to take account of the surface area of stems and other organs capable of transpiration and photosynthesis (2). The differentiation of green (active) and yellow (senescent) foliage is also relevant to the distribution of $CO_2$ sinks.

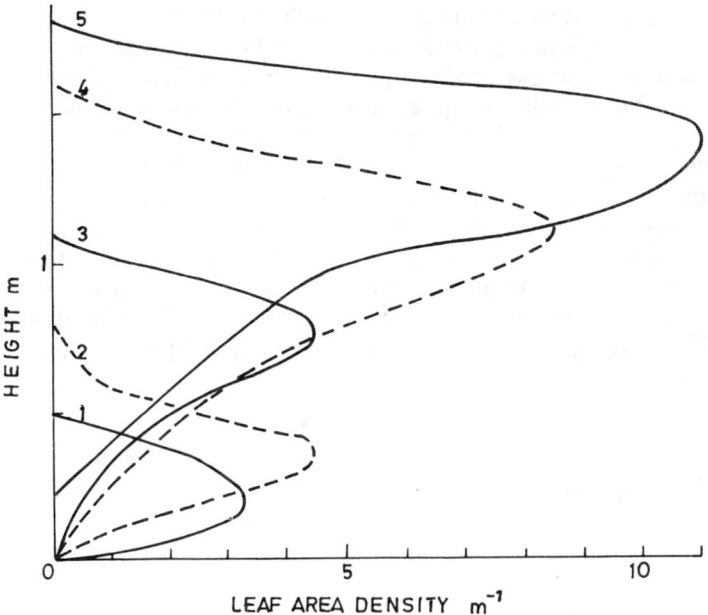

Figure 1. Vertical distribution of leaf area in a maize stand at Iygeva. 1963 (from Ross and Nilson (2)) 1, 12 July; 2, 25 July; 3, 3 August; 4, 13 August; 5, 20 August.

2) The spatial distribution of intercepted radiant energy which depends on the optical properties of the foliage and on its architecture (see contribution by Nilson).

3) The spatial distribution of potentials such as temperature, vapour pressure and $CO_2$ concentration. These potentials determine how the available radiant energy

is partitioned between convection, transpiration and photosynthesis. There is a strong element of feedback in the system, however, so that the existence of sources and sinks of heat and mass, whose magnitude and distribution depends on a set of potentials, is responsible in turn for the evolution of potential gradients within the foliage and for diurnal variation of the potentials about mean values which are set by weather and soil conditions.

4) Resistances to the transfer of heat and mass, i.e. the potential difference between any two points in the foliage required to maintain unit flux density of heat or of mass between these points.

The resistances are of two main types:

a) Physiological resistances to gaseous diffusion, in particular, the resistance of stomatal pores (discussed by Stigter). These resistances depend both on the physical environment of the foliage e.g. availability of light energy, temperature; and on biological factors such as age and genetic constitution.

b) aerodynamic resistances to heat and gaseous diffusion determined by the flow of air round individual foliage elements and through the canopy as a whole.

As the physiological resistances in a plant stand are often much larger than the corresponding aerodynamic resistances, they play a dominant role in determining the distribution of water vapour and carbon dioxide fluxes. However, as the main emphasis of this review is physical rather than biological, the following sections will be concerned mainly with the nature and specification of air flow and turbulence within canopies.

## 2.2. AERODYNAMIC REGIMES

A network of aerodynamic resistances extends from the surfaces of individual foliage elements to the free atmosphere above the canopy, and transfer across this network may be considered a three-stage process. In the first stage, transfer occurs across the laminar or turbulent boundary layer of air in immediate contact with the surface. In principle, the rate of transfer can be expressed as a difference of potential between the surface and the air divided by the diffusion resistance of the boundary layer. Numerous complications arise in practice: the effect of hairiness and of other surface irregularities; and the effect of externally imposed turbulence on boundary layer transport, a topic discussed by Clark and Wigley. Pressure fluctuations are likely to influence the behaviour of boundary layers within canopies but experimental evidence is lacking.

The third stage is transfer in the atmosphere, essentially a three-dimensional process. Over extensive, uniform vegetation, however, horizontal transfer may be neglected and most experimental studies have concentrated on the relation between vertical fluxes and concentration gradients of heat and mass in a one-dimensional boundary layer. The coefficient of turbulent transfer $K(z)$, representing the flux per unit gradient at height $z$ within this layer, can be estimated from a knowledge of windspeed and temperature profiles, and a corresponding resistance between two heights $z_1$ and $z_2$ can be found from $\int dz/K(z)$

The first and third stages of transfer are therefore related to the behaviour of specific boundary layers – the boundary layer of individual foliage surfaces and the boundary layer of the whole canopy acting as an extended surface. The intermediate second stage is much harder to identify and to define and it has no accepted name. At a conference in Canberra some years ago it was appropriately referred to as the Sargasso Sea where marine vegetation proliferates in the clear swirling waters!

## 2.3. HOMOGENEITY

At any given depth within the Sargasso Sea, the distribution of foliage depends partly on the spacing of individual plants within the community (e.g. in rows); and partly on the geometry of the foliage attached to each plant. Although in row crops and in cultivated forests the distribution of foliage may exhibit some degree of uniformity, the horizontal distribution of sources and sinks of heat and mass is usually very irregular (if not random). It follows that transport through the Sargasso Sea must depend on horizontal as well as on vertical gradients of temperature, water vapour concentration and other potentials. In all standard theoretical treatments, however, and in most experimental studies, the Sea, like the boundary layer above it, is treated as a one – dimensional system in which fluxes are exclusively vertical.

Agricultural crops can be regarded as horizontally homogeneous only on a scale that is larger than individual crop elements, and probably larger than the crop spacing. To some extent, it may be possible to overcome this problem by spatial averaging but severe difficulties are likely to arise in practice. Droppo and Hamilton (4) used three vertical arrays of instruments in an 18 $m$ tall stand of deciduous trees and they reported substantial horizontal variations in net radiation, temperature and humidity.

Non – uniformity on a larger scale may be responsible for horizontal fluxes for distances of several times the crop height. Byrne and Rose (5) released smoke beneath a crop of Townsville stylo that showed visible variations in the amount of water stress. Photographs show that it did not emerge uniformly, but in a few preferred places. Penman and Long (6) also reported hot spots in a wheat canopy with temperature anomalies of more than 1°C for sites only 8 $m$ apart and Legg (7) measured large horizontal $CO_2$ concentration gradients at a height of 70 $cm$ in a 125 $cm$ tall wheat crop. At times, the horizontal flux, calculated by multiplying the gradients by the wind speed, exceeded half the vertical flux. Although the crop looked perfectly uniform, random sampling revealed a variation of ± 22% in the fresh weight of plants per unit field area. There is an obvious need for more measurements of horizontal variability in crops to improve our understanding of what is meant by 'homogenous'.

## 3. PROPERTIES OF TURBULENCE

The ability of the Sargasso Sea to transport heat and mass between foliage and the atmosphere depends on the degree of turbulent mixing within the canopy. It is possible to measure or to estimate several relevant properties of turbulence, viz., ($i$) the input of energy to the turbulent flow; ($ii$) the intensity of turbulence; ($iii$) the scale of length of turbulent eddies; ($iv$) the frequency distribution of eddies. The specification and measurement of these properties will now be considered.

## 3.1. ENERGY INPUT

Turbulence within canopies is generated in three ways and there are three corresponding inputs of energy:

a) wind shear produces mechanical energy at a rate which is proportional to the shear stress $\tau$ and to the horizontal velocity $u$.

b) wind shear in the boundary layers of leaves and other foliage elements is also responsible for converting the kinetic energy of the mean motion into turbulent energy, e.g. in the wake flow.

c) buoyancy in the bulk flow or in the immediate neighbourhood of foliage surfaces may either promote or inhibit turbulence i.e. the action of buoyancy forces may increase or decrease the turbulent energy in the flow.

The two sources of mechanical energy can be considered together. The energy input at the top of the canopy is $\tau u$, and the energy converted into turbulent motion at each height is $d(\tau u)/dz = \tau(du/dz) + u(d\tau/dz)$. The first term is energy that comes from wind shear, and the second is the energy dissipated in the turbulent wakes behind crop elements. If $u_*$ above a canopy is $0.3\ m\ s^{-1}$, the total energy input is approximately $0.15 Wm^{-2}$ and for a crop 2 $m$ tall, say, the dissipation rates caused by shear and wake turbulence decrease exponentially from about 0.2 and $0.15 Wm^{-3}$ at the top of the canopy to less than a tenth of these values in the middle. Lower still, both terms would be less than 5 $m\ W\ m^{-3}$ and in some circumstances the shear component can be zero. Since both dissipation rates are roughly dependent on $u_*^3$ they decrease to less than 1 $mW\ m^{-3}$ on calm nights when $u_*$ is less than $0.3\ m\ s^{-1}$.

Buoyancy is caused by vertical density gradients and these arise from variations in temperature and humidity. The corresponding energy input to the turbulent motion is given by $(g/T)[(H/c_p) + 0.61\ ET]$ where $H$ and $E$ are vertical flux densities of heat and water vapour, $\lambda$ is the latent heat of vaporisation of water and $T$ is temperature ($K$). A modest sensible heat flux of 30 $Wm^{-2}$ which could occur at any level in a canopy, day or night, would contribute 1 $mWm^{-2}$ to turbulent energy. A latent heat flux of 300 $Wm^{-2}$, possible at the top of a canopy in bright sunshine, would achieve a similar input of energy. Vertical mixing in the free atmosphere is significantly affected by buoyancy when the ratio of buoyant to mechanical energy ($Ri$) exceeds 0.01 (8). It appears that this condition will often be satisfied within canopies, particularly at night when it is even possible for buoyant energy to exceed mechanical energy.

## 3.2. INTENSITY

The turbulent intensity associated with a mean horizontal windspeed is

$i = (\overline{(u - \overline{u})^2})^{\frac{1}{2}} / \overline{u}$ where $\overline{u}$ is an average and $u$ is an instantaneous horizontal

velocity. The relation of i to canopy scale and structure has been well summarised by Cionco (9) who reviewed a number of field and wind–tunnel studies. He concluded that in simple 'ideal' canopies where the foliage area density is nearly constant with height, the turbulent intensity is also nearly constant, but when the foliage is not uniformly distributed, i tends to be larger in layers where the foliage is most dense. In wind–tunnel experiments, it is possible to demonstrate a systematic dependence of i on the spacing of regular objects such as wooden pegs.

In general, i appears to be related to the scale of the canopy, increasing from about 0.4 for agricultural crops though values between 0.6 and 0.7 for temperate coniferous and deciduous forest to values greater than unity in dense tropical forests. Substantially smaller values of i have been recorded in the boundary layer above canopies.

In a study of deciduous forest, Houston showed that i tended to increase with thermal instability, presumably because turbulent mixing was enhanced by the ascent and circulation of warm air within the canopy (9). However, Cionco found no conclusive evidence for the effects of thermal stability in agricultural crops with dense uniform foliage.

### 3.3. SCALE LENGTHS

In the Eulerian system, the scale of flow at any instant of time is defined as $1 = \int R(y)dy$ where $R(y)$ is the correlation between the velocities of two particles separated by a distance y in the direction of the flow. In practice, 1 can be measured only when the wind direction is constant but if the turbulent flow is treated as a fixed pattern being swept past the observer with mean horizontal velocity $\overline{u}$, the correlation $R(y)$ for homogeneous turbulence is the same as the autocorrelation of the velocity at a point for the time interval $t = y/\overline{u}$, i.e. $1 = \overline{u} \int R(t)dt$. The scale length determind in this way is a measure of the average eddy size in the direction of the mean wind. A similar measure of eddy size in the vertical direction can be obtained from the autocorrelation of vertical velocity at a point. Scale lengths have been measured in a variety of stands, e.g. in a larch plantation (10) and in maize (11, 12). The general pattern emerging from these studies is that the longitudinal Eulerian scale length is of the order of $0.3\ h$, independent of wind speed in the lower half of crop canopies, increasing to $h$ near the top where it shows some increase with windspeed. The vertical scale length is an order of magnitude smaller and is less dependent on wind speed except at the very top of the canopy.

The scale length of eddies in the wake of individual leaves and other obstacles is given by $d/\sigma$ where $d$ is the characteristic dimension of the object and $\sigma$ is the Strouhal number, approximately 0.2 for Reynolds numbers exceeding 200 (13, 14). The nominal scale length of approximately $5d$ may be expected behind leaves depending on their orientation.

### 3.4. SPECTRAL COMPOSITION

For isotropic turbulence in which there is an equilibrium transfer of energy from larger to smaller eddies, the frequency distribution of the energy given by similarity theory is $F(n) :: n^{-5/3}$ (15). Shaw et al. (16) measured the spectral composition of all three wind components and of temperature within a mature maize canopy and found very accurate agreement with the $-5/3$ power law from 0.2 to 5Hz. Figure 2 illustrates the quality of their results. Uchijima and Wright (11) and Isobe (14) found similar results for the longitudional component, but in both cases the spectra were very ragged. Isobe noticed peaks at 6 and 16 Hz at all heights, though these were less pronounced for slow winds. These frequencies corespond to the expected frequencies in the wakes of leaves and stems. The fact that the same frequencies exist at all heights suggests that such eddies or vortices form mainly near the top of the canopy where the wind speed is greatest, and are then fed downwards. Allen (10) also noticed peaks at about 3 to 7 s in a larch plantation and associated these with eddies from individual trees whose spacing was 3 to 4 m. Near the floor of the forest, eddies with a wavelength of about 100 m were ascribed to pressure fluctuations.

When spectra are plotted as $nF(n)$ against $\log n$ to show the energy distribution, most of the energy is found at much lower frequencies. In a larch plantaion Allen (10) found most energy to be at frequencies of 0.04 to 0.1 Hz depending on wind speed, frequencies corresponding to eddy scale lengths of 20 to 100 m. In maize, Saito et al (12) and Isobe (14) found peaks in the u' and w' spectra at frequencies between 0.05 and 0.4 Hz corresponding to horizontal scale lengths of

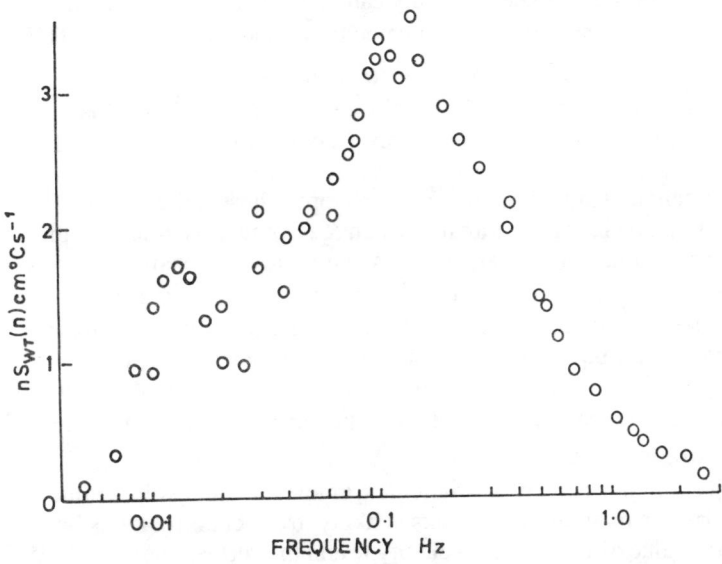

Figure 2. Cospectrum of $w$ and $T$ at a height of 1.8 m in a 2.9 m stand of maize at 1030 to 1215 hours on 8 October 1971. Mean wind speed $= 0.77$ m s$^{-1}$ (from Shaw et al (16)).

0.7 and 6 m and vertical scale lengths of 1 to 3 m in a crop that was only 3 m tall. These values are much larger than the Eulerian vertical scale length of only 0.15 to 0.3 m. When Shaw et al (16) measured the cospectrum of $u'$ and $w'$ within maize, the major contribution came from a frequency of 0.1 Hz corresponding to an eddy size of approximately 6 m.

Isobe (14) also measured the phase lag between the vertical velocity at heights of 1.7 and 3.0 m and found that w at 1.7 m had a phase lead over w at 3.0 m for the frequency range of 0.03 to 0.2 Hz, but a phase lag for frequencies outside this range. This result suggests that large eddies descend through the canopy, whereas the main energy–carrying frequencies from 0.03 to 0.3 Hz spread upward.

## 4. FLUX–RELATIONS

### 4.1. VALIDITY OF K

The measurements of turbulence within canopies which were reviewed in the last section cannot yet be placed in a general theoretical framework and therefore cannot be used directly to calculate rates of transport of heat and mass even when the relevant gradients are known. Micrometeorologists have therefore fallen back on the assumption that turbulent mixing can be expressed by a transfer coefficient $K$ which is defined as the flux of an entity per unit concentration gradient, i.e. $F(z) = -K(z) \partial \chi / \partial z$. The validity of this procedure may be challenged on at least three counts.

1. In the boundary layer which exists above uniform surfaces $K$ theory is well established. Values of $K$ determined experimentally can be related to a notional mixing length or eddy size $l$ and to a notional vertical velocity $\overline{w}$ such that $K = \overline{w} \, l$. This procedure is valid when the concentration gradient $\partial \chi / \partial z$ at a specific height is representative of the gradient within $\pm l$ of the height, i.e. if $(\partial \chi / \partial z)/(\partial \chi / \partial z^2)$ is comparable with or larger than 1. In the boundary layer above uniform plant stands, this condition appears to be satisfied, and provided gradients are not measured immediately above the canopy (say within $0.2 \, h$ of a canopy whose height is $h$), fluxes of heat and mass can be reliably estimated when $K$ is known. Within canopies, however, the presence of sources of heat and mass is commonly responsible for very large changes in $\partial \chi / \partial z$ within distances of the order of $0.2 \, h$, i.e. distances much smaller than the characteristic mixing length. It follows that fluxes of heat and mass at a specific level z cannot be uniquely related to a transfer coefficient which is a function of the velocity field in the layer $z \pm l$ and a concentration gradient measured at z. Either $\partial \chi / \partial z$ must be averaged in some way over the layer $z \pm l$, or $K$ must be regarded as a function of $\chi(z)$ as well as of $\overline{w} \, l$.

2. A second objection to the use of a simple flux/gradient relationship within canopies is that the existence of sources and sinks is likely to affect correlations between the instantaneous value of the vertical velocity $w$ and the corresponding values of the various potentials. For example, a small eddy which sweeps a parcel of air across a warm, sunlit leaf is likely to acquire a positive value of $w$. Corresponding deviations from the mean values of temperature and vapour pressure will also be positive (more sensible and latent heat loss) but the $CO_2$ deviation will be negative (more photosaynthesis). Above the canopy, similar correlations appear to be responsible for systematic differences between the turbulence transfer coefficients for different entities and these differences may well be greater within canopies. Very few experiments have tried to compare the values of $K$ for different entities (7, 17) and none has been accurate enough to associate systematic differences in $K$ with the physical structure of the foliage.

3. The third objection to calculating vertical fluxes from $K\, \partial\chi/\partial z$ is a matter of practice rather than principle. Most methods of estimating $K$ depend on the assumption that the divergence of horizontal flux for the appropriate entity plays no significant part in the continuity equation applied to the canopy, whereas the foliage in most canopies is inhomogeneous on a scale comparable with the mixing length. When the divergence of the horizontal flux $\partial(u\chi)/\partial z$ is compared with the vertical flux $K\,\partial\chi/\partial z$, estimates of $K$ may be seriously in error.

It is surprising that conventional $K$ theory has managed to stay afloat for so long in the Sargasso Sea but the plausible measurements which have appeared in the literature seem to have allayed fears of shipwreck. Some of these measurements will now be reviewed, paying special attention to the apparent variation of $K$ with height.

## 4.2. MEASUREMENTS OF $K(Z)$

The two methods most commonly used to determine $K$ as a function of height in plant stands depend on the application of the continuity equation to momentum and to sensible and latent heat. We shall consider the momentum balance method first.

The movement of wind through a crop imparts a drag on the crop elements, and on the ground beneath and at any level the downward momentum flux must equal the total drag beneath that level. It can be shown that the transfer coefficient for momentum is given by:

$$K(z) = \left[\int_0^z C_d a(z) u^2 dz\right] / (\partial u/\partial z) \qquad (1)$$

where $C_d$ is a drag coefficient for foliage elements $a(z)$ is a foliage area density. The drag on the soil surface can be added to the numerator if it is significant. Most workers have assumed that $C_d$ is independent of u and have estimated its mean value by analysing the wind profile above the canopy to estimate the total drag.

There are several theoretical and practical objections to this method. In the first place, the assumption that $C_d$ is constant is rarely valid. Within the range of Reynolds numbers representative of foliage in crops ($10^2$ to $10^4$) the drag coefficient of plane and cylindrical surfaces is expected to vary both with wind speed and with wind direction. Thom (18, 19) tried the more rigorous approach of applying values of $C_d$ measured in a wind tunnel to wind profiles in a field of beans, but the total drag obtained by integration through the canopy exceeded by a factor of four the drag calculated from the wind profile in the boundary layer. Thom attributed the discrepancy to mutual sheltering of the leaves but other factors may have been partly responsible e.g. a much greater turbulent intensity in the field than in the wind–tunnel which would decrease the true drag coefficient; and the use of hot–bulb anemometers in the canopy which gave a measure of the mean scalar wind instead of the vector mean wind required in drag calculation. The momentum balance method also needs mesurements of the distribution of leaf area with height which is often very difficult to determine in forest stands.

The profile of $K$ determined by this method is very sensitive to th precise shape assumed for the wind profile as well as to changes of $C_d$ with height. Conversely, it is not difficult to simulate realistic wind profiles from an estimated profile of $K$.

Applying the momentum balance method to a field of beans (Vicia faba), Thom found that the value of $K_M$ was approximately constant above $h/3$ (See Fig. 3) and suggested that the decrease of wind speed with depth below the surface was compensated to some extent by increased mixing as a result of wake turbulence in the layer of maximum foliage density (Fig. 1). Several workers however, including Uchijima and Wright (11) found that K decreased exponentially with depth in the top half of the canopy and expressed their result in the form

$$K_M(z) = K_M(h)\exp-\alpha(1-z/h) \qquad (2)$$

Values of $\alpha$ determined by the momentum method usually fall in the range 2 to 4 (Brown and Covey (36)).

The energy balance method for determining the transfer coefficient at any height $z$ within a canopy requires a measurement of the downward flux of net radiation $R R_n(z)$ and of the downward heat flux at the soil surface $G$. The upward flux of sensible and latent heat is $(\rho c_p)\partial\theta/\partial z$ where $\rho c_p$ is the volumetric specific heat of air and $\theta$ is the equivalent temperature, a simple linear function of air temperature and vapour pressure. Assuming there is no storage and that the transfer coefficients of heat and water vapour are both equal to $K_E$ in the layer, it can be shown that

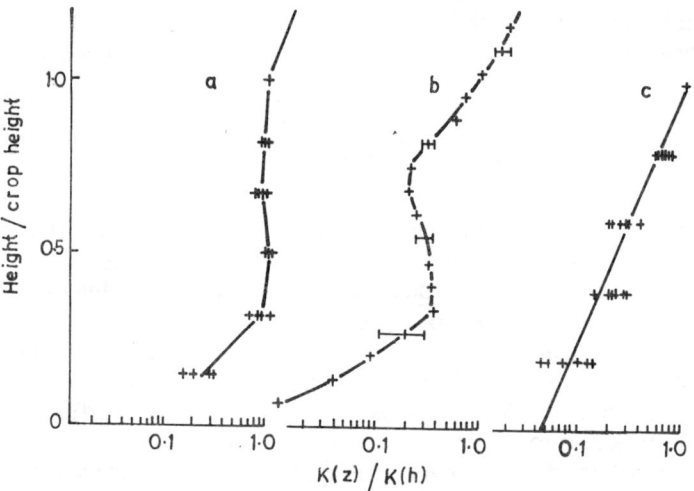

Figure 3. Profiles of $K(z)/K(h)$ a. from the momentum balance in a 1.18 m bean crop (Thom (19)); b. from the energy balance in a 2.5 m maize crop (Brown & Covey (36)); c. from the energy balance in a 2.2 m maize crop (Stewart & Lemon (44)).

$$K_E = -[R_n(z) - G]/\rho c_p \, \partial\theta/\partial z \qquad (3)$$

Compared with the momentum balance, this method has the advantage that laborious measurements of foliage density are avoided. In some conditions however, and particularly at night, both the numerator and denominator are small quantities with large (fractional) errors, making the determination of $K_E$ very inaccurate. Very few workers have been bold enough to estimate the error inherent in estimates of $K_E$ but analysis for a stand of maize presented by Lemon (20) showed standard errors ranging from about ±15% at the top of the canopy to about ±50% near the soil surface. The energy balance method, like the momentum balance method, has often yielded values of $K_E(z)$ which decrease almost exponentially with depth below the top of the canopy. However, a number of workers have found that the profile of $K_E$ is $S$ – shaped (see Fig. 3), possibly as a result of buoyancy induced by a temperature lapse in the lower part of the canopy (7).

Other methods of determining K can be summarised as follows: –

1) Ratio of water vapour flux to gradient of humidity. The flux may be calculated from an extended version of the Penman formula (21), applied layer by layer. Measurements of $R_n(z)$, $a(z)$ and stomatal resistance $r_s$ are needed and estimates of $K$ depend critically on $r_s$. In wheat, Legg (7) found reasonable agreement with $K_E$ for 4–hour averages. Gillespie and King (22) estimated the water vapour flux at night by absorbing dew with blotting paper from selected leaves. Profiles of $K$ for a 2.5 m tall maize canopy showed a maximum at 0.5 m where $\partial T/\partial z$ was negative and a minimum at 1.7 m where $\partial T/\partial z$ was positive, indicating thermal stability.

2) Ratio of flux of natural thoron from soil surface to thoron gradient. Druilhet et al (23) working in maize reported that the profile of $K$ was $S$ shaped during the day and the secondary maximum in the canopy was comparable with the value near the top of the stand. Thermal effects were invoked to explain qualitatively the dependence of $K$ on height.

3) Ratio of imposed constant flux to gradient. Legg (7) released $N_2O$ from a source with a diameter of 72 m in wheat. Profiles of $K$ at the centre of this area were often $S$ shaped on calm clear nights but no secondary maxima were found in the canopy during the day. Values of $K$ were strongly dependent on wind speed in the upper third of the canopy. There was no correlation in the lower two – trhirds except on clear nights when the maximum of $K$ was inversely proportional to $u^2$, suggesting that mechanical turbulence destroyed buoyant eddies. During the day, thermal stability influenced the value of $K$ at all heights but the precise form of the relationship was obscured by scatter ascribed to non–uniformity of the crop.

### 4.3. MODELS OF $K$ WITHIN CANOPIES

By making several assumptions about an 'ideal canopy' it is possible to derive various profiles for $K_M$ and wind speed. The basic equations are

$$\frac{d\tau}{dz} = \rho C_d a(z) u^2; \quad \tau = \rho \overline{w'u'} = \rho K \frac{du}{dz} \tag{4}$$

where $w'$ and $u'$ are deviations from mean vertical and horizontal velocities. By assuming isotropy and using a simple mixing length hypothesis it can be shown that

$$\tau = \rho l^2 \left(\frac{du}{dz}\right)^2 \tag{5}$$

By further assuming that $C_d$, $a(z)$ and $l$ are all constant with height, Inoue (24) and Cionco (25) both obtained profiles of $u$ and $K$ in the form

$$u = u(h)\exp[-\alpha(1-z/h)] \tag{6a}$$

and

$$K = (\rho^2 u(h)\alpha/h)\exp[-\alpha(1-z/h)] \tag{6b}$$

where

$$\alpha^3 = h^3 C_d\, a(z)/2l^2 \tag{6c}$$

Experimental evidence for an exponential relation between $K$ and height has been considered already. Cionco (25) used this type of analysis in conjunction with the wind profiles of Tan and Ling (26) measured in a maize canopy to show that l was almost constant with height above $0.1\,h$. However, values of $l$ or $K$ derived from the wind profile are very sensitive to the exact shape of the wind profile used, and evidence that $l$ is constant should be obtained independently. An exponential form for $u$ and $K$ can also be obtained by assuming that the shapes of the profiles of $u$ and $K$ are the same, i.e. $K(z)/u(z) = K(h)/u(h)$ (27) or that $u/u_*$ is constant with height. A variation on these models was suggested by Perrier (28) who assumed that the mixing length in the middle of a canopy is determined by the leaf density.

A major weakness in all attempts to model airflow and to derive values of $K$ within canopies is the way in which buoyancy effects are completely neglected. As in the boundary layer above the canopy, buoyancy will often be an important process in the transport of momentum and estimates of $K$ derived from wind shear are likely to be seriously in error when there are strong temperature gradients in the canopy.

## 5. MEASUREMENT OF HEAT AND MASS TRANSFER

### 5.1. CAUTIONARY NOTE

The momentum and energy balance methods of determining $K$ have been made by several groups to measure the vertical flux of heat, water vapour and $CO_2$ in agricultural crops and forests and to determine sources and sink strengths from the divergence of vertical flux, i.e. from the relation

$$\frac{\partial F}{\partial z} = -\frac{\partial}{\partial z}\left(\frac{\partial \chi}{\partial z}\right) = -\frac{\partial K}{\partial z}\frac{\partial \chi}{\partial z} + K\frac{\partial^2 \chi}{\partial z^2} \tag{7}$$

Whether this equation is used as it stands or in a more convenient finite difference form, it cannot be expected to yield accurate values of source and sink strength unless the height dependence of $K$ has been carefully established first and unless the profile of the potential $\chi$ is very exactly defined by measurements at an adequate number of levels preferably including horizontal integration. Despite the manifold sources of error already considered in this review, the full significance of these errors has rarely been admitted when studies of canopy flux have been reported in the literature.

The error in determining the gradient of a potential $\partial \chi/\partial z$ at a fixed height within a canopy is likely to be of the order of $\pm 10$ to $\pm 20\%$. If the error in $K$ is assumed to range from $\pm 20$ to $\pm 40\%$, the accuracy of single flux estimate is unlikely to be smaller than $\pm 22\%$ and may approach $\pm 50\%$.

The error in determining the flux convergence in any layer of the canopy, which is a measure of source and sink strengths, can be estimated from difference between the flux $F_1 \pm \delta F_1$ at level 1 and $F_2 \pm \delta F_2$ at level 2. In the unlikely event that $\delta F_1 = \delta F_2$, the fractional error in the flux divergence will be zero.

If, on the other hand, the two errors are treated as uncorrelated but of equal size, the fractional error in the flux divergence will be

$$\sqrt{2}\delta F_1/(F_1-F_2) = \sqrt{2}(\delta F_1/F_1)/(1-F_2/F_1)$$

Then if $\delta F_1/F_1$ is, say, 0.30 and $F_2 = 0.8F_1$, the error in estimating the flux divergence will exceed 200%.

In practice, some correlation will usually exist between the error in the flux at different levels so that the component error may be smaller than in the last example. Nevertheless, errors of the order of 100% are likely to be prevalent in the estimates of flux divergence which have been quoted in the literature.

It is also possible to measure vertical fluxes by eddy correlation. Shaw et al (16) measured the momentum and heat fluxes within a senescing maize canopy by using a modified three-dimensional hot-film anemometer and a finewire resistance thermometer. Their initial results look very promising: the horizontal variation of fluxes was only 10% in the upper part of the canopy and 25% in the centre. However, there are still several major difficulties: all sensors must be small (a few centimetres) and have very rapid response times (probably faster than 0.1 s), and such sensors are not yet available for measuring $CO_2$ and water vapour concentrations; horizontal averaging will usually be necessary and may require the use of several sensors at each height; and an on-line computer is an essential part of the equipment.

## 6. LITERATURE SOURCES

Examples of profiles, fluxes and estimates of flux divergence can be found in the following papers:

| Stand type | Approximate stand height (m) | Reference |
|---|---|---|
| Wheat | 0.4 | Denmead (29) |
| Red clover | 0.5 | Lemon (30) |
| Townsville stylo | 0.5 | Byrne and Rose (5) |
| Rice | 0.9 | Uchijima (31) |
| Beans | 1.2 | Thom (19) |
| Bulrush millet | 2.0 | Begg et al (32) |
| Sunflower | 2.2 | Impens (33) |
|  |  | Saugier (34) |
| Maize | 2.2 to 2.9 | Wright and Lemon (35) |
|  |  | Brown and Covey (36) |
|  |  | Inoue et al (37) |
|  |  | Lemon and Wright (38) |
|  |  | Gillespie and King (22) |
|  |  | Shaw et al (16) |
| Decidous forest | 18 | Droppo and Hamilton (4) |
| Coniferous forest | 5.5 | Denmead (39) |
|  | 30 | Baumgartner (40) |

Figure 4 illustrates the type of information available from these papers.

## 7. MODELS OF HEAT AND MASS TRANSFER

When a stand of vegetation is treated as a uniform plane with a single value for each surface property it is possible to express the partiton of energy into latent and sensible heat by a simple equation (41, 42). Even uniform crops, however, are very complex surfaces with sources of heat, water vapour and carbon dioxide at different heights. Several attempts have therefore been made to construct models of energy and $CO_2$ exchange in which transfer is assumed to occur in the vertical direction only.

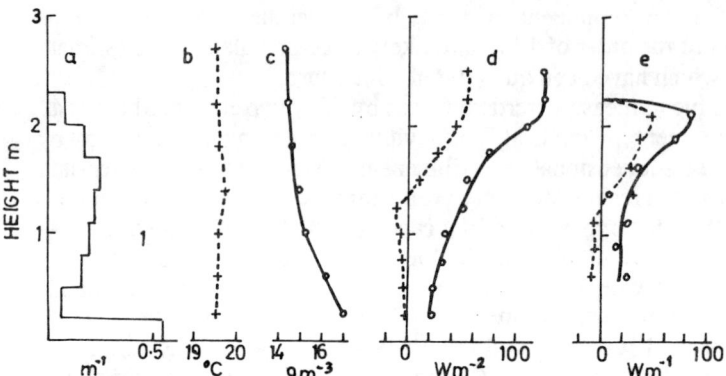

Figure 4. Energy balance of a sunflower canopy at Melle at 0830 hours 6 August 1969 (from Impens (33)). a. leaf area density; b. temperature; c. humidity; d. heat flux(——+——) and water vapour flux (——o——); e. sources and sinks of heat (——+——) and water vapour (——o——)

Such models are intended to reveal the distribution of sources and sinks, to improve the accuracy of fluxes estimated from field measurements, to simulate surface properties which can be measured on an artificial crop in the laboratory or wind tunnel, and to provide profiles of temperature, humidity and $CO_2$ concentration for comparison with the real world. Models have frequently been used to interrelate diverse field measurements and to show that they are consistent with some theory of energy exchange within the canopy. However, before models are used for prediction, they should be checked against independent measurements.

Most models work between boundary conditions at the soil surface and a height of about 1 m above the crop. This layer is convenient because its heat capacity is small and changes of heat content can often be neglected in relation to other components of the energy balance. The soil surface has been represented by a constant or zero flux of latent and sensible heat (43) or by a measured value for soil flux. Partition into sensible and latent heat can be represented by a surface resistance for sensible heat fransfer with a larger value for latent heat, or by a surface relative humidity below 100%, when the surface soil is not wet (44). However, the soil is not normally in thermal equilibrium, and a comprehensive model should allow the fluxes from the soil to depend on the previous history (45).The upper boundary conditions are specified by the macroclimate represented by the solar or net radiation, and a value for temperature, humidity, $CO_2$ concentration and wind speed.

The crop itself is normally represented by the leaf area density, a single valued function of height, and an associated average leaf dimension. The next step is to calculate the interception of radiation by the canopy, and the assumption that the net radiation decreases as an exponential function of cumulative leaf area index is used by Philip (43), Waggoner, Furnival and Reifsnyder (46) and Cowan (27). More complex functions are available (44).

One of the greatest uncertainties in model building has been the choice of values for stomatal resistance $r_s$. Philip (43) simply put $r_s = 0$ thereby dismissing the most important biological variable in the whole system, but in more realistic models $r_s$ is made a function of irradiance (44, 45, 46). Ideally $r_s$ should also depend on water stress, but the relationship is not known with any certainty.

The aerodynamic resistances of leaves can be related to a local wind speed, and then a profile of turbulent diffusivity is also needed. The models of Waggoner and Reifsnyder (47), Waggoner et al (46) and Goudriaan and Waggoner (45) all assume that wind and diffusivity decreases exponentially with height in the canopy, though Stewart and Lemon (44) used an exponential function of height in the upper parts of the canopy changing to an exponential function of cumulative leaf area index lower down. Paltridge (48) calculated $K$ by assuming that the kinetic energy of the wind is the source of turbulent energy $\varepsilon$, and that the scale size $S$ is set by the average distance between leaves. The turbulent diffusivity is then given by $K = \frac{3}{4} C^{1/2} \varepsilon^{1/3} S^{4/3}$ which applies to the inertial subrange of isotropic turbulence with $C$ a dimensionless constant. Not surprisingly, values given by this expression are numerically incorrect but the attempt to find an expression for $K$ that is not completely empirical, seems a step in the right direction.

There are several ways of solving the simultaneous differential equations for the partition of net radiation into heat and water vapour fluxes. Philip (41) assumed that leaf area density was constant with height and that $K(z) = 5z$, $r_a \alpha 1/K(z)$ giving a linear differential equation for air temperature which could be solved numerically. Cowan (27) also used a numerical solution after assuming the leaf density and stomatal resistance to be constant with height. An alternative approach is to divide the canopy into several layers giving a finite number of simultaneous linear equations. Originally Waggoner and Reifsnyder (47) solved these by successive approximation, but a later paper (46) used both a 6 layer model and a continuous integration using CSMP. Differences in total heat fluxes were only 3% of the net radiation, and errors of air temperature and humidity were never larger than 0.2°C or 0.2 mbar.

Canopy models used to predict the fluxes of heat and water vapour are not sensitive to values chosen for $K$, a fortunate circumstance noticed by Waggoner and Reifsnyder (47) and also demonstrated by Legg and Long (49). However, if it is necessary to predict accurate profiles of air or leaf temperature and humidity from the values above the crop, then the values for turbulent diffusion are very important indeed (45, 49). This dependence was not found by Waggoner and Reifsnyder (47) or Waggoner et al (46) because they used field measurements for soil temperature and humidity, and so constrained the profiles at both boundaries. It is, possible, of course, to get good temperature and humidity profiles if the $K$ profile used in the model is derived from neasured values of temperature and humidity. If completely independent values of $K$ are used it is not yet possible to predict $T(z) - T(h)$ with an accuracy of better than about 50%.

Crop models have also been used to predict the profile of $CO_2$ concentration (24, 50). The net photosynthesis is a function of solar radiation only, so the $CO_2$ fluxes, the height of the $CO_2$ minimum, and the height at which the vertical flux of $CO_2$ is constant are all determined by the radiation and leaf area distribution. In a more recent paper, Uchijima and Inoue (51) assumed that net photosynthesis at each height was proportional to the $CO_2$ concentration at that height. However, the total $CO_2$ flux was significantly affected only in conditions of intense solar radiation and unusually little turbulent mixing — an unlikely combination since turbulent diffusion is enhanced by buoyancy. This conclusion corresponds with the observation that the

$CO_2$ concentration in canopies rarely falls below 270 vpm or 85% of the mean value in the free atmosphere.

There have been very few attempts to model or measure mass fluxes within crop canopies, other than water vapour and carbon dioxide. One was by Parmele et al (52) who measured concentration profiles of dieldrin and heptachlor and used a value of $K$ which decreased exponentially to estimate the flux of vapour within the canopy. The results are very uncertain: some show the foliage to be a source of dieldrin while others show it to be a sink. At the moment it is not known how the accuracy could be improved enough to show the source distribution with height, but if the absorption coefficient of vapour on the foliage was known as a function of wind speed, a crop model could be used to show the relative amounts of vapour absorbed by leaves and lost in the air above the canopy.

## 8. CONCLUSIONS

We have considered the main features of plant communities which determine the pattern of heat and mass transfer, and have described how turbulence and turbulent mixing may be specified. Information about the statistical properties of turbulence in plant stands is not hard to find but is much more difficult to interpret in terms of the effectiveness of mixing processes and more fundamental work is needed. At the end of a comprehensive review of heat and mass transfer processes both within and above vegetation, Bradley and Finnigan (53) discussed the use of wind tunnel models to explore the air flow and turbulence in canopies. Most workers have fallen back on the conceptually simple $K$ theory applied in one dimension to vertical fluxes and gradients.

Rates of vertical transfer above vegetation can be successfully estimated from $K$ theory provided the divergence of horizontal flux is negligible and provided the vegetation can be regarded as 'uniform' in terms of the scale of turbulence. In practice, the first resriction requires an extensive level site and the second implies that the spacing of elements within the canopy must be small in relation to the height of measurements above the canopy. Within canopies, however, most methods used to measure or estimate $K$ have conspicuous shortcomings. The main objection is a matter of principle: substantial changes of potential are common within the scale length characteristic of turbulent eddies. It follows that $K$ will rarely be uniquely determined by the rate of turbulent mixing but will depend on a local potential gradient. The transfer rate of different entities will therefore be governed by different values of $K$. Interpretation of observed vertical profiles of $K$ is further obstructed by the complex way in which turbulence is generated by shear forces in the canopy as a whole, by the effects of buoyancy, and by the generation of wakes in the lee of canopy elements. Until the eddy correlation method (16) is developed to the point where fluxes of heat and mass can be measured reliably, the further study of flux/gradient relationships in canopies is likely to prove a sterile exercise. Meanwhile, published measurements of heat, water vapour, and $CO_2$ fluxes should be treated with caution and estimates of flux divergence with scepticism.

In the study of particulate transport however, particularly in relation to the distribution of spores, pollen and spray droplets, even approximate values of $K$ are useful in assessing the relative importance of gravity and turbulent diffusion as mechanisms of dispersal in canopies.

Lack of understanding about the nature of transfer processes in canopies has not inhibited the development of heuristic models. Some of these model are capable of generating realistic canopy microclimates. This success should be recognised not as a test of their validity but as the converse of the experimentally awkward fact that sources and sink distributions derived by differentiating potential gradients are extremely sensitive to the shape of measured profiles. Plausible profiles can therefore be obtained by integration from an approximate distribution of sources.

The usefulness of current models is severely limited by ignorance about the spatial distribution of physiological resistances to $CO_2$ and water vapour transfer in different types of canopy. In particular, when stomatal resistance is assumed constant (27) or zero (43) the distribution of sensible and latent heat fluxes cannot be confidently predicted. Although Philip chose to emphasise differences in the distribution of heat and water vapour sources, the experimental evidence suggests that these differences can often be neglected in practice, e.g. for the purpose of estimating a foliage resistance to water vapour diffusion analogous to the stomatal resistance of a single leaf (54, 55). Although simple parameters specifying the roughness or wetness of vegetation provide little insight into the nature of transfer processes in canopies they nevertheless have a useful role to play in models of water vapour and $CO_2$ exchange (42) and of sulphur dioxide uptake by vegetation (56).

Slow progress towards the physical description of transfer in canopies has fortunately not retarded the development of related ecological measurements and understanding. Several types of portable equipment have recently been developed to measure the gas exchange of rough leaves using an infra−red gas analyser or labelled $CO_2$. The availability of porometers for measuring stomatal resistance has already been referred to. In the current state of the art, these techniques are likely to yield much more information about the sources of water vapour and sinks of $CO_2$ in canopies than micrometeorological methods. Indeed, it is conceivable that we may soon be using the techniques of physical ecology to estimate fluxes and hence transfer coefficients in plant communities, thereby standing the conventional micrometeorological approach firmly on its head.

## REFERENCES

1. ROSS, Y.K. & NILSON, T. (1967a) The spatial orientation of leaves in crop stands and its determination. Photosynthesis of productive systems. ed. A.A. Nichiporovich, Israel Program for Scientific Translations. Jerusalem . 86–99.
2. ROSS, Y.K. & NILSON, T. (1967b) The vertical distribution of biomass in crop stands. Photosynthesis of productive systems. ed. A.A. Nichiporovich, Israel Program for Scientific Translations, Jerusalem. 75–85.
3. STEPHENS, G.R. (1969) Productivity of red pine, 1. Foliage distribution in tree crown and stand canopy. Agricultural Meteorology, $\underline{6}$, 275–282.
4. DROPPO, J.G.Jr. & HAMILTON, H.L.Jr. (1973) Experimental variability in the determination of the energy balance in a deciduous forest. Journal of Applied Meteorology, 12, 781–791.
5. BYRNE, G.F. & ROSE, C.W. (1972) On the determination of vertical fluxes in field crop studies. Gricultural Meteorology, 10, 13–17.
6. PENMAN, H.L. & LONG, I.F. (1960) Weather in wheat: an essay in micrometeorology. Quarterly Journal of the Royal Meteorological Society, 86, No. 367, 16–50.
7. LEGG, B.J. (1972) Turbulent transport in a crop canopy. Ph.D. Thesis, University of London.

8. DEACON, E.L. (1973) Transfer between surface and atmosphere. Proceedings of the first Australian conference on Heat and Mass Transfer. Monash University, Melbourne. Reviews pages 1 to 17.
9. CIONCO, R.M. (1972) Intensity of turbulence within canopies with simple and complex roughness elements. Boundary Layer Meteorology 2 (1972) 453–465.
10. ALLEN, L.H. Jr. (1968) Turbulence and wind speed spectra within a Japanese Larch plantation. Journal of Applied Meteorology, 7, No.1 73–78.
11. UCHIJIMA, Y. & WRIGHT, J.L. (1964) An experimental study of air flow in a corn plant-air layer. The Bulletin of the National Institute of Agricultural Sciences (Japan) Series A, No.11, 19–65.
12. SAITO, T., NAGAI, Y., ISOBE, S. & HORIBE, Y. (1970) An investigation of turbulence within a crop canopy. Journal of Agricultural Meteorology (Tokyo), 25, 205–214.
13. BAINES, G.B.K. (1972) Turbulence in a wheat crop. Agricultural Meteorology, 10, 93–105.
14. ISOBE, S. (1972) A spectral analysis of turbulence in a corn canopy. The Bulletin of the National Institute of Agricultural Sciences (Japan) Series A, No.19, 101–112.
15. LUMLEY, J.L. & PANOFSKY, H.A (1964) The structure of atmospheric turbulence. Monographs and texts in physics and astronomy XII Interscience Publishers, John Wiley & Sons.
16. SHAW, R.H., SILVERSIDES, R.H. & THURTELL, S.W. (1974) Some observations of turbulence and turbulent transport within and above plant canopies. Boundary Layer Meteorology, 5, 429–449.
17. WIGHT, J.L. & BROWN, K.W. (1967) Comparison of momentum and energy balance methods of computing vertical transfer within a crop. Agronomy Journal, 59, 427–432.
18. THOM, A.S. (1968) The exchange of momentum, mass and heat between an artifical leaf and the air flow in a wind tunnel. Quarterly Journal of the Royal Meteorological Society 94, No.399, 44–55.
19. THOM, A.S. (1971) Momentum absorbed by vegetation. Quarterly Journal of the Royal Meteorological Society, 97, No.414, 414–428.
20. LEMON, E.R. (1970) Mass and energy exchange between plant stands and environment. in Prediction and Measurement of Photosyntheic Productivity. ed. Setlik, Pudoc, Wageningen.
21. MONTEITH, J.L. (1964) Evaporation and environment. In "The state and movement of water in living organisms", the XIXth symposium of the Society for Experimental Biology, Swansea. Cambridge University Press, 205–233.
22. GILLESPIE, T.J. & KING, K.M. (1971) Night–time sink strengths and apparent diffusivities within a corn crop. Agricultural Meteorology, 8, (1971), 59–67.
23. DRUILHET, A., PERRIER, A., FONTAN, J. & LAURENT, J.L. (1971) Analysis of turbulent transfers in vegetation: use of thoron for measuring the diffusivity profiles. Boundary Layer Meteorology, 2, 173–187.
24. INOUE, E. (1965) On the concentration profiles within crop canopies. Journal of Agricultural Meteorology (Tokyo), 20, No.4, 137–140.
25. CIONCO, R.M. (1965) A mathematical model for air flow in a vegatative canopy. Journal of Applied Meteorology, 4, No.4, 517–522.
26. TAN, H.S. & LING, S.C. (1961) A study of atmospheric turbulence and canopy flow. Therm Incorporated and Department of Agriculture, Ithaca, N.Y., TAR–TR, 611, Cooperative Research Program.
27. COWAN, I.R. (1968) Mass, heat and momentum exchange between stands of plants and their atmospheric environment. Quarterly Journal of the Royal Meteorological Society, 94, No.402, 523–544.
28. PERRIER, A. (1967) Approche theorique de la microturbulence et des transferts dans les couverts vegetaus? en rue de Canalyse de la production vegetale. ? La Meteorologie I, 4, 527–550.

29. DENMEADM O.T. (1966) Carbon Dioxide exchange in the field: its measurement and interpretation. Proceedings of WMO Seminar on Agricultural Meteorology, Melbourne 1966, 445–482.
30. LEMON, E.R. (1965) Micrometeorology and the physiology of plants in their natural environment. Plant Physiology, Vol.4, Part A, 203–227. Academic Press Inc. New York.
31. UCHIJIMA, Z. (1962) Studies on the micro–climate within plant communites (1) On the turbulent transfer coefficient within plant layer. Journal of Agricultural Meteorology, Tokyo, 18, 1–9.
32. BEGG, J.E., BIERHUIZEN, J.F., LEMON, E.R., MISRA, D.K., SLATYER, R.O. & STERN, W.R. (1964) Diurnal energy and water exchanges in bulrush millet in an area of high solar radiation. Agricultural Meteorology 1, 294–312.
33. IMPENS, I.I. (1970) Daytime distribution of energy sinks and sources and transfer processes within a sunflower canopy. UNESCO Symposium on Plant Response to Climatic Factors, Upsala 1970.
34. SAUGIER, B. (1970) Transports turbulents de $CO_2$ et de vapeur d'eau audessus et a l'interieur de la vegetation. Methodes de mesure micrometeorologiques. Oecologia Plantarum. Gauthier Villars V, 179–223.
35. WIGHT, J.L. & LEMON, E.R. (1966) Photosynthesis under field conditions IX. Vertical distribution within a corn canopy. Agronomy Journal 58, 265–268.
36. BROWN, K.W. & COVEY, W. (1966) The energy budget evaluation of the micrometeorological transfer processes within a corn field. Agricultural Meteorology, 3, 73–96.
37. INOUE, E., UCHIJIMA, Z., UDAGAWA, T., HORIE, T. & KOBAYASHI, K. (1968) Studies of energy and gas exchange within crop canopies (2) $CO_2$ flux within and above a corn plant canopy. Journal of Agricultural Meteorology, Tokyo. 23, No. 4, 15–26.

International Seminar 1975

# HEAT AND MASS TRANSFER IN MODERN ENERGY SYSTEMS

## M. A. Styrikovich

*High Temperature Institute, USSR Academy of Sciences, Moscow, USSR*

The VIII ICHMT Symposium on Heat and Mass Transfer in Modern Energy Systems was held in the aftermath of the first energy crises of 1973-1974 and the pessimistic Roman Club forcasts, at a time when many of the prelevant ideas on energy development changed. It became apparent that the mostly inexpensive oil spurred exponential growth of energy consumption had come to an end. It also became clear that environmental protection had to be taken into account when considering the future of energy. All this increased the significance and contributed to the perfect timing of the ICHMT Symposium in August of 1975.

Of course, today, 13 years later, a considerable portion of the material presented, the assesment of the relative significance of various energy sources, in particular, seems pretty trivial. It should be noted, that the assumptions in the majority of the publications at that time, and long there after seem absurd today.

It is sufficient to recall that R. Gibrat stated in his presentation that "almost everyone agrees, that use of fossil fuel is doomed by the year 2000" and that the development of the power industries will be based on nuclear energy used to produce electricity and hydrogen. It is evident that J. Denton was much more perceptive in his paper "Perspectives of World Energy Production", an assessment of long-term energy source options. He laid emphasis on the more efficient use of traditional fossil and nuclear fuel and new techologies ensuring a higher level of environmental protection. Also of note is his discussion of heat resource quality (their energy potential) and the correctness of his assesment that the many energy sources then considered the energy sources of the future-solar energy, geothermal energy, marine temperature gradients are actually energy sources of low quality. It is significant that the author also attempted to evaluate energy sources not only on the basis of available reserves but also on the economics of their use. Today, we are of course used to this approach, though the economics are today somewhat different, the current price of oil still an indicator of the price of energy (incorporating inflation) being close to that in 1975.

If we consider the lectures and presentations at the Symposium, it is evident that during that "turbulent" time, the most exotic ideas on the perspectives of energy development were widespread. Nevertheless the Symposium as a whole, being devoted to fossil fuel and nuclear systems heat transfer in sufficient degree, had an impact on the technologies prerequisite to energy development in the forseeable future.

# ENVIRONMENT AND ENERGY PRODUCTION AFTER THE YEAR 2000

R.Gibrat

*Conseiller Scientifique SOLMER, Paris, France*

Gasoline crisis, energy crisis, civilization crisis, for several months we have been hearing all sorts of exclamations, lamentations, anguished questions, and apprehensive answers. Commissions on departmental and interdepartmental levels are succeeding one another and publishing bulky reports. Every day the press is full of opinions and contradictory judgments. The government steps in, explains, and makes laws.

We will avoid any commotion of this sort because we do not seek to define the future in a precise way by means of our long-term reflections. We will limit ourselves to examining the great lines of our present knowledge and if we quote as examples some figures to characterize the situation in the year 2000, our conclusions will nonetheless remain valid for energy situations that differ a great deal from each other in this period.

There is no contradiction with the difficulties that the futurologists foresee for the year 2000. Our viewpoint is different and takes into consideration the great inertia of decisions and the exceptionally long duration of scientific and technical research when energy is at stake. The choice of the year 2000 permits calm reflection and serves as an attraction not only because this round figure has been traditionally full of mystery but also for the following five reasons.

a) From now until the year 2000 no new energy source will be able to intervene in a significant fashion; but the quarter of a century that separates us from it, under the assumption that the necessary intellectual and financial efforts are made and the responsive structures are prepared, seems to be, from experience, an appropriate duration to assure some of them with sufficient technological maturity for their use on an industrial scale.

b) The two biggest technical energy ventures in the course of development are without question the breeder reactors and the HTR. But their success or failure will have no decisive influence on the unfolding of a period that separates us from the year 2000, because the various strategies of imaginable reactors lead to a utilization of uranium reserves of the same magnitude. (International Agency in Vienna, meeting of Nov. 5 - 9, 1973 of the study group for calculations on the strategy of reactors.)

c) The numerical size of the world population is a great factor in the energy problem, but a possible evolution in behaviour, and high handed measures that will perhaps be taken by certain countries, cannot seriously modify the level of the world's population in the year 2000, from a mathematical point of view. No serious uncertainty in this matter is expected and the population will amount at that time, if we disregard wars, epidemics and revolutions, to 6 or 7 billion people worldwide.

d) In the western world, at least, the natural resources (coal, natural gas, fuel oil ...) are slowly approaching their end with our present growth rate. We will cite just one example. A recent French study has shown that to meet the world's gasoline needs by the end of the century with moderate use of nuclear energy, it will be necessary to discover every year, in addition to the present deposits, 9 billion tons on the average which is the equivalent of the total reserves of Algeria and Libya, or three times the present petroleum reserves of the North Sea.

In addition, environmental problems make their use more and more difficult. The high sulfur content, especially in combination with ozone and nitrogen oxides, seems intolerable now. The pollution caused by the automobile or the airplane is the object of heavy attacks and one must not count on a significant reversal in public opinion. $CO_2$ or dust and their possible influence on the climate have also been attacked. One can, through its hot house effect, lead to a temperature increase on the earth's surface, the other can bring about the opposite effect by intercepting solar radiation.

It is therefore understandable why almost everyone agrees that use of fossil fuels is doomed by the year 2000. Their total disappearance seems unlikely but a considerable diminution of their role is desirable; at any rate, that is what we are assuming here.

e) Since 1968 we have been especially familiar with an extraordinary burst of ideas about these problems: contestation of our technical civilization, discussions on the industrial growth rate, a great awareness of the environmental problems and the quality of life, and industrialization of countries currently under development. The consequences of all this are beginning, due to the present crisis, to appear more clear, and it becomes possible to free oneself a little from the initial impassioned positions by imagining a reasonable long-term compromise.

As an average assumption we will suggest the following data to characterize the period up to the year 2000.

a) A moderate growth rate of the economy but a correspondingly great industrial development of the Third World, which will result in an energy consumption on a global scale in the year 2000 that is 3 to 5 times higher than the present level.

b) A high respect for the environment by continuous research on conditions for its improvement. One is talking about zero pollution growth; this is desirable but perhaps not possible.

c) Ever greater attention being given to a satisfactory solution of autonomy problems, i.e., a certain independence in energy resources for a given country.

It can be seen that our hypotheses remain pretty broad and therefore avoid any discussion of figures. The parameter of resources at acceptable prices (the only factor that was once important) is now complemented by two new factors just as important: environment and self-sufficiency, and this will suffice to define the big lines of the future.

How do energy problems appear from the year 2000 on? We are assuming here the more or less general view that electricity will be able by that time to satisfy 40 to 50 % of the total energy needs, with nuclear energy taking the greatest part in electricity production, 80 % for example. The extraordinary phenomenon of breeding will become more and more important from the year 2000, and in the first decades of the 21st century nuclear production will enjoy more or less complete self-sufficiency, with nuclear fuel becoming a byproduct of electricity production. This is a major fact and we will return to it.

Assuming this is true, the problems of resources and self-sufficiency will be extended to the portion of the energy consumption that is covered today by hydrocarbons. Electricity will perhaps be progressively substituted, but an all-out invasion is obviously impossible.

If a return to coal is dismissed, as will probably be the case except perhaps in the U.S., Poland and China, the only solution to the problem rests in the use of high temperature gas cooled reactors that allow the mass production of hydrogen from water; because numerous outlets that are inaccessible or accessible only with difficulty for electricity today will become feasible with nuclear energy through a hydrogen economy.

A totally new energy structure will come into existence due to the exceptional properties of hydrogen in matters of transportation, storage and its total absence of pollution - its combustion will produce only water as a byproduct.

Thus, breeding and hydrogen would be, thanks to nuclear reactors, the key items capable of changing the first decades of the 21st century. If they are successful, the resource and self-sufficiency problems will be practically solved. Thus this "all nuclear" solution seems to be the one that is best suited for the problems of the next century. It will be the subject of our first chapter but in studying its features against the background of the environment, we will find a limit to its extension because the profound nature of man cannot, in our opinion, stand the exceedingly high temptations that it will offer him.

Essentially for this reason, but also because failures and delays are possible (for example a serious nuclear accident would probably entail a temporary rejection of nuclear energy by the public), the resort to new sources will therefore be necessary in a short time.

We will therefore review in a chapter entitled "Long Term Solutions", the various natural energy flows surrounding us that we could hope to tap. Only three sources seem capable of playing a major role on a global scale: solar energy, geothermal energy, and nuclear fusion. These resources are, on our scale, practically unlimited and their self–sufficiency features are excellent.

We will mention in the second chapter certain other sources that are ready or almost ready, the influence of which, however, will be mostly local. We are thinking for example of tidal power plants for France. We will also have a few words to say of certain advanced techniques of conversion, storage, and transportation that can become important in all the solutions.

Solutions of any kind create meteorological problems and this, as we will see, will set limits for their use and therefore the production of energy and industrial growth. This major concern will be the topic of our fourth chapter.

We will not give technical descriptions and we will avoid floods of figures. But we will note that all the new sources or techniques have many varied properties that have had no equivalent up to now.

The energy world that is waiting for us will therefore not be a continuation of our present world; a profound change is certain and it is not easy to evaluate its "social and political" fall-out. Nevertheless, we will try to discuss it.

# 1. THE "ALL NUCLEAR" ECONOMY

"All nuclear" must not be confused with the present "Electricité de France" slogan "all nuclear, all electric" because "all nuclear" assumes here the introduction of a new energy carrier besides electricity: hydrogen. Three possibilities must be studied successively: breeding, high temperature reactors, and hydrogen.

The probability of solving the problems still unsolved before the year 2000 is great, because of these three items

a) There are already powerful industrial structures for research and development.

b) The problems all seem to be recorded, if not solved.

c) One can reasonably fix the schedule for their industrial introduction.

d) Their impact on the environment is well known.

For the "long-term solutions" more or less the opposite holds as far as these points are concerned.

## 1.1. BREEDING

This is an extraordinary phenomenon which in itself will bring about a revolutionary change of energy problems. Its understanding requires a few technical explanations. Natural uranium extracted from the earth contains two isotopes both of which have the chemical properties of uranium but differ in their nuclear properties. The one which represents 99.3 % of the total uranium is fertile. This is uranium 238, so named for the number of its nucleons; the other is fissionable and is uranium 235 (235 nucleons). Fissionable material in an operating reactor breaks into two pieces and thereby produces energy through mass loss. Some of the fertile material (0.3 % uranium 238) changes at the same time into fissionable material, plutonium.

In a reactor of the Anglo-French type (Magnox) for example, fissionable uranium (0.7 % of the total) and the plutonium produced (0.3 %) is consumed, but 99 % of the fertile portion remains unused.

In breeder reactors that use uranium and plutonium at the same time, the production of the fissionable material from the fertile material takes place with such success that after a first energy production one finds in the burnt fuel not only enough plutonium to replace the amount that underwent fission, but a surplus. We can, therefore, envisage transforming almost all fertile uranium into energy and thus can multiply the actual output by 100, in succession. The global reserves represents by this feat alone 100 times more energy than we have today. This is in itself extraordinary; but more and more it will be possible to exploit ores with much smaller content. There are, therefore, practically unlimited resources that for a long time to come will suppress any controversy on the availability of uranium reserves.

Today the breeder reactor is close to industrial realization. The French prototype got off to a good start. It was named "Pheonix", because like the legendary bird, the fissionable material is reborn of its ashes. However, the introduction of breeder reactors will change the uranium consumption little from now until the year

2000. Indeed, their massive industrial arrival will be late, and the lessons learned from the 1200 MWe reactor, that will follow Pheonix, will not be known until 1985. Furthermore, the breeding ratio that determines the plutonium surplus will, for technical reasons, be considerably lower than expected, at least in the beginning.

We have been able to demonstrate that breeder reactors will have more difficulty in taking an important place in the nuclear production as the total nuclear growth rate gets higher, a phenomenon that is little known, appears paradoxical, and to which we draw the attention of the economists. Indeed, the breeder reactors are in principle fed by plutonium and require large fissile investments. It is necessary in order to initiate a program of breeder reactors therefore to build both reactors that produce plutonium and reactors that consume plutonium.

The equation that expresses the absence of plutonium exchange with the outside is the dominating problem, and its solution shows the expected result. In the beginning it will be possible to modify this a little by feeding the reactors with uranium 235, if necessary, but this solution is not very economical.

It can be seen that breeding problems are rather complex. But it is clear that the fissionable material, the main nuclear energy source, will become a byproduct of industrial activity for a certain rate of growth of production, where the breeding of fissionable material practically cancels out all external requirements for raw materials. One could not ask for a more astonishing and perfect solution of the resource and self-sufficiency problem.

## 1.2. HIGH TEMPERATURE REACTORS (HTR)

The HTR is the only nuclear reactor type that allows heat production at high temperatures, due to its fuel (no metallic cladding) and its use of cooling gas that is chemically inert (helium). It exists in two types: in the U.S. through the atomic activities of two oil companies, Gulf and Royal Dutch-Shell, and in Germany through a group of industries that are supported by the State. In France, various agreements have been made with General Atomic, Industry the CEA, and Gaz de France. As of today, it is possible to achieve 800°C, but there is hope that with enough R&D, 1000°C or perhaps 1200°C can be reached in about 15 years. The main problems that are to be solved are structures, supports, exchangers, etc. Research and development methods on the behavior of metals at high temperature are well known. This heat will have valuable industrial applications in addition to the production of hydrogen (that we will look at further below). To mention only a few:

- production from fossil fuel of a reducing mixture for the total or partial reduction of iron ore, which would result in eliminating the principal pollution sources of steel industries,
- gasification of coal, which is today of the highest priority in countries rich in coal such as the U.S.,
- production of ethylene, etc.

All of this seems to be of a sufficient order of magnitude to justify a nuclear reactor that must necessarily be of very great power. It must be recalled that fuel, the large thermal capacity of the core, and the temperature coefficient that is largely negative, give the HTR an operational safety that is excellent. The liquid or gaseous wastes are insignificant, the solid waste will probably be on the order of a third of present-day reactors, and the thermal waste is the lowest of all nuclear sources. If we add their operational breeding capability, which means self – sufficiency, one must admit that without any doubt HTR technology has a future and is worthy of the greatest efforts of the coming 21st century.

### 1.3. HYDROGEN AS AN ENERGY CARRIER

In defining the "all nuclear" solution we have outlined an energy economy with two parallel structures: electricity and hydrogen, the latter mass – produced from water due to nuclear heat. Its introduction can be carried our progressively, because the transportation, storage and distribution installations of today can make use of some 10 to 15 % of hydrogen without any modifications. Its massive transportation resembles the present transportation of natural gas and poses no problem; there is a distribution network of 300 km in length installed in the Federal Republic of Germany at present time. It will be 5 to 10 times less expensive than that for electricity. Storage by means of subterraneous reservoirs of several billion $m^3$ will be possible, for instance, by using the presently exhausted deposits of natural gas.

The variety of ways in which hydrogen can be used is extraordinary. Nothing or almost nothing is lost. Its use as a reducing agent in steel making or in chemistry is well established. Today 20 million tons are consumed in the world.

Hydrogen promoters see no problem in its utilization in the automobile instead of gasoline. As early as 1927, Zeppelin had an internal combustion engine burning hydrogen; it was stored in his airship for altitude control. There are still certain questions concerning storage in a car, but for the optimists the hydrides offer a good solution. As far as the airplane is concerned, the same optimism applies because liquid hydrogen is 2.5 times lighter than kerosene at equal power. However, hydrogen takes up 3 times as much space. Highly advanced projects for a supersonic plane at Mach 8 use hydrides at the same time as a hydrogen source and as a heat sink that guarantees protection against thermal shock. For reasons of pollution, it will be used for take-off and landing at first.

On the domestic scene, hydrogen's features appear amazing. A group of gas companies in the U.S. financed the study of an "all hydrogen" house. Lighting will be affected by small amounts of hydrogen that come in contact with phosphorus in the presence of oxygen, which results in a strong luminiscence at low temperature without a flame. Heating will be achieved by means of hydrogen diffusion through decorative panels of porous plastic that are impregnated with an appropriate catalyzer. Heating plates without flame in the kitchen will be made of metals or porous ceramics. The electricity for household appliances will be available from hydrogen stored in fuel batteries, or, in large cities, from gas turbines. Such perfection sounds like a dream. We have read somewhere that the only unsolved problem was that of the electric razor, but we cannot remember why. What is missing for this paradise to become reality? Simply, it is inexpensive hydrogen production from water. The whole question rests on that. Electrolysis is the most simple procedure today but it is too costly. Experiments are being carried out with several procedures to obtain an acceptable price, electrolysis at high temperatures in the steam phase, for example. Certain experts are very optimistic.

The most significant research carried out today in the world has to do with the separation of water at approximately 1000°C through a number of chemical reactions between substances that produce oxygen at high temperatures, then become oxidized at a lower temperature to produce hydrogen, and cycle in endless repetition. The simultaneous production of oxygen is obviously a highly favorable factor. A great number of these reactions are currently being researched.

An international conference was held last year in May in Miami (U.S.A.) on this future hydrogen "civilization". Some one hundred papers, presented in 18 sessions, examined the various aspects of this exciting problem in today's energy world.

## 1.4. ENVIRONMENTAL PROBLEMS

This "all nuclear" solution is almost perfect as far as the problems of resources and self-sufficiency are concerned. It can obviously fail or succeed for technical and economical reasons, therefore the necessity for "long-term solutions". In case of success, however, the great problem is to find the limits that the environment might impose. We are not thinking here of the meteorological problems because they are common to all solutions. Also, we don't see the dangers of waste products radioactivity because they are not serious. Of course, we cannot guarantee that there will not be a nuclear accident one day, and we are well aware of the consequences which can be serious, but man has been living in danger since Adam and Eve and this danger seems much less probable and much weaker than the various "acts of God" - earthquakes, tidal waves, typhoons, etc. The problem for us is in the possible proliferation of nuclear energy, because as Alfven, the holder of the Nobel Prize for physics in 1970, in a paper at the second Pugwash Conference (Fall 1973), "one or several reactors under careful control cannot pose a serious ecological threat, but if they spread in almost unlimited fashion, everything changes."

It is no longer a question of simple multiplication, even by a high number, of small risks but a veritable new risk due to the profound nature of man.

Plutonium is a poison even if the amount is only one-thousandth of a milligram. The breeder reactors of a country such as France will be using tons of it every year in an "all nuclear" solution. All possible precautions will, of course, be taken in the recovery process or in unavoidable transportation accidents, but can we imagine any time in our world where violence was a greater temptation? All the problems of storage, fuel fabrication, and reprocessing will change in nature when they change in size.

This is not all. The safeguards to be taken in the matter of wastes will have to be continued perhaps for thousands of years. Man has never been confronted with responsibilities of such a duration, and no one can say today how he will assume them. Of course there is hope that the wastes will be changed into products of shorter life by means of radioactive transmutation. It will certainly be necessary to push research in this area with more vigor than has been the case up to now.

The "all nuclear" solution presupposes human beings that are civilized, disciplined, happy with their destiny, etc. This is impossible. Therefore its extension will certainly have a limit due to the nature of man, and we do not know this limit. We only know that today we are still far from it. But this -- we have to repeat it -- is the main reason to undertake research of new sources that are better suited to our own human nature at once. Let us listen again to Alfven: "Fusion, solar, or geothermal energy have not yet been developed enough to assure a solution to our energy problems ... But as the Manhattan and Apollo projects have shown, our science and technology are so powerful that if an intensive effort is made we will obtain what we want, let us say in 10 years, under the condition that we do not conflict with the laws of nature."

## 2. SECONDARY TECHNOLOGIES AND RESOURCES

These are energy sources that are technically mature and already capable of playing an interesting role, if only locally: tidal energy in first place, perhaps thermal energy of the seas or of the winds, with that of the swells seeming to be unusable for the time being. As far as the winds are concerned, we want, to mention here a serious Russian proposal that seems to border on science fiction: the installation of energy stations for capturing the wind at an altitude of 8 to 12 km, in the zone of subtropical jet currents, where their speed is 3 to 7 times greater than that on the surface of the earth. One may guess the problems that have to be resolved.

We must also mention certain reactor types, so far unexploited, that could in certain favorable cases find a market for environmental reasons, but will for the most part encounter the impossibility of obtaining the large investments that all nuclear research requires (molten salts, heavy water reactors cooled by an organic liquid...).

The present day significance of tidal power for France and Canada, and the time that we personally have spent on this energy source, require some explanations. The phenomenon of the tide has always been considered as one of the most powerful manifestations of nature but it is produced by very small forces. The greatest variation brought to bear by the moon on the earth's pull on the oceans is 6 million times smaller than the strength of gravity. It is equal to what we do to the weight of our watch band when we raise our fist 50 cm in an oratorical gesture. The sun is still less effective because, if it is bigger, it is also further away. Also, if the free surface of the seas would stay in equilibrium, the variation of the horizontal component of gravity would give a tide of a few centimeters and the variation of the vertical component, a tide of less than 1 mm. There would remain, therefore, no explanation for the 13 m tide at Mont Saint-Michel. But the surface of the seas cannot stay in equilibrium, the waters playing catch with each other and never coming to a stop. The tide is, therefore, essentially a resonance phenomenon and large amplitudes take place where the conditions are the most favorable, an astonishing energy source, we must admit.

The tidal power plant at la Rance was carried out on schedule, for the quoted price, and under remarkable technical conditions. It has already been in operation for eight years and its reliability is excellent: 23 out of 24 available turbosets on the average. However EDF is not studying any new projects of this type today; the high interest rate at present has increased the investment for a tidal power plant beyond the tolerable level.

Environment and self-sufficiency must one day bring back this energy source. The efforts of Mr. Caquot, one-time president of the Academy of Sciences, must be mentioned here. He continues to study and defend the Chausey project of Mont Saint Michel bay. With two basins of 1000 $km^2$ each (that of Rance has approximately 20), one would obtain 32 billion kWh. The manufacturing plant for the material would be built on solid ground obtained by fill-in, and this in turn could be used as a site for nuclear plants that would have at their disposal the necessary cooling water (150 MWe per $km^2$ of basin surface). Moreover, after Chausey, it would be possible to use the site at les Minquiers, which would double the energy produced.

It does not seem necessary here to demonstrate that tidal power plants look favorable from the point of view of the environment, which is not always the case with hydraulic plants. Indeed, the only modification brought about by the dam and the plant is to modify the phase lag between the variations of the level and the current without changing their magnitude. The absence of any thermal waste and therefore of any influence on the climate is particularly valuable. It would be possible to accept power plant concentrations in a tidal power plant that are 10 to 15 times higher than for nuclear plants.

The new technique of conversion, transportation, and storage must not be neglected. In about 2 to 3 years we will know whether large power industrial fuel cells (26 MWe) are feasible, because of the increased efforts of the U.S. utilities (by a factor of 4).

The use of topping cycles coupled with primary energy sources at high temperature (1000° and above) could cut thermal wastes in half, which seems to be essential. Certain new storage methods by phase change could also be used profitably.

The peaceful nuclear explosions presently in the course of industrial development must not be forgotten either.

## 3. LONG-TERM SOLUTIONS

One must not rely upon the energy flows that surround us (with the exception of solar energy), because if one calculates their size one finds that their power, if applied to the earth, is of the same order of magnitude as the one that is transformed today by man himself for his own use by means of fossil fuels, i.e., several billion kilowatts. We discussed some time ago cyclones, lightning, earthquakes, volcanic eruptions, the heat flow originating in the deep layers of the earth, etc. Making use of them would require a technological effort that would without any doubt be greater than the one man has just made for the atom and would do little to alleviate his problems. We must therefore concentrate essentially on the three great sources that are known today: solar energy, geothermal energy and nuclear fusion (as Alfven suggests).

### 3.1. SOLAR ENERGY

The amount of solar energy flux that is received continuously on the earth is very high: 2 calories per $cm^2$ per minute. This corresponds to an average power on the entire earth at the point of entrance into the earth's atmosphere of 170.000 billion kilowatts. To be sure, a portion is absorbed or reflected, another falls on the ocean, and only one-tenth reaches solid land, but this still amounts to several thousand times the present energy consumption of man.

This is fortunate because we have thus a measure of substantial growth possibilities of the energy released on the surface of the earth without seriously modifying the global thermal equilibrium. Soviet scientists believe that man could use up to 5% of the solar energy received. It is interesting to note here that the 1% efficiency of photosynthesis (forests, agriculture), on approximately one-tenth of the continents on the earth, corresponds to a power on the order of several billion kilowatts, which means that it is of the same order of magnitude as the world's present energy consumption.

One must not draw any hasty conclusions as to the inferiority of man's actions compared to the sun. When burning several hundred tons of leaves and brush for half an hour in an area of ten to twenty hectares (1 hectare equals 2.47 acres), farmers release energy that is ten times higher than the solar energy received on the same surface during the same time. The big industrial complexes or the large population centers release still more energy. Therefore man already influences the micro-climate.

Let us try to define the problems that arise from the use of solar energy on a large scale. The small efficiency (10%) known today, and that may hopefully increase in the future to perhaps 30%, indicate the character of the low density of solar energy. Japan, for example, would have to capture solar energy by means of mirrors over an area of 73.500 km$^2$ or 20% of its surface, to respond to its anticipated energy needs for the year 1985. This is impossible. If one wants to use solar energy it must be captured in other areas (for instance in the Sahara) and must be transformed into transportable fuel (hydrogen and oxygen). Desired self-sufficiency in this matter is still a long way off.

It has been said that solar energy would be free of any pollution. This is not exact, because the albedo is changed, i.e., the reflection coefficient from the ground. M. Peyches, member of the Academy of Sciences of Paris, has analyzed this problem very well. "The solar radiation on a desert consists of 40% that is reflected (albedo) and 60% that is absorbed by the lower atmosphere and the ground to be finally used by the environment. The presence of an absorbing surface reduces the portion that is reflected to 20% and if the 80% that is absorbed is transformed in a plant (close by) with 25% efficiency, 60% goes back to the environment and 20% is utilized. The useful energy is taken solely from the reflected portion."

This modification of the albedo and heat transfer from one place on the earth to another will certainly have meteorological consequences. But of all the heat sources with equal power, solar energy will cause the smallest disturbance.

Moreover, when it is used directly through photocells to produce electricity, even on a large scale, it is almost perfect as far as the environment is concerned: no $SO_2$ emission, no dust, no harmful liquid wastes or radioactive emissions. One can only hope that this will also apply in the case of direct transformation through photolysis of water into hydrogen and oxygen. All of this will be of great value. If solar energy is used through photosynthesis on immense forests and the wood then used as fuel, or if one envisages a production of algae with ensuing transformation into methane, one will again find the usual pollution.

Solar energy for domestic use without harm to the environment must also be mentioned. There is certainly an interesting possibility of alleviating foreign dependency problems due to the fraction of energy consumption that is supplied today by hydrocarbons. In this domain, technical solutions are well known and utilization conditions are well defined.

Development of solar energy on a large scale will depend a great deal on the solution of economic problems that are especially difficult because of the low density of energy. A solar power plant covers from 3 to 4 km$^2$ per 100 MWe for absorption systems using mirrors or photocells, i.e., an area of at least 10 times that of a nuclear plant. Furthermore, to give only one example, a cost reduction by 200 from the present price of photocells seems necessary to make the system competitive. In general, the mass production price is divided by 2 when the production rate is multiplied by 10. This gives an idea of the immense effort still required.

Finally, we want to recall an American proposal for a space energy station that would transmit the electricity obtained in an orbital station to the earth by means of microwaves (weight of the station 25.000 tons, cost 20 billion dollars). Today this is obviously still science fiction, but it may not be the case tomorrow.

The production of solar energy for medium or large power plants must first be introduced in the Mediterranean area where the best conditions are found for sunlight and available areas. For example, Central Spain, Southern Italy, Yugoslavia, Greece, etc. Perhaps its use can one day cross the borders and be of interest to the whole of Europe.

### 3.2. GEOTHERMAL ENERGY

In principle, three different heat sources of varying importance can be distinguished: the magnetic centers, the thermal water basins, and the heat that is stored in the rocks. The latter represents the essential one i.e., an energy of 7,5 kW/year/m$^2$ stored under several kilometers in depth. But one must not neglect subterraneous hot waters. For example, a fact that is given little attention, the area around Paris is very favored in this respect. The water carrying layer (Dogger) around Coulommiers represents the yearly equivalent of 10 to 15 million tons of oil, for about 30 years. Today in Melun, water that is gushing out at 75°C supplies 3000 homes with hot water and a fraction of their heating.

In certain cases, large rock masses have water circulation at temperatures higher than 200°C. In 1973 the total power in the world installed in this way reached 1000 MWe. Today it is planned to go much further by using the heat of the ground where the temperature increases rapidly with depth. Two bore holes, one for production and the other for reinjection, create a water circulation pattern and its reheating through the rocks. The United States, well situated on the earth's crust, is investing large sums of money in this field. Their predictions for use in this area are approximately 75.000 MWe (4% of the total electricity production in the U.S.) in the year 2000.

France has less ambitious plans and would like to realize, in the years to come, 2 or 3 small plants either at home or with its friends overseas.

This energy must not be confused with the flow that comes through the earth due to its natural conductivity (35 cal/year/cm$^2$), because this flow is very slow. In spite of the earth's age of several billion years, only the first hundred kilometers close to the surface have been capable of making their contribution to the flow through the surface since the beginning of time. Today we want to exploit the energy that is accumulated by going towards it. But once consumed, it will renew itself very slowly. For example, when operating over a period of 75 years, one has at one's disposal during this time a power of 0.1 kW/m$^2$: one finds again the order of magnitude per surface area of solar energy or wind energy. But this geothermal energy is not renewable.

Geothermal energy is not without harmful effects to the environment, some of which are similar to those of present plants (chemical or thermal wastes); other are different and not well known, such as local subsidence or creation of possible earthquakes, but the remedy is probably in the reinjection of the fluids that are extracted. For all these reasons, any development must be carried out with caution.

On the other hand, geothermal energy responds well to the self-sufficiency criterion because it is possible, at least in theory, to exploit it almost anywhere.

### 3.3. NUCLEAR FUSION

Fusion phenomena are very different from those of fission. The so-called DT reactor that is currently under investigation consists of a combination of two hydrogen isotopes (deuterium and tritium) that correspond under certain conditions to a mass loss, which results in energy production. Per unit mass involved, DT fusion provides eight times more energy than fission, but the reaction takes place only if these elements interact at very high speeds that allow shocks. The temperature must reach some one-hundred million degrees. By comparison, the highest temperature in a boiler does not reach 2000°C. How is it possible to obtain such a temperature and especially to maintain it without the mixture that is being heated coming into contact with the walls? No material can stand this heat. For some 20 years research in this area has used the principle of magnetic confinement; suitable fields force the DT mixture to stay inside magnetic tubes but these fields have been found to be very unstable and only now, after a great deal of effort, do we begin to see new signs of hope.

The first objective is to obtain a DT reaction that produces as much energy as that required for its heating (break even). The required conditions have been known for a long time. It is definitely necessary to have a temperature of 100 million degrees as we have already mentioned.

But it is also necessary to have a sufficiently high value of the product of particle density for the time during which it stays confined in the magnetic bottle. If the density is calculated in particles per $cm^3$ and time in seconds, this product must reach $10^{14}$.

With a project currently under way in Russia and in France by the name of Tokamak, it is possible to obtain 20 million degrees and a product of $5 \times 10^{12}$. This is still five times too low as far as temperature is concerned and twenty times too small as far as the product is concerned. This explains why it is advisable to obtain the desired values first, before going further and studying the industrial problems.

Approximately three years ago in Canada it was learned through a revelation of Teller, the father of the hydrogen bomb, that a second method was secretly being studied by the military - that of micronuclear explosions. The U. S., Russia, and France have begun to publish their results and it has been clear that the conditions of fusion are those of a thermonuclear bomb. The energy that corresponds to the disappearance of 1 milligram of material is of the magnitude of 80 kg of TNT. It is therefore possible to burn much smaller quantities ( a small fraction of a milligram) and hope to "contain" the micro explosion in a suitable vessel, and by repeating this process, obtain an energy production that is practically continuous just as in an internal combustion engine. To produce the micro explosion, it is necessary to heat a solid material up to more than 100 million degrees, more rapidly than it is going to destroy itself under the impact of the applied energy. The destruction takes place at the speed of sound at this temperature (1000 km/sec), i.e., for a fraction of a millimeter at a fraction of one – billionth of a second. Only lasers, coherent light sources, can deliver energy fluxes high enough for such a short duration on such a small target. One will understand the significance of one-billionth of a second if one knows the time to cast one's shadow takes several billionth of a second. This is like the story of the cowboy who wanted to draw faster than he could cast his shadow.

In the U. S., where this type of research is the most advanced, it is estimated that it is necessary to inject an energy of 20 to 30 g of TNT in one-billionth of a second on the target to bring about a reaction that is very small, to be sure, but sufficient to verify the numerical calculations. Today, one can only produce one hundred times less energy within this time, but the progress of lasers is such that we can hope to reach the desired results 2 or 3 years from now. This would then allow defining the necessary conditions (number of lasers for example) to realize an explosion that will produce as much energy as one will have consumed to produce it in the first place (the famous break even). At that moment, and only then, will one be able to think of industrial applications.

In one way or another, the U. S. is thinking of a first demonstration reactor around 1987 and the first industrial prototypes by the end of the century.

Careful reading of the reports that discuss the difficulties of industrial realization would certainly make the reader wonder, as we have done ourselves, as he looks at the number of recorded problems, the variety of technologies at issue, the exceptional size of difficulties that have to be surmounted and especially the seriousness of problems yet to come. But it must not be forgotten that around 1942, at the time of the Los Alamos explosion, the fission scientists and specialists did not know the necessary numerical values for the calculations of a reactor, the physical properties of the materials to be used, and the operating problems were not yet defined. Nonetheless, 30 years later the economic maturity of nuclear fission is uncontested and the reliability of nuclear energy is superior to that of conventional energy. Therefore an industrial realization of fusion in the year 2000 seems possible.

The environmental and safety problems arising from fusion are numerous but less difficult to solve than those of fission. For instance, the small mass of fusion material to be used is a very favorable factor. There are no serious problems of wastes or very long storage times. On the other hand, the trend towards concentration of very great power (5 or 10 GWe) will lead to considerable thermal waste. One will also find the weather problems that we mention below.

The deuterium-tritium reaction requires lithium, both to collect and transport the heat and to breed the tritium in a manner similar to breeder systems. It does not seem that we will have difficulties with supplies, but that would have to be verified. As far as deuterium is concerned, it exists in great quantity (1/5000) in all natural water.

The deuterium-deuterium reaction is also possible. It requires no lithium and is almost perfect for the environment but requires higher temperatures and raises additional problems. One will notice that the ultraviolet radiations inside such a reactor can photochemically decompose steam and thereby tie fusion energy to the hydrogen cycle, which is remarkable.

Fusion will certainly require a financial and intellectual effort in a few large laboratories that is much greater and much more concentrated than for solar or geothermal energy, but if it succeeds it will adapt itself without any problem to the present industrial or social structures. Its trend toward large quantities is obviously in line with the present development of the economy. Today, therefore, it is the long-term solution that must be considered first, to make sure that in case of a failure or rejection of the "all nuclear" solution the economic development of the 21st century does not stagnate. The U. S., Russia, and France through Euratom are working on this problem eagerly and with much hope.

## 4. METEOROLOGICAL PROBLEMS

For several years the influence of thermal wastes on the weather has been the subject of world-wide studies that have become more and more thorough without reaching any final results. The subject is very difficult because numerous counter reactions intervene and the climate is naturally unstable without man having to be there at all.

In the case assumed by the Russians of an energy produced by man which amounts to 5% of the solar energy that arrives on the continents, one would probably obtain a rise of the continents' average temperature of several degrees centigrade with consequences that are almost unforeseeable today.

The most exciting problem is that of the behavior of the ice in the Arctic Ocean. Present scientific studies cannot give the precise temperature increase that is necessary to melt the ice, but it is assumed that several degrees would suffice. It depends essentially on the more or less thorough mixing of the waters of the ocean with the surface layers under the Arctic ice that have less salinity and therefore freeze more easily. Actually, the thickness of the ice changes every year from 3 meters in winter to 2 meters in summer. Under the influence of wind, the disappearance of the Arctic ice even for a short time could entail the deletion of the surface layer of weak salinity and thus prevent refreezing. A recent study carried out by the Rand Corporation evaluated the influence on climate (wind, temperature) of the Northern Hemisphere by the heat source produced by the Arctic Ocean, assumed free of ice and with reduced albedo, but it was not possible to calculate the effect on rainfall.

Much has been said about the ice caps of Greenland and the Antarctic in case of an increase in temperature. The melting of Greenland ice caps in itself would bring about a rising of the level of the seas of 7 meters, thus drowning most of the inhabitants of the coastal cities. Nothing exact is known except that if anything like that should happen it would take hundreds of years. Indeed if heating of the polar regions caused the ice to melt slowly, the effect would probably be in the opposite direction, i.e., an increase in snowfall due to water vapor coming from the now ice-free Arctic Ocean.

To summarize, as far as the earth is concerned, there is nothing to fear for a long time to come. But the possible influence of local thermal wastes on the micro-climates seems much more serious and constitutes probably the greatest problem created by economic growth.

Let us give some order of magnitude figures. Solar heat, after passing through the atmosphere, reflection on the outside of the atmosphere and absorption on the ground, heats the earth's surface and the troposphere. The heat that is absorbed in this way is, on the average, 67 $W/m^2$ for the continents.

Today in the heavily industrialized areas of the world (500.000 $km^2$) the heat flow that is created by man is 12 $W/m^2$ on the average, but it reaches 630 $W/m^2$, that is 9 times the sun, for New York City. It is understandable that the heat rejection from a large concentration of nuclear power plants could cause serious local modifications of the winds and rainfall patterns, for example along the coast, since the earth could become warmer than the ocean.

The question has been raised whether because of the thermal wastes that go into the ocean local heating at a certain level would be capable of producing such localized peculiarities as waterspouts, tornadoes, etc., as are found in warm oceans or the desert. Preliminary comparison with the effects of forest fires seems to indicate that these phenomena would not occur for power concentrations of 5000 to 10.000 MWe. (All the projects of the big countries, France included, are on the order of 5000, for example four nuclear groups of 1200 MWe each.) These phenomena would probably appear at the next stage of, say 50.000 MWe. This phenomena would evaporate 30 m$^3$ per second into the atmosphere, which is the equivalent of a cumulo-nimbus of 1 km$^3$ in less than three minutes. The discharge heat would be equivalent to the solar radiation received in summer over an area of 3000 km$^2$. The climate would probably seriously deteriorate in an area surrounding the plant which could cover up to 1000 km$^2$ depending on the surrounding topography. Moreover, the operation of the plant could influence the overall weather patterns up to a distance of 100 km.

Russian engineers are attaching great significance to these problems; for this reason, they want to use the theoretical possibilities of heat absorption of the surrounding environment brought about by direct transformation of chemical energy into electricity (Gibbs formula). They feel sure that they will find, for instance for fuels extracted from coal, catalyzers that will allow using this procedure which are not feasible as yet. Thus, by working directly on thermal wastes, this difficult problem can change completely.

In concluding it can be said that we must be on our guard and allow only progressive installation of large power blocks on a given site. Micrometeorology, a science still in its infancy, will perhaps in the near future be the technique capable of defining the limits which will one day constrain that growth.

## 5. CONCLUSION

The great options in matters of energy in the 21st century have been well delineated. Breeding and hydrogen will allow establishment of an "all nuclear" solution for the year 2000, that will satisfactorily solve resource and independence problems and deliver us from the yoke of fossil fuels. In case of a failure or delay, the "long-term solutions" such as solar energy, geothermal energy, and especially nuclear fusion, should be ready to come to our rescue for an almost unlimited time we make a sufficient effort.

But the difficulties that have to be overcome are not small.

To begin with, as far as the technical problems are concerned, the issue is not simply one of substituting one fuel for another. We want to recall breeding, this extraordinary phenomenon that will make of nuclear fuel a simple byproduct of the industrial activity: hydrogen, the new energy carrier, that is drawn from water and is gaining ground slowly. These, together with electricity will slowly modify our domestic life and all our industrial structures: nuclear fusion will require bringing material to one hundred million degrees, and the corresponding industrial processes will be obtained by micro explosions of a duration that is shorter than one-billionth of a second.

Next, the problems facing man. Certainly, the new technologies will one day do away completely with the deficiencies of our present power plants in regard to the problems of environment, air and water pollution. They are, therefore, what man is demanding today. But man in turn has brought with nuclear energy the terror of the bomb and is freeing himself slowly of his nightmares. An accident would revive them and certainly create situations that would significantly slow down the "all nuclear" solution. Independent of this, the quasi – unlimited multiplication of nuclear reactors will create risks of a totally new type. The smallest amount of plutonium is a deadly poison, but breeding will one day introduce tons of it per year at hundreds of sites and it will be transported on highways and railroads. Also the storage of wastes will have to be assured for hundreds, perhaps thousands of years. Man has never thought in such long periods of time. Can he take on this responsibility? A certain limit must be set for the "all nuclear" solution, that leads to the "long-term" solutions.

Finally, the meteorological problems, with the earth dwindling to the size of man. The question of micro-climates can bring a limit to the power that may be installed on a given site for any of the energy solutions, and thus set a maximum to our economic growth. It seems hard for engineers to believe that a reasonable economic growth can be halted due to exhaustion of raw materials or smothering by pollution. We continue to believe, perhaps naively, in the power of science and technology to resolve the problems. But man is unquestionably arriving at the moment where, in certain places, he is going beyond nature with respect to energy production. For the first time he is encountering an obstacle and he does not know if he will be able to go around it or if one day he must get rid of it.

Before concluding, we must make the usual reservations. Thirty years from now discoveries that are absolutely unforeseen and that have not even been taken into consideration will certainly come about, which almost seems to make this paper superfluous. This is true, but it is also true that with the extraordinary inertia of new energy solutions, these discoveries will have little influence on the first years of the 21st century.

International Seminar 1976

# HEAT TRANSFER AND TURBULENT BUOYANT CONVECTION

**D.B. Spalding**

*Imperial College, London, U. K.*

It is well known that the turbulence models which engineers use in order to predict fluid flow and heat transfer provide only imperfect predictions of the real phenomena. For shear-dominated flows, experimental data on which were largely used in the selection of the turbulence-model constants, their agreement with reality is closest; however, for those turbulent flows which are strongly influenced by body forces resulting from gravity, rotation or other acceleration, the predictions leave much to be desired.

This was already apparent when the ICHMT decided to hold its 1976 symposium (then called 'seminar") on turbulent buoyant convection. Indeed it was the reason for the choice.

The practical importance of the ability to predict such flow phenomena accurately is very high, especially in relation to man's precarious control over the natural environment. Thus the spread of polluting smoke plumes through the atmosphere depends strongly on the extent to which thermal gradients stabilise or de-stabilise the air through which it spreads; and the warm-water effluent from power stations travels large distances from its point of entry, often with deleterious effects on wild-life, because the stabilising temperature gradient suppresses the turbulence which would cause mixing with the lower and cooler waters.

A related phenomena, fraught with even more hazardous consequences, is the movement of smoke from fires, especially those taking place in buildings from which people must be swiftly evacuated. The smoke rises from the fire; it then travels rapidly along the ceilings of rooms and escape corridors, often outpacing the retreating crowds. High-level building obstructions may then cause smoke to plunge downwards in frightening billows, confusing the escapers and often driving them back into the flames. Could the movement of smoke be better predicted, architects and fire-inspection authourities jointly would be able to devise more certain methods of evacuation; and the loss of life would be greatly diminished.

The 1976 ICHMT Symposium Proceedings contains papers related to the above-mentioned practical applications, and to others; and it also concerns more idealised situations, the study of which assist to promote understanding. Perusal of the appended titles of sessions and papers will confirm this. Some of the papers contain ideas and data which are still referred to by research workers; and others deserve to be re-examined in the light of more recent knowledge.

I am of course a highly partial scrutineer of these old papers; it is therefore rather apologetically that I report my gratification on finding that it was in the 1976 Symposium that I first described how computational-fluid-dynamics techniques could be extended to cover two-phase phenomena. The inclusion of the paper in the turbulent-buoyancy symposium may have seemed rather idiosyncratic at the time; yet the inkling that I had then has matured into the strong belief that there is an intimate connexion between buoyancy-influenced turbulence and two-phase flows. What is now called the "two-fluid model of turbulence" is the result, a subject about which there is alas no space to write more here.

It is however not (needless to say) the above paper, nor any concerned with environmental applications, that I single out as being the paper to be included in this volume. Instead I have selected the paper by Professor B.S. Petukhov of the Institute of High Temperatures of the USSR Academy of Sciences, entitled "Turbulent flow and heat transfer in pipes under considerable effect of thermogravitational forces". This author, now sadly deceased, devoted much of his life to the assembly of reliable experimental data in the field of heat and mass transfer. This paper is a very suitable one to remember him by.

# TURBULENT FLOW AND HEAT TRANSFER IN PIPES UNDER THE CONSIDERABLE INFLUENCE OF THERMOGRAVITATIONAL FORCES

B.S. Petukhov

*Institute of High Tempreatures, USSR Academy of Sciences, Moscow, USSR*

## 1. INTRODUCTION

At the present time great interest is being shown in the study of turbulent flows and heat transfer under the decisive influence of thermogravitational forces. The interest in this problem, including the case of such flows in pipes, is due both to scientific and practical considerations. Until recently, it was considered that in a developed turbulent liquid flow in tubes ($Re > 10^4$) the influence of thermogravitational forces on the heat transfer and resistance was insignificant. However, the latest experimental data show that this influence can be both significant and decisive under certain conditions. It has been found that sufficiently high Grashof numbers (or Rayleigh numbers) and moderate Reynolds numbers result in a change in the characteristics of turbulent transfer under the action of thermogravitational forces, deformation of the velocity and temperature profiles, and in a change in the heat transfer and resistance. The character of these changes depends on the relative direction of the forced flow and the gravity force vector. As applied to pipes, the following three typical situations should be distinguished:

(1) Flow in vertical pipes upward when the liquid is heated and downward when the liquid is cooled.

(2) Flow in vertical pipes upward when the liquid is cooled and downward when the liquid is heated.

(3) Flow in horizontal pipes with heating and cooling of the liquid.

In of each of these cases the character of the flow due to the interaction of forced and free convection is identical. Therefore, the dependencies for the heat transfer and resistance are also identical (provided that the properties of the liquid, except for the density, are practically constant).

## 2. THE RESULTS OF A THEORETICAL ANALYSIS

The system energy, momentum and mass conservation equations for a stationary (on the average) turbulent flow of incompressible fluid with constant physical properties (with the exception of the density in the expression for the thermogravitational force) in the absence of internal heat sources and energy dissipation is written as

$$\rho C_p \overline{W_k} \frac{\partial \overline{T}}{\partial X_k} = \frac{\partial}{\partial X_k}\left(\lambda \frac{\partial \overline{T}}{\partial X_k} - \rho C_p \overline{W'_k T'}\right) \quad k = 1,2,3$$

$$\rho \overline{W_k} \frac{\partial \overline{W}}{\partial X_k} = -\frac{\partial \overline{p}}{\partial X_i} + \rho g_i + \frac{\partial}{\partial X_k}\left(\mu \frac{\partial W_i}{\partial X_k} - \rho \overline{W'_i W'_k}\right) \quad i = 1,2,3 \qquad (1)$$

$$\partial \overline{W_k}/\partial X_k = 0$$

The nonuniform density distribution in the flow results in thermogravitational forces acting both on the averaged flow and on turbulent transfer. The latter can easily be seen from the equation of the pulse motion energy balance, which is written in a simplified form as

$$\rho \overline{W'_i W'_k}(\partial \overline{W_i}/\partial X_k) - \rho \overline{W'_i} g_i + \rho D = 0 \qquad (2)$$

Here only the generation of turbulent energy (first term), its dissipation (third term) and the work of thermogravitational forces (second term) is take into account.

As seen from (2), the Reynolds stresses $\rho \overline{W'_i W'_k}$ depend both on the averaged velocity field and on the work of thermogravitational forces. Therefore, the influence of thermogravitation on turbulent transfer is determined by two mutually interconnected effects. On the one hand, the thremogravitational forces change the averaged velocity and temperature fields resulting in a change in the turbulent transfer characteristics. On the other hand, the thermogravitational forces have a direct effect on the motion of the turbulent elements of the liquid thus strengthening or weakening the intensity of turbulent transfer.

In the case of stable density stratificatiion the vertical displacement of the turbulent elements is accompanied by consumption of energy for the work of Archimedean forces and leading to turbulence attenuation and, therefore, turbulent transfer. In the case of unstable density stratification Archimedean forces perform work of vertical motion of the turbulent elements and this raises the turbulence energy. The first case is observed, for example, in an ascending flow and the second in a descending flow in a heated vertical pipe. The thermogravitational forces differently affects not only turbulent pulse transfer but also heat transfer and, therefore, the turbulent Prandtl number [1]. The influence of thermogravitational forces on turbulent transfer is not small compared to their influence on the averaged flow. Depending on the conditions, both effects are commensurable or one of them is dominant.

The problem of turbulent flow and heat transfer in pipes with combined forced and free convection has been analyzed in a number of papers. These can be subdivided into two groups. The first group includes papers [2,3,4], in which the influence of thermogravitational forces is taken into account only when they act on an averaged flow; as for the turbulent transfer characteristics, they are accepted by the data purely forced convection. This approach does not allow the character of the $Nu$ number a $- Gr$ number dependence to be correctly ascertained qualitatively let alone quantitatively. Thus, for of an ascending flow in a heated vertical pipe the calculations show that Nu rises monotonically with an increase in $Gr$ at $Re$ = const. ; meanwhile, the experimental data indicate that the $Nu$ number first decreases with an increase in $Gr$ and only starts increasing after that.

The second group includes papers [5,6,7,8,9], in which the influence of thermograviational forces on turbulent transfer is taken into account to a certain extent. Thus, in the L.E.Ber papers [5,6] an attempt is made to take into account the effect of thermogravitational forces on turbulent transfer of a pulse in a flow in a vertical pipe. The thermogravitational corrections of the turbulent transfer coefficient for purely forced convection specified on the basis of experimental heat transfer data. Thus the corrections were not adequately justified non accurate. The Ber calculations are approximate and were not followed up by numerical results, making their analysis difficult.

The A.F. Polyakov papers [7,8,9] are interesting in that they present an approximate analysis of turbulent flow and heat transfer in vertical and horizontal pipes at a distance from the take with a weak influence of the thermogravitational forces on the forced flow and the heat transfer. The influence of thermogravitation on both the averaged flow and on the turbulent transfer is studied. The latter effect is taken into account using the turbulent energy balance equation.

The analysis given in [7] shows that for an ascending flow in heated vertical pipes at $Pr > 0.5$ and low $Gr$ numbers the effect of thermogravitation on the turbulent transfer is dominant* . Therefore, the influence of thermogravitation on the averaged flow is not taken into account. Owing to a decrease in the intensity of turbulent transfer with stable stratification the Nu number decreases with an increase in $Gr$. This is in qualitative agreement with the experimental data (see Fig. 1). After the $Gr$ number reachs a certain value, thermogravitation becomes predominant and the $Nu$ number increases with the $Gr$ number. However, the latter situation was not quantitatively analyzed in the paper [7].

In the case of a descending flow in heated vertical pipes the influence of thermogravitation on the turbulent transfer considerably exceeds its effect on the averaged flow.The intensity of turbulent transfer under conditions of unstable stratificaiton increases, and the $Nu$ number rises monotonically with an increase in the $Gr$ number in complete agreement with the experimental data (see Fig.1).

Approximate analysis of a flow in heated horizontal pipes [8,9] shows that at low values of the $Gr$ number thermogravitational forces act, first of all, on the averaged flow inducing secondary flows in a plane normal to the axis. The effect of thermogravitational forces on the turbulent transfer apparently manifests itself at higher $Gr$ numbers. The secondary flows break the axial symmetry of the velocity and temperature fields, resultat in a nonuniform heat transfer distribution over the pipe perimeter. As the number increases, the heat transfer increases near the lower generatrix and decreases near the upper generatrix.

The approximate flow and heat transfer calculation at weak thermogravitational influence conducted in [7,9] yield equations describing the boundaries of the influence of thermogravitation on heat transfer. A 1% deviation of the $Nu$ number under the effect of thermogravitation from its purely forced convection value is observed at values of the number determined from the relations:
for vertical pipes

---

*At $Pr \ll 1$ the predominance of any effect depends on the relation between the $Re$ and $Pr$ numbers.

$$Gr_q = \frac{1.3 \times 10^{-4} \times Re^{2.75} \times Pr[Re^{1/8} + 2.4(Pr^{2/3} - 1)]}{lgRe + 1.15 lg(5Pr + 1) + 0.5Pr - 1.8} \quad (3)$$

for horizontal pipes

$$Gr_q = 3 \cdot 10^{-5} \cdot Re^{2.75} \cdot Pr^{0.5}[1 + 2.4(Pr^{2/3} - 1)Re^{-1/8}] \quad (4)$$

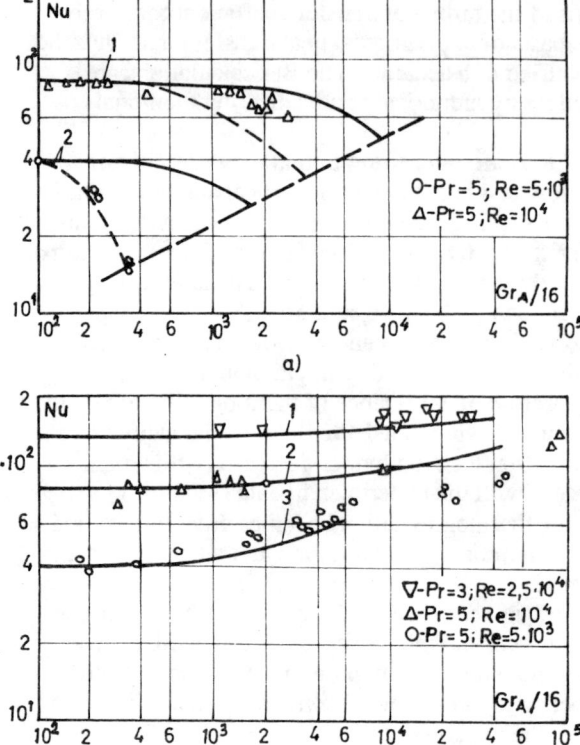

Fig. 1 The Nu number at ascending (a) and descending (b) flows in a heated vertical pipe. Full lines – theoretical calculations
Dashed lines – empirical equations (5) and (6);
Dotted lines – experimental data.

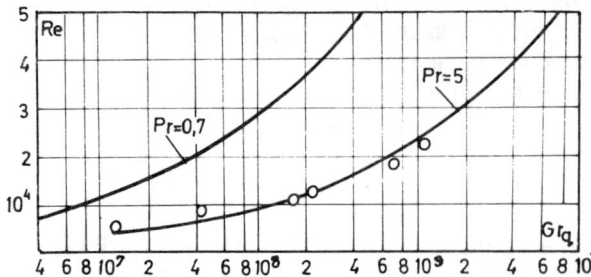

Fig.2 Boundaries of the effect of thermogravitation on the heat transfer with turbulent flow in vertical pipes.

Equations (3) and (4) are validat a distance from the inlet at $q_w$ = const and at $Pr$ 0.5. The limiting curves constructed for equations (3) and (4) are in agreement with measurement results. Figure 2 shows the limiting curves for vertical tubes.

In conclusion, it should be recognized that at the present time an adequately complete analysis of problems of turblulent flows and heat transfer in pipes under the considerable influence of thermogravitational forces does not exist. In order to fill this gap, it is necessary to obtain experimental data on flow structure and heat transfer under the given conditions.

## 3. EXPERIMENTAL DATA FOR VERTICAL PIPES

When a liquid moves through a vertical pipe in a gravitational field, the ascending flow with heating and the descending flow with cooling are practically equivalent; this also applies for an ascending flow with cooling and a descending flow with heating. In accordance with the experimental data available, we will further consider the ascending and descending flows when the liquid is heated.

### 3.1. ASCENDING FLOW WITH HEATING OF THE LIQUID

In the near – wall region of such a flow the directions of the thermogravitational forces and of the averaged flow velocity coincide. Therefore, with an increase in the $Gr$ number the velocity of the liquid increases near the wall and decreases near the axis. At a certain value of $Gr$ minimum velocity occurs at the tube axis, while

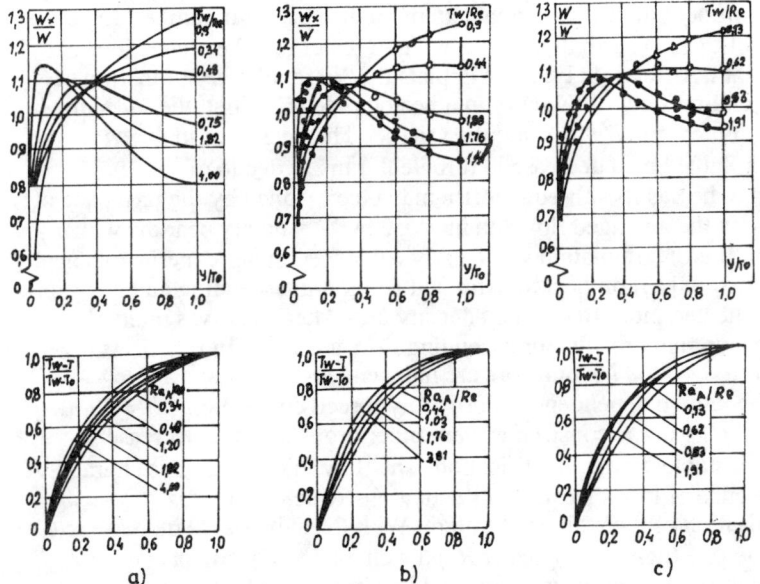

Fig. 3. The velocity and temperature profiles for ascending flow in a heated vertical pipe:
a) $Re \cong 2 \cdot 10^4$, b) $Re \cong 3 \cdot 10^4$, c) $Re \cong 6 \cdot 10^4$.

maximum velocity is observed between the axis and the wall. The velocity maximum approaches the wall with an increase in $Gr$. This velocity profile behaviour under the effect of thermogravitation has been observed experimentally in air and mercury flows [10,11,12]. As an example, Fig. 3 depicts the velocity and temperature profiles measured by Buhr, Horsten and Carr for mercury for $Re \cong 2 \cdot 10^4$, $3 \times 10^4$ and $6 \times 10^4$.

There is scant data on the influence of thermogravitation on turbulent transfer in vertical pipes. This problem was not studied systematically. Paper [11] gives the results of measurements of Reynolds stresses in air moving upwards in a heated pipe, i.e. under conditions of stable density height stratification. It has been found that at $Re \cong 5000$ the Reynolds stresses decrease with an increase in the $Gr$ number and at $Gr_A \cong 2.5 \cdot 10^4$ they approach zero over the entire cross section of the pipe.

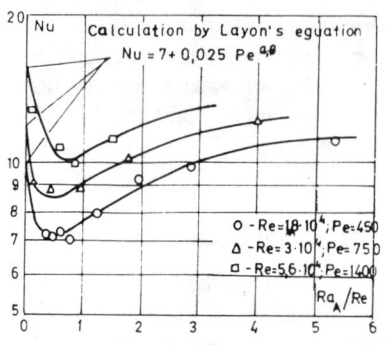

Fig. 4. Heat tranfer in an ascending mercury flow in a heated vertical pipe.

The coefficients of turbulent momentum and heat transfer, calculated in paper [10] by measuring the velocity and temperature profiles in a mercury flow, depend not only on the Reynolds number but also to a great extent on the Rayleigh number. Insigh into the influence of thermogravitation on turbulent transfer may be gained indirectly by studying the heat transfer data. Figure 4 depicts the $Nu$ number as a function of $Ra_A/Re$ for an ascending flow of mercury in a heated pipe /10/. First, the $Nu$ number decreases with an increase in $Ra_A/Re$ and then rises*. The decreases in the $Nu$ number may be explained by a decrease in turbulent transfer due to of thermogravitation, whereas its subsequent rise may be explained by the influence of thermogravitation on the averaged flow (an increase in the velocity near the wall).

The effect of thermogravitation was clearly evidenced in our experiments made jointly with B.K.Strigin [13] on heat transfer and friction in ascending and descending water flows in heated pipes (a constant density of the heat flux was maintained on the wall). The measurement results for ascending flow are given in Fig. 5. At low values of $Cr_A$ and $Re > 5000$ $Nu$ numbers are independent of $Gr_A$ and are in agreement with the known dependencies for purely forced convection ($Nu = Nu_0$). With an increase in $Gr_A$ ($Re$ = const.) the $Nu$ number drops due to a decrease in turbulent transfer at stable density stratification. The flow in this region is characterized by the presence of low – frequency fluctuations which result in a corresponding pulsation of the wall temperature. With the subsequent increase in $Gr_A$ over a certain value (the higher, the greater $Re$) the character of the dependence changes: the $Nu$ number increases with $Gr_A$ and becomes weakly dependent on $Re$. This section of the curve reflects the dominant role of free convection in the formation of the averaged flow. In the is case $Gr_A$ is more "responsible" for heat transfer than $Re$. The role of forced convection is reduced essentially to the maintenance of the specified flow rate. This, in particular, is confirmed by the fact that the points for low $Re$ values (about 500) also lie on this section of the curve.

---

* In this connection, it should be noted that the discrepancy of the avilable experimental data on heat transfer to liquid metals, particularly at low values of the $Re$ number, is probably due to a different degree of the influence of free convection, which was not taken into account during the analysis.

It should be noted that the scatter of the experimental points in Fig.5 is not due to experimental errors but is caused by the fact that the dependence of the $Nu$ number on the $Pr$ and $Re$ numbers is not taken into account in the given system of coordinates

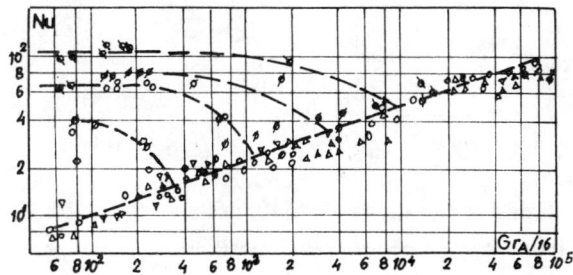

Fig. 5. Heat transfer in ascending water flow in a heated vertical pipe.

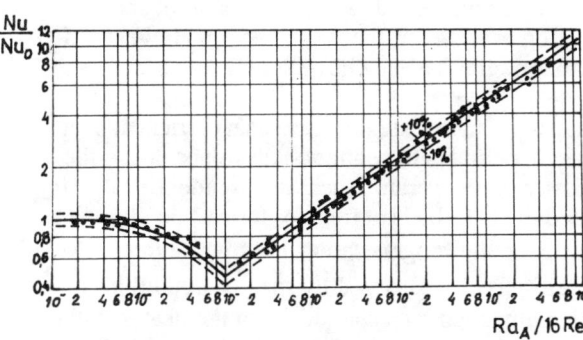

Fig. 6. Generalization of the experimental data on the heat transfer for ascending water flow in a heated vertical pipe. Continuous lines — equations (5) and (6), dotted lines — experimental data.

A generalization of experimental data on heat transfer for an ascending flow of water in heated vertical pipes (part of the data is given in Fig. 5) was made in [13] and is given in Fig. 6. In the region of developed flow and heat transfer ($x/d > 40$) the experimental data are described with $\pm 10\%$ accuracy by the equations:

for $Ra_A/Re^2 < 10^{-4}$

$$Nu/Nu_0 = [1 + 720(Ra_A/Re^2)]^{-1} \qquad (5)$$

for $Ra_A/Re^2 > 10^{-4}$

$$Nu/Nu_0 = 3.97\,(Ra_A/Re^2)^{1/3} \qquad (6)$$

Here $Nu_0$ is calculated according the formula [14]

$$Nu_0 = (\xi/8)\,Re \cdot Pr/K + 12.7\sqrt{\xi/8}(Pr^{2/3} - 1) \qquad (7)$$

where

$$\xi = (1.82\,lg\,Re - 1.64)^{-2} \qquad (8)$$

$$K = 1 + 900/Re \qquad (9)$$

Equations (5) and (6) are valid for the experimental data in the range of $300 \leq Re \leq 3 \cdot 10^4$; $5 \cdot 10^3 \leq Ra_A \leq 8 \cdot 10^6$ and $2 \leq Pr \leq 6$.

In [13] the friction factor, averaged over the pipe length is measured also. These data given in Fig. 7 are satisfactorily described by the following equation:

$$\xi/\xi_0 = [1 + 3.5(Ra_A/Re^{3/2})]^{0.4} \qquad (10)$$

where $\xi_0$ is the value of $\xi$ according to (8).

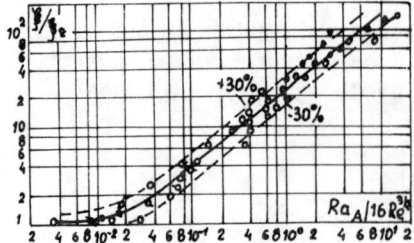

Fig. 7 Friction factor for ascending water flow in a heated vertical pipe. Continuous line – equation (10), dotted line – experimental data.

Equation (10) holds for values $300 \leq Re \leq 3 \cdot 10^4$, $8 \cdot 10^4 \leq Ra_A \leq 6.5 \cdot 10^6$ and $2 \leq Pr \leq 6$.

The physical properties of the liquid in equations (5) to (9) are chosen at a bulk temperature of the liquid in a given pipe cross-section, while in equation (10) at a length temperature average.

The above experimental data were obtained at low Reynolds numbers. Of course, the effect of thermogravitation is stronger, the lower the $Re$. However, if $Gr$ is suffiiciently large, the effect of thermogravitation will be significant at high Reynolds numbers as well.

This is confirmed by heat transfer data for single phase heat carriers at near-critical parameters. In this case the $Gr$ numbers may attain high values due to the pronounced dependence of the density on temperature near the pseudocritical point. Thus, according to the data obtained in [15], the heat transfer for carbon dioxide at near-critical parameters increases due to thermogravitation 1.5 times at $Re \cong 2.5 \cdot 10^5$ and 2 times at $Re \cong 10^5$. Nearcriticallity is characterized by specific conditions of flow and heat transfer which are accompanied by a sharp drop of the heat transfer coefficients (up to five times) at some sections of the pipe. According to modern concepts, these conditions are also associated with the action of thermogravitation and thermal acceleration of the flow.

### 3.2. A DESCENDING FLOW WITH HEATING OF THE LIQUID

In the near-wall region of such a flow the thermogravitational forces act in the direction opposite to the direction of the averaged flow velocity. In this case we may expect some decrease of the velocity near the wall at small $Gr/Re$ ratios, as it is observed in a laminar flow. Unfortunately, the flow velocity near the wall was not measured; therefore, this effect may only be considered on the basis of some heat transfer data (see below). At higher values of $Gr/Re$ we have favourable conditions for intensive generation of turbulent energy due to the opposite directions of the forced and free convection near the wall and due to the unstable density distribution. In this case the velocity profiles fill out, as seen in Fig.8, where the measurement results are given. However, the given flow may in principle be characterized by the heat transfer data.

Figure 9 presents the results of the heat transfer measurements for an ascending and descending flow of water in a heated pipe [12]. The measurements were conducted at a distance from the heated section inlet (at $x/d \cong 50$). The $Gr/Re$ ratios for each experimental point are given in the figure. The striking difference in the character of the dependence of the $Nu$ number on the $Re$ number to descending flow (upper curve) and for ascending flow (lower curve) is immediately aparrent.

The character of the dependence for the ascending flow confirms the considerations expressed above. In the case of a descending flow, owing to the action of thermogravitation, much more intensive heat transfer is observed, therefore, here we

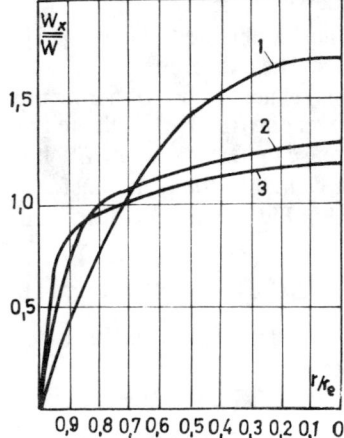

Fig. 8. The velocity profiles for descending air flow in a heated vertical pipe.
1 – $Re = 2650$; $Gr_q/Re = 178$;
2 – $Re = 5780$; $Gr_q/Re = 206$;
3 – $Re = 16890$; $Gr_q/Re = 15500$.

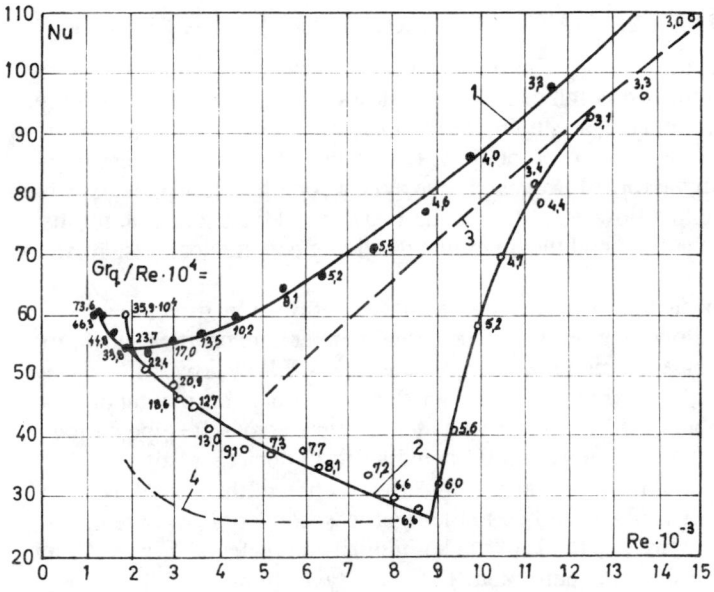

Fig. 9. Heat transfer in a turbulent water flow in a heated vertical pipe.
1 – descending flow with combined forced and free convection;
2 – ascending flow with combined convection;
3 – purely forced turbulent flow;
4 – ascending laminar motion with combined convection.

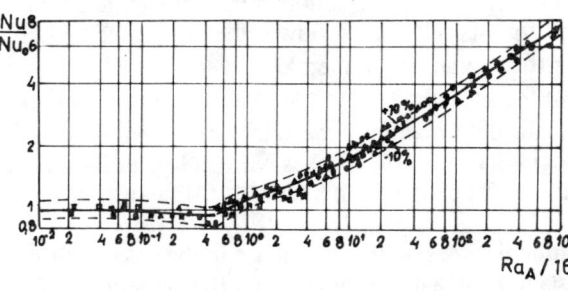

Fig. 10. Generalization of experimental data on heat transfer for a descending water flow in a heated vertical pipe. Continuous line – equation (11), dotted line – experimental data.

have more intensive turbulent transfer not only in comparison to the ascending flow in the presence of thermogravitation but also in comparison to purely forced convection. It is also interesting to note that intensive mixing in the flow under the effect of thermogravitation is observed already at Reynolds numbers of $Re \cong 10^3$.

Figure 10 presents the experimental data on local heat transfer obtained in paper [13] for the case of a descending flow of water in heated pipes. The data are for a developed flow pipe section. The measured $Nu$ number is related to the $Nu$ number at purely forced convection and presented as a value dependent on $Ra_A/Re$. With the increase of $Ra_A/Re$, we first observe an insignificant decrease in heat transfer, which is probably due to a decrease of the velocity gradient at the wall; then the heat transfer increases continuously. The experimental data presented in Fig. 10 may be described by:

$$Nu/Nu_0 = [1 + 0.031(Ra_A/Re)]^{1/3} - 0.15\exp\{-2[(Ra_A/Re) - 8]^2\} \qquad (11)$$

At $Ra_A/Re \geq 16$ the second term on the right-hand side of equation (11) becomes negligible compared to the first term and it may be neglected.

Equation (11) is valid for $300 \leq Re \leq 2.5 \cdot 10^4$, $5 \cdot 10^3 \leq Ra_A \leq 13 \cdot 10^6$ and $2 \leq Pr \leq 6$.

## 4. EXPERIMENTAL DATA FOR HORIZONTAL PIPES

As noted above, the action of thermogravitation on the averaged flow in horizontal pipes results in the formation of secondary flows in a plane normal to the pipe axis. When the liquid is heated, ascending flows appear near the walls and descending flows appear in the central part of the flow. Due to the interaction of these free-convective flows with the forced flow along the axis, a complex flow appears, which may be decomposed into a flow ascending along two vertical lines, one line for the clockwise rotation of the fluid and the other line providing counterclockwise rotation of the same.

Such averaged flow behaviour is rather convincingly confirmed by direct measurement of averaged flow longitudinal velocity and temperature profiles shown in Figs. 11 and 12. The measurements were conducted by A.F.Polyakov, V.A.Kuleshov and Yu.L.Shekhter [17] in an air flow in a heated horizontal pipe. A clear idea of the distribution of the velocity and temperature across the pipe during simultaneuse free and forced convection is given by Fig.13 depicting isotachs (equal-velocity lines) and isotherms across the cross section of the heated pipe through which air flows [16]. As shown in the figures, secondary flows break the axial symmetry of the flow. The fields of the logitudinal components of the velocity and temperature are considerably deformed. The velocity maximum and the temperature minimum are shifted downward. The deviation of the velocity and temperature profiles from those typical for a purely forced flow is the larger, the higher the Grashof number and the lower the Reynolds number. However, even at $Re \cong 5 \cdot 10^4$ the deviation is noticeable (see Figs. 11 and 12). We may assume that if the $Gr$ number is sufficiently large, the deformation of the velocity and temperature profiles under the action of thermogravitational forces takes place at higher Reynolds numbers. Such a situation is observed, for example, in a turbulent flow of heat carriers near critical parameters.

The action of thermogravitational forces on the turbulence in the case of a flow in heated horizontal pipes manifests itself in a different way in different regions of the

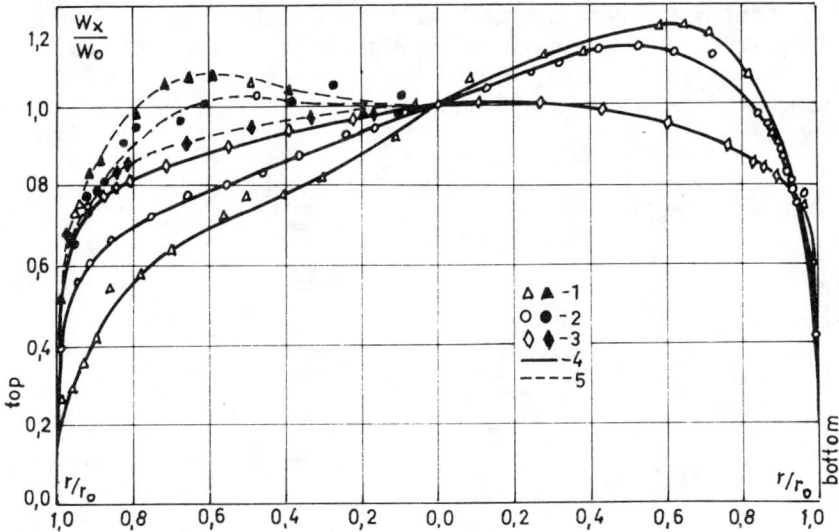

Fig. 11. Velocity profiles. 1) $Re = 1.3 \cdot 10^4$, $Gr_q = 4.2 \cdot 10^8$; 2) $Re = 2.6 \cdot 10^4$; $Gr_q = 7.7 \cdot 10^8$; 3) $Re = 5.2 \cdot 10^4$; $Gr_q = 1.03 \cdot 10^9$; 4) in the vertical diametral plane; 5) in the horizontal diametral plane.

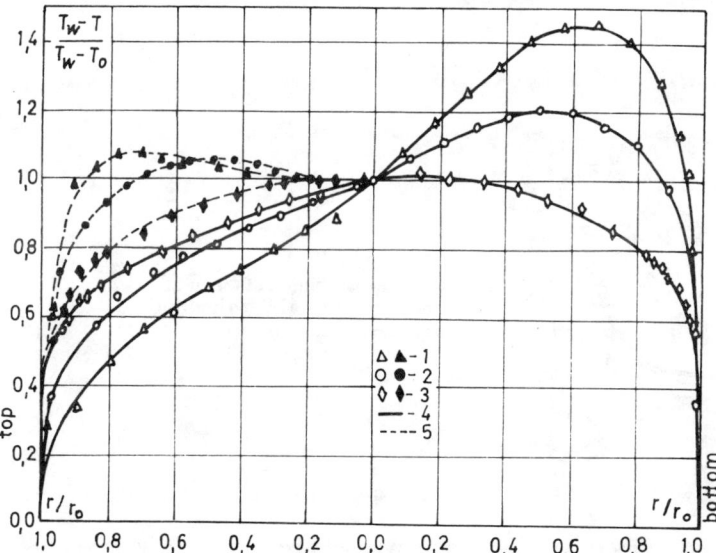

Fig. 12. The temperature profiles for an air flow in a heated horizontal pipe. The designations are the same as in Fig. 11.

flow. The gradients of the averaged velocity are reduced near the upper generatrix and this leads to a decrease in the generation of energy of the pulsatory motion. Furthermore, in this case the density stratification is stable and this also induces a decrease in the turbulent energy. Therefore, considerable reduction of the intensity of the pulsatory motion is to be expected near the upper generatrix. On the contrary, the velocity gradients increase near the lower generatrix due to thermogravitation while the density stratification is ustable. Both these factors increase turbulent

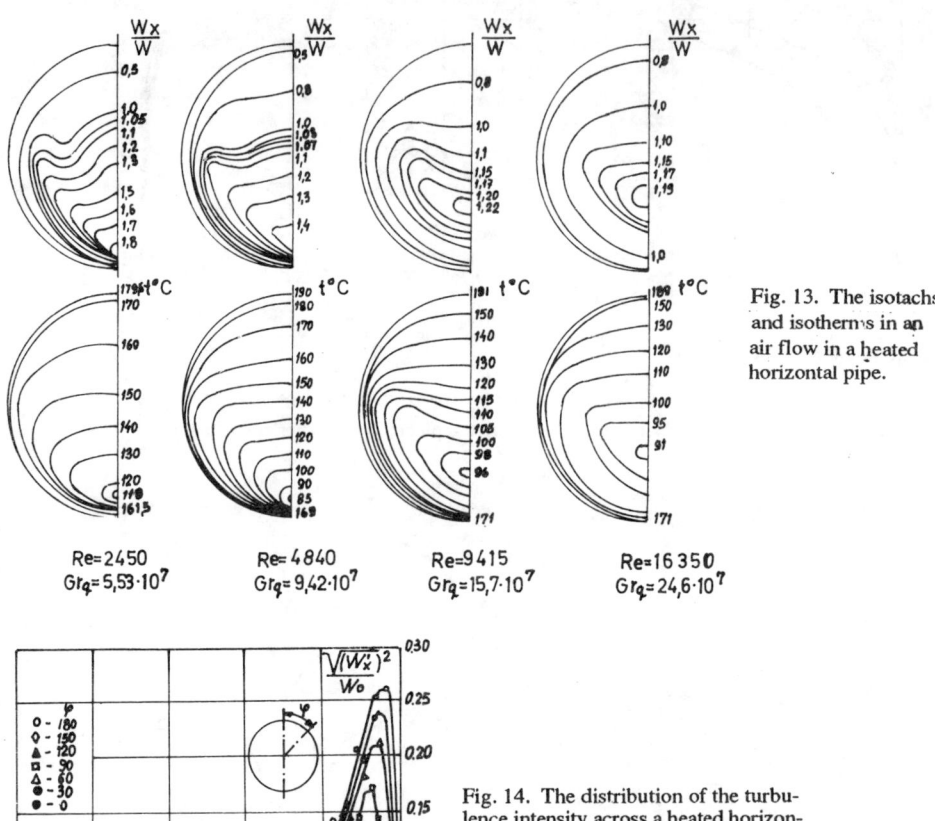

Fig. 13. The isotachs and isotherms in an air flow in a heated horizontal pipe.

Fig. 14. The distribution of the turbulence intensity across a heated horizontal pipe at the Reynolds numbers of $Re = 1.2 \cdot 10^4$ and $Gr_q = 4 \cdot 10^8$.

energy; therefore, the intensity of pulsatory motion should be high near the lower generatrix. This is confirmed by the experimental data [17] on the turbulence intensity distribution in an air flow at $Re = 1.2 \cdot 10^4$ and $Gr_q = 4 \cdot 10^8$ shown in Fig. 14. The intensity is highest near the lower generatrix ($\varphi = 180°$, $r/r_0 = 0.95$), drops with a decrease in the angle $\varphi$ and near the upper generatrix ($\varphi = 0$, $r/r_0 = 0.95$) is about $\sim$ 1/20 of the value near the lower generatrix.

The above–mentioned characteristic features of the flow also determine the nature of the heat transfer. During combined forced and free covection in horizontal pipes the heat transfer varies over the pipe perimeter the stronger, the higher the $Gr_q$ number. On the upper genratrix the $Nu$ number is minimal and decreases with an increase in $Gr_q$; near the lower generatrix the $Nu$ number is at maximum and increases with $Gr_q$.

Only secondary flows develop near the input to the heated part of the pipe; therefore, the $Nu$ number is practically independent of the $Gr_q$ number along a certain section of the pipe having a length or $X<X_{\lim}$, and the $Nu$ number remains constant over the perimeter. $X_{\lim}$ is the distance from the input at which $(Nu_{\varphi=\pi}-Nu_{\varphi=0})/Nu_{\varphi=\pi}=0.5$, and this may be regarded as an indication of a start of the influence of thermograviation on heat transfer. Let $Gr_{q\lim}$ stand for the number $Gr_q$ at which we observe the beginning of such an effect over the distance $X=X_{\lim}$. According to the analysis of the experimental data, there is an unambiguous connection between the ratios $Gr_{q\lim}/Gr^{\infty}_{q\lim}$ and $X_{\lim}/d$. As $X_{\lim}/d$ increases, the number $Gr_{g\lim}$ decreases and assumes a constant value equal to $Gr^{\infty}_{q\lim}$ at $X_{\lim}/d \geq 45$. The values of $Gr^{\infty}_{q\lim}$ are found using equation (4).

Figure 15 gives $Nu/Nu_0$ as a function of $Gr_g/Gr_{g\lim}$ on the upper and lower generatrices of the pipe. The experimental data given this are obtained for the motion of air [17] and water [18] through heated pipes and cover the range of values

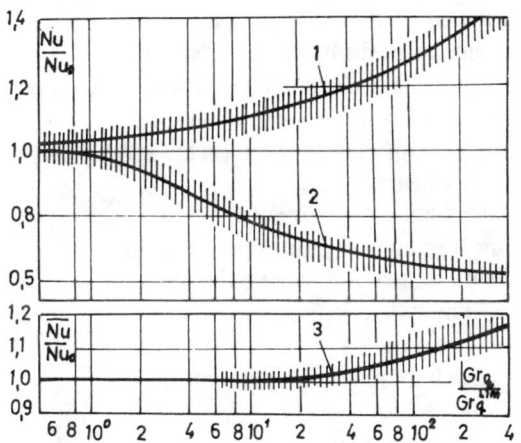

Fig. 15. The Nusselt number on the lower generatrix (1), upper generatrix (2) and averaged – over – perimeter (3) with air and water flows in heated horizontal pipes.

$8 \cdot 10^3 \leq Re \leq 10^5$, $10^7 \leq Gr_q \leq 10^{10}$ and $0.7 \leq Pr \leq 3.3$. Shown in the same drawing is dependence of the $Nu$ number average over the perimeter (related to $Nu_0$) on $Gr_q/Gr_{q\lim}$. As seen from the drawing, the $Nu$ number increases on the lower generatrix and decreases on the upper generatrix with an increase in $Gr_q$. Deviation to either side from $Nu_0$ is as high as 40%. The $Nu$ number, which is averaged over the perimeter, slightly rises with an increase in the $Gr_q$ number changing approximately by 15%.

## 5. CONCLUSIONS

Further progress in the study of turbulent flows and heat transfer at combined forced and free convection requires systematic experimental study of the structure of stratified flows: velocity and temperature fields, turbulent stresses, turbulent heat fluxes, and other correlation functions. In addition to the study of the flow structure, it is necessary to obtain more detailed experimental data on heat transfer and friction in a wide range of Grashof, Reynolds and Prandtl numbers with different orientations of the system in the gravitation field. general methods of calculation of turbulent stratified flows and heat transfer in pipes can be developed from the experimental data and methods used in semi−empirical turbulence theory.

## NOMENCLATURE

| | | |
|---|---|---|
| $T$ | — | temperature |
| $T_b$ | — | bulk temperature |
| $T_w$ | — | wall temperature |
| $P$ | — | pressure |
| $W_i, W_k$ | — | velocity vector components |
| $W$ | — | average velocity of liquid across the pipe section |
| $X_i, X_k$ | — | Cartesian coordinates |
| $\rho$ | — | density |
| $C_p$ | — | specific heat |
| $\lambda$ | — | thermal conductivity coefficient |
| $a$ | — | temperature diffusivity coefficient |
| $\mu, \nu$ | — | dynamic and kinematic viscosity coefficients |
| $\beta$ | — | thermal expansion coefficient |
| $g, g_i$ | — | modulus of gravity force vector and its component |
| $d$ | — | pipe diameter ($d = 2r_0$) |
| $\varphi$ | — | angle read from pipe upper generatix |
| $q_w$ | — | heat flux per unit area on the wall |
| $\alpha$ | — | heat transfer coefficients |
| $Nu = \alpha \cdot d/\lambda$ | — | Nusselt number |
| $Nu_0$ | — | Nusselt number with purely turbulent forced flow, see equation (7) |
| $Re = \overline{w} \cdot d/\nu$ | — | Reynolds number |
| $Pr = \nu/d$ | — | Prandtl number |
| $Gr_q \dfrac{g\beta q_w \cdot d^4}{\nu^2 \cdot \lambda}, Gr_A \dfrac{g\beta d^4 \cdot A}{\nu^2}$ | — | Grashof numbers |
| $Pe = \overline{w} \cdot d/k$ | — | Peclet number |
| $Ra_A = Gr_A \cdot Pr$ | — | Rayleigh number |
| $\xi$ | — | frition factor |
| $\xi_0$ | — | friction factor at pulery turbulent forced flow, see equation (8) |
| $A = dT_b/d_x$ | — | longitudinal temperatture gradient |

# REFERENCES

1. MONIN A.S., YAGLOM A.S., Statistical Hydromechanics, Part I, "Nauka" Publishers, 1965.
2. BER L.E. Bulletin of the USSR Academy of Sciences, No.11, 1957.
3. BER L.E. Bulletin of the USSR Academy of Sciences, No.6, 1962.
4. OJALVO M.S., ANAND D.K., DUNBAR R.P., Journal of Heat Transfer, vol. 89. No.4, 1967.
5. BER L.E. PMTF, Journal of Applied Mechanics and Technical Physics, No.4, 1967.
6. BER L.E. Bulletin of the USSR Academy of Sciences, Fluid and Gas Mechanics, No. 3, 1974.
7. POLYAKOV A.F.Teplofizika vysokikh temperatur (High Temperatures), vol. 11, No.1, 1973.
8. POLYAKOV A.F. In collected paper on "Teploobmen i fizicheskaia gazodinamika" (Heat Transfer and Physical Gas Dynamics", Nauka Publishers, Moscow, 1974.
9. POLYAKOV A.F. Journal of Applied Mechanics and Technical Physics, PMTF, No.5, 1974.
10. BUHR H.O., HORSTEN E.A., CARR A.D. Journal of Heat Transfer, vol. 96, No.2, 1974.
11. CARR A.D., CONNOR M.A., BUHR H.O. Journal of Heat Transfer, vol. 95, No.4, 1973.
12. SARABI A.R. Theses de docteur–ingenier, A L'Universite Paris, 1971.
13. PETUKHOV B.S., STRIGIN B.K. Teplofizika Vysokikh Temperatur, (High Temperatures), vol. 6, No.5, 1968.
14. PETUKHOV B.S., GENIN L.G., KOVALEV S.A., Heat Transfer in Nuclear Power Plants. "Atomizdat" Publishers, 1974.
15. IKRYNNIKOV N.P., PETUKHOV B.S., PROTOPOPOV V.S. Teplofizika Vysokikh Temperatur (High Temperatures) vol. 10, No. 1, 1972.
16. MREIDEN A. Theses de docteur –ingenier, A L;Universite Paris, 1968.
17. PETUKHOV B.S., POLYAKOV A.F., SHEKHTER YU.L., KULESHOV V.A. Report at the International Seminar "Heat Transfer at Turbulent Free Convection", Yugoslavia, 1976.
18. PETUKHOV B.S., POLYAKOV A.F. 4th International Heat Transfer Conference, Versailles, September, 1970.

International Seminar 1977

# ENERGY CONSERVATION IN HEATING, COOLING AND VENTILATING OF BUILDINGS

**C.J. Hoogendorn**

*Delft University of Technology, Delft, The Netherlands.*

World wide concern about the depletion of our energy resurces led in 1975 to the decision by the ICHMT-executive to organize a Symposium on Energy conservation in Dubrovnik in 1977. The energy crises of 1973 initiated many studies on energy use in the industrialized, high energy consuming countries. In retrospect the crisis in 1973, followed by a smaller one in 1980 were political crises. Essentially there was no real shortage in fossil fuel supply. Declining oil prices in the `80`s have shown this. However there is no doubt that the fossil fuel reserves we are using are time – limited. The need for energy conservation is still undisputable. It is interesting to note that since 1973 energy consumption per unit of GNP has been reduced: by 33% in Japan, 25% in the USA and 15% in Europe. This is the result of the energy conservation measures in these countries.

The topic chosen for the 1977 ICHMT Symposium was energy use for heating and cooling buildings: both residential and commercial. This was done because, about one third of primary energy consumption is in that area. The main topics of the Symposium were:
- Heat and moisture transport in building materials.
- Ventilation and air movements.
- Modelling of energy requirements for buildings.
- New techniques.
- Solar energy

A few advances since 1977 will be noted here.

## 1. VENTILATION AND AIR MOVEMENTS

Since 1977, numerical modelling of 3-dimensional turbulent flows has greatly improved. Finite volume element methods combined with $k$-$\varepsilon$ modelling have led to new possibilities. Essentially, one can now calculate air movements and temperature distributions in rooms and other spaces inside buildings. The effect of ventilation, cold windows, a heated ceiling or floor can be simulated using these mathematical models. A difficulty is that experimental validation at high Rayleigh numbers (Ra from $10^{11}$ to $10^{14}$) is still difficult. The prediction of the fully turbulent natural convection layer is possible but experimental data are lacking. In many countries, such as France the USA and Holland, research efforts are concentrated on this, ref. [1] and [2]. The interaction with radiation between wall elements is a complicating factor in the heat transfer tests. For a review of the topic of natural convection in spaces see ref. [3].

## 2. MODELLING OF ENERGY REQUIREMENTS

Much attention has been given to accurately predict the dynamic behaviour of heated (or cooled) rooms. This includes the transient response of walls, and the in – time fluotuating outdoor meteorological conditions; temperature, humidity, solar radiaiton and wind speed. The advance of computer techniques makes it possible to make simplified dynamic calculations on a $P.C.$ For the design of a new large building one uses large models to find the optimum energy building design. Here heating and coolings loads, including the energy for lighting and solar radiation through windows are calculated. The total annual minimum energy requirement is then optimized. Especially in regions with hot summers and relatively cold winters this can lead to a design which saves up to a third of the energy consumption of a normal building.

However in addition computer studies have advanced more than the necessary experimental validation of the modelling of the physical phenomena. A review of this area can be found in ref. [4].

## 3. NEW TECHNIQUES AND INSULATION

Since 1977 many new techniques energy conservation have been introduced. Heat pumps, short term energy storage as well as seasonal storage using aquifers have been developed. Also much attention is given to lighting and solar radiaiton as well as to insulation. Better window design, double or triple glazing and new infrared reflecting coatings have been well advanced. This has led to energy-efficient windows. Lighting by means of light pipes and fibre optics is a new technique that uses solar radiation for lighting without the need of windows [5].

The related development of transparent insulating materials is of great interest [6,7]. They prevent heat losses but still allow outside light to enter reducing lighting energy requirements.

## 4. SOLAR ENERGY

In 1977 there were great expectations for the use of renewable energy sources: the sun and wind. It was realized that the economics were only attractive if a further strong increase of oil prices ensued. The research and development initiated in the USA, Europa and Asia have led to extensive heat transfer knowhow in thermal solar engineering: collectors, thermal storage, heating installations and systems modelling studies. It seems that solar hot water systems are now attractive and are economically attractive in countries with a sunny climates. Active solar heating and colling has reached a mature state, however it is still too expensive for general use.

The development of passive solar energy has been most successful. Essentially the use of glazing on facades to capture winter solar energy: window, Trombe walls or transparent insulating materials. With the right kind of internal storage in the building construction, many architects cooperating with solar engineers were able to introduce passive systems. It has always been combined with good thermal insulation of the envelope of the building.

A good review of the state of the art in solar energy can be found in the Proceedings of the most recent International Solar Energy Society (ISES) Conference in Hamburg, [8].

## REFERENCES

1. KIRKPATRICK, A.T.and BOHN, A.M., An Experimental Investigaiton of Mixed Cavity Convection in the High Rayleigh Number Regime, Int. J. Heat Mass Tr., vol. 29, pp. 69-8, 1986.
2. LANKHORST, A.M.. YGUEL, F., BRETON, J.L., ALLARD, F. and BENARD, C., Natural Convection, in Book of Extended Abstracts of Euromech. Coll. 207 ed. C.J. HOOGENDOORN, publ. Dept. Appl. Physics, Delft Un. Techn.,The Netherlands, 1986.
3. HOOGENDOORN, C.J., Natural Convection in Enclosures, Proc. 8th Int. Heat Tr. Confer., San Francisco, Vol. 1, pp. 111-120, 1986.
4. CLARKE, J.A., Energy Simulation in Building Design, Adam Hilger Ltd., Bristol, Boston, 1985.
5. LEARN, M., A New Economics for Solar Energy: Core Daylighting with heliostat and light pipes, Paper 8.7.17, Proc. ISES World Congress, Hamburg, 1987, Pergamon Press 1988.
6. PLATZER, W.J., Solar Transmission of Transpsarent Insulation Materials, Solar En. Materials, vol. 16, pp. 275-287, 1987.
7. PFLÜGER, A., Minumum Thermal Conductivity of Transparent Insulation Materials, Solar En. Materials, vol. 16, pp. 255-265, 1987.
8. Proceedings *ISES* World Congress, Hamburg, 1987, Pergamon Press, 1988.

# THERMAL PROCESSES OF SOLAR HOUSES

K.-I. Kimura

Department of Architecture, Waseda University Shinhuku,
Tokyo 160, Japan

## 1. INTRODUCTION

The construction of solar houses is expected to grow quite rapidly almost everywhere in the world as many people have come to recognize that the solar heating of building spaces and domestic hot water is very effecient in saving fossil fuels and thus conserves energy for our future generations. The reason for the high effectiveness of solar heating is that solar heat can be utilized at lower temperatures than in other solar applications. Though the total amoung of solar energy falling on the earth is enormous, the net amount of solar energy that can be utilized for various modes of human activity in our modern society is rather limited, if one takes into account the coexistence with plants, animals and other living creatures on this planet. It is reasonable to contemplate the use of the surface area already disrupted by human beings by utilizing roofs and walls of existing buildings and those areas to be inevitably disrupted such as surfaces of buildings to be built in the future, thus preserving meadows, fields, forests and water bodies essential to our daily lives as well as for the maintenance of the global energy balance.

Heat and mass transfer are thermal processess occuring in solar houses throughout from solar energy collection to the end use of energy in the form of space and hot water heating. It is often said that there are two basic types of solar houses: passive solar houses and active solar houses. A passive solar house is a house where space heating or cooling can be achieved imperfectly but satisfactorily without relying on mechanical means, whereas in active solar houses different mechanical components are combined to form a solar heating, cooling and hot water supply system to be installed in the house.

Thermal characteristics of the building structure stongly affect the performance of solar heating and cooling both in passive and active solar houses, but solar hot water can only be supplied through mechanical means. It would be interesting, therefore, to deal with this important problem using current theory and incorporating the state–of–the–art techhnology of heat and mass transfer in a concerted effort by architects, and the scientists and engineers of various fields.

## 2. BASIC SYSTEM OF A SOLAR HOUSE

It is well known that the collection and storage of solar heat are necessary if we are to make effective use of solar energy to save energy in buildings, regardless of the type of system, active or passive. Figure 1 depicts the schematic diagram of the energy flow pattern of a solar house. Heat storage is required to store the collected solar energy for a certain period of time because the period of collection and the period of energy demand usually do not coincide. In order to ensure comfort it is necessary to provide an auxiliary energy source to supplement solar energy when the collected solar energy can not meet the energy demand.

Strictly speaking, there is no clear distinction between passive and active systems. Some say that a house without moving parts may be defined as a passive solar house, but this definition fails to recognize movable panels as the components of a passive system. Since natural environmental conditions differ during the night and day and between summer and winter, no optimal stationary device can be designed from the viewpoint of the efficient use of solar energy.

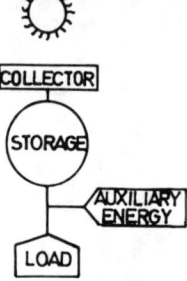

Fig. 1. Energy flow diagram of a solar house

## 3. GLASS WINDOWS AS COLLECTORS

In the northern hemisphere the amount of solar radiation transmitted through glass windows on the south side is enormous on clear winter days and contributes significantly to the heating the interior spaces. People think of it as a kind of a natural blessing, but the south window in itself shold be recognized as one of the most excellent devices for collecting solar energy as well as a means of day lighting and a port for the enjoyment of views.

From the thermal point of view, treating the south window as a collector, one can easily reach the following conclusion: the net heat gain from the window is the transmitted solar radiation minus heat loss across the glass pane in proportion to the indoor-to-outdoor temperature difference as expressed by:

$$q(t) = \tau I(t) - K_G[\theta_r(t) - \theta_a(t)] \qquad (1)$$

where:
- $q(t)$ – net heat gain from glass window unit area at time $t$ [W/m$^2$]
- $\tau$ – transmissivity of glass pane as a function of incident angle of solar radiation
- $I(t)$ – solar radiation incident on the window surface at time $t$ [W/m$^2$]
- $K_G$ – overall heat transfer coeficient of glass window [W/m$^2$]
- $\theta_r(t)$ – room air temperature at time $t$ [°C]
- $\theta_a(t)$ – outdoor air temperature at time $t$ [°C]

With a view to greater precision, the following equation must supercede Eq. (1):

$$q(t)A_G = \tau I(t)A_{GC} - K_G A_G[\theta_r(t) - \theta_a(t)] \qquad (2)$$

where:

$A_G$ – glass window area including framing [m²]
$A_{GC}$ – unshaded glass area [m²]

As $q(t)A_G$ is the instantaneous heat gain from windows, it often happens that no heating is required during the sunny daytime in winter in the milder climatic regions when solar radiation through windows exceeds the heat loss from opaque walls and roofs. It is possible, however, that the daily cumulative value of $q(t)$ will fall below zero, because the first term of Eq. (2) is zero during the night and the second term may be even larger than the daytime value. In this case the glass window cannot be considered a solar collector. It is wise, therefore, to install insulating panels inside or outside the window to minimize heat loss during the night, unless the owner feels that the operation of moving the panels twice a day is troublesome, as shown in
Fig. 2.

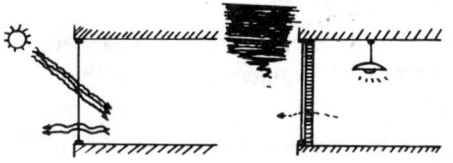

Fig. 2. Window insulated during the night.

The next question is when should the windows be covered with insulation. If artificial lighting must be used as soon as the windows are covered, use of electricity for lighting must lead to the reconsideration of the time to cover the windows. It then follows that the windows may be kept uncovered a little longer depending on the balance of heating and lighting requirements, ignoring the fact that one would like to see the sunset.

Furthermore we must take into account of the effect of room air temperature on the second term of Eq. (2). As a result of solar radiation being transmitted through glass windows, the room air temperature may become higher than the level required for thermal comfort, when the excess amount of heat, which is a part of the collected solar energy, must be lost to the ambient across the glass pane. Use of multiple glazing reduces such losses but cannot make them zero.

## 4. ROOM AIR TEMPERATURE VARIATION DUE TO SOLAR RADIATION

When we calculate the room air temperature variation due to the solar radiation transmitted through glass windows, the thermal storage effect of the building structure must be taken into account. It is convenient to use the weighting factor technique and convert all time-dependent continuous variables into time series variables. This approach is currently widely used in the computerized calculation of heating and cooling loads of a space as recommended by the American Society of

Heating, Refrigerating and Air Conditioning Engineers [1] and that of the Society of Heating, Air Conditioning and Sanitary Engineers of Japan [2] and possibly the procedures of other societies.

The basic equation of room air temperature variation is expressed by the following convolution, taking into accounting for all of the heat flow per unit floor area of the space concerned:

$$H_G(n) = H_L(n) + \sum_{j=0}^{\infty} W_z(j) T_r(n-j) + K_v(n) T_r(n) \qquad (3)$$

where:

$H_G(n)$ – Heat gain from glass windows and plus internal heat generation by lighting and home appliances per unit floor area at time n [W/m$^2$]. The sampling time interval is usually one a standard hour. Time $n$ is an integer representing every hour, $t = n\Delta t$.

$H_L(n)$ – Heat loss from indoors to outdoors through walls and roof and due to ventilation based on the constant reference room air temperature per unit floor area at time n [W/m$^2$].

$T_r(n)$ – Room air temperature deviation from the reference room air temperature at time $n$ [deg]. The reference room air temperature is usually occupant the comfort temperature.

$W_z(j)$ – Weighting factors relating the room air temperature deviation to the heat loss per unit floor area [W/m$^2$deg].

$K_v(n)$ – Variable coefficient of heat loss due to ventilation depending on the variable rate of ventilation per unit floor area at time $n$ [W/m$^2$deg], namely,

$$K_v(n) = C_p \gamma V(n) \qquad (4)$$

where

$V(n)$ – Rate of ventilation per unit floor area at time $n$ [m$^3$/m$^2$s].
$C_p$ – Specific heat of ventilation air [J/kg deg].
$\gamma$ – Specific wieght of ventilation air [kg/m$^2$].

Then the room air temperature deviation from the reference temperature at time n may be derived from Eq. (3) as follows:

$$T_r(n) = \frac{-1}{W_z(0) + K_v(n)} \left[ H_L(n) - H_G(n) + \sum_{j=1}^{\infty} W_z(j) T_r(n-j) \right] \qquad (5)$$

To simplify the calculation procedure, weighting factors $W_z(j)$ may be approximated so as to be represented by the first three terms and the common ratio, viz.,

$$W = W_z(0) + W_z(1) + W_z(2) + c W_z(2) + c^2 W_z(2) + \ldots = W_z(0) + W_z(1) + \frac{1}{1-c} W_z(2) \qquad (6)$$

where:

W – sum of the product of the overall heat transfer coefficient and the area of the room enclosure components per unit floor area [W/m²deg].

$W_z(0)$, $W_z(1)$, $W_z(2)$ – the first three terms of $W_z(j)$ [W/m²deg].

$c$ – common ratio of weighting factors $W_z(j)$ as defined by Eq. (6).

The second term on the right hand side of Eq. (3) represents the increase in heat loss due to the room air temperature deviation from the reference temperature on which the calculation of $H_G(n)$ and $H_L(n)$ are based. The third term represents the increase in the heat loss by ventilation due to the room air temperature deviation at time n.

Then Eq. (5) can be rewritten in simpler form:

$$T_r(n) = \frac{-1}{W_z(0)+K_v(n)} \{[H_L(n)-H_G(n)] - c[H_L(n-1)-H_G(n-1)] + [W_z(1)-cW_z(0)]T_r(n-1) + [W_z(2)-cW_z(1)]T_r(n-2)\} \qquad (7)$$

The author has attempted to obtain the values of $W_z(0)$, $W_z(1)$, $W_z(2)$ and $c$ appropriate to different types of the rooms for easy, approximate calculations and proposes a temporary procedure as follows. Figure 3 shows the chart used to obatain $W_z(0)$ on a per unit floor area basis as a function of the ratio of the total interior surface area $S_s$[m²] to the floor area $S_p$[m²] of the space concerned. For concrete structures:

$W_z(1) = -7$ W/m²deg
$c = 0.9$

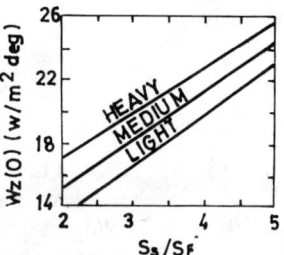

Fig. 3 Chart to obtain $W_z(0)$ [3]

The values of $W_z(2)$ can be obtained from Eq. (6)

For example, when a small house has the following exterior enclosure:
Floor area – $S_F = 8 \times 12 = 96$ m²
Room height – 2.5 m
Total interior surface area – $S_s = 2.5 \times (8+12) \times 2 + 96 \times 2 = 292$ m²
Total glass window area – $A_g = 20$ m²
Total exterior wall area – $A_w = 80$ m²
Total roof area – $A_r = 96$ m²
Overall heat transfer coefficient of glass window – $K_G = 2.5$ W/m² deg
Overall heat transfer coefficient of exterior wall – $K_W = 0.8$ W/m² deg
Overall heat transfer coefficient of roof – $K_R = 0.6$ W/m²deg
the values of the weighting factors $W_z(j)$ and c can be obtained by the following procedure:

$S_S/S_F = 292/96 = 3.04$

$W = (2.5 \times 20 + 0.8 \times 80 + 0.6 \times 96)/96 = 171.6/96 = 1.79$
$W_z(0) = 16.5$ from Fig. 3 for light weight constructions.
$W_z(1) = -7$ and $c = 0.9$ by assumption as in the above.

Substitung the above values into Eq. (6)

$$1.79 = 16.5 - 7 + \frac{1}{1-0.9} \times W_z(2) \text{ and}$$

$W_z(2) = -0.77$ can be obtained.

It may be concluded that according to the theory presented above as long as $T_r(2)$ is positive the glass window acts as solar collector.

## 5. SOLAR RADIATION HEAT STORAGE IN THE BUILDING STRUCTURE

Solar radiation transmitted through glass windows is partially stored in the building structure and is reintroduced into the occupied spaces later on by convection and radiation from the floor and wall surfaces directly irradiated by solar radiation. In this sense $H_G(n)$ in the previous equations must be substituted by the following convolution form:

$$H_{GG}(n) = \sum_{j=0}^{\infty} W_G(n) \, H_G(n-j) \qquad (8)$$

where

$H_{GG}(n)$ — Actual heat input to the space due to solar radiation from windows per unit floor area at time $n$ [W/m$^2$].

$W_G(j)$ — Weighting factors relating the solar heat gain to the actual heat input to the space at reference room air temperature due to solar radiation from windows [dimensionless].

Table 1 shows the values of $W_G(0)$, $W_G(1)$ and common ratio $c_G$ for different structure weight is recommended for approximate calculation [3].

Table 1 Weighting Factors $W_G(j)$ recommended by SHASE [3]

|  | STRUCTURE | $W_G(0)$ | $W_G(1)$ | $c_G$ |
|---|---|---|---|---|
| WITHOUT VENETIAN BLINDS | HEAVY | 0.3774 | 0.0551 | 0.9086 |
|  | MEDIUM | 0.3990 | 0.0629 | 0.8593 |
|  | LIGHT | 0.3728 | 0.0991 | 0.8420 |
| WITH VENETIAN BLINDS | HEAVY | 0.4828 | 0.0473 | 0.9086 |
|  | MEDIUM | 0.4849 | 0.0539 | 0.8953 |
|  | LIGHT | 0.4765 | 0.0827 | 0.8420 |

Using the values in Table 1, Eq. (8) can be rewritten as follows:

$$H_{GG}(n) = c_G H_{GG}(n-1) + W_G(0) H_G(n) + [W_G(1) - c_G W_G(0)] H_G(n-1) \qquad (9)$$

It is certain that the thermal capacity of the building structure helps to spread the peak heat gain by solar radiation from glass windows over a longer period of time, thus reducing the heat loss due to the temperature difference between indoors and outdoors. In most cases, however, the room air temperature tends to become very much higher than required for many hours during the daytime in winter, causing an excessive heat loss through the windows. This means that the thermal capacity of the floor slab and principal building structures is less than is desirable.

Taking this effect into consideration, several investigators have attempted to install a very massive body intercepting and absorbing the excess amount of solar radiation in order to store it for a period of several hours and to use it for heating later on in the evening.

For example, M.I.T. Solar House II [4] had a water tank inside of the glass window with a drape hung between the glass panes and the tank during the night as shown in Fig. 4(a). Odeillo Solar House [5] as shown in Fig. 4(b) has a 60 cm thick wall inside the vertical glass pane with a provision for holes in the upper and lower parts for natural circulation of warm air. The stored heat is naturally dissipated to the interior space toward the evening without any mechanical means. It has been proven, however, that this arrangment loses much of the heat from the outer surface of the wall to the outside from sunset to nighttime. Another disadvantage of this system is that a larger portion of the south facade is occupied by an opaque wall, though this can be overcome by proper design of the house. In order to avoid these two drawbacks, a solar house proposed in Adelaide, Australia [6] provides storage for the solar radiation transmitted directly through glass windows in winter as shown in Fig. 4(c). Theoretically this should have an excellent thermal effect, but practically there may arise such problems as the storage might not be able to absorb solar radiation at different incident angles or furniture might partially intercept it.

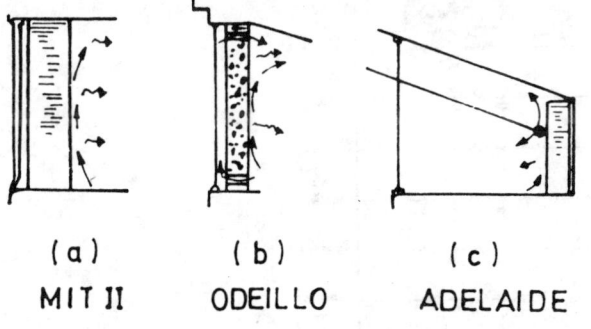

Fig. 4. Typical passive solar houses

(a) MIT II    (b) ODEILLO    (c) ADELAIDE

All of these solar houses may be called passive solar houses, as only natural means are employed to harness solar energy; the water tank of M.I.T. Solar House II, the thick concrete wall of Odeillo Solar House and the interior tank of the proposed Adelaide House, all act as collector, storage and room heater on their own. It is important to note, however, that provision of thermal insulation between the window and the storage is desirable for the efficient use of solar energy.

The Pheonix Solar House [7] designed by Hay and Yellott known as a passive solar house has a water pond on the roof. Thermal insulation panels

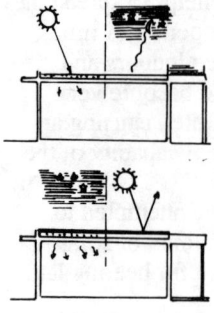

WINTER   SUMMER

cover the roof pond during a winter night and summer daytime and are removed during winter daytime and summer night as shown in Fig. 5. The idea was developed later by Hay and realized as Atascadero Solar House [8], where insulation panelsare moved by motor drive; some say that this is an active system. Nevertheless it is reported that the natural air conditioning is successful.

Fig. 5. Pheonix Solar House.

## 6. THERMAL EVALUATION OF A PASSIVE SOLAR HOUSE

It is quite difficult to assess the effectiveness of passive solar houses based on a common criterion, because the thermal systems are all so very different and in many cases non-linear. The finite difference method may be a straightforward approach to the solution of the problem. The thermal system involves components among which heat conduction, radiation and convection take place. Thermal diffusion and heat transfer by natural convection also take place in the water tank and thus increase complexity if rigorous solution is attempted.

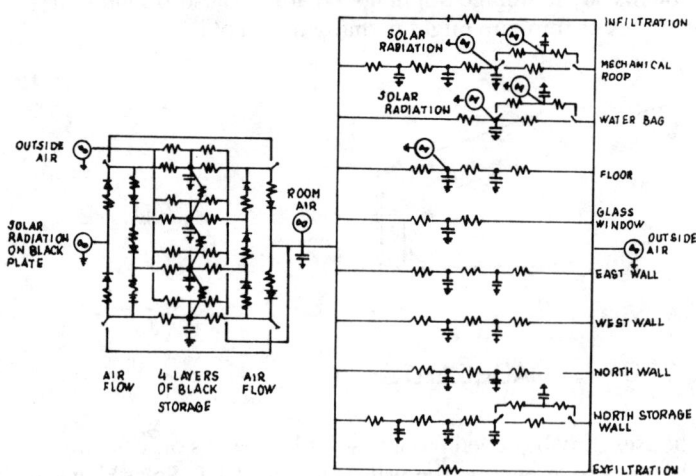

Fig. 6. Thermal network of a semi-passive solar house

The thermal network for the system consisting of all of the building components among which heat transfer and heat accumulation occur, may be modelled by a set of simultaneous equations. Figure 6 shows an example of the thermal network of a passive system as illustrated in Fig. 7 [9]. This may be better called a semi-passive system.

Here the brick storage is treated as four masses each of which has its own thermal capacity and thermal resistance in respect to adjacent masses. There are two kinds of semi-passive solar systems on the roof: water bags with movable insulation as used in the Atascadero House and a mechanical roof consistng of an array of pivoted insulation boards that conduct solar energy between them to be directly

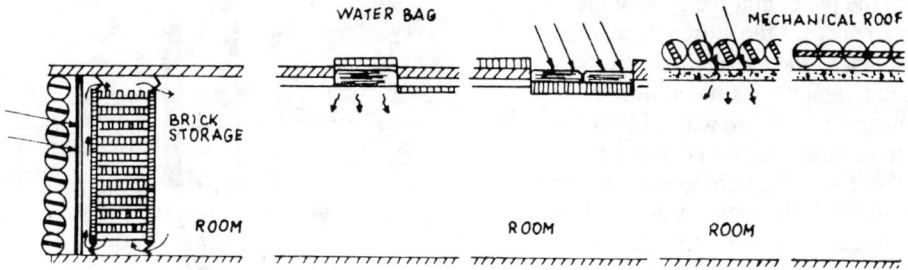

Fig. 7. A semi-passive system proposed for case analysis

stored in the concrete roof slab. There are ordinary glass windows, too. The results of a simulation under actual Tokyo weather conditions for several consecutive days are depicted in Fig. 8, where discharge of stored heat into the occupied space is started at 2 p.m. It is found that the room air temperature can be kept considerably higher (8-10 degrees) than the outdoors air temperature without auxiliary energy. Alternative designs can be tested by modifying the thermal network.

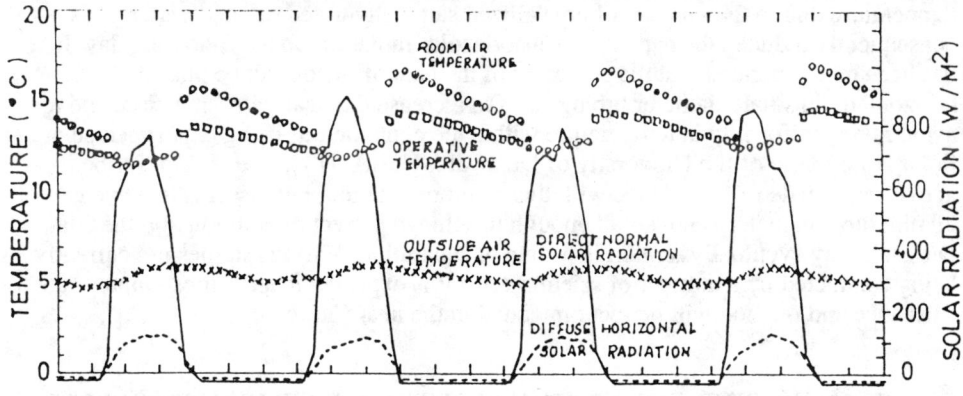

Fig. 8. Resuls of simulation of a semi-passive system

Another approach may be attempted experimentally. The author has made a special curtain wall that acts as collector, storage and heater as illusrated in Fig. 9. Solar radiation through glass is absorbed by black painted glass tubes which contain Glauber's salt ($Na_2SO_4 \cdot 10H_2O$). The crystalized Glauber's salt gradually changes to liquid as it absorbs solar energy while the temperature stays the same, around 32°C. The temperature gradually rises after the entire substance has dissolved and equilibrium is reached. During the sunshine hours the insulation panel is placed inside the curtain wall to store as much solar energy as possible. Before sunset the insulation panel should be placed between the panel and the glass pane so as to prevent the stored heat from being lost to the ambient air. In the evening, heat is dissipated into the room space from the warm tubes and thus solar heating is affected Glauber's salt then changing back to crystals, and the cycle resumes. The use of heat of fusion of Glauber's salt helps to reduce the heat storage volume, which is very important and convenient for the architectural design of solar houses.

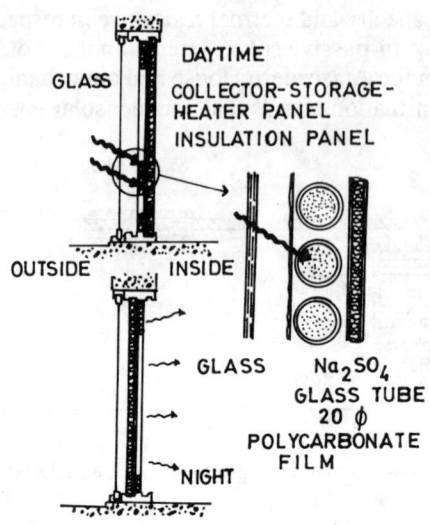

Fig. 9. Collector-storage-heater integrated curtain wall.

Figure 10 shows an example of the experimental results [10] from which the two phase changes described above can be recognized. It was not successful, however, in that the salt seperated after a number of cycles. It was seen that the dissolved salt did not entirely crystallize and the temperature dropped below the phase change temperature due to the mixture of crystallized salt, dehydrated salt and water. This consequently reduced the capacity to absorb solar radiation on the following day. It has been recommended that the container of the heat of fusion salt be placed horizontally in small diameter tubing so as to increase the heat transfer surface and to avoid non – uniformity in temperature within the container, as being experimented at Solar House One of the University of Delaware, U.S.A. [11]. The tests described above were made in accordance with this recommendation, but it seems that even a 20 mm tube diameter is not small enough to achieve perfect dissolution and the Gluber's salt may eventually remain in a state of separation. Various studies are currently being conducted by a number of scientists and it is expected that the most appropriate substance and method will be recommended in the near future.

## 7. AN ACTIVE SOLAR HOUSE WITH A HOT WATER SUPPLY SYSTEM

Hot water is needed throughout the year while space heating is necessary only on cold days for a limited period of the year in many countries. This means that the cost effectiveness is higher in solar hot water supply than space heating. An active solar house may be defined as a house equipped with mechanical components and systems for solar energy collection and storage together with an auxiliary heating system and, in the sense described above, the system includes a solar hot water supply system. Solar cooling is not referred to in this article the discussion concentrating on solar space heating and hot water supply.

There are many different solar heating and hot water supply systems proposed and actually built. None of the active systems built so far seem economically feasible, if the future oil price is taken into account, in terms of the pay back period of the initial cost of solar equipment versus the expected annual energy savings within the expected life of solar equipment. Energy conserving building design makes the pay back period of solar equipment even longer. It should be stressed here that it is more important to use conventional fuels as little as possible than to live on economics in order to maintain our future generations in a clean and safe environment. In this sense it is worthwhile to discuss the performance of economically unjustifiable solar heating systems because it saves energy once it is installed.

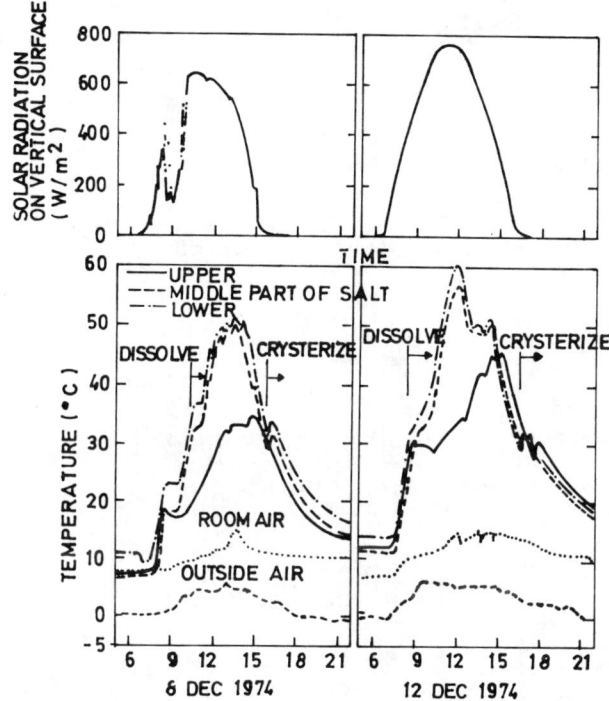

Fig. 10. Experimental results with collector-storage-heater integrated curtain wall

There are two basic kinds of active solar heated houses with regard to the combination of space heating and hot water heating: separate systems and integrated systems. Separate systems are easy to design, to install and operate, as everything can be handled separately, but are considered wasteful since the collectors for space heating cannot be used for hot water supply during the nonheating season when auxiliary heat may be called for on cloudy days. The integrated system, on the other hand, can be designed so that maximum use of solar energy can be achieved for space heating and hot water supply, but integration of the two systems requires a heat exchanger and related control devices which tend to complicate the system and result in unexpectedly higher costs than initially predicted.

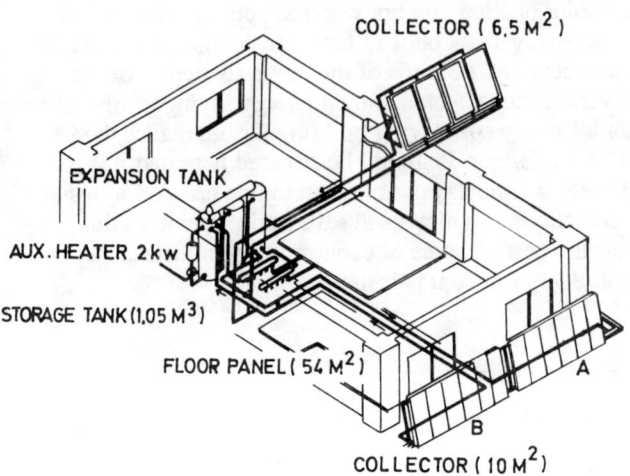

Fig. 11. Direct solar heating system - KEP Solar Experimental House.

(a) Daily variation of energy     (b) Hourly variation of temperature and heat flow

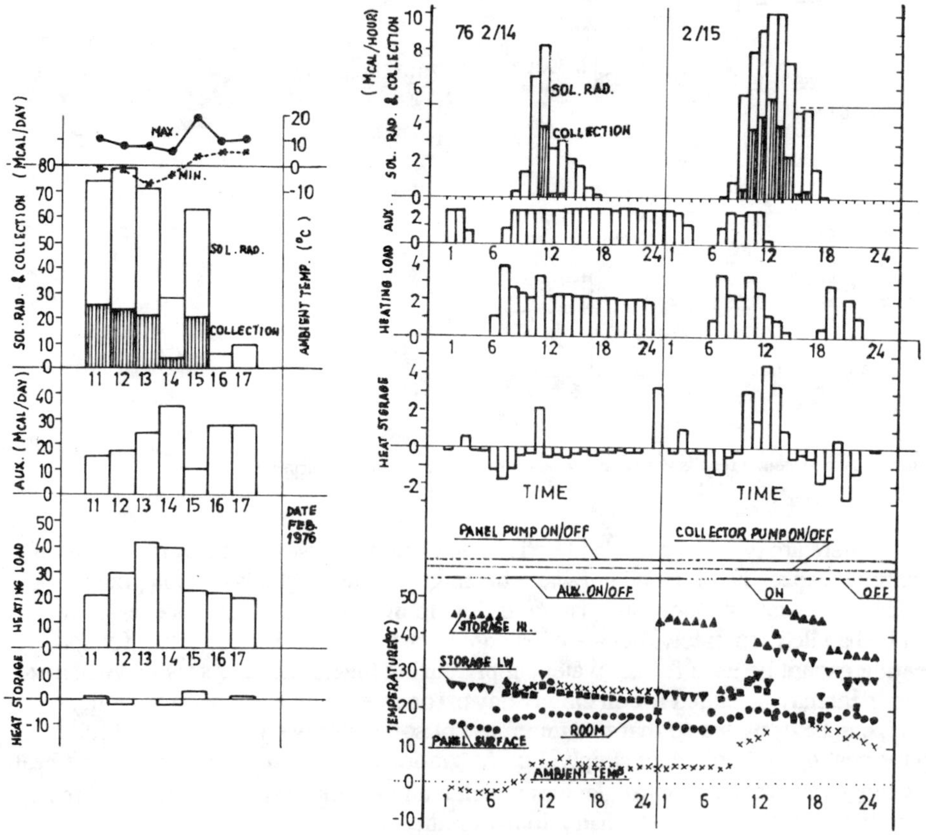

Fig. 12. Experimental results of solar space heating at KEP Solar House.

Space heating systems can be classified into two groups from the viewpoint of thermal processes: direct heating systems and heat pump systems. The direct heating system is simple and has been used in many solar houses such as the M.I.T. Solar House [12] and Löf House in Denver [13]. It is necessary for the system to use a heating media at as low a temperature as possible to avoid the use of auxiliary energy. Panel heating is particularly suitable for this purpose as the liquid temperature can be as low as 35 - 40°C.

Figure 11 is a diagram of the space heating system at the KEP Solar House of the Japan Housing Corporation, where an auxiliary heater has been installed to heat only one of three water tanks connected in series and a floor panel heating system in used. Figure 12 depicts the results of an experiment at the KEP Solar House. It was found that about 50 - 60% of the total heating energy durng the period of one week in February 1976 was solar energy using a 16.5m² collector for a floor area of 89m² under no – occupancy conditions. The operation time of the circulating pump turned out to be much longer than expected becuse of the heat delivery from the storage to the heating panels at such a low temperature [14].

A heat exchanger must be provided somewhere so that solar heat reaches the hot water from the solar tank directly connected to the heating system or to the heating circuit from the hot water tank directly connected to the hot water sevice system. Integration of an auxiliary system is also very difficult as the required hot water temperature is usually higher than that of space heating.

## 8. HEAT PUMP IN SOLAR HEATING

In the case of a heat assisted solar heating system the collector temperature can be lower than for a direct heating system. This increases collector efficiency and the utilizability of solar radiation over the long term higher as lower intensity solar radiation can be made use of. This was the topic of extensive studies made earlier by Liu and Jordan [15]. Figure 13 depicts the approximate linear relationship between collector efficiency and the temperature difference between collector inlet and ambient air over the intensity of solar radiation on the collector surface based on the so-called Hottel-Bliss-Whilier equation that states:

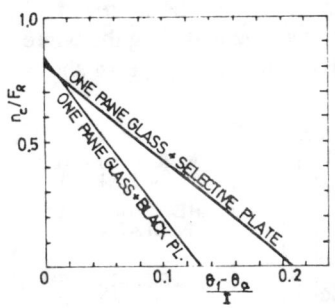

Fig. 13. Collector efficiency of flat-plate collector.

$$\eta_c = \frac{q_c}{I} = \frac{F_R}{I}[\tau\alpha - K\frac{\theta_1 - \theta_a}{I}] \qquad (10)$$

where

- $\eta_c$    – collector efficiency
- $q_c$    – collection per unit collector area [W/m²]
- $I$    – solar radiation on collector surface of unit area [W/m²]
- $F_R$    – heat removal efficiency associated with fin efficiency of collector plate, bond conductance between tube and plate and flow factor taking

account of collector plate temperature distribution from inlet to outlet [16]

$\tau\alpha$ – product of glass covering transmissivity and collector plate absorptivity

$K$ – overall heat transfer coefficient from collector plate to ambient air [W/m²deg]

$\theta_1$ – collector inlet temperature [°C]

$\theta_a$ – ambient air temperature [°C]

As the heat pump is a tool of energy transfer from the lower temperature side to the higher temperature side, the advantage of the heat pump taking heat out of the solar heated water over an outside air source heat pump cannot be evaluated only on the basis of the increase in the value of the coefficient of performance (COP) if heating energy were delivered through the heat pump all the time. We have to take into account of course that the air source heat pump does not work when the outside temperature falls too low, though a type of heat pump working at an air source of –7°C is currently available. It is advantageous, therefore, to design the system so that direct heating can be effected whenever the tank temperature is high enough to be suited for direct heating.

Experience with the authot's solar house [17] shows that the switch-over to pump operation from direct heating operation is not as simple as conceived from the standpoint of manual operation. Moreover, direct heatng may fail perform efficiently because the heat in the tank supplied by an auxiliarty electric heater during the night may end up in the collector the next morning. On account of insufficient collector area ( 24 m² for floor area of 150 m²) the heat exchange from the solar heated water was very weak during the winter as the tank temperature remained as low as 0 - 20°C in the evening. Figure 14 shows the power consumption of Kimura Solar House on a

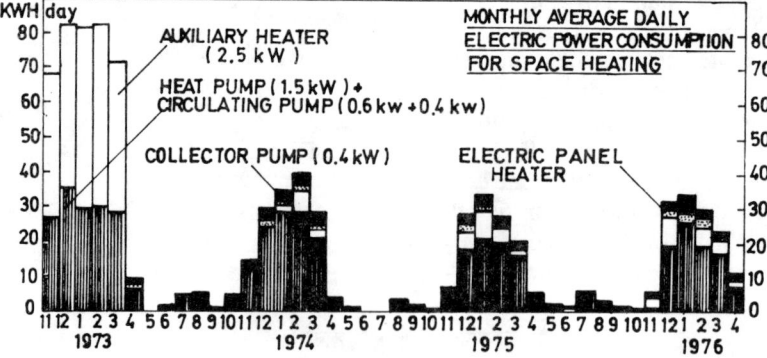

Fig. 14. Monthly average daily electric power consumption of Kimura Solar House.

monthly average basis of daily electric consumption for space heating [18]. Measurements were made without the collector in the first winter and it has been proved that about 60% of the energy for space heating was solar energy.

The obvious advantage of a solar heat pump system has been utilized recently by two Japanese manufacturers: One is a dual purpose heat pump which can pick up heat either from solar water or from outside air depending on the conditions by automatic switch-over and the other system uses the air source heat pump as an auxiliary to recover heat from the outside air and supply it to a rock bed storage. These are reported to be functioning quite well [19], [20]. There are other schemes solar house schemes using the heat pump in Japan on the grounds that it also can serve to cool in the summertime.

It must be stressed here that the advantage of the heat pump does not lie solely in the dual effect of working as a heater in winter and cooling in summer but in the fact that the heat pump is simply a heating apparatus for making effecient use of low temperature energy, because the use of electricity for cooling in the summer daytime must be avoided as much as possible. It is justified to assume, that the heat pump will be more effeciently and widely used in combination with solar collection systems of various designs utilizing low temperature energy in the colder regions for heating purposes.

## 9. LONG-TERM HEAT STORAGE

As long as the collector is installed at a certain tilt angle suited to the most effecient use of solar energy for year – round hot water supply and winter space heating, it naturally follows that there should be an excess amount of solar energy unused during from summer to autumn. In this respect long term heat storage is considered worthwhile to be attempted.

In the past the annual heating load has been much greater than the annual hot water load in most Japanese single family houses and long-term storage has never been justified. In recent years, however, owing to the fact that various energy coservation in centives are making the space heating load less and less, whereas the use of hot water has tended to increase in the mentioned period, the central hot water supply system is becoming popular in many households.

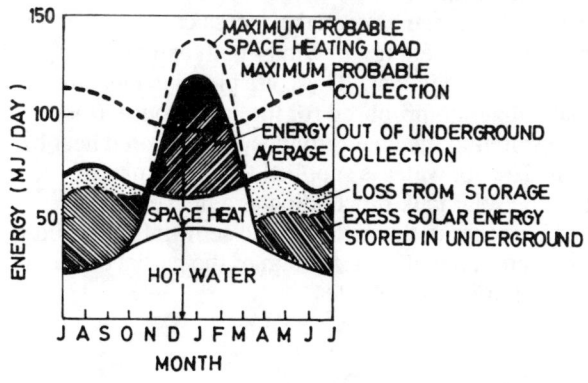

Fig. 15. Annual variation of heating requirements and collected energy for a proposed house with long-term storage.

It is roughly estimated that the annual space heating load of a single family house in Japan could be decreased to 3 MWh in the Tokyo area if proper insulation were used the annual hot water heating load of the average family in Japan being about 3.5 MWh and might increase in the future. With this order of magnitude in the load variation pattern, storage of summer heat underground may be regarded possible.

A water storage tank of 30 - 40 m$^3$ is considered capable of keeping the heat for winter heating in northern Europe and is being experimented with at the Zero Energy House [21] in Denmark and Aachen Solar House [22] in West Germany. Mathew Solar House [23] in Oregon, U.S.A. has an underground water storage tank below which an air void layer is provided so that the heat can be stored in the soil beneath the tank

The use of summer heat for winter heating would of course reduce not only the annual total heating load by auxiliary energy use but also reduce the collector area. Underground heat storage is being tested by Nakajima at his house [24] and at the Building Research Institute [25] experimental house in Japan. The success of long term storage would certainly improv of the economic feasibility of solar space heating.

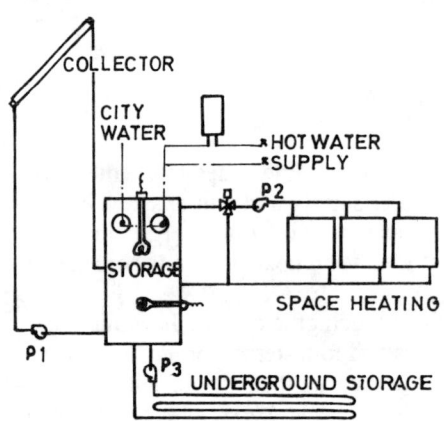

Fig. 16. Schematic diagram of solar space heating and hot-water supply system proposed.

Figure 15 shows the annual variation pattern of the space heating and hot water loads of a single family house versus solar radiation on the collector surface and collection for weather conditions near Tokyo. It can be seen the excess collected energy in summer is comparable with the shortage of energy in winter both expressed in hatched area, allowing for loss from the storage tank. These are predicted features.

Figure 16 shows the schematic diagram proposed for the solar house. It is planned that the excess heat may be stored in the underground and the stored heat be reclaimed if the house calls for heat. The hot water is supplied through a pair of cylindrical small tanks immersed in the solar tank functioning as a heat exchanger. The unsteady state thermal diffusion process in the underground semi-infinite solid is an interesting problem and may be approximated by a solution of the partial differential equation of heat conduction in cylindrical coordinates.

## 10. SUMMARY

The thermal processes in solar houses vary considerably from case to case as seen in the examples presented in this paper. It is important to note that the thermal processes in solar houses must be understood not only in terms of the behaviour of the mechanical sub-systems but also in terms of the nature of heat transfer in building structures. The optimality of solar house designs depends on the logical estimation of the variation of temperature and heat flow at various points so that the environment of the house can be comfortably maintained and the hot water be properly supplied with minimal consumption of auxiliary energy owing to the sound harnessing of solar energy.

## REFERENCES

1. Procedure for Determining Heating and Cooling Loads for Computerized Energy Calculations, ASHRAE Task Group on Energy Requirements for Heating and Cooling 1961.
2. Computerized Calculation Procedures of Dynamic Air Conditioning Load Developed by SHASE of Japan, The Second Sub-committee of the Air Conditioning Standarcds Committee of SHASE of Japan represented by H. SAITO & K. KIMURA, Second jSymposium on the Use of Computers for Envornmental Engineerig Related to Buildings, Paris 1974.
3. SHASE Handbook of Air-Conditioning and Sanitary Engineerig, 1975, p. II-90.
4. DIETZ, A.G.H. & CZAPEK, E. L., Solar Heating of Houses by Vertical South Wall Storage Panels, Paper Presented to 56th Annual Meeting of the American Society of Heating and Ventilating Engineers, Dallas, Texas, Jan. 1950.
5. TROMBE, F., Heating by Solar Radiation, CNRS Report B-1-73-100, 1973.
6. MORSE, R. N., CSIRO, Australia, personal communication.
7. HAY, H.R. & YELLOTT, J.I., Natural Air Conditioning with Roof Ponds and Movable Insulation, ASHRAE Transactions, Vol. 75 Part I, 1968.
8. NILES, P.W.B., Thermal Evaluation of a House Using a Movable-Insulation Heating and Cooling System, paper presented to ISES Congress, Los Angeles, July, 1975.
9. KIMURA, K. & OKYUMA, H., Study on Natural Air Conditioning with Numerical Solution Method of Thermal Network, Transactions of Architectural Institute of Japan, Oct. 1976, p. 333
10. KIMURA, K. , UDAGAWA, M. et al, Experimental Study on Collection-Storage-Heater Type of Curtain Wall Using Latent Heat Storage, paper presented to the Technical Meeting of the Society of Heating, Air-Conditioning and Sanitary Engineers of Japan, Oct. 1975, p.41.
11. BÖER, K.W., A Combined Solar Thermal and Electrical House System, International Congress-The Sun in the Service of Mankind, Paris, July 1973.
12. ENGEBRETSON, C.D., The Use of Solar Energy for Space Heating M.I.T. Solar House IV, U.N. Conference on New Sources of Energy, E/CONF. 35/S/67, Apr. 1961.
13. LÖF, G.O.G., El-WAKIL, M.M. & CHIOU, J.P., Residential Heating with Solar Heated Air - Colorado Solar House -, ASHRAE Journal, Oct. 1963.
14. KIMURA, K. UDAGAWA, M. et al. Operation Results of Space Heating and Hot Water Sypply Systems of Solar Experimental Multi-family Housing, Transactions of A.I.J., Oct. 1976.
15. LIU, B.Y.H. and JORDAN, R.C., A Rational Procedure for Predicting the Long-Term Average Performance of Flat-Plate Solar Energy Collectors, Solar Energy, 7 (2), 1963.
16. WHILLIER, A., Design Factors Influencing Solar Collector Performance, "Low Temperature Engineering Application of Solar Energy" edited by Jordan, R.C., Technical Committee on Solar Energy Utilization of ASHRAE, 1967.
17. KIMURA, K.,Design and Yearround Performance of Kimura Solar House, ISES Los Angeles Meeting, 1975, Extended Abstract 42-4.
18. KIMURA, K., Investigation on the Annual Energy Consumption of All Electric Solar House, Transactions of A.I.J., Oct. 1976.
19. SAKAI, J. et al, Solar Space Heating and Cooling with Bi-Heat Source Heat Pump and Hot Water Supply System, Extended Abstract 45-1, ISES, Los Angeles, 1975.

20. KOIZUMI, H., Kawada, Z. et al, Space Heating Performance of Toshiba Solar House I, Second Technical Meeting, Japan Solar Energy Society, Dec. 1976.
21. ESBENSEN, T.V. & KORSGAARD, V., Dimentioning of the Heat Balance and the Solar Heating System in the Zero Energy House in Denmark, Report Meddelelse NR, 31A, Thermal Insulation Laboratory, Technical University of Denmark DK-2800, Denmark.
22. The Experimental House, brochure issued by Philips, Aachen, West Germany.
23. REYNOLDS, J.S. et al, The Atypical Mathew Solar House at Coos Bay, Oregon, ISES Los Angeles Meeting, 1975.
24. NAKAJIMA, Y. & OHASHI, K., On the Underground Heat Storage with Solar Energy, Second Technical Meeting of Japan Solar Energy Society, Tokyo, Dec. 1976.
25. TSUCHIYA, T. & SETO, H., A Model Experiment on the Utilization of Summer Solar Energy for Heating with the Underground Heat Storage, Paper Presented to Architectural Institute of Japan Meeting, Nagoya, Oct. 1976.

# TWO-PHASE-MOMENTUM, HEAT AND MASS TRANSFER IN CHEMICAL, PROCESS AND ENERGY ENGINEERING SYSTEMS

F. Durst

*Fridrich – Alexander Universität, Erlangen, FR Germany*

From the 4th to the 9th September 1978, a seminar of the International Centre for Heat and Mass Transfer was held in Dubrovnik, concentrating on Two-Phase Flow Momentum, Heat and Mass Transfer in Engineering Systems. The seminar provided a platform for presentations of the results of research work from laboratories all over the world. One hundred fifty participants from 18 countries took part in the meeting to present and listen to presentations on the following subjects:

– Two-Phase Flow Fundamentals
– Two-Phase Flows of Rigid and Deformable Particles
– Non-Equilibrium Phenomena
– Interface Transport in Liquid Films
– Numerical Studies of Two-Phase Flows
– Heat Transfer and Pressure Drop in Power Generator
– Mist Flows, Sprays and Dispersed Bubble Flows
– Two-Phase Flows and Reactor Safety
– Isothermal Two-Phase Flows in Chemical Systems
– Heat and Mass Transfer in Two-Phase Flow Chemical Systems

Selected papers given at the meeting were published in past conference proceedings presenting a lasting contribution to the subject of the symposium. The above given subtitles of the proceedings characterize the grouping of the papers each being introduced by an introductory article by a well known scientist in the field of two-phase flow momentum heat and mass transfer. The introductory and review papers were followed by new contributions to the subject and demonstrate the state-of-the art at the time of the meeting.

At the meeting, presented papers dealt with two-phase flows in chemical reactors. The atmosphere of this strong part of the seminar is captured by one of the review lectures of the meeting, namely

– Heat and Mass Transfer at Liquid Gas Interphase, by F. Mayinger
– Design of Gas-Liquid-Reactors: Mass Transfer Area and Input of Energy, by O. Nagel, H. Kürten and B.Hegner

A further strong part of the meeting concentrated on two-phase flows in nuclear reactors and this part is well dealt with in the review lecture:

– Liquid Mass Transport in Annular Two-Phase Flow, by G.F. Hewitt

There were many good papers at the seminar, among them quite a number that should be presented in this summary of the meeting. However, limited space does not permit this.

The conference was sponsored by the United Nations Educational, Scientific and Cultural Organisation and generously prepared by the members of the Organizing committee.

The personal participation of the members of the Organizing Committee in the meeting stressed its importance and attracted good scientists from all over the world. Participants from Eastern and Western countries were present to introduce new results of their research and to discuss the outcome of the research of other contributors to the meeting. During the entire seminar, lively discussions took place and good ideas for future research were exchanged.

# HEAT AND MASS TRANSFER AT THE LIQUID-GAS INTERPHASE

F. Mayinger

*Technische Universität, Hannover, West Germany*

## 1. INTRODUCTION

Heat and mass transfer in two or multiphase systems is a function of the interfacial area, of the fluid dynamic behaviour in each phase near the phase boundary, of the equilibrium conditions between the phases and in some cases also of interface active effects.

Generally, one can distinguish two flow conditions in two-phase systems, namely:

– Dispersed particles (bubbles or droplets) in a continuous phase

– Two continuous phases (falling film-flow)

The flow conditions - dispersed or continuous - are a function of the forces acting on the phases such as pressure drop, interfacial shear stress, buoyancy and surface tension.

Tehnical apparati for heat and mass transfer in two-phase gas liquid systems, depending on their design features, work in a wide range of possible flow conditions whereby the interfacial area, its rearrangement and the turbulence near the phase boundaries are mostly artificially promoted. Theoretical deliberations and fundamental experiments very often consider a single particle in an infinite fluid and then in practicle application, the difficulty is how to take into account the influence of neighbouring particles and interaction phenomena. Therefore, for the layout of heat and mass exchanging apparati very often pure empirical correlations are used. Especially in complicated fluid dynamic systems where the phases and the interfacial area continuously change, mostly only phenomenological descriptions of the heat and mass transfer processes are available.

## 2. FUNDAMENTAL ASPECTS

There are several models in the literature that describe the mechanism of heat and mass transfer between phases. A detailed discussion of these models is given in the papers by Sideman and Pinczewski [1] as well as by Johannisbauer [2]. Generally the mocels can be divided into three groups:

1. Two film theory.
2. Penetration hypothesis.
3. Turbulent diffusivity assumptions.

The two film model assumes steady-state diffusion in the boundary layers of each phase near the interface. Assuming a boundary layer thickness $\delta$, the mass transfer coefficient $\beta$ in each phase can be correlated with the diffusion coefficient $D$ by the simple equation $\beta = D/\delta$. For most practical cases this model gives mass transfer coefficients that are far too low.

Higbie [3] proposed the penetration model, according to which fluid parts having their origin in the turbulent area of the flow penetrate the thin laminar layer at the phase boundary. During their residence at the interface there is unsteady mass transfer due to molecular diffusion with the other phase. Dankwerts [4] developed a surface rearrangement model from the penetration model abandoning the idea that all fluid elements stay the same time period at the phase boundary but giving a distribution function for the resting time. With Higbie's assumption of constant resting time t the mass transfer coefficient reads:

$$\beta = 2 \cdot \sqrt{\frac{D}{t}} \tag{1}$$

and using Dankwerts theory with the area rearranging factor S which is the ratio of the produced new area per time unit and the total interfacial area one obtains:

$$\beta = \sqrt{D_l \cdot S} \tag{2}$$

The two film model and Higbie's penetration model were combined by Toor and Marchello [5] in the so-called film penetration model.

The models mentioned so far are based on the concept, that the fluid elements behave like stiff bodies and that the heat and mass transfer occurs through rigid phase boundaries which does not correspond to physical reality. Therefore, numerous new models were developed describing turbulence in the neighbourhood of the phase boundary. The models of Lamont [6], Fortescue [7], Ruckenstein [8] or Barnejee [9] consider the inner motion of the fluid elements near the phase boundary and take into account a velocity distribution in the flow. While these models consider the microstructure of the turbulence near the phase boundary there are also other models, working with eddy-diffusivity and describing the macro-structure of the turbulence. They mainly deal with turbulent falling film flow and define turbulent diffusion coefficients for momentum-mass-and heat flux.The correlations for these coefficients are coupled according to Bousinesq [10] via the mean transport velocity. The method based on the eddy – diffusivity is well-known in single phase boundary layer theory. For mass transfer at the phase boundary the turbulent diffusion coefficient $\varepsilon_m$ usually is assumed to be proportional to the square of the distance from the phase boundary.

$$\varepsilon_m \sim y^2 \tag{3}$$

Lamourelle and Sandal [11] basing on the correlation of King [12]

$$\varepsilon_m = b \cdot y^n \qquad (4)$$

found for falling film flow

$$b \sim Re^{1.789} \quad n = 2 \qquad (5)$$

which is only valid for large values of $Re \cdot Sc$. For wavy falling film flow Javdani [13] obtained an expression for $\varepsilon_m$ by solving the momentum equation for wavy film flow. Limberg [14] used a correlation for the turbulent momentum exchange coefficient from the literature and by assuming analogy between momentum-heat- and mass transfer he calculated the mass transfer via the velocity profile. Spalding's [15] theory for the turbulent momentum exchange coefficient is the basis for Iribane's et al. [16] correlation of the mass transfer through a solid phase boundary for short inlet lengths.

All these theories are primarily only valid for regular wave motion in a falling film flow at large values of $Re \cdot Sc$. For irregular wave motion ($12 \leq Re \leq 400$) there is no theoretical solution for the heat and mass transfer which gives satisfactory agreement with measured data. Also in turbulent falling film flow, the theoretical predictions are not too good. Recently Carrubba [17] found that the velocity profile has no influence on the mass transfer and that in agreement with the usual assumption in the literature, the total mass flow density in the immediate neighbourhood of the solid phase boundary is approximately constant and equal to that at the gas liquid phase boundary in wavy falling film flow.

In an empirical manner, the mass transfer is often correlated by

$$Sh = C \cdot Re^m \cdot Sc^n \cdot Ga^\circ \qquad (6)$$

where the dimensionless numbers are defined as follows:

$$Sh = \frac{\beta \cdot \delta}{D} \qquad (7)$$

$$Re = \frac{\overline{w} \cdot \delta}{\nu} \qquad (8)$$

$$Sc = \frac{\nu}{D} \qquad (9)$$

$$Ga = x^3 \cdot g/\nu^2 \qquad (10)$$

Equation (6) is mainly valid for separated flow such as falling film flow. In the Galilei – number Ga, the distance from the flow inlet is denoted by $x$.

Surface active solutes such as tensides can affect the stability of the phase boundary. Recently Palmer [18] demonstrated, that these solutes have a profound stabilizing influence on convection induced by differential vapor recoil but no notable effect on the criteria for instability via the fluid inertial mechanism. Extensive work into interfacial resistance was performed by Nitsch [19], however, with liquid-liquid two-phase systems. Gradients in the surface tension due to concentration – or temperature differences can also produce micro-convection near the phase boundary via the Marongoni – effect as shown by Beer [47] in bubble boiling.

## 3. DISPERSED DROPLET SYSTEMS

The transport phenomena from droplets to a gas flow and vice versa are mainly of interest for spray dryers and spray evaporators, but also for absorption apparati used for example to purify air or other gases. With small freely falling spray droplets the flow conditions around each droplet can be assumed purely laminar which results in a constant value of the Nusselt - or Sherwood-number (Nu = 2.0). The main

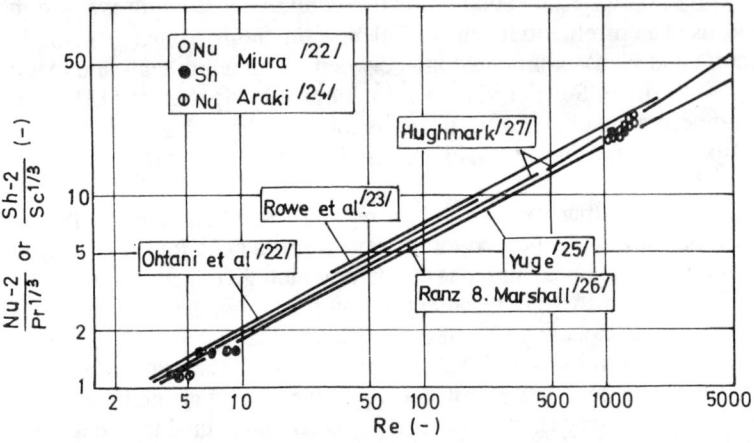

Fig. 1. Relationship between $Re$ and $(Nu-2)/Pr^{1/3}$ or $(Sh-2)Sc^{1/3}$ [21].

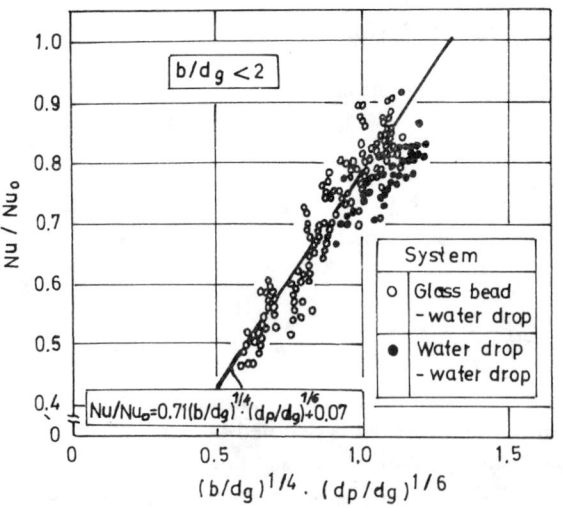

Fig. 2. Relationship between $Nu/Nu_0$ and $(b/d_g)^{1/4} \cdot (d_p/d_g)^{1/6}$ [21].

transport resistance is on the gaseous side and the conduction or diffusion distances in the droplet are small. Mullin and Treleaven [20] considered the effect of neighbouring particles on the mass transfer rate. Recently this effect of drop to drop interaction on the rate of drop-to-fluid heat ant mass transfer was studied by Miura et. al. [21] by measuring the transfer rate from drops. In their measurements for single drops they found good agreement with older experimental and theoretical data [22-27] as shown in Fig. 1. From this figure it also can be seen, that the Nußelt - or the Sherwood – number respectively can be expressed as a function of the Reynolds - and the Prandtl - or Schmidt – number. In a continuation of their research work, the influence of neighbouring droplets was imitated by Miura using glass beads.

The effect on heat transfer depends on the glass bead – to – droplet diameter ratio and on the distance between glass bead and droplet. In Fig. 2 the reduction of the heat transfer due to the presence of other particles - glass beads or water droplets - is shown as a function of the mentioned geometrical conditions. In this figure $Nu_o$ is the Nußelt – number around a single droplet without neighbouring particles.

Miura et al. [21] gave the following correlations for neighbouring effects on the heat or mass transfer from droplets.

$$b/d_g \leq 2 \quad Nu/Nu_o \text{ or } Sh/Sh_o = 0.71(b/d_g)^{1/4}(dp/d_g)^{1/6} + 0.07 \qquad (11)$$

$$b/d_g > 2 \quad Nu/Nu_o \text{ or } Sh/Sh_o = 0.42(b/d_g)^{1/8} + 0.41 \quad Nu/Nu_o \text{ or } Sh/Sh_o \cong 1.0 \qquad (12)$$

$$db/B \leq 2 \quad Nu/Nu_o \text{ or } Sh/Sh_o \cong 1.0 \qquad (13)$$

$$db/B > 4 \quad Nu/Nu_o \text{ or } Sh/Sh_o \cong 0.57$$

For larger droplets the heat and mass transport inside the droplet has to be taken into account also. This transport is agitated by circulation effects as has been known for many years [28]. These circulations are due to the shear stress resulting from the gas velocity around the droplet.

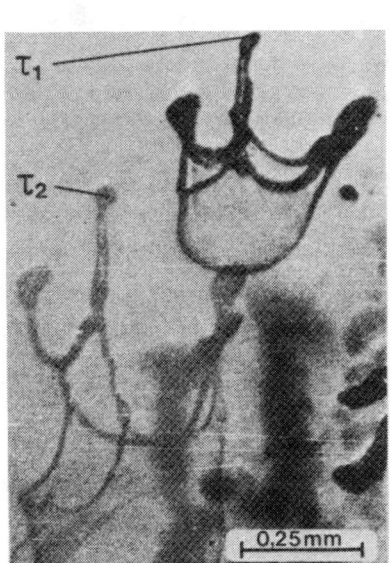

Fig. 3. Liquid membrane in Venturi-nozzles formed as parachute [29].

At very high gas velocity, when liquid dispersion is not performed by high pressure spray nozzles, but by adding the liquid in a continuous jet which is fragmentated by the shear stress of the gas, filigrane lamina-like particles may be formed. This is for example the case in Venturi scrubbers which are used for the absorption of aerosols and the cleaning of air of micro particles. Neumann [29] studied liquid fragmentation in Venturi-nozzles and found extremely high interfacial areas.

Figure 3, obtained by high-speed cinematography, shows an example of such a liquid membrane formed in a Venturi-nozzle. The air-flow fragmenting the liquid had a velocity of 80 m/s. In Fig.3 the liquid membrane was photographed twice with a time difference between exposures of $10^{-5}$ s. From a comparison of exposures one can see that the membrane expands due to the flow forces of the gas. Downward

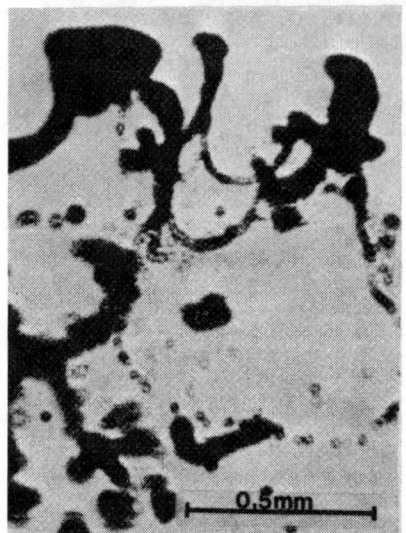

Fig.4. Disruption of parachutes [29].

from the Venturi-nozzle, the velocity difference between the liquid and the gas decreases, which results in a reduction of the gas forces and so a few inches after the membrane formation starts out of the liquid jet and the surface tension can prevail again and reduce the membrane into droplet form. Sometimes, the flow forces are high enough to disrupt the membrane as shown in Fig. 4 which results in the formation of numerous very small droplets.

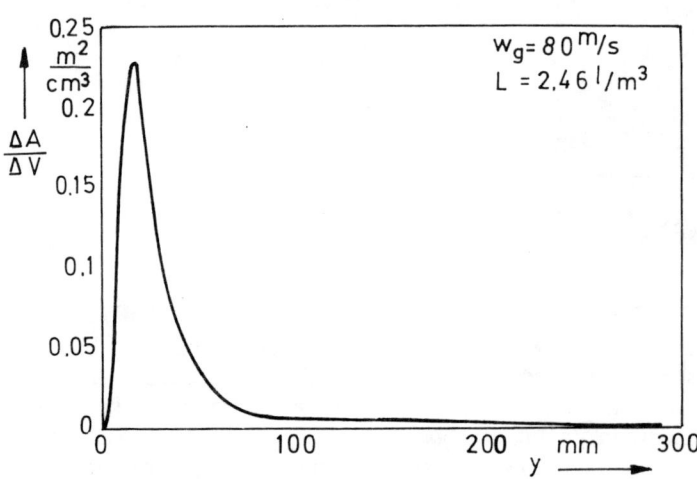

Fig. 5. Interfacial area as a function of flow path [29].

These mechanisms guarentee not only a high area for heat and mass transfer between the phases but also a violent rearrangement of the phase boundaries. The interfacial area is a strong function of the flow history as shown in Fig. 5 and reaches a maximum a few inches behind the point of water injection in the Venture-nozzles. From there it decreases rapidly again due to the deceleration of the gas flow in the expanding part of the nozzle and the decrease in velocity difference between liquid and gas. A theoretical description of this mechanism to predict the mass transfer and the separation efficiency of a Venturi-scrubber seems to be without prospects and therefore Neumann described the product of mass transfer coefficient and superficial interfacial area A by an empirical correlation.

$$\frac{\beta \cdot A}{\dot{V}_g} = 94{,}9 \cdot Re^{0.1} \cdot Eu^{-0.2} \cdot We^{0.24} \cdot \frac{1-\dot{x}}{\dot{x}} \cdot \left(\frac{w_s}{w_g}\right)^{0.2} \quad (14)$$

with

$$We = \frac{d_d \cdot \rho_g \cdot w_g}{\sigma_l}$$

$$Re = \frac{w_g \cdot d_d}{v_g}$$

$$Eu = \frac{\Delta p}{\rho_g \cdot w_g^2}$$

$w_g$ – gas velocity in venture nozzle
$d_p$ – mean particle or aerosol diameter
$\dot{V}_g$ – volumetric gas flow
$d_d$ = mean liquid particle diameter

$$ws = \frac{d_g^2 \cdot \rho_p \cdot g}{18 v_g \cdot \rho_g}$$

$$\dot{x} = \frac{\dot{m}_g}{\dot{m}_g + \dot{m}_l}$$

## 4. BUBBLE SYSTEMS

The momentum and mass transfer through the phase boundary of a bubble with or without a change of shape was discussed in detail by Glaeser and Brauer [30]. Only small bubbles are purely spherical during rising in a liquid. Surface active materials such as tensides can stabilize the spherical form of a bubble. With increasing diameter the bubble tends to oscillate between a spherical and an ellipsoidal shape which results in a circulating flow inside the bubble. The further increase of the diameter of the bubbles finally results in complete deviation from the spherical, they become mushroom or umbrella-like or flat discs.

An attempt to give a theoretical explanation for the effect of tensides on bubble behaviour was made by Levich [31]. According to his theory, surface active molecules are transported to the lee-side of the bubble resulting in a concentration gradient of the tensides over the phase boundary. This produces a gradient in the surface tension, which acts against the fluid motion in the neighbourhood of the phase boundary.

The mass transfer through the interface of a spherical stable bubble can be correlated by the equation [32].

$$Sh = 2 + \frac{0,651(Re \cdot Sc)^{1.72}}{1+(Re \cdot Sc)^{1.22}} \tag{15}$$

A more detailed description of the mass transfer from or into spherical bubbles needs the numerical solution [33] of the mass transfer-differential equation. This then gives also information about the local distribution of the Sherwood – number around the bubble and one finds, that on the front side of the bubble Sherwood – numbers are much higher than on the back.

When the bubble changes its form the mass transfer coefficient is highly dependent on the local conditions around the bubble. A detailed theoretical analysis by Glaeser and Brauer [30] showed a tremendous variation in the Sherwood – number over the perimeter of the bubble as pointed out in Fig. 6. Glaeser and Brauer also considered the influence of turbulent diffusivity and in Fig. 6 results with and without this correction factor are presented. The differences in the local Sherwood – number around the bubble become greater with increasing Reynolds – number as demonstrated in Fig. 7.

Interesting information for a better understanding of the heat transfer from condensing vapour bubbles to a subcooled liquid can be drawn from holographic observations performed by Nordmann [34]. The inclinations of the interference fringes in Fig. 8 can be regarded as temperature gradients in a first rough approximation. For a quantitative evaluation, due to the three-dimensional almost spherical form of the bubbles, the so-called Abel-correction has to be made with which one gets the temperature distribution in the immediate neighbourhood of the phase boundary as also demonstrated in Fig. 8. Asssuming a laminar sublayer on the liquid side of the interface the heat transfer coefficient can be immediately deduced from the temperature profile.

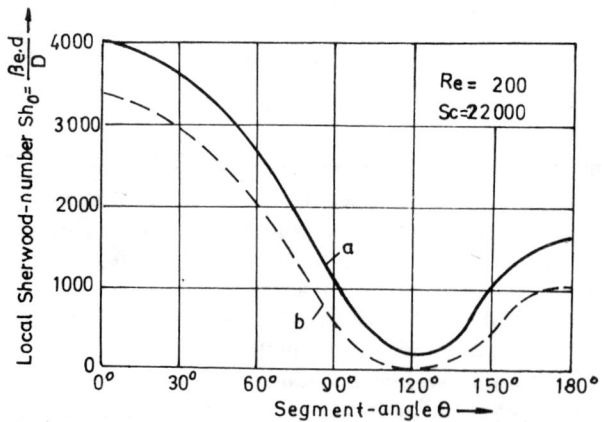

Fig. 6. Local Sherwood – number vs. segment-angle θ with (a) and without (b) turbulent mass transfer [30].

A condensing bubble has a fast moving phase boundary due to its volumetric reduction. This results in high turbulence in the surrounding liquid. Figure 9 allows

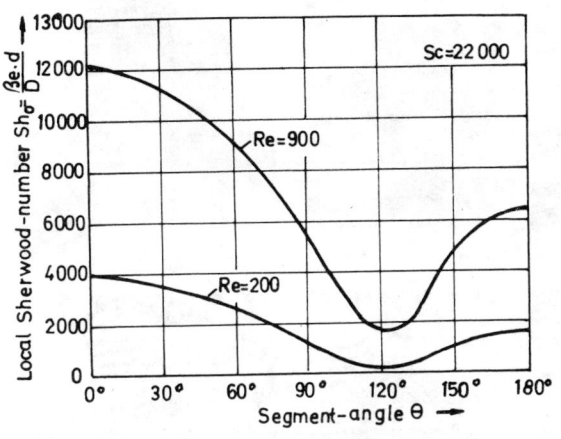

Fig. 7. Local Sherwood-number vs. segment-angle θ influence of Reynolds – number [30].

us to compare the interfacial conditions at slow condensation rates with low subcooling of the liquid and at high condensation velocities due to large heat transfer rates and large subcooling with violent interface movement. A comprehensive theoretical description of the heat and mass transfer from and to bubbles with unsteady phase boundaries, as is the case during recondensation, is not yet available in the literature.

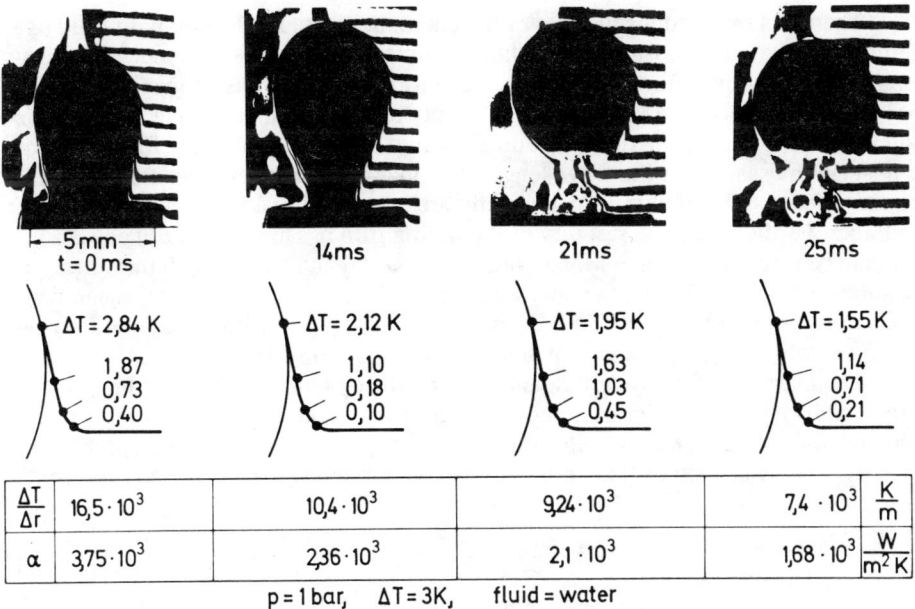

Fig. 8. Temperautre gradients during bubble growth and condensation [34].

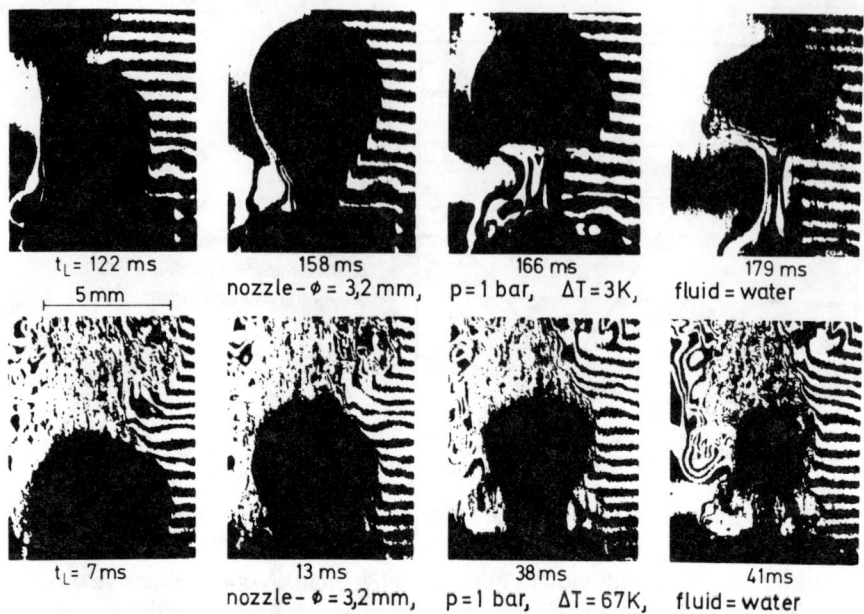

Fig. 9. Growth and condensation of bubbles produced with nozzles influence of subcooling [34].

## 5. LIQUID FILMS

There is a tremendous number of papers in the literature dealing with fluid dynamics, and heat and mass transfer in falling film systems. In most practicle applications the falling liquid film is thick enough to form waves and as shown in Fig. 10 [35] these waves produce vortices which finally result even in a reversed flow in the valleys of the waves. For a detailed fluid dynamic description of wavy film flow and its heat and mass transfer the wave length, the phase velocities and the amplitude of the waves have to be known, which usually are expressed as a function of the Weber – number. Furthermore, the stability of the falling film has to be carefully checked and may be influenced by the Marangoni-effect and by interaction with the surface of the solid wall. The stability of evaporating falling films for example was researched by Coulon [36] and Damman [37]. On both sides of the phase boundary vortices can be formed which influence the heat and mass transfer, too.

The mass transfer through the moving interface of a falling liquid film was measured by a great number of authors. In Fig. 11 [17] the experimental results of Kamei and Oishi [38], Lamourelle and Sandall [39], Hikita [40], Emmert and Pigford [41] and Malewski [42] are compared and equations (16), (17), (18) represent these data.

$$Sh_\infty = 2.24 \cdot 10^{-2} \cdot Re^{0.8} \cdot Sc^{.5} \tag{16}$$

for $12 \leq Re \leq 70$ and $Sc \geq \dfrac{2.32 \cdot 10^4}{Re^{1.6}}$

$$Sh_\infty = 8.0 \cdot 10^{-2} \cdot Re^{0.5} \cdot Sc^{.5}$$

$$\text{for } 70 \leq Re \leq 400 \text{ and } Sc \geq \frac{1.82 \cdot 10^3}{Re} \qquad (17)$$

$$Sh_\infty = 8.9 \cdot 10^{-4} \cdot Re^{1.25} \cdot Sc^{0.5}$$

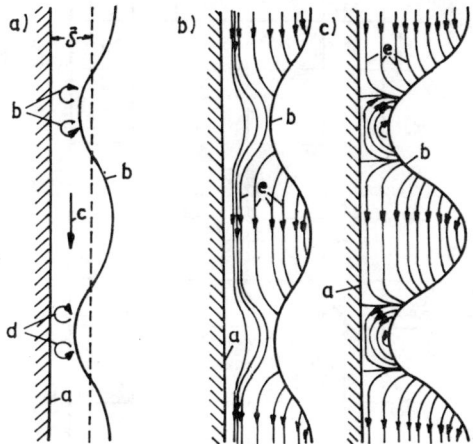

a. wall, b. film surface, c. direction of flow, d. formation of vortex pairs, e. flow line, $\overline{\delta}$ average film thickness

a) turbulence through move motion by S. Portalski
b) and c) flow lines in wavy films by C. Massot with and without return current

Fig. 10. Characteristics of wavy films [35].

$$\text{for } 400 \leq Re \text{ and } Sc \geq \frac{1.47 \cdot 10^7}{Re^{2.5}} \qquad (18)$$

Carrubba [17] developed a theory for predicting heat and mass transfer coefficients in wavy film flow, when the transport resistance is on the liquid side. He used the eddy-diffusivity model and in the diffusion equation added a turbulent diffusivity coefficient $\varepsilon_m$ to the molecular diffusion coefficient. An example of the variation of the reduced turbulent coefficient with respect to the distance from the interphase is shown in Fig. 12. As well-known from single phase boundary layer theory the eddy-diffusivity is a strong function of the Reynolds – number.

Recently, interfacial heat and mass transfer in turbulent flows was also researched by Kolar [43]. He mainly considers the structure of the turbulence near the interface and comes to a general expression for the mass transfer based on the local degree of turbulence and the thickness of the laminar layer.

## 6. TRANSPORT BEHAVIOUR IN PRACTICAL AND INDUSTRIAL EQUIPMENT

Liquid/gas interphase heat transfer phenomena occur in a wide variety of apparati in the chemical industry as for example in bubble columns, packed and fluidized beds, spray columns and jet – mixers. For their optimal layout, a reliable prediction of the heat and mass transfer exchanging area and of the transfer coefficients is needed. Systematic criteria for selecting gas/liquid contact apparati with respect to their interfacial area were elaborated by Nagel et al. [44]. He correlates the interfacial area A per unit of volume V to the dissipated energy E which is due to the pressure drop in the apparatus. The interfacial area is a function of the gas flow rate $\dot{V}_g$ as shown in Fig. 13 for a pebble bed reactor with co-current gas/liquid flow.

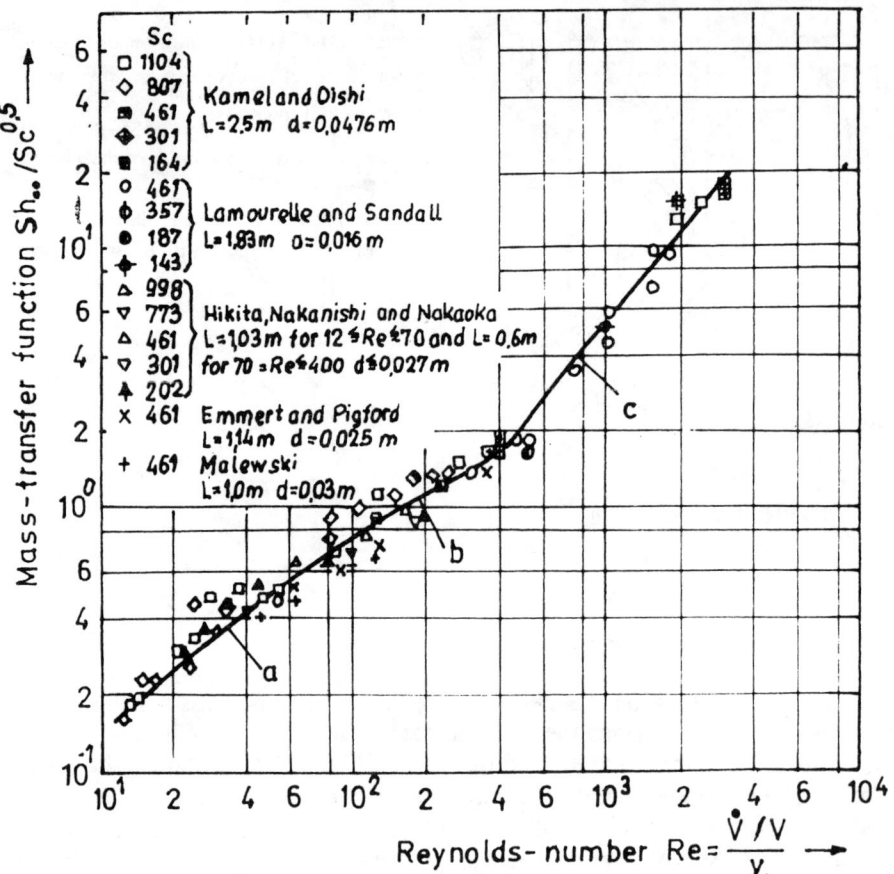

Fig. 11. Mass transfer-function $Sh_\infty / Sc^{0.5}$ vs. Reynolds-number Measurements recalculated by /17/ with equation (16) for a, (17) for b, (18) for c.

the desired heat and mass transfer area can be deduced. As expected high interfacial areas and thus also high mass transfer is linked to high energy consumption by the circulation pump or the blower. The comparison between the Venturi-scrubber and the tubular reactor/jet nozzle is interesting as evidenced in Fig. 14 where it is obvious

This unique relationship between the interfacial area and the energy dissipation, both referred to the unit of volume, is valid for such different heat and mass transfer apparati as jet nozzles, bubble columns, packed beds, Venturi-scrubbers and stirred mixers. From Fig. 14 [45] the demand for a special apparatus design depending on

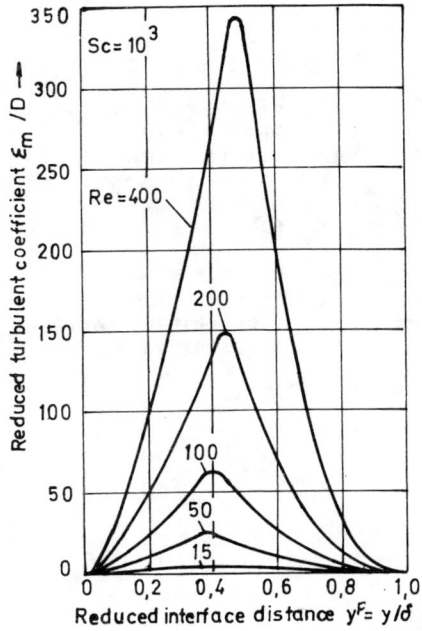

Fig. 12. Reduced turbulence mass-transfer coefficient $\varepsilon_m/D$ vs. reduced coordinate $y' = y/\delta$ [17].

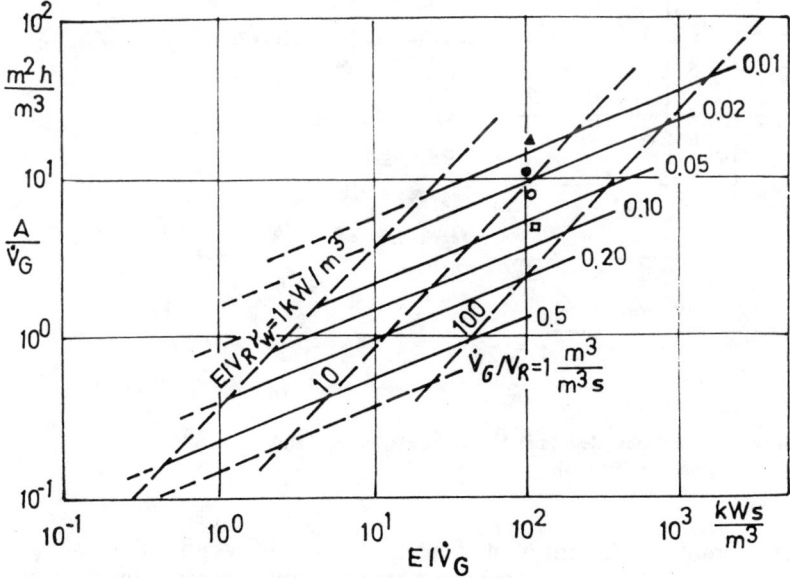

Fig. 13. Reduced interfacial area as a function of the reduced gas-flow rate [44].

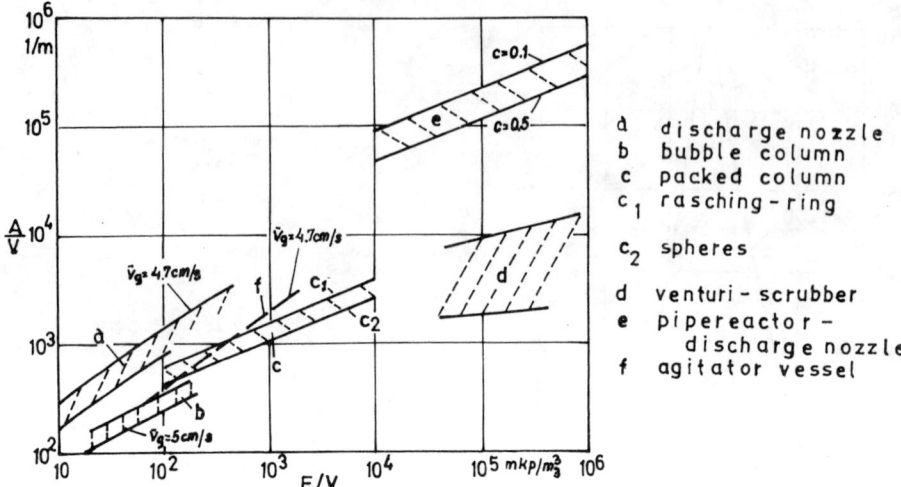

Fig. 14. Comparison of gas/liquid-reactors [45].

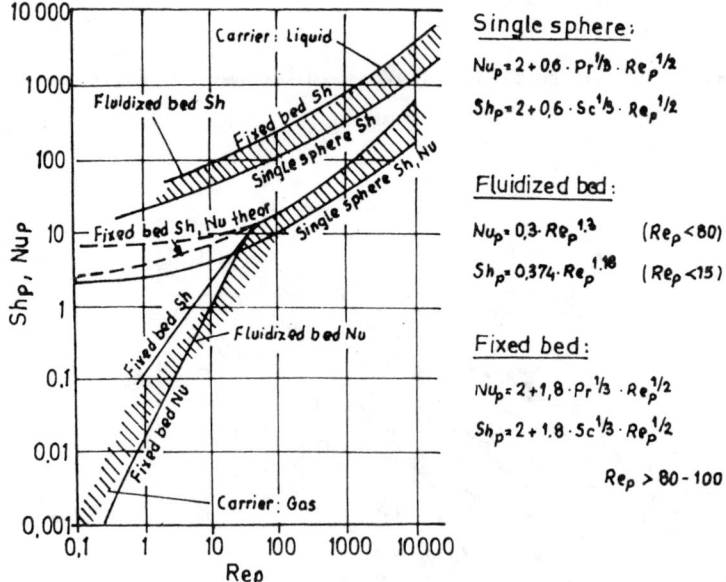

Fig. 15. Deviation of fluid – to – particle heat and mass transfer in fixed and fluidized beds from single sphere data [46].

that the Venturi – scrubber – needing a lot of energy - is not always an optimal design for mass transfer processes. Here, however, one has to distinguish carefully whether there is mass transfer resistance in both phases as is the case in the example in Fig. 14 or whether the mass transfer is only controlled by fluid dynamic behaviour of the gaseous phase as for example with the separation of submicron particles or aerosols from gas flow. Under the later mentioned conditions the high turbulence and the large shear stress in the Venturi-scrubber is beneficial to the separation effect.

A critical review of the trends in research and the industrial application of fluidization including general remarks on the interfacial mass transfer in these apparati was recently presented by Reh [46]. In his comparison of heat and mass transfer behaviour he unfortunately does not consider gas-liquid systems, but only liquid/solid and gas/solid systems. In his comparison he notices - as shown in Fig. 15 - a remarkable drop of fixed bed and fluidized bed transfer data for gas/solid systems at low Reynolds – numbers compared with those of single sphere transfer values.

## 7. CONCLUDING REMARKS

There is such a large number of theoretical and experimental papers dealing with the fundamental questions of heat and mass transfer at gas/liquid interfaces, that it is completely impossible to discuss even a small fraction of them in a representative way within a short review paper. However, faced with the task of making a layout for an industrial plant operating in two-phase conditions, one realizes that there is still a great lack of experimental data and theoretical correlations which are valid under the complex conditions of industrial plants. Therefore, large scale experiments based on the experience of fundamental research results are needed to get a better prediction of two-phase flow interfacial transfer behaviour.

This, however, does not mean, that there is no further need for fundamental research. We still do not have sufficient understanding of the interfacial phenomena even in simple phase geometries for developing theoretical models and correlations of general validity. Optical methods like interferometry or Laser-doppler-anemometry should help to give more detailed information.

### NOMENCLATURE

| | | |
|---|---|---|
| $A$ | test section area | $m^2$ |
| $b$ | width | m |
| $B$ | width in special type by /22/ | m |
| $d$ | diameter | m |
| $D$ | diffusion coefficient | $m^2/s$ |
| $Eu$ | Euler – number | |
| $g$ | acceleration due to gravity | $m/s^2$ |
| $Ga$ | Galilei – number | |
| $m$ | exponent in equaiton (6) | |
| $\dot{m}$ | mass flux | $kg/m^2s$ |
| $n$ | exponent in equation (4) | |
| $Nu$ | Nußelt – number | |
| $o$ | exponent in equation (6) | |
| $\Delta p$ | pressure difference | $N/m^2$ |
| $Pr$ | Prandtl – number | |
| $Re$ | Reynolds – number | |
| $S$ | area rearranging factor def. by /4/ | |

| | | |
|---|---|---|
| $Sc$ | Schmidt – number | |
| $Sh$ | Sherwood – number | |
| $t$ | time | s |
| $V$ | volume | m³ |
| $\dot{V}$ | volumetric flow rate | m³/s |
| $w$ | velocity | m/s |
| $We$ | Weber – number | |
| $x$ | length | m |
| $\dot{x}$ | quality | |
| $y$ | phase boundary distance | m |
| $\beta$ | mass transfer coefficient | m/s |
| $\delta$ | boundary layer thickness | m |
| $\varepsilon$ | turbulent diffusion coefficient | m²/s |
| $\nu$ | kinematic viscosity | m²/s |
| $\rho$ | density | kg/m³ |
| $\tau$ | surface tension | kg/s |

*Subscripts*

| | |
|---|---|
| $g$ | gas |
| $l$ | liquid |
| $p$ | particle or aerosol |
| $d$ | mean liquid particle |
| $o$ | without particle |

## REFERENCES

1. SIDEMAN, S., and PINCZEWSKI, W.V., Turbulent heat and mass transfer at interfaces, Transport models and mechanisms, Technion R. & Found. Ltd. Chem. Engng. Res. Center Haifa, Israel.
2. JOHANNISBAURE, W., Stoffaustauschmodelle für den Rieselfilm, Diplomarbeit, Institut für Verfahrenstechnik, Technische Hochschule Aachen, 1971.
3. HIGBIE, R., The rate of absorption of a pure gas into a still liquid during short period of exposure, Trans. Amer. Inst. Chem. Engng. 31, 1935, 365-389.
4. DANKWERTS, P.V., Significance of liquid-film coefficients in gas absorption, Ind. Engng. Chem. 43, 1951, 6, 1460-1467.
5. TOOR, H.L., and MARCHELLO, J.M., Film-penetration model for mass and heat transfer, AIChE Journ. 4, 1958, 1, 97-101.
6. LAMONT, J.C., and SCOTT, D.S., An eddy cell model of mass transfer into the surface of turbulent liquid, AIChE Journ. 16, 1970, 4, 513-519.
7. FORTESCUE, G.E. and PEARSON, J.R.A., On gas absorption into a turbulent liquid, Chem. Engng. Sci. 22, 1967, 1163-1176.
8. RUCKENSTIN, E., On turbulent mass transfer near a liquid fluid interface, The Chem. Engng. J. 2, 1971, 1-8.
9. BANERJEE, S., SCOTT, D.S., and RHODES, E., Mass transfer to falling wavy liquid films in turbulent flow, Ind. Engng. Chem. Fundamentals 7, 1968, 1, 22-27.
10. BOUSINESQ, J., Theorie de l'ecoulement tourbillant, Mem. Pres. Acad. Sci. XXIII 46, Paris, 1877.

11. LAMOURELLE, A.P., and SANDALL, O.C., Gas absorption into a turbulent liquid, Chem. Engng. Sci. 27, 1972, 1035-1043.
12. KING, C.J., Turbulent liquid phase mass transfer at a free gas-liquid interface, Ind. Engng. Chem. Fundamentals 5, 1966, 1, 1-8.
13. JAVDANI, K., Mass transfer in wavy liquid films, Chem. Engng. Sci. 29, 1974, 61-69.
14. LIMBERG, H., Über die turbulente Strömung in einem Rieselfilm, Ber. Dt. Akad. Wiss. Berlin 12, 1970, 5, 333-341.
15. SPALDING, D.B., A single formula for the law of the wall, J. Appl. Mech. 83, 1961, 455-458.
16. IRIBANE, A., GOSMAN, A.D., and SPALDING, D.B., A theoretical and experimental investigation of diffusion-controled electorlytic mass transfer between a falling liquid film and a wall, Int. J. Heat Mass Transfer 10, 1967, 1661-1667.
17. CARRUBBA, G., Stoff- und Wärmetransport in welligen Rieses-filmen, Diss. Techn. Univ. Berlin, 1976.
18. PALMER, H.J., The effect of surfactants on the stability of fluid interfaces during phase transformation, AIChE J. 23, 1977, 6, 831-839.
19. NITSCH, W., and MATSCHKE, K., Einfluß des Grenzflächenwider standes auf die Stoffübertragung zwischen flüssigen Phasen bei verschiedenen Strömungszuständen, Chem. Ing. Techn. 40, 1968, 13, 625.
20. MULLIN, J.W., and TRELEAVEN, C.R., Proceeding of the Symposium of the Interaction between Fluids and Particles (London), p. 203, 1962.
21. MIURA, K., MIURA, T., and OHTANI, S., Heat and mass transfer to and from droplets, AIChE Sympos. Ser. 73, 1977, 163, 95-102.
22. MIURA, K., OUCHI, I., ATARASHIYA, K., and OHTANI, S., Kagaku Kogaku, 35, 643, 1971.
23. ROWE, P.N., CLAXTON, K.T., and LEWIS, J.B., Trans. Inst. Chem. Engrs. (London) 43, T14, 1965.
24. ARAKI, N., Trans. Japan Soc. Mech. Engrs., 37, 1178, 1971.
25. YUGE, T., Trans. A. Soc. Mech Engrs., Series C, 82, 214, 1960.
26. RANZ, W.E., and MARSHALL, W.R., Jr., Chem. Eng. Prog., 48, 141, 174, 1952.
27. HUGHMARK, G.A., AIChE J. 13, 1219, 1967.
28. BRAUER, H., Stoffaustausch einschließlich chemischer Reaktionen, Verlag Sauerländer, Aarau, 1971, Kap. 14.
29. NEUMAN, M., Ein beitrag zum Abscheidemechanismus im Ventureiwäscher, Diss. Techn. Univ. Hannover, 1978.
30. GLAESER, H., and BRAUER, H., Berechnung des Impuls- und Stoff-transports durch die Grenzfläche einer formveränderlichen Blase, VDI-Forschungsheft Nr. 581, 1977.,
31. LEVICH, V.G., Physicochemical hydrodynamics, Englewood Cliffs, N.J.: Prentice Hall, 1962.
32. SCHMIDT-TRAUB, H., Impuls- und Stoffaustausch an umströmten Kugeln unter Berücksichtigung der inneren Zirkulation und der freien Konvektion, Diss. TU Berlin, 1970.
33. OELLRICH, L., SCHMIDT-TRAUB, H., and BRAUER, H., Theoretische Berechnung des Stofftransports in der Umgebung einer Einzelblase, Chem. Engng. Sci. Bd. 28, 1973. 3, 711-21.
34. MAYINGER, F., and NORDMANN, D., Temperauture, Pressure and Heat Transfer near Condensing Bubbles, Paper presented at ICHMT-Meeting, 1978, Dubrovnik.
35. STRUVE, H., Der Wärmeübergang an einem verdampfenden Rieselfilm, VDI-Forschungsheft 534, 1969.
36. COULON, H., Wärmeüberg und Stabilitätsverhältnisse bei Rieselfilmen, Diss. TU Braunschweig, Braunschweig 1971.
37. DAMMANN, J., Wärmeübergang und Stabilitätsverhältnisse bie Rieselfilmen aus wasserigen Losungen, Diss. TU Braunschweig, 1973.

38. KAMEI, S., and OISHI, I., Mass and heat transfer in a falling film of wetted wall tower, Mem. Fac. Engng. Kyoto Univ. 17, 1955, 227-239.
39. LAMOURELLE, A.P., and SANDALL, D.C., Gas absorption into a turbulent liquid, Chem Engng. Sci. 27, 1972, 1034-1043.
40. HIKITA, H., NAKANISHI, K., and NAKAOKA, T., Liquid phase mass transfer in wetted wall columns, Chem. Engng. (Japan) 23, 1959, 459-466.
41. EMMERT, R.E. and PIGFORD, R.L., A study of gas absorption in falling liquid films, Chem. Engng. Prog. 50, 1954, 2, 87-93.
42. MALEWSKI, W., Zusammenhang zwischen Stoffübergang un Wellenstruktur beim willigen Rieselfilm, Chem.-Ing.-Techn. 37, 1965, 8, 815-825.
43. KOLAR, V., Interfacial heat and mass transfer under the turbulent motion of fluids, Coll. Czech. Chem. Com. 42, 1977, 4, 1310-1324.
44. NAGEL, O., KÜRTEN, H., and HEGENER, B., Die Stofaustauschfläche in Gas/Flüssigkeits-Kontaktapparaten - Auswahlkriterien und Unterlagen zur Vergrößerung, Chem.-Ing.-Techn. 45, 1973, 14, 913-920.
45. NAGEL, O., KÜRTEN, H., and SINN, R., Stoffaustauschfläche un Energiedissipationsdichte als Auswahlkriterien für Gas/Flüssigkeits-Reaktoren, Chem.-Ing.-Techn. 44, 1972, 14, 899-903.
46. REH, L., Trends in Research and Industrial Application of Fluidization, Part 1: Research, Verfahrenstechn. 11, 1977, 6, 381-384.
47. BEER, H., RONNENBERG. M., Einige Besonderheiten technischer Siedevorgänge, Verfahrenstechnik 11, 1977, 10, 614-619.

International Seminar 1979

# HEAT AND MASS TRANSFER IN METALLURGICAL SYSTEMS

D.B. Spalding

*Imperial College, London, U. K.*

The materials men work with have characterised their societies since the beginning of pre-history. Historians indeed speak of the Stone, Bronze and Iron Ages; and to call our present times the Plastic Age is to make a significant, if exaggerated, comment. The comment is too extreme because metals are still among the most important of our structural materials, whether embedded within concrete for the construction of buildings and bridges, or thinly covered by paint when used for automobiles and ships, or accesible to the eye in the fuselage of an aircraft or in a domestic cooking utensil.

It is therefore not surprising that, in 1980, the International Centre for Heat and Mass Transfer devoted a symposium to the study of the role of its science in the manufacture, processing and use of metals.

Typically, metals are found in nature only in chemical combination with oxygen, chlorine or some other element. In order to separate them, heat must be applied, as in the blast furnace which separates iron from oxygen; or electrical power must be supplied, as when aluminium is electrolysed from its molten oxide, or sodium from its chloride. In all cases, transfer of mass must occur, for example in order that the oxygen can forsake its metallic partner in order to combine with carbon. Often the heat and mass transfer processes are complicated by the presence of phase change, as when bubbles of gas are formed at an electrode immersed in molten ore.

Once converted to the metallic state, metals are repeatedly subjected to further thermal and mass-exchange processes. Molten iron is poured into a Bessemer Converter, through which oxygen gas is blown, in order to effect the reduction of carbon content that turns it into steel; and the mechanical properties of the steel can be significantly improved by carefully timed heating and cooling processes such as "quenching" and "tempering".

The finished metal products must subsequently often resist exposure to extremely high and extremely low temperatures, usually in rapid succession; and the magnitudes of stresses which sometimes cause failure of the material depend on the manner in which heat is transferred internally.

The proceedings of the 1980 Symposium reflect the above -mentioned relationships between the science of heat and mass transfer and technology of metallurgical processing. The blast furnace was the subject of 7 papers, and other iron and steel processes the subject of 6 more; then 5 papers were directly related to processes for winning non-ferrous metals from their ores. Further contributions concerned such more specialised topics as mathematical modelling, crystallization and heat and diffusion treatment. Lastly, and not inappropriately, a session was devoted to that process by which metals, having served their purpose, revert to something like their original chemically-bound state. The process is corrosion, a mass-transfer phenomenon of which the prevention or control could save mankind much expenditure.

To select a paper of especial excellence from so many valuable ones is not easy. My choice falls on that of Professor Julian Szekely, entitled "Transport processes in agitated ladles: problems, solutions and experimental techiques." Its author has made it his special concern to apply systematically to metallurgical processes the techniques and insights which have been regarded with much success; and he can truly be regarded as a significant benefactor to the metallurgical industry, as well as to research workers in heat and mass transfer.

# TRANSPORT PROCESSES IN AGITATED LADLES: PROBLEMS, SOLUTIONS, AND EXPERIMENTAL TECHNIQUES

J. Szekely

Massachusetts Institute of Technology
Cambridge, USA

## 1. INTRODUCTION

In recent years there has been a growing interest in the "ladle refining process" as a means of obtaining molten steel of given temperature and composition [1,2].

For the present purpose the term "ladle refining" will be broadly interpreted, to include all the processing operations in the molten state subsequent to the actual steelmaking operation, such as argon stirred ladles, induction stirred vessels, vacuum degassing, injection of powders and the like.

Two main factors have been responsible for this marked growth of ladle metallurgy. One of these has been the need for steel of higher quality, meeting more stringent specifications regarding composition, which would have been difficult to attain in conventional steelmaking furnaces.

The second has been the need for higher productivity, particularly relevant to electric furnace operations, which motivated operators to restrict the function of the electric furnace to the actual melting (and possibly the carbon oxidation) and carry out the subsequent refining and deoxidation steps in another processing unit.

Table 1. Listing of Ladle Metallurgical Operations by Function

| Function | Argon Stirring | Vacuum Degassing | Induction Stirring | Powder Injection |
|---|---|---|---|---|
| Bath homogenization | XX | X | X | |
| Alloy additions | X | | XX | XX |
| Deoxidation | X | X | X | X |
| Degassing | X | XXX | X | |
| Removal of inclusions | X | XX | XX | X |
| Desulfurization | X | X | | XX |

The main functions of ladle metallurgy operations have been usefully summarized by Ohman and Lehner [3] and these include:
- bath homogenization, both regarding temperature and composition
- alloy additions
- degassing
- removal of inclusions
- desulfurization

The principal systems that may be used to perform these functions are given as:
- argon stirring
- vacuum degassing
- induction stirring
- powder injection using inert gas carriers

While in principal all these systems may be used to perform the above mentioned functions, one may construct a matrix, indicating the preferred techniques for certain applications, as shown in Table 1. A great deal of experience is being accumulated regarding the application of these techniques, from both the fundamental and the practical viewpoints.

The purpose of the present paper is to review briefly the techniques that are available for studying the heat transfer, mass transfer and fluid flow phenomena in these systems and then summarize our current state of knowledge. The paper will then be concluded with a discussion of the areas of uncertainty that still remain and where further research needs to be done.

## 2. EXPERIMENTAL AND MATHEMATICAL TECHNIQUES AVAILABLE FOR STUDYING RATE PHENOMENA IN LADLE METALLURGICAL OPERATIONS

The principal systems to be discussed are sketched in Fig. 1. It is seen that these would include:

(a) induction stirring
(b) agitation by injected gas streams at atmospheric pressure
(c) steady state or pulsed vacuum degassing systems
(d) the injection of powders

Ideally in describing these systems we would wish to have information on the overall performance characteristics, such as the time required for homogenization, the desulfurization rate, the deoxidation rate and the like for a given set of circumstances. However, in addition, it would be desirable to have information on the detailed behavior of the systems, such as the velocity fields, turbulence levels, concentration profiles, local mass transfer rates, local rate of agglomeration of the inclusion particles, and so on.

This latter information would, of course, be very valuable in identifying the actual mechanism of the process and thus in developing process improvements, or a wider range of applications.

The techniques that are available for studying rate phenomena in ladle metallurgical operations are conveniently divided into:

2.1 Experimental

## 2.2 Mathematical

while recognizing the need for the close connection between these two usually complementary approaches.

## 2.1 EXPERIMENTAL TOOLS

### 2.1.1 Models

Because of the difficulties inherent in taking velocity, turbulence, etc., measurements in molten steel, frequently recourse is made to the use of physical modelling. A very good description of physical modelling techniques is available in a

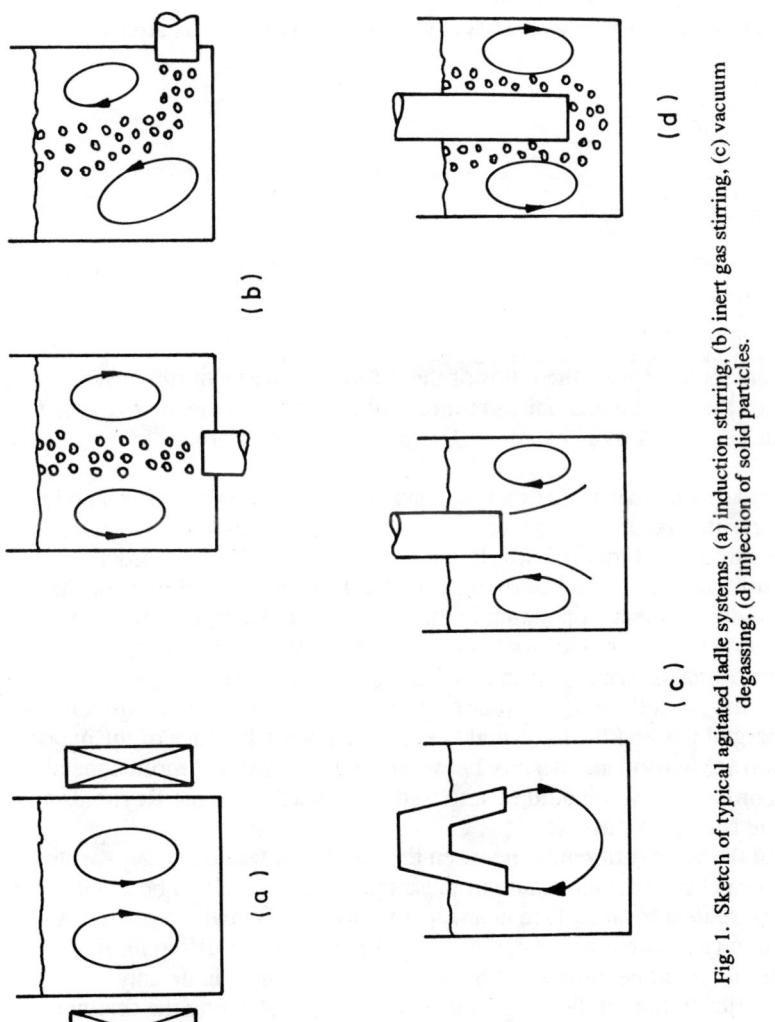

Fig. 1. Sketch of typical agitated ladle systems. (a) induction stirring, (b) inert gas stirring, (c) vacuum degassing, (d) injection of solid particles.

paper by Hlinka, [4] and also in several texts. [5,6]. In physical modelling work steel tends to be replaced by water or by other fluids, which are much easier to handle.

However, in order that the physical model does indeed correspond to the actual system it seeks to represent, certain quite strict rules have to be observed, namely the law of geometric similarity and the law of dynamic similarity.

*Geometric Similarity* is the similarity of shape. Systems are geometrically similar when the ratio of any length in one system to the corresponding length in the other system is everywhere the same. This ratio is usually termed the scale factor. While geometric similarity is one of the most obvious requirements in modelling, often it may not be possible to attain perfect geometric similarity. Examples include the modelling of lakes, rivers, particulate systems and the like.

*Dynamic Similarity* represents the similarity of forces. Dynamic similarity is observed between two systems, when the magnitude of forces at corresponding locations in each system are in a fixed ratio.

The principal forces that one encounters in metallurgical systems are listed in the following:

inertial forces

pressure forces

viscous forces

gravity (also buoyancy)

surface tension

elastic forces

electromagnetic forces

It is convenient to express the ratios of these forces in terms of dimensionless numbers; some of the key dimensionless numbers of relevance to the present discussion are listed in Table 2. A more extensive list of dimensionless groups is available in references [5,7].

It is stressed that the behavior of a one-phase system is quite readily modelled; typical examples being the natural convection driven flow of steel held in a ladle, the pouring of steel through submerged nozzles and the like [4,5]. Here, in order to satisfy dynamic similarity, one would have to match the Reynolds, the Froude or the Grashof and Prandtl numbers, depending on the application. However, even under these conditions one may encounter some problems, because it would be quite difficult to model electromagnetically driven flows in sequency systems.

In contrast the modelling of two phase systems, such as gas bubble driven circulations, submerged jets and the like would be quite difficult, because of the major differences in surface tension and density between aqueous - gas and molten metal - gas systems. In order words, it would be impossible to match both the Reynolds and the Weber, or the Morton numbers.

Because of the great differences between the interfacial tension of gas - water and gas - molten steel systems the nature of these interfaces is likely to be different. It is much easier to create a fresh surface in an aqueous system than in a molten metal, thus the nature of any dispersions created is also likely to be quite different. Additional complications could be introduced by increased differences in density.

A dramatic illustration of the dangers involved in extraplotating the findings from two phase aqueous systems to those involving gases and molten metals is

Table 2. Some key dimensionless numbers

| | | | | |
|---|---|---|---|---|
| Froude number $U_{Fr}$ | $\dfrac{u^2}{gL}$ | $L$ = Charestic length dimension of system | $\dfrac{\text{inertial force}}{\text{gravitational force}}$ | wave and surface behavior pouring streams |
| Froude number, modified, $N'_{Fr}$ | $\dfrac{\rho_g u^2}{(\rho_l-\rho_g)gL}$ | $\rho_g$ = gas density $\rho_l$ = liquid density | $\dfrac{\text{inertial force}}{\text{gravitational force}}$ | fluid behavior of gas liquid systems |
| Galileo number $N_{Ga}$ | $\dfrac{gP^2L^3}{\mu^2}$ | $\mu$ = viscosity | $\dfrac{(\text{inertial})\times(\text{gravity})}{(\text{viscous force})}$ | flow in baths of viscous liquids |
| Mach number $N_{Ma}$ | $\dfrac{u}{u_s}$ | $u$ = fluid velocity $u_s$ = velocity of sound in fluid | | high speed flow |
| Morton number $N_{Mo}$ | $\dfrac{g\mu_L^4}{\rho_L(\sigma)^3}$ | $\mu_l$ = liquid viscosity | $\dfrac{(\text{gravity})\times(\text{viscous})}{\text{surface tension force}}$ | velocity of bubbles in liquids |
| Power number $N_P$ | $\dfrac{P'}{\rho n^3 L^5}$ | $L$ = sharacteristic dimensionr of agitator paddle $n$ = angular sped of rotation $L$ = characteristic length dimension of system | $\dfrac{\text{drag force on paddle}}{\text{inertial force}}$ | power consumption in agitated vessels |
| Reynolds number $N_{Re}$ | $\dfrac{Lu\rho}{\mu}$ | $\rho$ = fluid density $\mu$ = fluid viscosity | $\dfrac{\text{inertial force}}{\text{viscous force}}$ | fluid flow |
| Weber number $N_{We}$ | $\dfrac{\rho Lu^2}{c}$ | $\sigma$ = surface tension $L$ = characeristic length dimension $\rho$ = density of fluid | | |

| | | | | |
|---|---|---|---|---|
| Grashov number $N_{Gr}$ | $\dfrac{g\rho^2 L\beta \Delta T}{\mu^2}$ | $\beta$ = coefficient of thermal volume expansion $\beta = -\dfrac{1}{\rho}\left(\dfrac{\partial \rho}{\partial T}\right)_p$ $\Delta T$ = temperature difference | $(N_{Re})\left(\dfrac{\text{buoyancy force}}{\text{vicous force}}\right)$ $=(N_{Ga})\beta \Delta T$ | free convection |
| Péclet number $N_{Pi}$ | $\dfrac{L u \rho C_p}{k} = \dfrac{L u}{\alpha}$ | $u$ = fluid velocity $C_p$ = specific head $\alpha$ = thermal diffusivity | $\dfrac{\text{heat transver by bulk motin}}{\text{conductive heat transver}}$ $=N_{Re}N_{Pr}$ | forced covection |
| Prandtl number $N_{Pr}$ | | $C_p$ = specific heat of fluid $\mu$ = viscosity $k$ = thermal conductivity | $\dfrac{\text{momentum diffusivity}}{\text{thermal diffusivity}}$ | forced and free convection |
| Rayleigh number | $\dfrac{\rho^2 L^3 C_p \beta \Delta T}{k}$ | $\beta$ = coefficient of thermal $\Delta T$ = temperature difference aerous film $L$ = characteristic length dimension | $\dfrac{\text{heat transfer by convection}}{\text{heat transfer by conduction}}$ | free convection |
| Hartman number | $LB\sqrt{\dfrac{\sigma_e}{\mu}}$ | $L$ = characteristic length $B$ = characteristic magnetic flux density $e$ = electric conductivity | $\dfrac{(\text{electromagnetic force})}{\text{viscous force}}$ | |

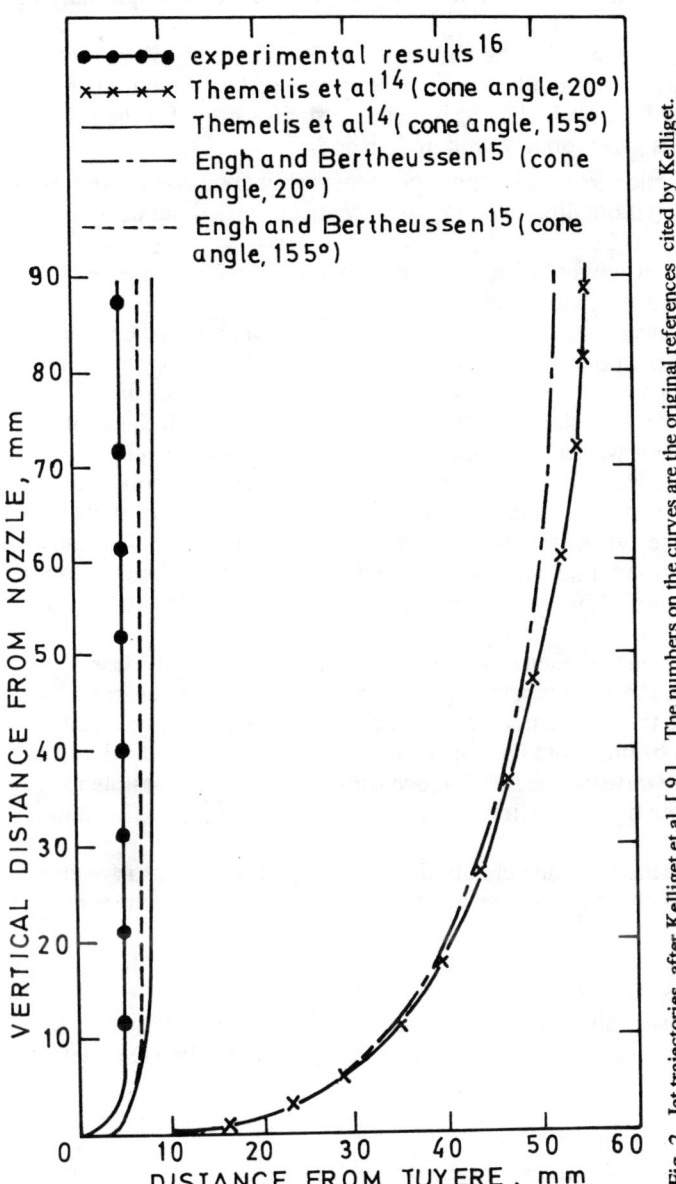

Fig. 2. Jet trajectories, after Kelliget et al. [9]. The numbers on the curves are the original references cited by Kelliget.

shown in Fig. 2 where it is seen that an expression deduced for the trajectory of an air jet in water (the lines on the right hand side) [8] cannot be applied to the nitrogen - mercury system, unless a correction is made for the difference between the cone angles, [9,10].

This statement is not intended as a general condemnation of modelling work relating to multi-phase systems, but rather as caution, that when the similarity criteria are not met, the models may not faithfully reproduce the behavior of the molten steel systems. Nonetheless, even under these conditions, useful qualitative insight may be gained.

It should be noted, however that the bulk of the melt, in virtually all of the systems sketched in Fig. 1 is essentially a one phase system and so the fluid flow field in these regions may be modelled using aqueous systems - provided the nature of the gas liquid interfaces is reasonably well understood.

When electromagnetic forces are important a reasonably good conductor has to be used. In this area useful modelling experiments have been carried out using mercury [10] and Woods Metal [11].

The principal attractiveness of using models is that measurements are much more readily made under controlled conditions.

When transparent systems are used (e.g. aqueous solutions) both the velocity fields and the turbulence parameters (turbulent kinetic energy, Reynolds stresses etc.) are readily measured, using either *hot film anemometry* or *laser anemometry*.

The hot film anemometer, sketched in Fig. 3 is in essence a small filament heated by an electric current which is immersed in the moving fluid. The electric resistance of the filament depends on its temperature, which in turn is determined by a balance between the resistive heat supplied by the current and the heat lost to the fluid. This latter, of course depends on the velocity of the fluid. By accurately measuring the current needed to maintain the filament at a constant temperature we can obtain the fluid velocity, while the fluctuations in the current provide information on the turbulence characteristics of the system.

A somewhat more sophisticated device used for measurement is the laser anemometer, sketched in Fig. 4. It is seen that a laser beam is split into beams of equal strength, which are then made to cross at the point where the measurement is to be made. Where the two beams cross an interference fringe pattern is created and by tracking particles, as they traverse the interference fringe pattern, it is possible to obtain information on both the mean (time smoothed) velocity and on the fluctuating velocity components.

A qualitative assessment of the velocity field in transparent systems is readily made by using tracer particles and time lapse photography, to obtain the streamline pattern.

The trajectories of the buoyant particles introduced into transparent liquids is also readily measured, as reported by Guthrie and co-workers [12,13].

Heat and mass transfer studies are also readily carried out in these systems, although as noted earlier, it is questionable whether the appropriate dimensionless criteria can be met.

### 2.1.2. Modelling work using metallic melts

Modelling work using metallic melts may be necessary when we wish to study electromagnetically driven flows, or particular gas - melt interactions. Under these conditions the principal attractiveness is that the system may be operated at lower temperatures (e.g. mercury or Woods Metal) - although in many instances it is difficult to meet all the criteria for dynamic similarity.

The optical methods (flow visualization and laser anemometry) cannot be applied to opaque systems, so that the measurement techniques are necessarily more restricted.

Regarding electromagnetically driven flows, Evans and co-workers have done useful work by measuring the surface velocities of mercury melts, using a photographic technique [10].

Szekely, Chang and Ryan [11] used a mechanical pressure transducer to determine the velocity fields in electromagnetically driven melts of Woods Metal.

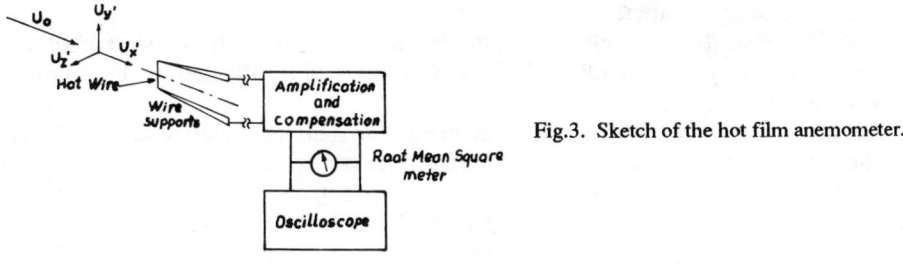

Fig.3. Sketch of the hot film anemometer.

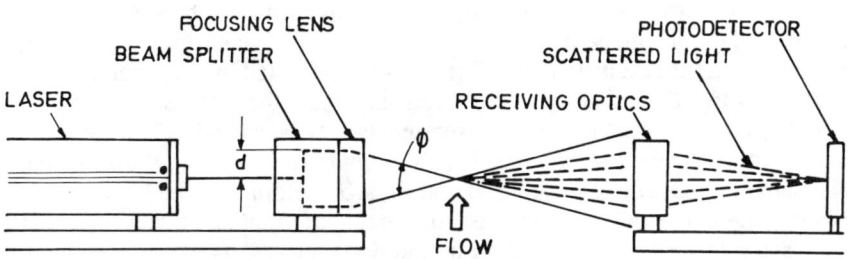

Fig. 4. Sketch of the optics for a laser anemometer.

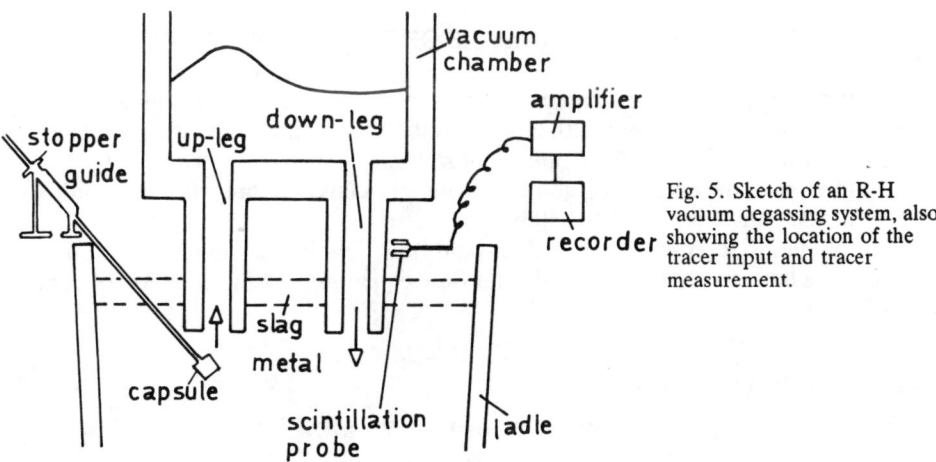

Fig. 5. Sketch of an R-H vacuum degassing system, also showing the location of the tracer input and tracer measurement.

The behavior of gas bubbles or jets in molten metals may be studied using X-ray photography and useful work in this area has been reported by Davies [14] and Brimacombe [15]. Finally, mass transfer phenomena between mercury and aqueous solutions may be used to simulate slag-metal reactions, as pioneered by

Richardson and his co-workers, although it is not quite clear, whether in th absence of meeting the similarity criteria, such work can be more than a qualitative guide.

### 2.1.3. Plant Scale Measurements

The tools available for carrying out measurements, using molten steel on tonnage scale installations are rather more limited.

Estimates of the melt velocities at the free surface may be made using motion picture photography [10,16] although these readings may be affected by the conditions at the melt surface.

Radioactive tracers may be introduced into the melt, as sketched in Fig. 5 [17] and the melt either analyzed continuously for radioactivity at a given location, e.g. by using a Geiger counter, or samples may be taken, which can then be analyzed. These measurements provide information that is immediately usable for estimating the mixing time, i. e. the time required for the homogenization of the bath.

Furthermore, through the interpretation of these measurements it may be possible to deduce experimental values of the eddy diffusivity within the system. It is noted, that since both the melt velocities and the eddy diffusivities tend to be strongly spatially dependent, the results may be quite markedly affected by the location of both the points where the tracer was introduced and where the sample was taken.

In addition to tracer dispersion, measurements may be taken to characterize the bath composition, such as its oxygen, sulfur, aluminum, etc. content, thus one may follow the path of deoxidation, desulfurization decarburization and the like.

Finally, one may study the local dissolution rate of graphite or iron rods, that were immersed into steel baths, and these dissolution patterns, may in turn be interpreted to provide information on the local values of the turbulence intensity in these systems.

## 2.2. MATHEMATICAL ANALYSIS OF TRANSPORT PROCESSES

In recent years major advances have been made in our understanding of turbulent recirculating flows [18,19,20,21,22] and the application of these concepts to metallurgical systems has also made major strides.

In view of the extensive published material in this area only the basic outlines of the mathematical approaches will be given here.

In order to represent the turbulent fluid flow field in the one phase region of the system, we have to use the standard fluid flow equations, viz the equation of continuity:

$$\nabla \cdot \vec{U} = 0 \tag{1}$$

and the equation of motion:

$$\rho(\vec{U} \cdot \nabla)\vec{U} = -\nabla P + \nabla \cdot \underset{\sim}{\tau} + \vec{F}_b \tag{2}$$

where  $\vec{U}$  is the velocity vector
- $\rho$  is the density
- $P$  is the pressure
- $\underset{\sim}{\tau}$  is the stress tensor, which includes both the laminar and the turbulent components

$\vec{F}_b$ is the body force field vector, due to electromagnetic or buoyancy forces

The components of the stress tensor are related to the velocity gradients through the effective viscosity ($\mu_t$) components:

$$\mu_e = \mu_t + \mu \qquad (3)$$

The molecular viscosity is a property of the material, however, the turbulent viscosity which is a multiple of the molecular value depends on location within the system and on the nature of the flow.

Several models have been proposed to represent $\mu_t$ in terms of the properties of the flow, the most commonly used of the being the $K-W$ and the $K-\varepsilon$ models [23].

The $K-W$ model postulates that

$$\mu_t = K/W^{1/2} \qquad (4)$$

while the $K-\varepsilon$ model proposed that

$$\mu_t = \rho K^2/\varepsilon \qquad (5)$$

where $K = \frac{1}{2}\left(\overline{U'_x}^2 + \overline{U'_y}^2 + \overline{U'_z}^2\right)$ is the kinetic energy of turbulence

$W$ is the square of the frequency of the turbulent fluctuations

$\varepsilon$ is the turbulent energy dissipation.

As a matter of fact separate differential equations have to be solved, to obtain experssions for $K$, $W$ or $\varepsilon$, simultaneously with the equation of motion. The procedures that have to be employed are complex but well documented for mathematically two dimensional systems, viz axial cylindrical symmetry, or a two dimensional slice in the cartesian coordinate system. The tackling of mathematically three dimensional flows is feasible, but requires large amounts of computer time (1000 – 3000 seconds on modern machines).

When the flow is driven by electromagnetic forces, $F_b$ is given by:

$$\vec{F}_b = \vec{J} \times \vec{B} \qquad (6)$$

where $\vec{J}$ is the current density in the melt and

$\vec{B}$ is the magnetic flux intensity

The quantities $\vec{J}$ and $\vec{B}$ have to obtained by solving Maxwell's equations [19].

Heat or mass transfer in the bulk of the melt may be represented by the standard conservation equations, which take the following form:

$$\nabla k_e \nabla T - \rho C_p \vec{U} \cdot \nabla T = \dot{q}''' \qquad (7)$$

for heat transfer and

$$\nabla D_e \nabla C - \vec{U} \cdot \nabla C = \dot{r}''' \qquad (8)$$

for mass transfer
where
    $k_e$ is the effective thermal conductivity
    $T$ is the temperature
    $\dot{q}'''$ is the local rate of heat generation

$D_e$ is the effective diffusivity
$C$ is concentration of the transferred species
$\dot{r}''''$ is the rate at which the species is being produced by chemical reaction.

The quantities $D_e$ and $k_e$ may be estimated from the values of the effective diffusivity by postulating that:

$$N_{Pr}^{(t)} = \frac{C_p \mu_t}{k_t} \approx 1 \quad \text{and} \quad N_{Sc}^{(t)} = \frac{\mu_t}{\rho D_e} \approx 1$$

where $N_{Pr}^{(t)}$ and $N_{Sc}^{(t)}$ are the turbulent Prandtl and Schmidt numbers respectively.

As will be discussed subsequently, numerous solutions have been presented in the literature for the system of equations (1-8); however, most of this work has been restricted to systems where axial symmetry may be assumed.

## 3. SUMMARY OF INFORMATION AVAILABLE ON THE MODELLING OF LADLE METALLURGY SYSTEMS

In the preceeding section we discussed the experimental and the analytical tools that are available for the study of ladle metallurgy systems.

In this section we shall present a brief review of the information that is available at present, with emphasis on the fundamental understanding of the relevant transport processes.

In general induction stirred systems, which essentially involve just one melt phase, are fairly well understood and the theoretical predictions are well supported by experimental measurements. Useful work has also been done regarding the application of these concepts to heat and mass transfer and to deoxidation kinetics.

Gas bubble driven circulation systems are understood in a qualitative sense, although a great deal more work needs to be done to characterize the nature of the phase boundaries between the bubble streams and the bulk of the melt.

Regarding injection systems, useful data have been collected concerning the overall kinetics, but the fundamental basis for the interpretation of these measurements is not yet fully available.

### 3.1. INDUCTIVELY STIRRED SYSTEMS

As a result of extensive work during the past five years the behavior of inductively stirred systems is well understood. One may readily calculate the electromagnetic force field generated by a variety of coil designs and then use the resultant values of $F_b$ to solve the fluid flow, heat flow and mass transfer problems.

As in illustration, Figs. 6 (a) and (b) show a comparison between the theoretically predicted and the experimentally measured velocity fields in an inductively stirred low melting alloy [11].

Figure 7 shows a comparison between the experimentally measured and the theoretically predicted tracer dispersion in a 50 ton ASEA–SKF furnace [24]. It is seen that the agreement is very good in both instances.

Regarding mass transfer phenomena Fig. 7 shows a comparison between experimentally measured and theoretically predicted Mg contents of a 140 kg ferrous melt. It is seen that the predictions, based on the solution of the appropriate electromagnetically driven fluid flow equations and the coupled diffusion equation provide a reasonably accurate representation of the experimental measurements.

Finally Fig. 8 shows a comparison between the experimentally measured deoxidation rates in an ASEA - SKF furnace with the theoretical predictions, based on a coalescence model, which took into account the turbulent dissipation in the inductively stirred melt ]26]. It is seen that here again the theory does provide a rational framework for the interpretation of the measurements.

It is suggested, therefore, that at present we do have a good, quantitative understanding of fluid flow in turbulent electromagnetically stirred melts and that useful applications have been developed to represent heat and mass transfer phenomena in these systems.

Further work could be usefully done to model truly three dimensional situations (such as found in the use of ASEA's linear stirrer) on the use of spatially non-uniform fields (e. g. the use of coil design to generate turbulence at some locations and quiescent melts at other locations, etc) and on the more broad application of this technique to heat and mass transfer problems.

### 3.2. ARGON - STIRRED LADLES

The behavior of argon – stirred ladles has been extensively studied, both experimentally and through physical and mathematical modelling. As a result of this

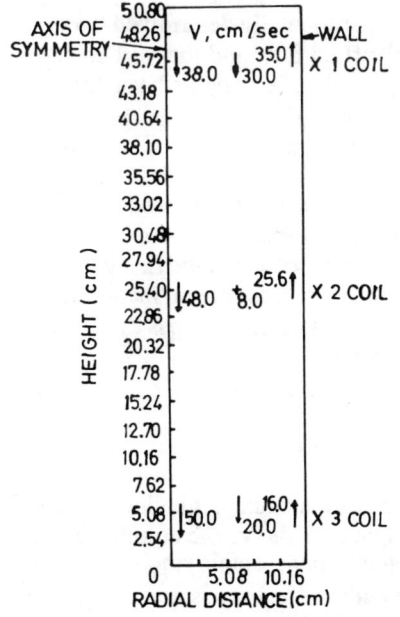

Fig. 6(a). Theoretically predicted velocity fields in an electromagnetically stirred low alloy melt.

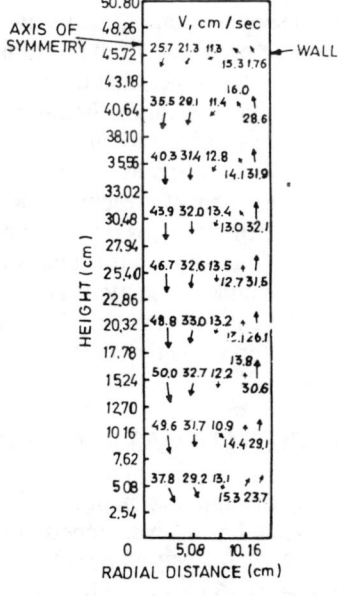

Fig. 6(b). Experimentally measured velocity fields in an electromagnetically stirred low alloy melt.

work there is a good qualitative understanding of both the flow patterns and the mixing processes in these systems [27].

The spatially distributed eddy diffusivity for a 60 ton argon-stirred vessel is shown in Fig. 9, where it is seen that as expected, the eddy diffusivity is the largest in the vicinity of the central core and decreases radially outward toward the wall.

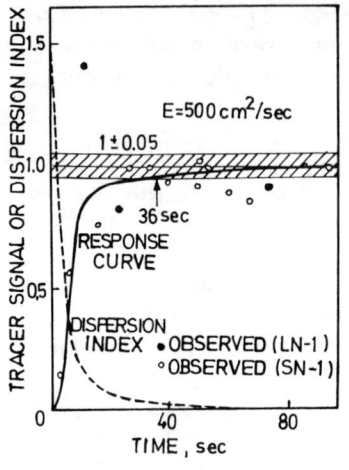

Fig. 7. Comparison of the experimentally measured and the theoretically predicted tracer dispersion in an ASEA-SKF Furnace.

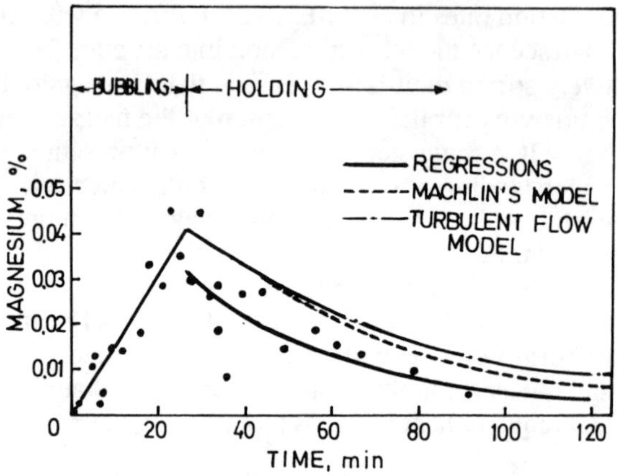

Fig.8. Comparison between experimentally measured and theoretically predicted Mg content of 140 kg inductively stirred ferrous melt after Irons et al. Met. Trans. 9B 151

Figure 10 shows a comparison between the experimentally determined tracer dispersion and that predicted from the solution of the fluid flow and the (turbulent) diffusion equations. It is seen that the model does provide a reasonable basis for estimating the time required for the homogenization of the bath.

A fuller discussion of measurements on argon - stirred ladles is available in the MEFOS reports by Lehner and coworkers [27].

It is noted here that at present the principal area of uncertainty regarding the behavior of argon stirred ladles is the definition and characterization of the interface that separates the "bubble column" that is the region which contains the rising stream of bubbles and the bulk of the melt. The actual definition of this region, including its geometry, and the characterization of heat and momentum transfer across this (rather poorly defined) interface is the major remaining problem.

### 3.3. VACUUM DEGASSING SYSTEMS

Regarding the application of turbulence theories to vacuum degassing the R-H system, [17] which has been sketched in Fig. 2. Figure 11 shows the computed contours of the eddy diffusivity $E = D_{e'}$ for a 150 ton system, employing a recirculation rate of 36 tons/min. It is seen that the eddy diffusivity is quite large, having an average value of about 100 cm$^2$/s, much larger than the values obtained in the argon - stirred ladle.

Figure 12 and 13 show a comparison between the experimentally determined tracer distribution and that predicted from the solution of the turbulent Navier - Stokes equations. It is seen that the agreement is quite reasonable, essentially quantitative, for both the case when the tracer was introduced into a very active region of the ladle (Fig. 12) and when the tracer was introduced into a relatively quiescent region (Fig. 13),

It is noted, that while the $R-H$ system is in essence a gas bubble driven circulation system, the boundary conditions are much more readily established than in case of an argon - stirred ladle, because of the essential absence of a two phase region in the bulk of the ladle. Some useful work has been done on the modelling of the $D-H$ system, however, the inherent complexity of these flow fields has rendered these early results somewhat preliminary [29].

## 4. DISCUSSION

In the paper we discussed the tools that are available for the study of fluid flow, heat and mass transfer phenomena in agitated ladle systems of importance in ladle metallurgical operations, together with the experimental results that are available.

It can be stated in general that industrial experience is being rapidly accumulated with these systems and that the application of quite sophisticated mathematical

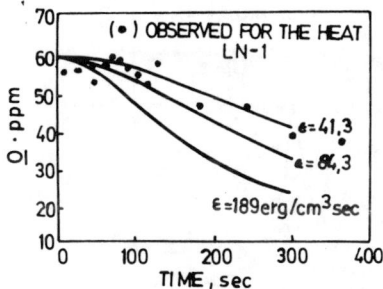

Fig.9. Experimentally measured and theoretically predicted deoxidation rates in an inductively stirred melt, after Nakanishi and Szekely.

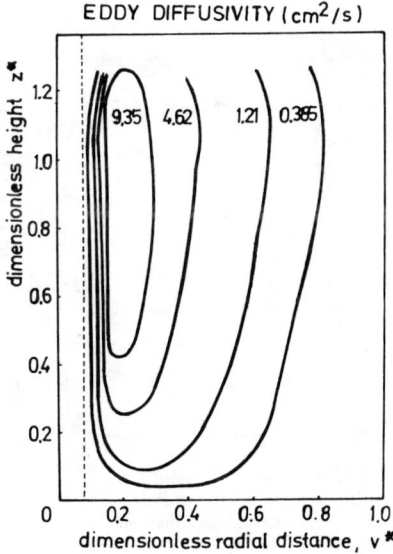

Fig.10. Computed profiles of the eddy diffusivity in a 60 ton argon-stirred ladle, after Szekely, Lehner and Chang.

modelling and laboratory-scale investigations have provided us with a greatly improved insight regarding the mechanism of agitation and that of the transport processes.

Induction stirred systems appear to be very well understood, because here we have to deal with just one, continuous molten phase. Gas bubble - driven circulation

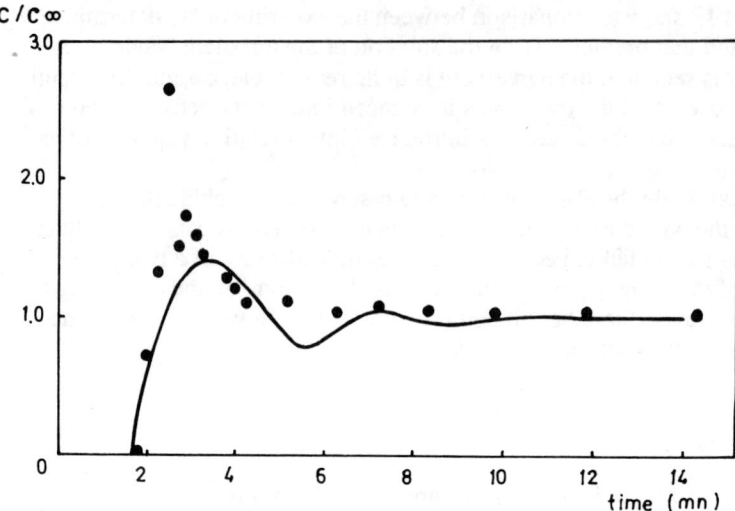

Fig 11. Comparison between the experimentally measured and the theoretically predicted tracer dispersion in a 60 ton argon-stirred ladle, after Szekely, Lehner and Chang.

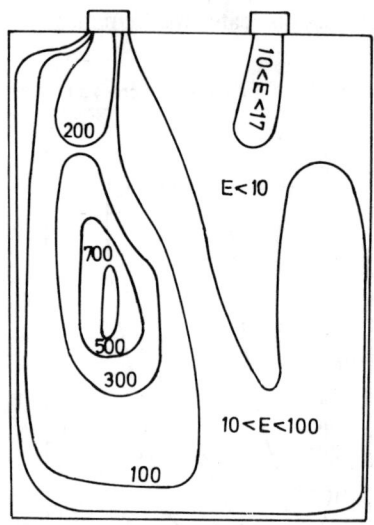

Fig.12. Computed contours of the eddy diffusivity in a 150 ton R-H vessel, after Szekely, Nakanishi and Chang, [17].

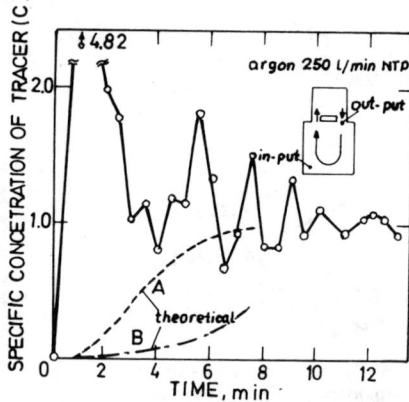

Fig.13. Comparison between the experimentally measured and the theoretically predicted tracer dispersion in a 150 ton R-H vessel. Tracer introduced near the bottom of the vessel [17].

systems are understood in a semi-quantitative sense, in that the overall flow patterns are known, and the mixing rates may be evaluated, either from turbulence models, or from empirical correlations. However, in this area a great deal more work needs to be done regarding the definition of the interface between the gas bubble stream and the bulk of the melt.

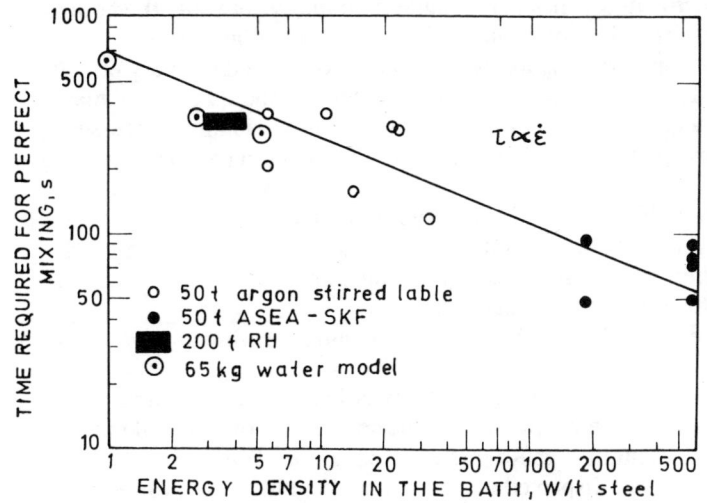

Fig.14. Plot of the mixing time against the rate of dissipation of the energy density in various agitated melts [30].

From a basic standpoint the understanding of fluid flow and mixing necessarily has to precede the development of models for heat and mass transfer processes and for deoxidation. Understandably this effort has lagged somewhat behind the fluid flow work.

All commercially used agitated ladle systems have the common characteristic that the turbulence intensity, which also corresponds to the eddy diffusivity is the highest in the central portion of the vessel and the bottom of the ladle. This is particularly the case for gas bubble driven circulation systems and for melts operating under vacuum.

In induction stirred systems there is the possibility, at least in principle, to modify the coil design so as to produce a wider range of patterns for the turbulent kinetic energy distribution, to suit particular purposes. Such an arrangement is much less likely for gas bubble driven systems because the upper part of the ladle will always receive better agitation.

Figure 14 provides an interesting summary of our empirical knowledge pertaining to agitation and energy dissipation in a variety of systems. It is seen that the mixing time in the vessel may be related, albeit in an approximate fashion, to the rate of energy density dissipation. The higher the rate of energy density dissipation, the better is the agitation and hence the lower is the mixing time.

There is an appreciable scatter in the data, which is to be expected because clearly one system may be more effective in using the energy dissipated than the other.

Nonetheless this finding seems to be in general accord with the chemical engineering experience which postulates that the rate of mixing may be related to the energy input into agitated systems.

Inspection of Fig.12 shows that induction stirring, with higher specific energy dissipation allows more rapid mixing than agitation provided by bubbles. In comparing the relative effectiveness of gas bubble and induction stirring, it should be noted that argon stirring is relatively inexpensive and it can provide effective agitation to promote slag metal reactions. Its disadvantage is lack of flexibility, i.e. one cannot impose a desired turbulent energy dissipation pattern on the system, and the bottom of the ladle, especially in the vicinity of the walls, will be rather quiescent.

In contrast, induction stirring is likely to be more costly, and not very effective in promoting slag metal reactions; however induction stirring should be particularly attractive when high energy dissipation is required (certain deoxidation processes) and when a specific flow pattern and turbulent energy dissipation has to be imposed on the system.

In conclusion it should be stated that a very impressive array of research tools both experimental and analytical is at our disposal for the further study of agitated ladle systems, and in view of the useful results obtained up to the present it would be definitely worthwhile to continue with this effort.

It may be convenient to classify the future work into the following two categories:

(1) *Extension of existing efforts* by straightforward application of currently available tools. Work of this type would include the application of heat and mass transfer theory to predict vacuum degassing rates, mixing rates and the rate at which alloying additions melt and are dispersed in systems, and the like, for a variety of ladle metallurgical applications. Such calculations are quite readily done and comparison with measurements would be a worthwhile exercise.

In the same vein useful work could be done in the study of pulsed systems, such as the D-H degassing operation, e.g. by using water models.

(2) *New research areas* that would be very valuable include the detailed study of the nature of gas - melt interfaces in gas agitated systems, both regarding their geometry and the transport processes. Such work would be essential for a basic understanding of many kinetic processes in injection metallurgy. Another potentially very fruitfully area would be the study of pulsed systems, because these could offer special advantages for certain circumstances.

Finally, there should be some quite exciting opportunities for a new generation of induction stirring systems, where a spatially non-uniform turbulence field is imposed on the system, possibly in conjunction with bubble or with vacuum applications.

All in all a great deal of progress has been made, but the use of transport fundamentals should be of considerable help in maintaining the momentum of advances in ladle metallurgy.

## REFERENCES

1. International Conference on Injection Metallurgy, Lulea, 1977.
2. Secondary Steelmaking Conference, Ondon, 1977.
3. OHMAN, B., and ., T., Proceedings of Scaniject, Lulea, 1977.
4. HLINKA, J.S., in Symposium on Mathematical Process Models in Iron and Steelmaking, Amsterdam, 1973.
5. SZEKELY, J., and THEMELIS, N.J., Rate Phenomena in Process Metallurgy, Wiley, New York, 1971.
6. SZEKELY, J., Fluid flow Aspects of Metals Processing, Academic Press, New York, 1979.
7. KNUDSEN, J.G., and KATZ, D.L., Fluid Mechanics and Heat Transfer, McGraw Hill, New York 1958.

8. THEMELIS, N.J., TARASOFF P., and SZEKELY, J., Trans. A.I.M.E. 245 2452, (1969)
9. MCKELLIGET, J.W. CROSS, M., and GIBSON, R.D., Ironmaking and Steelmaking, 5, 282 (1978).
10. TARRAPORT, E.D., and EVANS J.W., Met. Trans, 7B (1976) 343.
11. SZEKELY, J., CHANG, C.W., and RYAN, R.E., "The Measurement and Prediction of the Melt Velocities in a Turbulent, Electromagnetically Driven Recirculating Low Melting Alloy System", Met. Trans. 8B, p 333-338, June 1977.
12. GOURTSOYANIS, L., HEINEIN, H., and GUTHRIE, R.I.L., in Proc.Physical Chemistry of the Production and Use of Ferroalloys (j. Farrell, ed. ), AIME, 1974.
13. GUTHRIE, R.I.L., CLIFT, R., and HEINEIN, H.,Met. Trans. 6B (1975) 321.
14. DAVIES, M.W., HAZELDEAN, G.S.F., and SMITH, P.N. n Proceedings of the Richardson Conference J.H.E. Jeffes and R.J> Tait, Eds. London, 1974.
15. BRIMACOMBE, K., paper presented at 1978 AIME Annual Meeting, Denver.
16. SZEKELY, J., CHANG, C.W. , and JOHNSON, W.E., Experimental Measurement and Prediction of Melt Surface Velocities in a 30,000 lb Inductively Stirred Melt, Met. Trans. Vol.8B p 514-517.
17. NAKANISHI, K., SZEKELY, J.,and CHANG, C.W., Experimental and Theoretical Investigation of Mixing Phenomena in the R-H Vacuum Process, Ironmaking Steelmaking 3, 115 (1975)
18. SZEKELY, J., Turbulent Recirculating Flow in Ladles, Proc. Int. Conf. on Injection Metallurgy, Lulea, Sweden, 1977.
19. SZEKELY, J. and CHANG, C.W., Turbulent Electromagnetically Driven Flow in Metals Processing, Part I: Formulation and Part II: Practical Applications, Ironmaking and Steelmaking 1977, No. 3, 190-204.
20. SZEKELY, J., Proc Int. Conference on Chemical Metallurgy of Iron and Steel, Versailles, 1978.
21. SZEKELY, J. and ASAI, S., On Some Turbulent Fluid Flow Problems in Steel Processing Operations: Part I: The General Mathematical Statement of Turbulent Recirculatory Flow, Trans. J.I.S.I., 15 No 5, 276 (1975)
22. SZEKELY, J., ASAI, S., On some Turbulent Fluid Flow Problems in Steel Processing Operations, Part II: Applications, Trans. J.I.S.I., 15 No. 5276 (1975)
23. LAUNDER, E. and SPALDING, D.B., Mathematical Models of Turbulence, Academic Press, New York, 1972.
24. NAKANISHI, K., SZEKELY, J., FUJII, P., and MIHARA, y., Stirring and its Effects on Aluminum Deoxidation in the ASEA-SKF Furnace, Met. Trans. 6B, 111 (1975).
25. SZEKELY, J., and NAKANISHI, K., Stirring and its Effects on Aluminum Deoxidation in the ASEA-SKF Furnace, Part II: Mathematical Representation of the Turbulent Flow Field and of Tracer Dispersion, Met. Trans. 6B, ;245 (1975).
26. NAKANISHI, K., SZEKELY, J., Deoxidation Kinetics in a Turbulent Fluid Flow Field, Trans. J.I.S.I. 15, 522, (1975).
27. LEHNER, T., MEFOS reports.
28. SZEKELY, J., LEHNER, T., and CHANG. C.W.
29. EBNETH, G., RUTTIGER, K., and ZAHS, G., in Proceedings of the 5th International Conference on Vacuum Metallurgy Munich, 1976.
30. NAKANISHI, K., FUJII, T., and SZEKELY, J., Possible Relationships Between Energy Dissipation and Agitation in Steel Processing Operations, I. and S. Quart., No. 2 193,(1975)/

International Seminar 1980

# HEAT TRANSFER IN NUCLEAR REACTOR SAFETY

G. Bankoff

*Northwestern University, Evanston, USA*

The topic of the 1980 seminar was "Heat Transfer in Nuclear Reactor Safety". There were eleven invited lectures of a survey nature. Most contributed papers dealt with light water reactor safety. These included fourteen papers on basic heat transfer studies, ten papers on critical heat flux, and sixteen papers on blowdown, rewet and reflood heat transfer. In addition, there were seven papers on liquid metal fast breeder reactor safety and one paper on gas cooled reactor safety heat transfer. Prior to the Three Mile Island accident in 1978, activity along these lines in the light water reactor field was confined to the early stages of the accident, before any melting could occur. The objective in most countries in the world which carried out this type of research was to demonstrate that, under the worst conceivable initiating accident circumstances, the peak clad temperature would not exceed a dangerous limit, which was still well below the clad melting temperature. The worst initiator was conceived to be the 200% guillotine break in the cold leg of a pressurized water reactor. In this imaginary loss-of-coolant accident one of the cold legs was instantaneously slashed through, and the two cut ends were displaced far enough so that they could both discharge at full force into the surroundings. This highly unlikely (and probably impossible) accident initiator was considered to be sufficiently severe that there was adequate safety margin against severe initiators. Research, therefore, focused on such questions as the rate of loss of coolant inventory during the blowdown process, critical heat flux and dryout in large tube bundles, the behaviour of the emergency core coolant on injection into a downcomer in which there was an ascending flow of steam (the so-called "steam binding" problem), the re-wetting of the dried-out fuel bundles by coolant injected from below and reflooding by coolant entering from the top of the tube bundle. The subjects of the research papers of this conference reflect this earlier emphasis. One finds papers on transient flow boiling from vessels, thermodynamics of critical two-phase flow from long pipes of initially subcooled water, pressure waves and oscillatory flow instabilities in gas-liquid flows, studies of flooding in tubes and annuli and in parallel channels, and predictions of subchannel two-phase flow in interconnected channels. In view of the transient nature and high fluid velocities, there were papers on critical heat flux in dryout and two-phase shear flow, and in tube bundles under quasi-steady conditions, and on film boiling at various pressures under transient conditions. Rewet and reflood heat transfer was studied using Refrigerant-12 and water in single tube experiments under a variety of boundary conditions. There was also considerable activity in generating computer codes, using the results of experiments to predict the two dimensional fluid mechanics and heat transfer within the reactor core during the reflood and rewet process. One problem which is unique to the BWR is the phenomenon of steam chugging, where steam coming from the reactor enters the pressure suppression pool and condenses in an oscillatory faschion. This was reflected in several analytical papers. Reflood heat transfer involves cooling of the hot cladding surfaces by dispersed flow of steam and droplets. This also received attention. The LOFT experiments and the TRAC code were both large efforts aimed at quantitative demonstration of the safety margin of a PWR or BWR under large-break conditions. These efforts were well summarized in two review papers.

However, the Three Mile Island Accident changed the emphasis in the light water program completely. It was realized that small break accidents were much more probable, and that a combination of events, including human error, could produce disastrous results. Althouth not enough time had elapsed for the new directions of research to be reported, there were already some indications of this shift in the papers of this conference. A paper by Zuber on problems in the modelling of small break LOCA and some tests reported by McPherson on small break tests in LOFT facility were the chief among these. Also noteworthy was a paper by Lahey, based upon purely analytical methods, predicting the transient behavior or the two-phase level in the PWR core during conditions of severely reduced liquid mass inventory. The importance of overall systems variables, such as interactions with the steam generators, was emphasized in another paper.

In the liquid metal fast breeder reactor (LFMBR) area, te possibility of a severe accident, in which melting and fuel coolant interactions might take place, had long been recognized, because of the very hight power densities in a typical fast breeder core. Thus, there were papers on sodium boiling, clad melting and relocation, and natural convection heat transfer from downwards facing surfaces in sodium for fast reactor internal core catchers. Other papers dealing with the advanced phases of a severe accident were studies of boiling fragmentation and the propagation of fuel-coolant interactions, the interaction of a molten $UO_2$ pool with the support structure and post-accident heat removal.

Since that time emphasis in the light water reactor fields has shifted practically entirely to the severe accident case, while research on LMFBR safety has considerably declined. Current interest is in advanced reactor designs which would be inherently safe, if possible, owing to thermal expansion or other effects. Nevertheless, the subjects and the distribution of papers in this seminar gives a fairly accurate picture of the status of nuclear reactor heat transfer safety research at the time.

# THE ANALISYS OF TWO-PHASE LEVEL IN A PWR CORE DURING CONDITIONS OF SEVERELY REDUCED LIQUID MASS INVENTORY

R. T. Lahey, JR.

*Rensselaer Polytechnic Institute*
*Troy, New York, USA*

## 1. INTRODUCTION

The recent incident at the Three Mile Island Nuclear Power Station, Unit #2 (TMI-2), has focused world wide attention on non-design-basis-accident (DBA) loss-of-coolant-accident (LOCA) phenomena. In particular, the need to be able to rapidly and accurately calculate the position of the two-phase level in a PWR core was clearly demonstrated at TMI-2. Indeed, as has been shown experimentally [1,2,3] if the core is submerged in a two-phase mixture, the resultant boiling heat transfer is normally sufficient to remove the decay heat from the fuel rods, thus preventing overheating of the core.

While an evaluation of the position of the two-phase level is possible using existing digital computer codes (eg: RELAP/4, TRAC), this approach is quite costly and, in general, the information of interest is not available when needed by the operators experiencing a "small break" LOCA incident.

In this paper a closed form, analytical model for prediction of the two-phase level in a PWR core, during conditions of severely reduced liquid mass inventory, is derived. This model is based on drift-flux (4) techhiques and assumes that thermodynamic equilibrium exists during the quasistatic process involved. It is easily shown that these assumptions are valid for many situations of interest.

Figure 1 is a schematic of the primary sistem of a typical PWR. For clarity, the elevation of the hot and cold legs are shown at different elevations. In actuality, they are at the same elevation, but at different azimuthal positions. It can be seen that for a hypothetical event, in which considerable liquid mass inventory has been lost from the primary system, the two-phase level in the core is determined by s manometer-like balance between the core and downcomer region. The higher the collapsed liquid level ($\xi$) in the downcomer, the higher will be the two-phase level ($L_{2\emptyset}$) in the core.

While the situation shown schematically in Fig. 1 is somewhat contrived (i. e.: not necessarily typical of TMI-2 conditions) it does serve to focus attention on the essential features of the thermal-hydraulics involved, and thus will be used as the basis for model development.

Figure 2 indicates the control volumes of interest in the PWR core region during conditions of severely reduced liquid mass inventory in the primary system. The manometer-like balance is evident. It can be noted that the collapsed liquid level in the downcomer forces water up into the core region. If the water which enters the core is subcooled, there will be a non-boiling length, $\lambda$, after which the water boils and swells-up into the core. If the core power level were zero, the system would have an equal liquid height in each leg Normally however, the core is at a decay heat power level, and thus $L_{2\emptyset} > \xi$.

If coolant supplied to the primary system from the emergency core cooling system (ECCS) is at a rate which exceeds the discharge rate to the containment, then the level in the downcomer ($\xi$) will increase. If not, this level may decrease with time, resulting in lower two-phase levels ($L_{2\emptyset}$) in the core. For purposes of the analysis done herein, it is assumed that the process is quasi-static and thus the time rate of

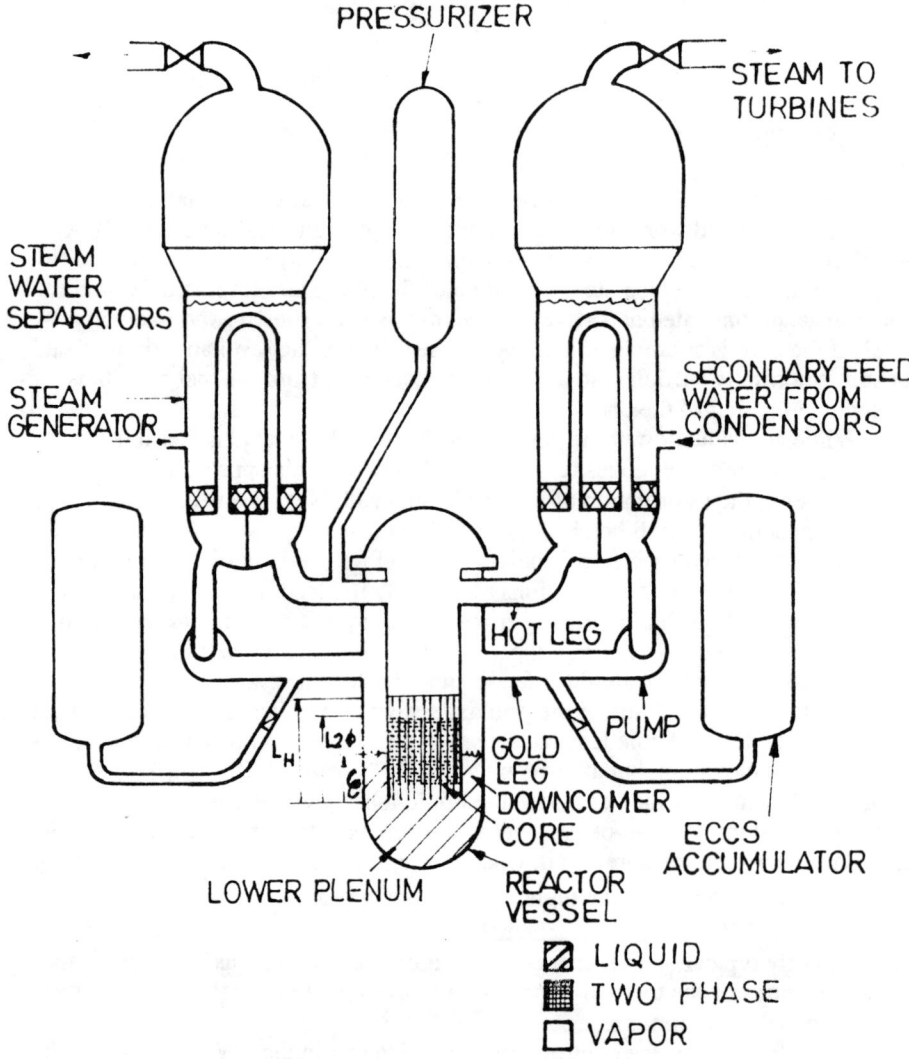

Figure 1. Core cooling in a PWR during conditions of severely reduced liquid mass inventory.

change of mass inventory in the primary system is negligible. This situation approximates the conditions which might have existed at TMI subsequent to system blowdown when the operators attempted to cool the core with the decay heat removal system.

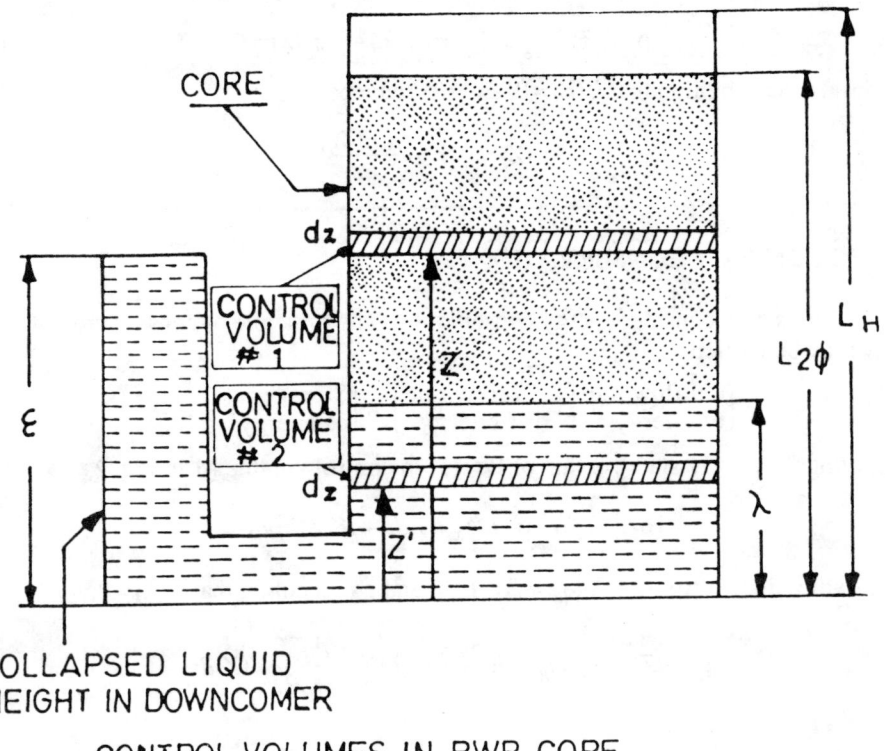

Figure 2. Control volumes in PWR core.

## 2. DERIVATION OF THE ANALYTICAL MODEL

A closed-form analytical expression for the two-phase level ($L_{2\emptyset}$) in a PWR core, with severely reduced liquid mass inventory, will now be derived. The approach taken is to derive and integrate the appropriate conservation equations of mass, energy and momentum.

### 2.1. MASS CONSERVATION

The principle of conservation of mass may be written as:

$$\begin{bmatrix} \text{Outflow rate} \\ \text{of mass} \end{bmatrix} - \begin{bmatrix} \text{Inflow rate} \\ \text{of mass} \end{bmatrix} + \begin{bmatrix} \text{Storage rate} \\ \text{of mass} \end{bmatrix} = 0 \qquad (1)$$

Applying the steady-state version of Eq. (1) to control volume #1 in Fig. 2,

$$\frac{d}{dz}(\rho_f j_f + \rho_g j_g) = 0 \tag{2}$$

Integrating Eq. (2) between $z = \lambda$ and $z = Z < L_{2\emptyset}$,

$$\rho_f j_f(Z) + \rho_g j_g(Z) - \rho_f j_f(\lambda) - \rho_g j_g(\lambda) = 0 \tag{3}$$

Imposing the condition that $j_g(\lambda) = 0$, Eq. (3) reduces to,

$$j_f(Z) = j_f(\lambda) - \frac{\rho_g j_g(Z)}{\rho_f} \tag{4}$$

## 2.2 ENERGY CONSERVATION

Let us now consider energy conservation. The principle of conservation of energy may be written as,

$$\begin{bmatrix}\text{Outflow rate} \\ \text{of energy}\end{bmatrix} - \begin{bmatrix}\text{Inflow rate} \\ \text{of energy}\end{bmatrix} + \begin{bmatrix}\text{Storage rate} \\ \text{of energy}\end{bmatrix} = 0 \tag{5}$$

Applying the steady-state version of Eq. (5) to control volume #1 in Fig. 2,

$$\frac{d}{dz}[\rho_f j_f A_{xs} h_f + \rho_g j_g A_{xs} h_g] dz - q''(Z) P_H dz = 0 \tag{6}$$

Integrating Eq. (6) from $z = \lambda$ to $z = Z \leq L_{2\emptyset}$, and again noting that $j_g(\lambda) = 0$,

$$\rho_f j_f(Z) h_f + \rho_g j_g(Z) h_g - \rho_f j_f(\lambda) h_f = \frac{P_H}{A_{xs}} \int_\lambda^Z q''(Z) dz$$

Solving for $j_g(Z)$,

$$j_g(Z) = \frac{\rho_f h_f}{\rho_g h_g}[j_f(\lambda) - j_f(Z)] + \frac{P_H}{A_{xs} \rho_g h_g} \int_\lambda^Z q''(Z) dz \tag{7}$$

Combining Eqs. (4) and (7)

$$j_g(Z) = \frac{P_H}{A_{xs} \rho_g h_{fg}} \int_\lambda^Z q''(Z) dz \tag{8}$$

Equations (4) and (8) give the superficial velocities of the liquid and vapor phases as a function of axial position, Z, for $\lambda \leq Z \leq L_{2\emptyset}$. The void fraction may now be determined using these superficial velocities and the standard drift-flux (4) relationship,

$$<\alpha(Z)> = \frac{j_g(Z)}{C_0[j_f(Z) + j_g(Z)] + V_{gj}} \tag{9}$$

where the notation, $< >$, indicates averaging over the cross-sectional flow area.

It should be noted that continuity implies,

$$\rho_f j_f(\lambda) = \rho_g j_g(L_{2\emptyset}) = \frac{w}{A_{xs}} \tag{10}$$

where w is the flow rate through the heated section.

Now, using Eqs. (4) and (10), Eq. (9) may be expressed as,

$$<\alpha(Z)> = \frac{j_g(Z)}{C_0[\frac{w}{\rho_f A_{xs}} + j_g(Z)(1-\rho_g/\rho_f)] + V_{gj}} \quad (11)$$

The radial void concentration parameter, $C_0$, well approximated by a constant and the drift velocity, $V_{gj}$, may be obtained from the empirical correlation (5),

$$V_{gj} = k_3 \left[\frac{(\rho_l - \rho_g)\sigma g g_c}{\rho_l^2}\right]^{1/4} \quad (12)$$

Let us now determine the unheated length, $\lambda$, from an energy balance on the non-boiling portion of the core.

Applying the principle of conservation of energy, Eq. (5), to control volume #2 in Fig. 2,

$$\frac{d}{dz}(\rho_f j_f A_{xs} h_l)dz = q''(Z)P_H dz \quad (13)$$

Integrating between $z=0$ and $z=\lambda$,

$$w\Delta h_{sub} = P_H \int_0^\lambda q''(Z)dz \quad (14)$$

where we have used Eq. (10) and the definition,

$$\Delta h_{sub} = h_f - h_i$$

The axial heat flux profile in a PWR core is well approximated by a chopped cosine and thus the decay heat is taken to be of the form,

$$q''(Z) = K\overline{q''}\cos[k(Z - L_H/2)] \quad (15)$$

where $K$ is the peak-to-average axial heat flux ratio, and $k$ depends on $K$ and the heated length, $L_H$.

Substituting Eq. (15) into (14), and integrating,

$$w\Delta h_{sub} = \frac{P_H K \overline{q''}}{k}\left[\sin k(\lambda - L_H/2) + \sin\left(\frac{kL_H}{2}\right)\right] \quad (16)$$

Noting that the power, $Q$, is given by,

$$Q = \overline{q''} P_H L_H \quad (17)$$

Equations (16) and (17) yield $\lambda$,

$$\lambda = \frac{1}{k}\sin^{-1}\left[\frac{w\Delta h_{sub} L_H k}{KQ} - \sin\frac{kL_H}{2}\right] + \frac{L_H}{2} \quad (18)$$

Since the axial heat flux profile has been specified as an explicit expression, Eq. (8) may also be integrated,

$$j_g(Z) = \frac{KQ}{A_{xs}\rho_g h_{fg} L_H k}[\sin k(Z - L_H/2) - \sin k(\lambda - L_H/2)] \quad (19)$$

Eqaution (19) can now be substituted into Eq. (11) to obtain the global void fraction explicitly as a function of axial position, Z. The resulting expression is,

$$<\alpha(Z)> = \frac{\frac{KQ}{A_{xs}\rho_g h_{fg} L_H k}[\sin k(Z-L_H/2)-\sin k(\lambda-L_H/2)]}{C_0\left[\frac{w}{\rho_f A_{xs}}+\frac{KQ(1-\rho_g/\rho_f)}{\rho_g A_{xs} h_{fg} L_H k}\{\sin k(Z-L_H/2)-\sin k(\lambda-L_H/2)\}\right]+V_{gj}}$$

which can be written more compactly as,

$$<\alpha(Z)> = \frac{K_1 \sin[k(Z-L_H/2)]+K_2}{K_3 \sin[k(Z-L_H/2)]+K_4} \quad (20)$$

where,

$$K_1 \triangleq \frac{KQ}{A_{xs}\rho_g h_{fg} L_H k}, \quad K_2 \triangleq -K_1 \sin k(\lambda-L_H/2)$$

$$K_3 \triangleq C_0\left[\frac{KQ(1-\rho_g/\rho_f)}{\rho_g A_{xs} h_{fg} L_H k}\right] = C_0(1-\rho_g/\rho_f)K_1$$

$$K_4 \triangleq V_{gj}+\frac{C_0 w}{\rho_f A_{xs}}-K_3 \sin k(\lambda-L_H/2)$$

Before proceeding, it is convenient to eliminate the ratio $w/Q$ in Eq. (18) to avoid singularities when $Q=0$. If we combine Eqs. (10) and (19),

$$\frac{w}{\rho_g A_{xs}} = j_g(L_{2\emptyset}) = \frac{KQ}{A_{x-s}\rho_g h_{fg} L_H k}[\sin k(L_{2\emptyset}-L_H/2)-\sin k(\lambda-L_H/2)] \quad (21)$$

Rearranging Eq. (16) and using Eq. (17),

$$\sin k(\lambda-L_H/2) = \frac{w \Delta h_{sub} L_H k}{KQ} - \sin\frac{kL_H}{2} \quad (22)$$

combining Eqs. (21) and (22) and rearranging,

$$w/Q = \frac{K}{h_{fg} L_H k(1+\frac{\Delta h_{sub}}{h_{fg}})}\left\{\sin k(L_{2\emptyset}-L_H/2)-\sin k\left(\frac{L_H}{2}\right)\right\} \quad (23)$$

Now Eqs. (18) and (23) can be combined to yield,

$$\lambda = \frac{1}{k}\sin^{-1}\left\{\frac{1}{(1+h_{fg}/\Delta h_{sub})}[\sin k(L_{2\emptyset}-L_H/2)+\sin(kL_H/2)]-\sin(kL_H/2)\right\}+L_H/2 \quad (24)$$

Equation (24) is the preferred form for numerical evaluation of the boiling boundary ($\lambda$).

## 2.3 MOMENTUM CONSERVATION

Let us now consider the principle of conservation of linear momentum, which may be written as:

$$\begin{bmatrix} \text{Outflow rate} \\ \text{of axial} \\ \text{momentum} \end{bmatrix} - \begin{bmatrix} \text{Inflow rate} \\ \text{of axial} \\ \text{momentum} \end{bmatrix} + \begin{bmatrix} \text{Storage rate} \\ \text{of axial} \\ \text{momentum} \end{bmatrix} = \begin{bmatrix} \text{Sum of the axial} \\ \text{forces acting on} \\ \text{the control volume} \end{bmatrix} \quad (25)$$

Applying the steady-state version of Eq. (25) to control Volume #1 in Fig. 2,

$$\frac{1}{g_c}\frac{d}{dz}\left[\frac{\rho_f j_f^2}{(1-\langle\alpha\rangle)} + \frac{\rho_g j_g^2}{\langle\alpha\rangle}\right] = -\frac{dp}{dz} - \frac{g}{g_c}\overline{\rho} - \frac{\tau_f P_f}{A_{xs}} \quad (26)$$

Equation (26) can be integrated, term-by-term, from $z=0$ to $z=L_{2\emptyset}$ (see Fig. 2).

The spatial acceleration term in Eq. (26) is zero from $z=0$ to $z=\lambda$ since, neglecting the effect of subcooling on liquid density, the momentum flux does not change with position in this region. Hence,

$$\frac{1}{g_c}\int_0^{L_{2\emptyset}}\frac{d}{dz}\left[\frac{\rho_f j_f^2}{(1-\langle\alpha\rangle)} + \frac{\rho_g j_g^2}{\langle\alpha\rangle}\right]dz = \frac{1}{g_c}\int_\lambda^{L_{2\emptyset}}\frac{d}{dz}\left[\frac{\rho_f j_f^2}{(1-\langle\alpha\rangle)} + \frac{\rho_g j_g^2}{\langle\alpha\rangle}\right]dz =$$

$$= \frac{1}{g_c}[\rho_g j_g^2(L_{2\emptyset})/\langle\alpha(L_{2\emptyset})\rangle - \rho_f j_f^2(\lambda)] \quad (27)$$

where we have made use of the fact that $j_g(\lambda)=0$, and have asssumed no carryover, i.e.: $j_f(L_{2\emptyset})=0$. Equation (27) may be written in terms of the mass flow rate, $w$, in the form,

$$\frac{1}{g_c}\int_0^{L_{2\emptyset}}\frac{d}{dz}\left[\frac{\rho_f j_f^2}{(1-\langle\alpha\rangle)} + \frac{\rho_g j_g^2}{\langle\alpha\rangle}\right]dz = \frac{w^2}{g_c A_{xs}^2}\left[\frac{1}{\rho_g\langle\alpha(L_{2\emptyset})\rangle} - \frac{1}{\rho_f}\right] \quad (28)$$

Now, integrating the first term on the right-hand side of Eq. (26), from $z=0$ to $z=L_{2\emptyset}$,

$$-\int_0^{L_{2\emptyset}}\frac{dp}{dz}dz = -p(L_{2\emptyset}) + p(0) \triangleq \Delta p$$

We can also note from Fig. 2 that, if we neglect the effect of subcooling on the liquid density,

$$\Delta p = \frac{g}{g_c}\rho_f \xi$$

Hence,

$$-\int_0^{L_{2\emptyset}}\frac{dp}{dz}dz = \Delta p = \frac{g}{g_c}\rho_f \xi \quad (29)$$

Let us now integrate the density head term in Eq. (26),

$$\int_0^{L_{2\emptyset}} \bar{\rho}(z)dz = \int_0^\lambda \rho_f dz + \int_\lambda^{L_{2\emptyset}} [\rho_f + (\rho_g - \rho_f)<\alpha>]dz =$$

$$= \rho_f \lambda + (L_{2\emptyset} - \lambda)\rho_f + (\rho_g - \rho_f) \int_\lambda^{L_{2\emptyset}} <\alpha(z)> dz \qquad (30)$$

Using Eq. (20) for the void fraction,

$$\int_\lambda^{L_{2\emptyset}} <\alpha(z)> dz = \int_\lambda^{L_{2\emptyset}} \left[\frac{K_1 \sin[k(Z-L_H/2)]+K_2}{K_3 \sin[k(Z-L_H/2)]+K_4}\right] dz$$

which can be rewritten as,

$$\int_\lambda^{L_{2\emptyset}} <\alpha(z)> dz = \frac{K_1}{K_3}(L_{2\emptyset} - \lambda) + \left(K_2 - \frac{K_1 K_4}{K_3}\right) \int_\lambda^{L_{2\emptyset}} \frac{dz}{K_3 \sin[k(Z-L_H/2)]+K_4} \qquad (31)$$

The integral in Eq. (31) may be reduced to a more convenient form by using the transformation, $\chi = k(z - L_H/2)$.
Hence,

$$\int_\lambda^{L_{2\emptyset}} \frac{dz}{K_3 \sin[k(Z-L_H/2)]+K_4} = \frac{1}{k} \int_{k(\lambda - L_H/2)}^{k(L_{2\emptyset} - L_H/2)} \frac{d\chi}{(K_3 \sin\chi + K_4)} \qquad (32)$$

From standard integral tables,

$$\int \frac{dz}{[K_3 \sin\chi + K_4]} = \begin{cases} \dfrac{2}{\sqrt{K_4^2 - K_3^2}} \tan^{-1}\left[\dfrac{K_4 \tan\left(\frac{\chi}{2}\right) + K_3}{\sqrt{K_4^2 - K_3^2}}\right], & K_4^2 > K_3^2 \\[2ex] \dfrac{1}{\sqrt{K_3^2 - K_4^2}} \ln\left[\dfrac{K_4 \tan\left(\frac{\chi}{2}\right) + K_3 - \sqrt{K_3^2 - K_4^2}}{K_4 \tan\left(\frac{\chi}{2}\right) + K_3 + \sqrt{K_3^2 - K_4^2}}\right], & K_3^2 > K_4^2 \end{cases} \qquad (33)$$

If we let,

$$\left.\begin{aligned} X &\triangleq K_4 \tan\left(\frac{\chi}{2}\right) + K_3 \\ \text{and,} & \\ Y &\triangleq \sqrt{|K_3^2 - K_4^2|} \end{aligned}\right\} \qquad (34)$$

Then Eq. (33) may be written more compactly as,

$$\int \frac{d\chi}{[K_3 \sin\chi + K_4]} = \begin{cases} I_1(X,Y) = \frac{2}{Y} \tan^{-1}\left[\frac{X}{Y}\right], & K_4^2 > K_3^2 \\ I_2(X,Y) = \frac{1}{Y} \ln\left(\frac{X-Y}{X+Y}\right), & K_3^2 > K_4^2 \end{cases} \quad (35)$$

Substituting Eqs. (31), (32) and (35) into (30), the integral of the density head tearm becomes,

$$\int_0^{L_{2\emptyset}} \bar{\rho}(z)dz = \rho_f \lambda + (L_{2\emptyset} - \lambda)\rho_f + (\rho_g - \rho_f) \cdot$$

$$\left[\frac{K_1}{K_3}(L_{2\emptyset} - \lambda) + (K_2 - \frac{K_1 K_4}{K_3}) \frac{1}{k}\{I_{1,2}[X(L_{2\emptyset}),Y] - I_{1,2}[X(\lambda),Y]\}\right] \quad (36)$$

where

$$X(z) = K_4 \tan\left[\frac{k}{2}(z - L_H/2)\right] + K_3 \quad (37)$$

and $I_{1,2}$ are defined in Eq. (35).

Let us not consider the wall friction term in Eq. (26),

$$\frac{\tau_w P_f}{A_{xs}} = \frac{4\tau_w}{D_H} = \frac{fw^2}{2g_c A_{xs}^2 \rho_f D_H} \quad (38)$$

Equation (38) is directly applicable to single-phase flows. For two-phase flows, a two-phase multiplier, $\emptyset_{lo}^2$, can be used, such that the total frictional pressure drop can be written.

$$\int_0^{L_{2\emptyset}} \frac{\tau_w P_f}{A_{xs}} dz = \int_0^{\lambda} \frac{fw^2}{2g_c A_{xs}^2 \rho_f D_H} dz + \frac{fw^2}{2g_c A_{xs}^2 \rho_f D_H} \int_\lambda^{L_{2\emptyset}} \emptyset_{lo}^2 dz \quad (39)$$

For simplicity, let us assume that the appropriate two-phase multiplier is that for homogeneuous flow (5),

$$\emptyset_{lo}^2 = 1 + \frac{v_{fg}}{v_f} x \quad (40)$$

where the flow quality, $x$, is defined as,

$$X(z) \triangleq \frac{\rho_g j_g(z)}{[\rho_g j_g(z) + \rho_f j_f(z)]} = \frac{\rho_g[K_1 \sin k(z - L_H/2) + K_2]}{(w/A_{xs})} \quad (41)$$

Hence, Eqs. (39), (40) and (41), imply,

$$\int_\lambda^{L_{2\emptyset}} \emptyset_{lo}^2(z)dz = \int_\lambda^{L_{2\emptyset}} \left\{1 + \frac{v_{fg}}{v_f}\left[\frac{\rho_g[K_1 \sin k(z - L_H/2) + K_2]}{(w/A_{xs})}\right]\right\} dz$$

which integrates to,

$$\int_\lambda^{L_{2\emptyset}} \emptyset_{lo}^2(z)dz = (L_{2\emptyset}-\lambda)\left\{1+\frac{v_{fg}}{v_f}\frac{\rho_g K_2}{(w/A_{xs})}\right\} - \frac{v_{fg}}{v_f}\frac{\rho_g K_1}{(w/A_{xs})}J(\lambda,L_{2\emptyset}) \qquad (42)$$

where,

$$J(\lambda,L_{2\emptyset}) \triangleq \frac{1}{k}[\text{Cos}k(L_{2\emptyset}-L_H/2) - \text{Cos}k(\lambda-L_H/2)] \qquad (43)$$

Substituting Eq. (42) into (39), the total frictional pressure drop is found to be,

$$\int_0^{L_{2\emptyset}} \frac{\tau_w P_f}{A_{xs}} dz = \frac{fw^2\lambda}{2g_c A_{xs}^2 \rho_f D_H} +$$

$$+ \frac{fw^2}{2g_c A_{xs}^2 \rho_f D_H}\left[(L_{2\emptyset}-\lambda)\left\{1+\frac{v_{fg}}{v_f}\frac{\rho_g K_2}{(w/A_{xs})}\right\} - \frac{v_{fg}}{v_f}\frac{\rho_g K_1}{(w/A_{xs})}J(\lambda,L_{2\emptyset})\right] \qquad (44)$$

We have now integrated all pressure drop components in Eq. (26). The final result is obained by combining Eqs. (28), (29), (36), and (44) with the integrated form of Eq. (26),

$$\Delta p = \frac{g}{g_c}\rho_f \xi = \frac{w^2}{g_c A_{xs}^2}\left[\frac{1}{\rho_g <\alpha(L_{2\emptyset})>} - \frac{1}{\rho_f}\right] + \frac{g}{g_c}\rho_f\lambda +$$

$$+ (L_{2\emptyset}-\lambda)\left[\frac{g}{g_c}\left\{\rho_f + \frac{K_1}{K_3}(\rho_g - \rho_f)\right\} + \frac{fw^2}{2g_c A_{xs}^2 \rho_f D_H}\left\{1+\frac{v_{fg}}{v_f}\frac{\rho_g K_2}{(w/A_{xs})}\right\}\right] +$$

$$+ \frac{fw^2\lambda}{2g_c A_{xs}^2 \rho_f D_H} - \frac{fw^2}{2g_c A_{xs}^2 \rho_f D_H}\frac{v_{fg}}{v_f}\frac{\rho_g K_1}{(w/A_{xs})}J(\lambda,L_{2\emptyset}) +$$

$$+ \frac{g}{g_c}(\rho_g - \rho_f)\left(K_2 - \frac{K_1 K_4}{K_3}\right)\frac{1}{k}\{I_{1,2}[X(L_{2\emptyset}),Y] - I_{1,2}[X(\lambda),Y]\} \qquad (45)$$

For a given power level, $Q$, and inlet subcooling, $\Delta h_{sub}$, Eqs. (23), (24) and (45) may be solved for the liquid head in the jet pump ($\xi$) required for a given two-phase level ($L_{2\emptyset}$), or vice versa. The former evaluation procedure is straigtforward and is thus preferred to the latter, which involves solving a transcendental equation for $L_{2\emptyset}$.

## 3. DISCUSSION OF THE MODEL

The analytical model was evaluated for a 49-rod BWR fuel bundle having an axial heat flux profile given by Eq. (15), where $K = 1.37$ and $k = 0.221784$ ft.$^{-1}$. These parameters were chosen so that the model could be compared with existing experimental data [1,2,3] on BWR long-term cooling (no comparable PWR exists).

As expected, the calculated level, $L_{2\emptyset}$, was found to be quite sensitive to the drift velocity, $V_{gj}$, and the concentration parameter, $C_0$. As can be seen in Fig. 3, both $C_0$ and $V_{gj}$, have a similar effect on the calculated results. An increase in either parameter causes the void fraction to decrease, hence increasing the collapsed liquid level ($\xi$) necessary to submerge the core (i.e., to achieve $L_{2\emptyset}=L_H$). As can be seen in Fig. 4, there is considerabel "scatter" between the various data sets [1,2,3], however, due to the method of subcooling control, the German data [1] is considered to be the most reliable. It can be seen that values of $C_0 = 1.1$, and a $V_{gj}$, coefficient of $k_3 = 2.5$, give the best agreement with these data, and follow the observed trends with core power level. It should be noted that these valuse for the drift-flux parameters ($C_0$ and $V_{gj}$) may not be appropriate for higher system pressures, however adequate data does not currently exist to quantify the pressure dependence.

We are now in a positioh to evaluate the model for the two-phase level in a PWR core during conditions of severely reduced liquid mass inventory. For purposes of this evaluation, a typical large (1000 Mw$_e$) PWR was chosen.

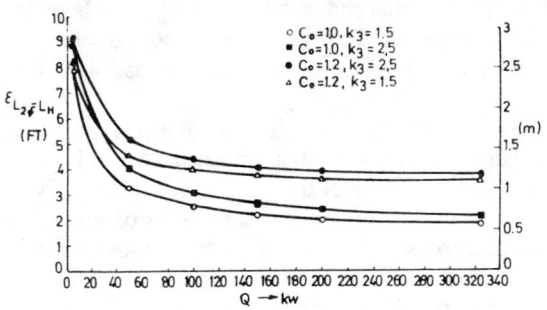

Figure 3. Effect of drift-flux parameter variation

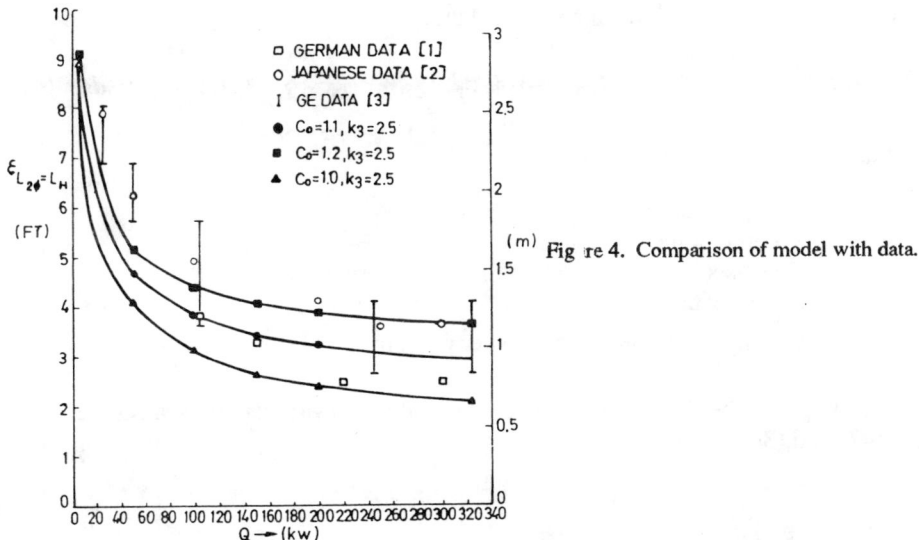

Figure 4. Comparison of model with data.

The primary system pressure used in this evaluation was containment backpressure (0.21 MPa). It should be noted that reduced two-phase levels would be predicted for higher pressure in the primary system.

Figures 5 and 6 show the effect of the collapsed liquid level in the downcomer ($\xi$) on the normalized two-phase level in the core ($L_{2\emptyset}/L_H$), for various power levels and subcoolings. As expected, the effect of inlet subcooling is to reduce the two-phase level in the core. If, due to heat transfer from the pressure vessel wall and components, the fluid in the downcomer region is a two-phase mixture, then the curves shown in Fig. 5 are appropriate when the collapsed liquid level ($\xi$) is evaluated by,

$$\xi = \frac{1}{\rho_{f}} \int_{0}^{\xi_{2\emptyset}} [\rho_f(1-<\alpha_{dc}>) + \rho_g<\alpha_{dc}>]dz \qquad (46)$$

Using curves such as Figs. 5 and 6, one can easily cross-plot the collapsed liquid level in the downcomer required to completely submerge the core ($\xi_{L_{2\emptyset}=L_H}$), versus the normalized decay heat power level ($Q/Q_{max}$). Figure 7 indicates that, for a given inlet subcooling, the core will be submerged as long as the power level is sufficient to produce the necessary froth level. To determine the normalized power level which exist at a given time after a SCRAM it is convenient to use standard decay heat curves [6], such as tose shown in Fig. 8.

It should be obvious to the reader that plant-specific calculations can be readily done using the model presented herein. Curves, such as those shown in Figs. 7 and 8, would then be available to operational personnel for a rapid, yet reasonably accurate, evaluation of the two-phase level in the core during conditions of severely reduced liquid mass inventory. In order to use these curves, however, one must have direct or indirect information on the collapsed liquid level in the downcomer. If appropriate liquid level detectors are installed in the downcomer, then the evaluation of $\xi$ is straightforward. On the other hand, if the residual mass (M) in the system is known, one can evaluate $\xi$ from a mass balance,

$$M = \rho_g[V_p - V_{Lp} - \xi A_{dc} - L_{2\emptyset}A_{xs}] + \rho_f \xi A_{dc} + V_{Lp}\,\bar{\rho}\,_{Lp} + A_{xs}\int_{0}^{L_{2\emptyset}} \bar{\rho}\,(z)dz \qquad (47)$$

where,

$V_p$ — Total volume of the primary system
$V_{Lp}$ — Volume of the lower plenum
$A_{dc}$ — Cross-sectional flow area of the downcomer
$A_{xs}$ — Cross-sectional flow area of the core
$\bar{\rho}\,_{Lp}$ — Density of the fluid in the lower plenum

As an example, if the fluid in the downcomer and the lower plenum is subcooled, Eqs. (47) and (36) yield,

$$\xi = \frac{1}{(\rho_f - \rho_g)A_{dc}} \left\{ M - (\rho_f - \rho_g)A_{x-s} \times \right.$$

$$\left. \times \left[ L_{2\emptyset} + \frac{K_1}{K_3}(L_{2\emptyset}-\lambda) + \left(K_2 - \frac{K_1 K_4}{K_3}\right)\frac{1}{k}\left[I_{1,2}(X(L_{2\emptyset}),Y) - I_{1,2}(X(\lambda),Y)\right]\right] - \right.$$

$$\left. - \rho_g V_p - (\bar{\rho}\,_{Lp} - \rho_g)V_{Lp} \right\} \qquad (48)$$

For this specific case, Eqs. (48), (45), (23) and (24) can be solved simultaneously to give the position of the two-phase level in the core ($L_{2\emptyset}$).

While the analytical model presented here–in is a powerful tool for evaluating certain phenomena of safety significance it should be realized by the reader that it has

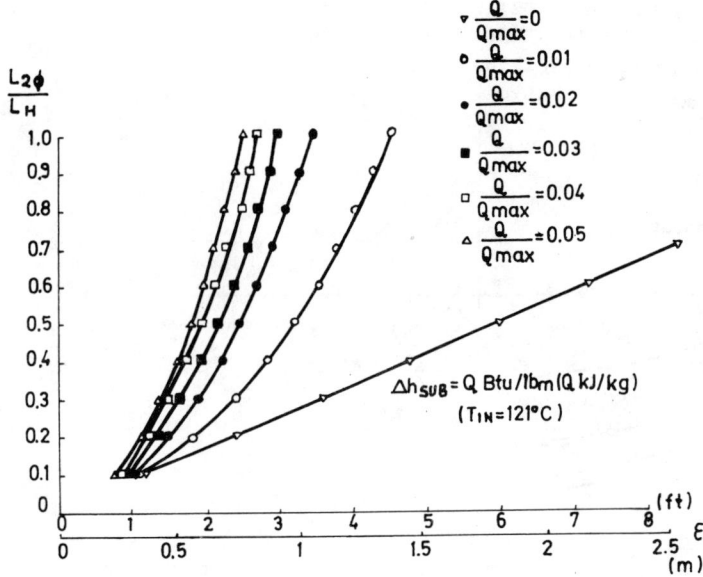

Figure 5. Two-phase levels for zero subcooling.

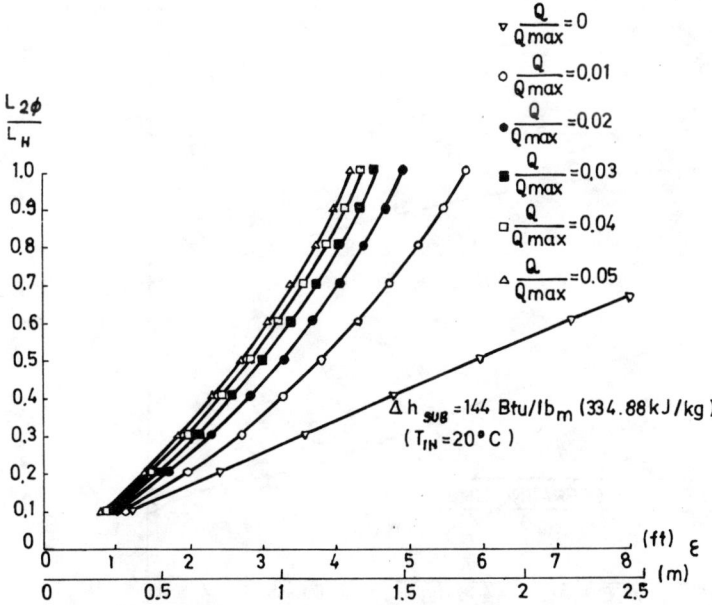

Figure 6. Two-phase levels for high subcooling.

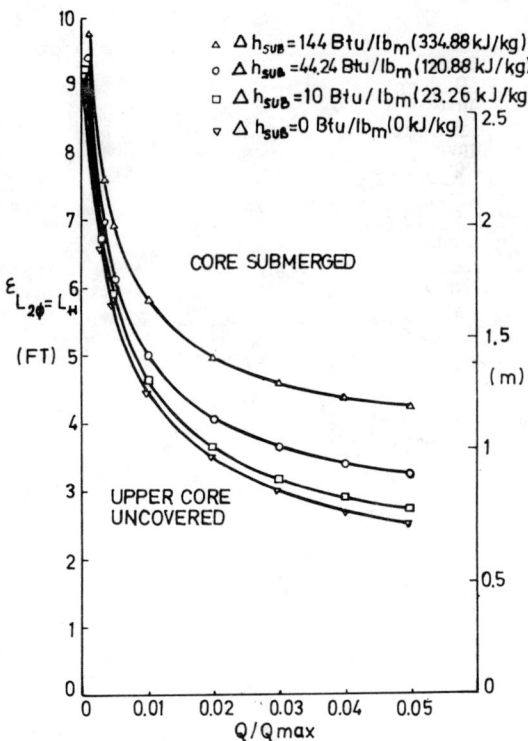

Figure 7. Core submergence map.

Figure 8. Decay heat curves.

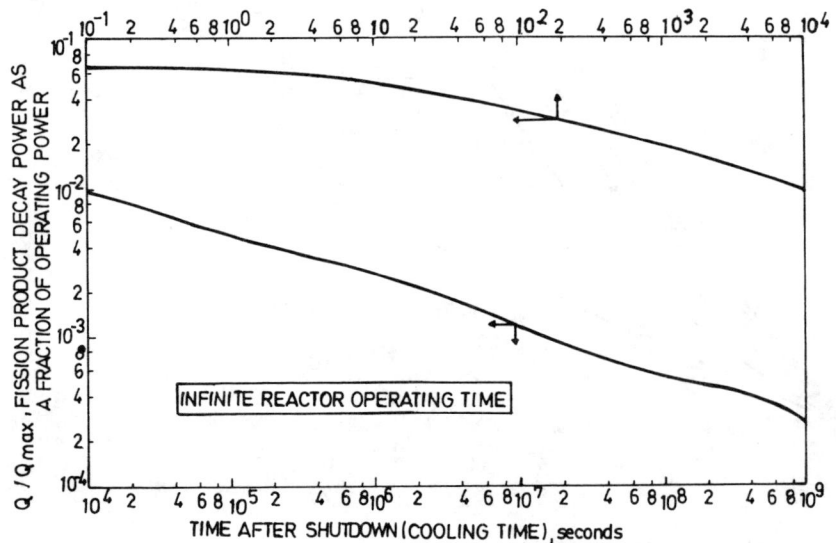

some important limitations. Specifically, transient phenomena have not been considered. In addition, situations such as preferential core voiding (which may have oc cured at some point during the TMI-2 incident) and the effect of "steam binding" induced back-pressures have not been considered in the present model. Nevertheless, the model presented herein is potentially valuable for a whole host of problems concerning two-phase levels in LWRs (e.g., the analysis of BWR long term cooling (7)).

## 4. SUMMARY

A closed-form analytical model for the position of the two-phase level in a PWR core during conditions of severely reduced mass inventory, has been derived from first principles.

The model is based on drift-flux techniques. The drift-flux parameters ($C_0$ and $V_{gj}$) have been optimized to produce agreement with the existing data and the model has been applied to a typical large PWR. It is shown that core submergence will occur as long as the collapsed liquid level in the downcomer ($\xi$) and the normalized core decay heat power level ($Q/Q_{max}$) are sufficiently high.

## 5. ACKNOWLEDGEMETS

The numerical evaluations and helpful comments of Mr. P. Kamath (RPI) are gratefully acknowledged.

## NOMENCLATURE

| | | |
|---|---|---|
| $A_{xs}$ | — | Cross-sectional flow area in the core. |
| $A_{dc}$ | — | Cross-sectional flow area in downcomer annulus. |
| $C_0$ | — | $\frac{<\alpha j>}{<\alpha><j>}$, concentration parameter. |
| $D_H$ | — | Hydraulic diameter. |
| $f$ | — | Friction factor. |
| $g$ | — | Acceleration of gravity. |
| $g_c$ | — | Gravitational conversion factor. |
| $h_k$ | — | Enthalpy of phase-$k$. |
| $h_{fg}$ | — | Latent heat ($h_g - h_f$). |
| $h_i$ | — | Inlet enthalpy. |
| $\Delta h_{sub}$ | — | $h_f - h_i$, inlet subcooling. |
| $j_k$ | — | Superficial velocity of phase-$k$. |
| $j$ | — | $j_g + j_f$, velocity of the center-of-volume of the two-phase flow. |
| $L_H$ | — | Heated length of core. |
| $L_{2\emptyset}$ | — | Two-phase level in the core. |
| $M$ | — | Mass of fluid in primary system. |
| $P$ | — | Primary system pressure. |
| $P_H$ | — | Heated perimeter. |
| $P_f$ | — | Friction perimeter. |
| $q''$ | — | Local heat flux. |

$\overline{q''}$ = $\dfrac{1}{L_H}\int_0^{L_H} q''(z)dz$, axial-average heat flux.

$Q$ — $\overline{q''}P_H L_H$, core power level.

$Q_{max}$ — Rated power level of core.

$u_k$ — Velocity of phase-$k$.

$v_k$ — Specific volume of phase-$k$.

$v_{fg}$ — $v_g$-$v_f$

$V_{gj}$ — $\dfrac{<\alpha(u_g-j)>}{<\alpha>}$, drift velocity.

$V_n$ — Volume of region-$n$.

$w$ — Flow rate.

$x$ — Flow quality.

$z$ — Axial position.

## GREEK SYMBOLS

$<\alpha_{dc}>$ — One-dimensional void fraction in the downcomer annulus.

$<\alpha(Z)>$ — Local one-dimensional void fraction in the core.

$\rho_k$ — Density of phase-$k$.

$\overline{\rho_k}$ — $\rho_f(1-<\alpha_n>) + \rho_g<\alpha_n>$, two-phase density in region-$n$.

$\overline{\rho}$ — $\rho_f(1-<\alpha>) + \rho_g<\alpha>$, two-phase density in the core.

$\xi$ — Collapsed liquid level in the downcomer annulus.

$\xi_{2\emptyset}$ — Two-phase level in the downcomer annulus.

$\sigma$ — Surface tension.

$\lambda$ — Axial position of the boiling boundary.

$\tau_w$ — Wall shear.

$\emptyset_{lo}^2$ — Two-phase friction miltiplier.

$<\psi>$ — $\dfrac{1}{A_{xs}}\iint \psi da$, cross-sectional average of $\psi$.

## SUBSCRIPTS

$dc$ — Downcomer

$f$ — Saturated liquid

$g$ — Saturated vapor

$i$ — Inlet

$l$ — Subcooled liquid

$2\emptyset$ — Two-phase

$p$ — Primary system

$Lp$ — Lower plenum

# REFERENCES

1. RIEDLE, K., GAUL, H.P., RUTHROF, K. AND SARKAR, J., *"Reflood and Spray Cooling Heat Transfer in PWR and BWR Bundles"*, ASME Preprint 76-HT-10, 1976.
2. OGASAWARA, H. AND TAKASHIMA, Y., *"Cooling Mechanism of the Low Pressure Coolant Injection System of Boiling Water Reactors and Other Studies on the Loss-of-Coolant accident Phenomena"*, ANS Topical Meeting on Water Reactor Safety, Salt Lake City, 1973, p. 351.
3. SCHRAUG, F.A. AND LEONARD, J.E., *"Thermal Response of a Recator Fuel Assembly Cooled by Flooding Under Loss-Coolant Conditions"*, ASME Preprint 68-WA/NE-9, 1968.
4. ZUBER, N., AND FINDLAY, J., *"Average Volumetric Concentration in Two-Phase Flow Systems"*, J. Heat Transfer, 1965.
5. LAHEY, R.T., JR. AND MOODY, F.J., *"Thermal-Hydraulics of a Boiling Water Nuclear Reactor"*, ANS Monograph, 1977.
6. ANS-5.1, *"Decay Energy Release Rates Following Shutdown of Uranium-Fueled Thermal Reactors"*, 1971.
7. LAHEY, R.T., JR. AND KAMATH, P.S., *"The Analysis of BWR Long-Term Cooling"*, ANS Journal of Nuclear Technology, 1980.

International Seminar 1981

# HEAT EXCHANGERS – THEORY AND PRACTICE

**J. Taborek**

*University of Texas, Austin, USA*

The topic of the Seminar was very timely, as great attention was given to heat exchanger design as related to the effectiveness of thermal energy use, re–use and conservation, as these became an imperative during the energy crisis. Selected papers (52) and 11 Invited Lectures in major subject groups were published in [1]. Space limitations permit to include in this review only such contributions which became lasting references and/or influenced subsequent research.

*Boiling* is one of the areas of heat transfer where new developments, even of rather fundamental nature, are required. A group of four papers presented at the Seminar by researchers from Prof. Schluender's Institute at the University of Karlsruhe laid a foundation to better understanding of the flow boiling phenomena. Subsequent work by Steiner resulted in formulation of a new approach to flow boiling which may be loosely termed the "method of predominant coefficients", as published 1986 in [2].

It is imperative that the three methods now recognized in the literature, namely a) Chen's "suppression" factor method; b) M.M.Schah's method and c) Steiner's "redominant coefficient" method, should be compared and evaluated.

Enhanced boiling surfaces were represented at the Seminar by two papers which are frequently quoted. Work on a variety of enhancement techniques for boiling, especially at low temperature differences, continues to be one of the prime subjects of research.

*Condensation* is a much more mature science (described by Nusselts in 1916) compared to boiling, which was defined in gross fundamentals only in 1932. Nevertheless, many outstanding problems in condensation remain. These were represented at the Seminar by several significant papers in the area of reflux condensers, pressure drop inside horizontal tubes and partial condensation. The paper by J.M.McNaught "An Assessment of Design Methods for Condensation of Vapors from a Noncondensing Gas" is an excellent survey on this very important subject, useful to the heat exchanger designer as well as to the researcher, and was therefore selected for inclusion into this Volume as a reference of fundamental importance.

The paper evaluates a representative set of data on multi-component vapor mixtures with noncondensible gases against the three presently most popular methods, namely a) the Silver or Bell and Ghally "resistance proration" method; b) the same method with a mass transfer correction factor; and c) the classical Colburn-

Hougen method and its subsequent extension termed the "matrix method". It is of interest to notice that with the exception of some extreme compositions, the resistance proration approach with mass transfer correction emerged very favorably, compared to the much more demanding mass transfer models.

*Heat transfer and pressure drop in tube banks* was represented by three rather definitive papers on specialized topics: augmentation of heat transfer by surface roughness; yawed tube banks; flow at high Reynolds numbers ($< 10^7$). While all the papers were largely oriented toward nuclear energy interests, the flow in yawed tube banks - essentially present in all baffled shell and tube heat exchangers - should be noted by workers in that field for advanced model development.

*Compact heat exchangers* and their applications were covered by two well organized survey lectures, first dealing with compact heat exchangers in general (R.K.Shah and R.L.Webb) and second their application to cryogenic process industry (J.M.Robertson). Another paper of interest dealt with compact exchangers fabricated from ceramic materials.

It should be noted here that the use of compact heat exchangers is gaining steadily acceptance in a variety of applications and a voluminous literature was published since the Seminar.

*Fludized bed systems* were represented by the first publication in English of a new approach to this subject by H.Martin. Several subsequent publications of this method since the Seminar represent later developments.

*Heat exchanger design* session was noted by the lecture by R.V.Macbeth. He presented a large set of data on shell and tube heat exchanger with six types of baffles and concluded that a given pressure drop produces the same value of the heat transfer coefficient, virtually regardless of the baffle configuration (some extremes excepted). However, the heat transfer data were obtained by electrochemical methods and thus reflect the effects of flow velocity only, and the very crucial effects of the mean temperature difference are neglected. Thus the validity of the results (and the measurement technique for that matter), is restricted to investigation of the effect of a single variable in otherwise same geometries. Unfortunately there was no follow-up by the author.

A recent study of the same subject based on experimental data shows great differences in the effectiveness of pressure drop to heat transfer conversion between various baffle types [3]. Hopefully, this very important subject will be clarified in subsequent discussions and papers.

*Fouling* session introduced several excellent papers which became widely quoted, namely "Water Quality Effects on Fouling" by A.P.Watkinson; "Fouling of Cooling Water" by Lahm and Knudsen; and "Fouling in Plate Exchangers" by L.Novak.

The survey of "Fouling in Crude Oil Preheater Trains" by Lambourne and Durrieu is probably the best work in this area.

It is of general interest to notice that while the Art (or Science ?) of fouling keeps accumulating data and better understanding of the mechanisms involved by small but systematic steps, no "real" breakthrough is imminent in the way fouling is actually handled in design practice. There is great need for some summary of the past work and how it could reflect on improved practices.

## REFERENCES

1. J.TABOREK, G.F.HEWITT and N.AFGAN edd. "Heat Exchanger - Theory and Practice", Hemisphere Publ. Corp., 1982.
2. D.STEINER, Section "Boiling" in VDI Waermeatlas, VDI Verlag 1986.
3. TABOREK, J. and SHARIF, A., "Effectiveness of Heat Transfer to Pressure Drop Conversion in Various Shell Side Flow Arrangements"; paper presented at the ASME/AIChE Heat Tr.Conference, Pittsburgh, PA, 1987. To be published in Heat Tr.Eng.

# FOULING IN CRUDE OIL PREHEAT TRAINS

## G. A. Lambourn

TOTAL – C.F.R. Paris, France

## M. Durrieu

TOTAL – C.F.R. Harfleur, France

## 1. MONITORING PREHEAT EXCHANGER PERFORMANCE

### 1.1. THE NECESSITY FOR GOOD HARDWARE AND SOFTWARE

Typically the preheat temperature at the inlet of the crude unit heater decreases by 30–50 °C over a 12 month period (2); this loss of preheat temperature can be particularly serious for units which have elaborate heat recovery systems where preheat temperatures may approach 300°C.

As shown later some fouling phenomena are transient in nature, this fact together with the need to monitor preheat train performance efficiently convinced TOTAL–C.F.R. of the need to improve the frequency and accuracy of its procedures for calculating fouling resistances.

All exchangers downstream of the desalter are now equiped with thermocouples, replacing the previous thermowells; in some cases the additional thermocouples are directly linked to online process control computers. In order to provide a uniform basis of calculation HTRI's ST–4 program was selected and used on an off–line basis; this type of advanced program is sufficiently accurate to check the efficiency of cleaning ($R_s \neq 0$ after cleaning) which is not always the case for methods in the accessible literature.

Data input has now been simplified by using the ST–4 program to generate simple power type relationships based on flow–rates and physical properties such as viscosity. These are then input into the on–line computer for daily use by operating personnel.

### 1.2. EXCHANGER FOULING RATES MEASURED BY RIGOROUS METHODS

Figure 1 below shows the measured fouling rates for the hottest crude/residue exchangers of two different crude distillation units of relatively modern design. The operating characteristics and structural details of the two sets of exchangers are shown in Table 1. The fouling rates shown compare well with those reported by Haluska for similar exchangers.

Figure 2 shows the measured fouling rates for crude/hot reflux exchangers in the same two units, the evolution of fouling is here much more erratic being due to excessive break through of salt from the desalter.

Fouling curves of the same form as those shown were reported by Van der Wee and Tritsmans (3).

## TABLE 1

### CRUDE/HOT RESIDUE EXCHANGER DESIGN CHARACTERISTICS (BY HTRI ST-4 PROGRAM)

|  | UNITS | UNIT A | | UNIT B | |
|---|---|---|---|---|---|
|  |  | CRUDE | RESIDUE | CRUDE | RESIDUE |
|  | tonnes/d | 30000 |  | 20000 |  |
| Flowrate | kg/s | 323.5 | 117.1 | 217.7 | 75.7 |
| Temperature IN | °C | 229 | 374 | 199 | 361 |
| OUT | °C | 269 | 265 | 247 | 224 |
| Duty | MW | 37 |  | 29 |  |
| Shell type |  |  | AES |  | AES |
| Tube length | m | 6.1 |  | 6.1 |  |
| Tube diam | mm | 25.4 |  | 19.4 |  |
| N. of passes |  | 2 |  | 2 |  |
| Baffle spacing/cut | mm/% |  | 250/15.0 |  | 229/10.0 |
| Surface/Shell | m$^2$ | 536 |  | 484 |  |
| N Shells in series |  | 2 | 2 | 2 | 2 |
| in parallel |  | 2 | 2 | 2 | 2 |
| Tube side velocity | m/s | 1.34 |  | 1.32 |  |
| Reynolds No. in |  | 30000 | 13800 | 25420 | 6870 |
| out |  | 37000 | 6420 | 35000 | 1720 |
| Shell side velocity |  |  |  |  |  |
| Bundle/Window | m/s |  | 0.94/0.98 |  | 0.57/1.22 |
| Clean heat trans coeff. | W/m$^2$K |  | 490 |  | 468 |
| Service heat trans coeff. | W/m$^2$K |  | 280 |  | 290 |
| Fouling factor (Design) | m$^2$/WK | .0005 | .0009 | .00035 | .0009 |

In each case it is to be noted that the fouling resistances following 6 months of operation considerably exceed the recommended TEMA values.

In addition it is clear that the rate of fouling increase in the crude/residue exchangers of UNIT B is considerably greater than in the same exchangers of UNIT A. A very careful review of all the pertinent operating conditions has failed to pin-point the exact reason for this difference; however as will be explained later fouling on the residue side appears important for UNIT B. As can be seen from Table 1 UNIT B's exchangers suffer from excessively low velocities on the residue (shell) side.

### 1.3. FOULING RESISTANCES ADOPTED FOR DESIGN PURPOSES

As a result of these repeated observations TOTAL – C. F. R. now specifies a fouling resistance of $40 \times 10^{-4}$ m$^2$/WK for average quality crude oils between 180

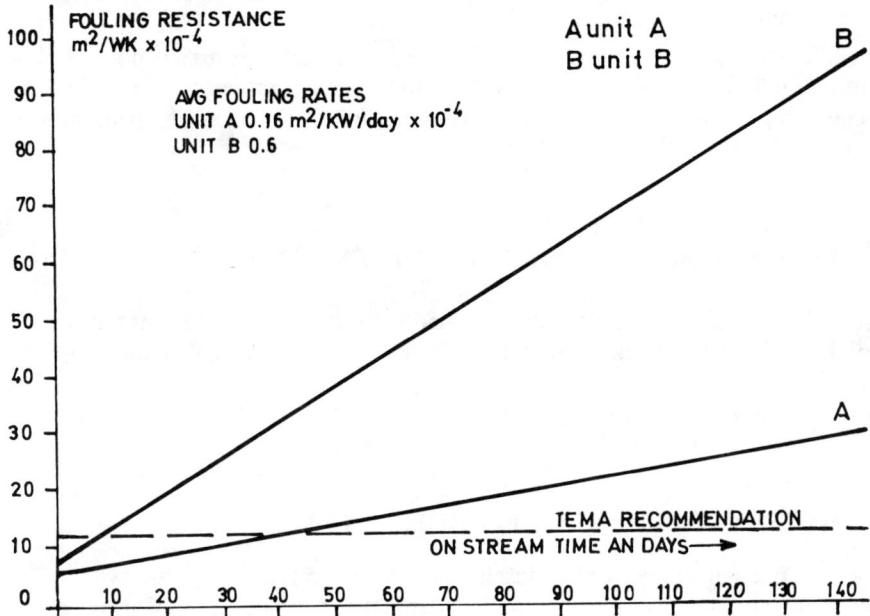

Figure 1. Evolution of fouling resistances for crude/residue exchangers.

Figure 2. Evolution of fouling resistances for crude/hot reflux exchangers.

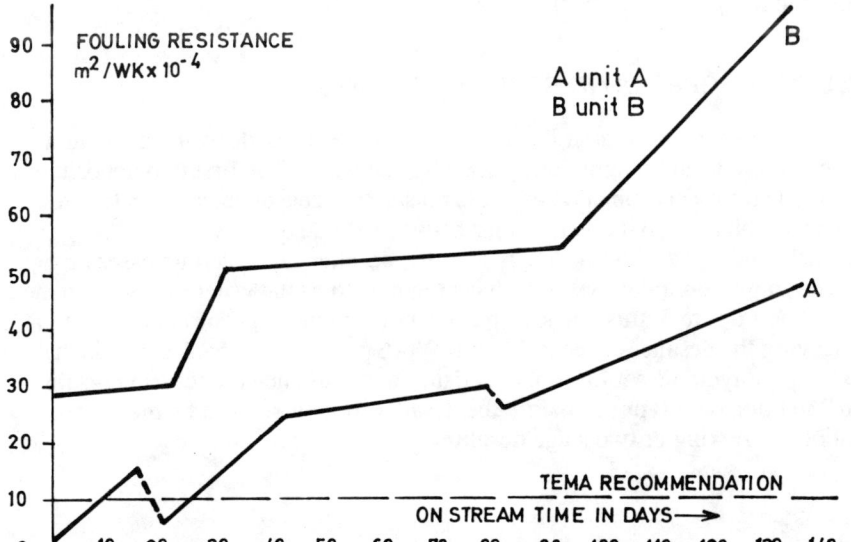

and 300°C, this increased fouling resistance is aimed at a cycle length of at least six months betwen cleanings.

The effect of this greater value on the required heat exchange area is not as large as might first be expected as many preheat improvement projects are based either on recovering additional low level heat from products or by exporting heat from pumparound refluxes to other services.

## 2. MECHANISMS OF FOULING FOR CRUDE OILS

The reduction of the levels of fouling found in crude oil preheat trains requires first of all that valid mechanisms are available for the different types of fouling encountered.

The examination and analysis of the performance of different preheat trains enables the following conclusions to be drawn.

### 2.1. EXCHANGERS UP STREAM OF DESALTER

The cold exchangers before the desalter are subject to fouling by salts which are easily eliminated by water injection to the charge pump, an injection of stripped process water (2 to 6 % vol) is sufficient to cure this problem.

It is worthwhile to point that although desalter temperatures are normally below 140°C the use of split–stream preheat trains before the desalter can exacerbate this problem. Such a configuration has recently become popular as a means at improving preheat train low level heat recovery by a better match of heat capacity flowrates (1).

It is relatively easy to allow the temperature in one branch of such a system to exceed 180°C while still maintaining the specified desalter temperature. Thus the monitoring and control of crude oil temperatures in each branch must be borne in mind when designing such a system.

### 2.2. EXCHANGERS DOWNSTREAM OF DESALTER

The exchangers situated after the desalter can suffer from deposits of calcium sulfate, of bicarbonate and magnesium salts from sea water. The first two deposits can be precipitated by a combination of crude oil salinity, caustic injection rate and desalting water which is too hard and occur between 180 and 200°C.

The exchangers placed immediately after the desalter can in certain cases be subject to very rapid fouling by salt which is precipitated as the water dissolves in the crude as shown in figure 3; this phenomena occurs when the concentration of salt in the crude leaving the desalter exceeds 25 ppm Wt (6 pt b). Heavy crudes of less than 30 API exhibit unfavorable settling characteristics and thus such crudes can contain more than 250 ppm Wt (60 pt b)* before the desalter. Recourse must be made in such situations to fluxing or two stage desalting.

---

* pt b pounds per thousand barrels.

## 2.3. CRUDE/HOT RESIDUE EXCHANGERS

The hottest exchangers (generally crude/residue exchangers) are subject to heavy fouling, generally on the tube side. The deposits are formed from a mixture of asphaltenes and iron sulfide strongly bound together, see Table 2.

The composition of these deposits varies with the crude oil treated; in the case of light North African crudes (API gravity above 40°) it is composed essentially of iron salts (up to 80%) and can be easily removed by acid washing. The treatment of heavier crudes (API gravity below 37°) results in the appearance of asphaltic deposits which cannot be easily removed by chemical treatment.

### TABLE 2
AVERAGE COMPOSITION OF THE CRUDE SIDE DEPOSITS IN CRUDE/RESID EXCHANGERS

|  |  | Various crudes API Gravity ≅ 34° | Light crudes API Gravity ≥ 40° |
|---|---|---|---|
| Asphaltenes + Carbenes + Condensation Products | Wt % | 60-75 | 3-10 |
| Water soluble Salts | Wt % | 1-5 | 1-5 |
| Iron salts (Oxydes and/or sulfides) | Wt % | 20-35 | 75-90 |

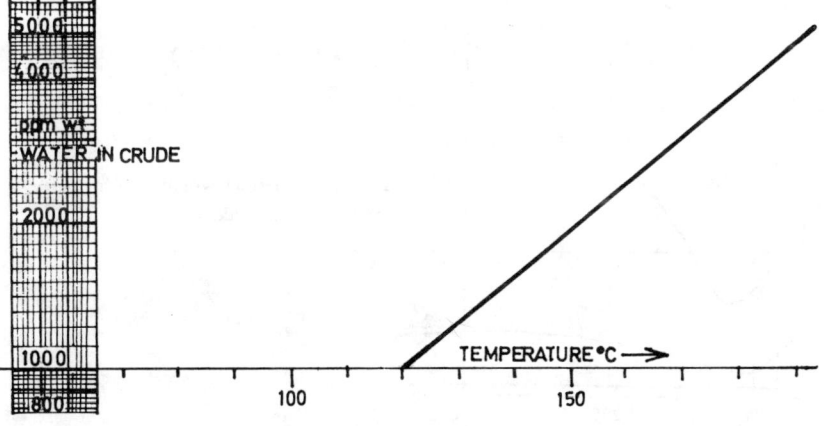

Figure 3. Solubility of water in Middle East crudes (30 API - 880kg/m$^3$).

## 3. A NEW METHOD FOR CONTROLLING CRUDE OIL FOULING

### 3.1. NATURE OF STABLE DESALTER EMULSIONS

About 4 years ago work was started by TOTAL–C.F.R. on a new approach to anti–fouling control which aims at limiting asphaltene – iron sulfide deposits in the hottest crude/residue exchangers and has formed the subject of a systematic study in our laboratories. (5) (6)

Van der Wee & Tritsmans emphasised some time ago the importance of the emulsion which is after formed at the water–interface level in a crude oil desalter. It is surprising that their pioneer work has not been followed up since desalter demulsifiers are selected and injected solely on the basis of maintaining an oil free desalter effluent water while anti–foulants are injected into the crude downstream of the desalter.

Evaluations of common crudes (density below 840 kg/m$^3$ or 37 API) show the following characteristics upstream of the desalter:

| | | |
|---|---|---|
| Water soluble salts (as NaCl) | ppm Wt | 50–300 (15–100 ptb) |
| Water content | % Wt | 0.2–0.8 |
| Suspended solids content: | ppm Wt | 100–530 |
| Iron salts | ppm | 5–30 |
| Insoluble asphaltenes | " | 100–500 |
| Sand etc | " | 5 |

Meanwhile we have found out that the insoluble asphaltene content varies significantly with temperature (Fig. 4) as suspended asphaltenes are partly redissolved near 130°C and then reprecipitated by thermal deasphalting above 200°C.

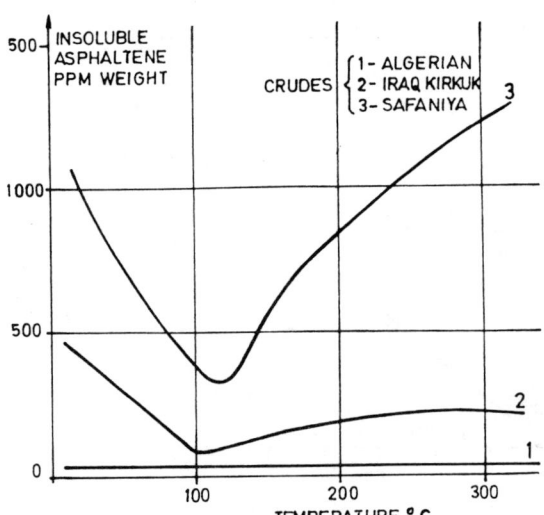

Figure 4. Thermal evolution of insoluble asphaltenes in crude oils.

The usual water content of the desalted crude ranges from 0.2 to 0.6 % Wt and a typical single stage desalting efficiency is 90%. Nevertheless certain situations can arise which result in the following composition:

| | | |
|---|---|---|
| Water soluble salts (as NaCl) | ppm Wt | 10–20 (3–6 ptb) |
| Water | Wt % | 0.7–1.0 |
| Suspended solids content | ppm Wt | 250–800 |
| Iron salts | " | 100–300 |
| Insoluble asphaltenes | " | 150–500 |

During such an upset a thick "stable emulsion" layer gradually forms above the normal oil/water interface, having the following composition:

| | | |
|---|---|---|
| Desalting water | % Wt on emulsion | 50 – 75 |
| Crude | "         "         " | 50 – 25 |
| Solid particles | % Wt on crude | 2 – 15 |
| Viscosity | cPs 60°C | 1000 – 10000 |
| Waterdrop diameter | m | 10 – 100 |

The "stable emulsion" is immiscible with water and oil at desalter temperatures.

In addition a microscopic examination shows a shell of colloidal asphaltenes and solid particies which coats the water droplets and effectively impedes any natural coalescence.

The solid particies have the following typical composition:

| | | |
|---|---|---|
| Iron oxides/sulfides | % Wt | 50 – 70 |
| Insoluble asphaltenes | " | 50 – 30 |

This composition closely approximates that of the fouling deposits found in exchangers operating above 180°C on the crude side. The stability of the emulsion is promoted by the presence of solid particles that are equally well wetted by the oil and water phases.

Photo 1. Tube side fouling crude hot residue exchanger UNIT A.

The stable emulsion builds up and is then removed in an irregular fashion by the desalted crude. The severity of the fouling deposit thus caused can be seen from the thickness of the deposit shown in Photo 1.

It should be emphasised that this mechanism is applicable to crude oils having an asphaltene content above 1.3 % Wt, the practice of recycling recovered slop oils to the fresh crude can lower this threshold level to about 1.0 % Wt.

## 3.2. PILOT PLANT STUDY OF FOULING CAUSED BY STABLE EMULSIONS

The mechanism proposed by Van der Wee & Tritsmans was confirmed in a series of pilot plant experiments previously described (4) in which a variety of clean crudes were "doped" with a stable emulsion and heated to 320°C under laminar flow conditions (velocity 20 mm/s $N_{Re}=250$).

A test rig was used to develop a dual purpose additive designed to:

Break the stable emulsion in the desalter
Prevent solids deposition by dispersing solids
and preferentially wetting exchanger tube walls.

In the pilot plant the additive formulation finally retained resulted in a six – fold reduction in the fouling rates measured under standardised conditions . (Fig. 5) (6)

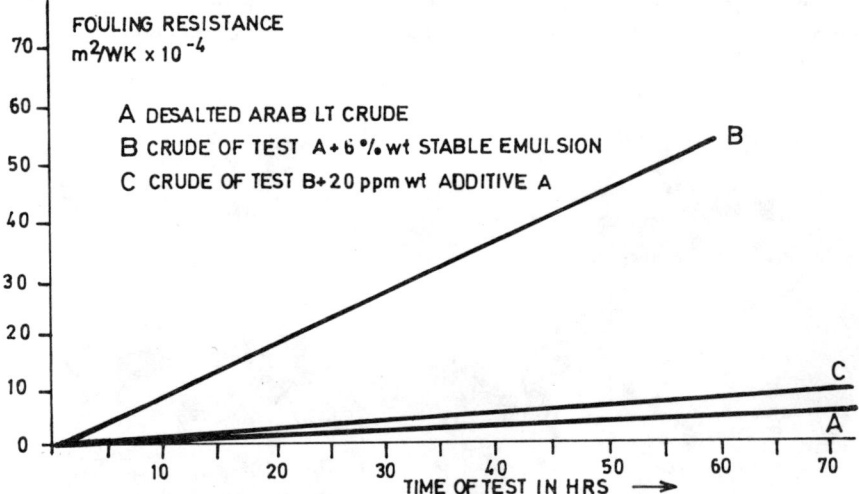

Figure 5. Fouling resistance build-up measured on pilot plant.

## 3.3. AN IMPROVED ADDITIVE INJECTION SYSTEM

A novel injection system has also been developed for the injection of this additive by which the surfactant is injected directly into the "stable emulsion" layer thus increasing the effective concentration of the additive as compared to simple injection into the bulk of the crude. (5).

## 4. INDUSTRIAL RESULTS WITH ADDITIVE INJECTION

### 4.1. FOULING RATES WITH ADDITIVE INJECTION

To date two industrial trials have been carried out using an additive developed by TOTAL and marketed by Universal Matthey Products.

The first trial was made on the crude/hot residue exchangers of UNIT A referred to in para. 1.2.; the injection rate was 13 ppm Wt. This trial was followed by a similar application on Unit B previously referred to.

The results are shown in Figure 6 from which it can be concluded that in each case the fouling rate has been halved as compared with the reference cases. As remarked on earlier it would appear that residue side fouling is a significant factor for the exchangers of Unit B.

Confirmation of the additives' efficiency can be seen in Photos 2 and 3 which show the tube side conditions for Unit A's crude residue exchangers.

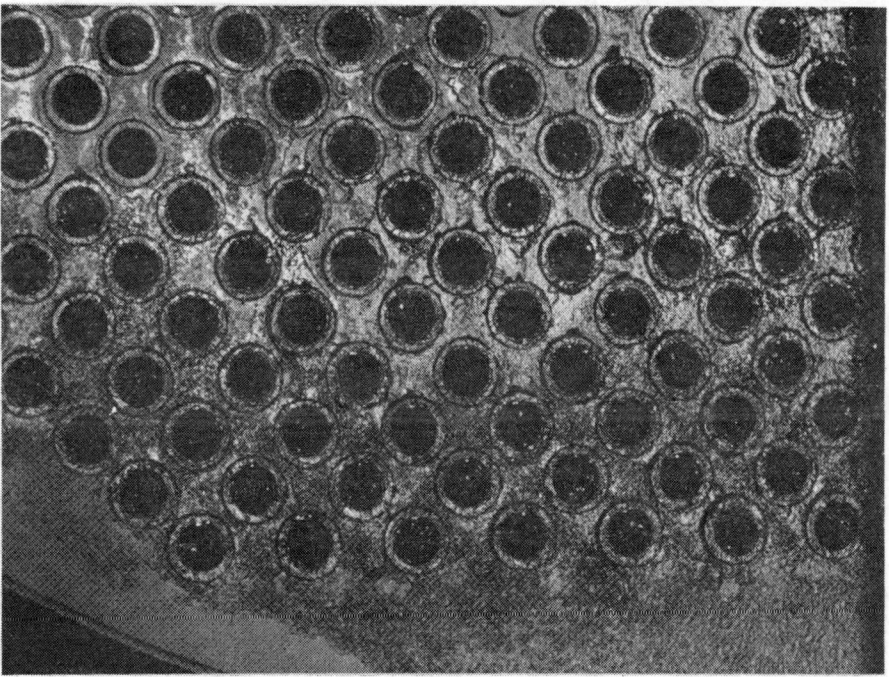

Photo 2. Crude side crude/resid exchanger UNIT A without additive.

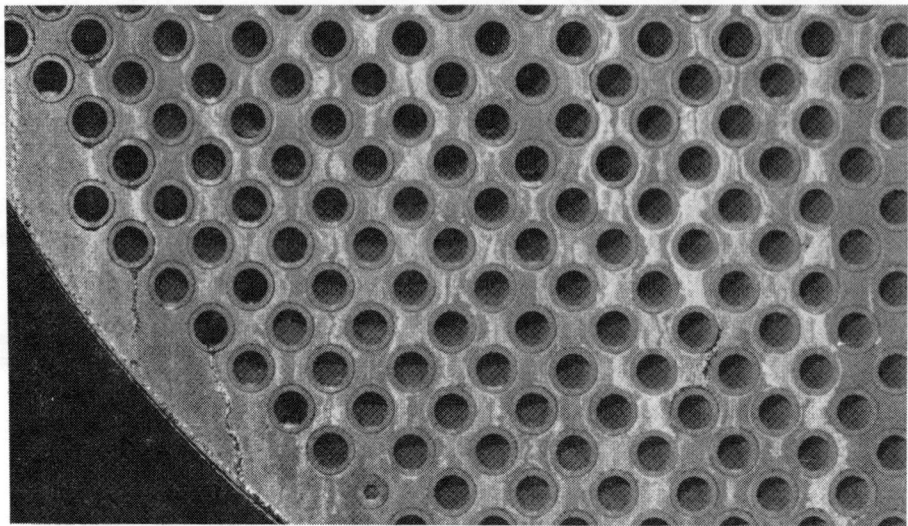

Photo 3. Same exchanger as photo 2 with additive.

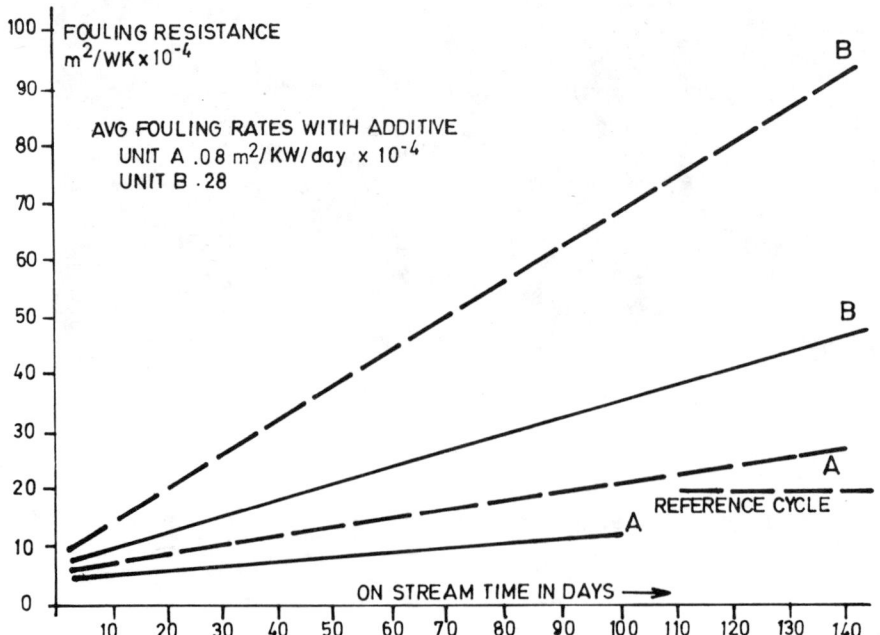

Figure 6. Evolution of fouling resistances for crude/residue exchangers using TOTAL/UMP additive injection.

## 4.2. ECONOMIES OF ADDITIVE INJECTION

The efficiency of the additive has been defined with respect to a reference cycle as:

$$E = 100 \times \left(\frac{\Delta R - \Delta R'}{\Delta R}\right)$$

R: Evolution of the fouling resistance over a cycle without additive.
R': Evolution of the fouling resistance over a cycle with additive.

for each of the cycles shown in Fig. 6 the efficiency is around 50 %.

In order to optimise the use of the additive we developed an empirical relationship for the cost of fouling as a function of the time since on – stream cleaning can be accomplished on both of the units in question, the cost of this operation is also taken into account.

$$C = K_1 K_2 t^{1.2} + K_3 \left(\frac{12}{t} - 1\right) + K_4 + K_5$$

$t$ = length of cycle in months
$C$ = cost of fouling
$K_1$ = constant for the unit
$K_2$ = cost of fuel
$K_3$ = cost of one on – stream cleaning
$K_4$ = cost of annual cleaning
$K_5$ = annual cost of additive

For unit B (20 000 t/day) in our example the following cost table has been drawn up for end of 1980 economics using an additive injection rate of 10 ppm Wt.

### TABLE 3

### ANNUAL COST OF FOULING K $ (1 $ = 5 F)

| Annual cost | Annual cost with additive at an efficiency of: | | | Cost of fuel | 140 $/tonne |
|---|---|---|---|---|---|
| | 75 % | 50 % | 20 % | | |
| without additive | | | | Cost of additive | 2 $/kg |
| 4 month cycle | | | | Cost of on – stream | |
| 1 040 | 580 | 780 | 1 000 | cleaning | 102 000 $ |
| 6 month cycle | | | | Cost of off – line | 52 000 $ |
| 1 440 | 600 | 920 | 1 300 | cleaning | |
| 12 month cycle | | | | | |
| 3 000 | 920 | 1 680 | 2 540 | On – stream time | 8 000 hours |

The optimum cycle length is in fact of 4 months in duration while the borderline efficiency which justifies use of the additive is an improvement of 20 % over the reference conditons.

Two other units are now being equipped with the necessary equipment to permit additive injection to the desalter.

## 5. CONCLUSIONS

This work which is still continuing has enabled a number of important conclusions to be drawn:

- The fouling of crude oil preheat trains is a complex phenomenon which above all is not caused primarily by the crude oil itself but by some constituents of the crude, by various impurities and by synergistic action between different impurities and the crude.
- The values of fouling factor currently admitted by TEMA for crude above 180°C appear much too low if an economical cycle length between exchanger cleaning is to be attained.
- The use of dispersants within the desalter emulsion phase offers a major hope for decreasing the fouling of the hottest exchangers which recover the most valuable heat.
- Fouling on the shell side can be aggravated by inadequate turbulence and velocity.
- Care should be focussed on the temperature control of preheat trains using parallel circuits to increase heat recovery.

The authors wish to express their thanks to M.J.Febvay, Director of TOTAL TECHNIQUE, for permission to publish this paper and to all C.FR & UMP personnel associated with this project over the last three years.

## REFERENCES

1 PLATT, G., Oil and Gas Journal 78, 159–173, Oct 13 1980
    HUANG F., ELSHOUT R., Chemical Engineering Progress, 72 (7), 68–74, July 1976.
2 HALUSKA, J.L., Hydrocarbon Processing, 55 (7) 153–156 (1976).
3 VAN DER WEE, P., TRITSMANS, P.A., Hydrocarbon Processing, 45 (8), 141–144 (1966).
4 SCHERRER, C., DURRIEU, M., RICHMOND, J.R., 1979 NPRA Meeting, Paper AM–79–50.
5 FRENCH PATENT No. 2388037, US.PATENT No. 4200550
6 FRENCH PATENT No. 2421958, US.PATENT No. 4222853

# International Symposium 1982

## HEAT AND MASS TRANSFER IN ROTATING MACHINERY

D. Metzger

*Arizona State University, Tempe, USA*

The XIV Symposium of the International Centre of Heat and Mass Transfer held at Dubrovnik in September 1982 was devoted to Heat and Mass Transfer in Rotating Machinery. The successful meeting brought together an outstanding worldwide group of researchers and practitioners with extensive background in the manyfacted subject of transport phenomena in rotating machinery.

Professor Darryl Metzger served as Chairman of the Symposium Committee, which consisted of twelve members from eight countries world-wide. The overall purpose of the meeting was to create a forum for the exchange of information on both research topics and on design problems and strategies. To this end, the contributed papers were organized into sessions on generic, research-oriented subjects and into sessions on specific types of machinery. In retrospect, this session organization helped to promote information exchange and resulted in a good balance between contributions emphasizing research and those primarily concerned with problems and design needs in actual machines.

On the research-oriented side, contributions and presentations were grouped into a sessions on experimental techniques, into sessions on transport phenomena in rotating tubes and channels and on rotating surfaces and enclosures. Both of these latter two areas were, in 1982, beginning to receive intense worldwide attention, particularly because of their applications in cooling of new generations of high temperature, high preformance gas turbine aero-engines. The XIV Symposium volume provides and excellent summary of the 1982 research status of these areas, and a good starting point for study of the extensive work published from 1982 to the present. This is especially true in the area of rotating tubes and channels, where the keynote paper by Y. Mori and W. Nakayama from Japan provides and excellent overview of the phenomena involved. In this area in particular there has been and continues to be very extensive national and industrial support of rotating cooling channel research in all of the aero-engine producing countries since 1982.

Similar comments are applicable to the area of transport phenomena on rotating surfaces and in enclosures, although the research activity level subsqunt to 1982 has not yet been as extensive as that experienced in the rotating tube area. Nevertheless, research in this area is now, in 1987, rapidly expanding, again stimulated by the need for basic understanding and design information related to the development of advanced gas turbine engines and liquid recket turbopumps. At the XIV Symposium, the keynote paper by J.M. Owen from England gives comprehensive background and summary information on the 1982 state-of-the art, and describes areas where much additional theroretical and experimanetal work is needed.

At the sessions on gas turbine engines, an excellent overview of heat transfer problems in aero-engines was given in a keynote paper by D.K. Hennecke from F. R. Germany. The paper contains an especially lucid description of the thermal and material responses associated with the transient conditions existing within aero-engines during their operating cycles. Cooling and thermal management problems associated with both steady state and transient engine operation and flight are described, and critical areas requiring additional research are identified. In the years following the XIV Symposium, this keynote paper has been widely referenced in the research literature on gas turbine heat transfer, and remains one of the best and most up-to-date introductions to the field.

An overview of some heat and mass problems in steam turbines was provided by a keynote paper by M. Majcen form Yugoslavia, and emphasized two-phase flow and heat transfer research as areas important to the development of higher efficiency machines. This theme was carried throughout the sessions on steam turbines, with several papers addressing topics associated with wet streams in tubines, especially condensate droplet evolution and blade erosion.

Finally, a session specifically on two -phase phenomena as experienced in rotating heat pipes and thermosyphons was organized around an excellent keynote paper on rotating heat paper provides and excellent physical description of the operating principles of rotating heat pipes, including a description of various applications. The current heat transfer literature (as of 1982) pertinent to such devices is reviewed and summarized, and areas in need of further research are identified.

In summary, the XIV Symposium must be judged an unqualified success in satisfying the objective of bringing together a broad spectrum of researchers and practitioners for a concentrated exchange of information and ideas related to heat and mass transfer in rotating machinery. A direct line age from this Dubrovnik meeting can be found in a number of subsequent meetings held under various sponsoring agencies and societies, in continued technical communication between many of the participants, and in several instances of international cooperative research programs and personnel exchanges. For individuals working in this area of technology, the meeting remains a unique forum that truly captured and expressed the spirit of international technical cooperation.

# HEAT TRANSFER PROBLEMS IN AERO-ENGINES

D. K. Hennecke

*MTU Motoren- und Turbinen-Union Munchen GmbH,
Munich, F. R. Germany*

## 1. INTRODUCTION

The purpose of this paper is to describe the major heat transfer areas that have to be dealt within the development of a modern aero-engine. The focus will be on gas turbine engines; piston and rocket engines will be excluded. Although the focus will be on aero-engines, many statements will also be relevant to stationary gas turbines.

Gas turbines, being heat engines, have always received the attention of heat transfer specialistics. The present paper will show that, owing to the trends in aircraft gas turbine design, which will be described briefly, the scope of heat transfer problems has greatly increased. By looking at each component separately it will be demostrated that many new challenging problems have arisen besides the hitherto classical areas of combustor and turbine cooling.

## 2. SOME TRENDS IN AERO-ENGINE DESIGN

Modern aero-engines may be characterized by their extremely high power concentration and low specific fuel consumption leading to relatively low wieght, small size, a high thrust-to-weight ratio and low mission fuel consumption. To illustrate this trend, two fighter engines with about the same thrust are shown in Fig. 1. The top one, the RB 199, was developed two decades later than the bottom one, the J 79. The dramatic reduction in engine size should be noted.

For civil engines the fuel consumption is of paramount importance. With and advanced technology engine such as the PW 2037, shown in Fig. 2, which is currently under development, the Boeing 757 will consume about 40% less fuel per passenger seat than the Boeing 727.

It is well known that two of the main factors contributing toward this goal are the high compression ratio and a high turbine entry gas temperature, providing for high thermal efficiency of the thermodynamic cycle and a high power density.

Figure 3 depicts the development of the turbine entry temperature of actual aero-enginee since 1950 (Ref. 1). Also shown is the increase in the maximum allowable blade material temperature. It may be seen that the turbine blades operate in and environment in which the gas temperature is well above the material temperature even for ceramics. And the tendency is towards increasing. Today's aero-engines have a turbine entry temperature of 1600 to 1700 K and above. It is obvious that intensive cooling is required.

The reduction in engine mass achieved to date is presented in Fig. 4 (Ref. 2). The trust-to-mass ratio has increased from about 3:3 to 8:1 for military engines, with a similar gain for commercial engines.

The following chapter will show that the tendency to light weight parts, high temperatures, highly loaded compressors and turbines combined with the extremely severe requirement for reliability has greatly increased the importance of the heat transfer engineer in the design of aero-engines. In addition, a large number of challenging new heat transfer problems has emerged, calling for intensified research, which will be identified.

## 3. HEAT TRANSFER PROBLEMS IN THE MAJOR ENGINE COMPONENTS

Figure 5 shows the cross-section of a modern aero-engine with the major components indicated. The heat transfer areas will be described for these components.

### 3.1. INLET, FAN AND COMPRESSOR

*Inlet and fan.* In certain altitude, flight and atmospheric conditions the engine inlet and fan may be subject to severe icing problems. Tha task of the heat transfer engineer is to identify and assess these conditions and to design appropriate means to prevent the build-up of harmful ice. This may be achieved by internally heating the bullet nose, the fan vanes and blades, etc. by warm air form higher stages. Even film heating may be employed in severe cases. Surface areas that are prone to icing may also be covered with a thin electrically heated mat. The problem of icing is gone into in detail in Ref. 3.

*Compressor.* A typical compressor is sketched in Fig. 6. The trend in compressor design is toward higher pressure ratios (now up to 30 -40) resulting in high exit temperatures (800 - 900 K), increased rotor speed, lower weight, as well as higher aerodynamic efficiency and loading. This has called for much more detailed analytical design, including thermal analysis, bringing the heat transfer engineer into the picture. His task is to predict accurately the transient temperature distribution of the compressor disk and casing for the whole flight cycle form start to landing with two objectives:

- To facilitate computation of the life of the disks and casing and, on this basis, to design for the required life at minimum weight, and

Fig. 1  Progress in aero-engine design
        Top: RB 199 engine; bottom: J79 engine

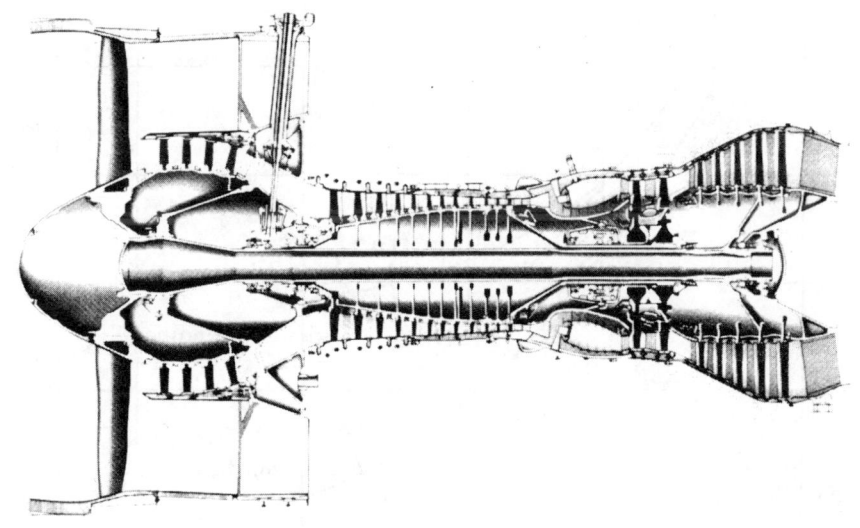

Fig. 2  Modern civil high bypass ratio engine: The PW 2037

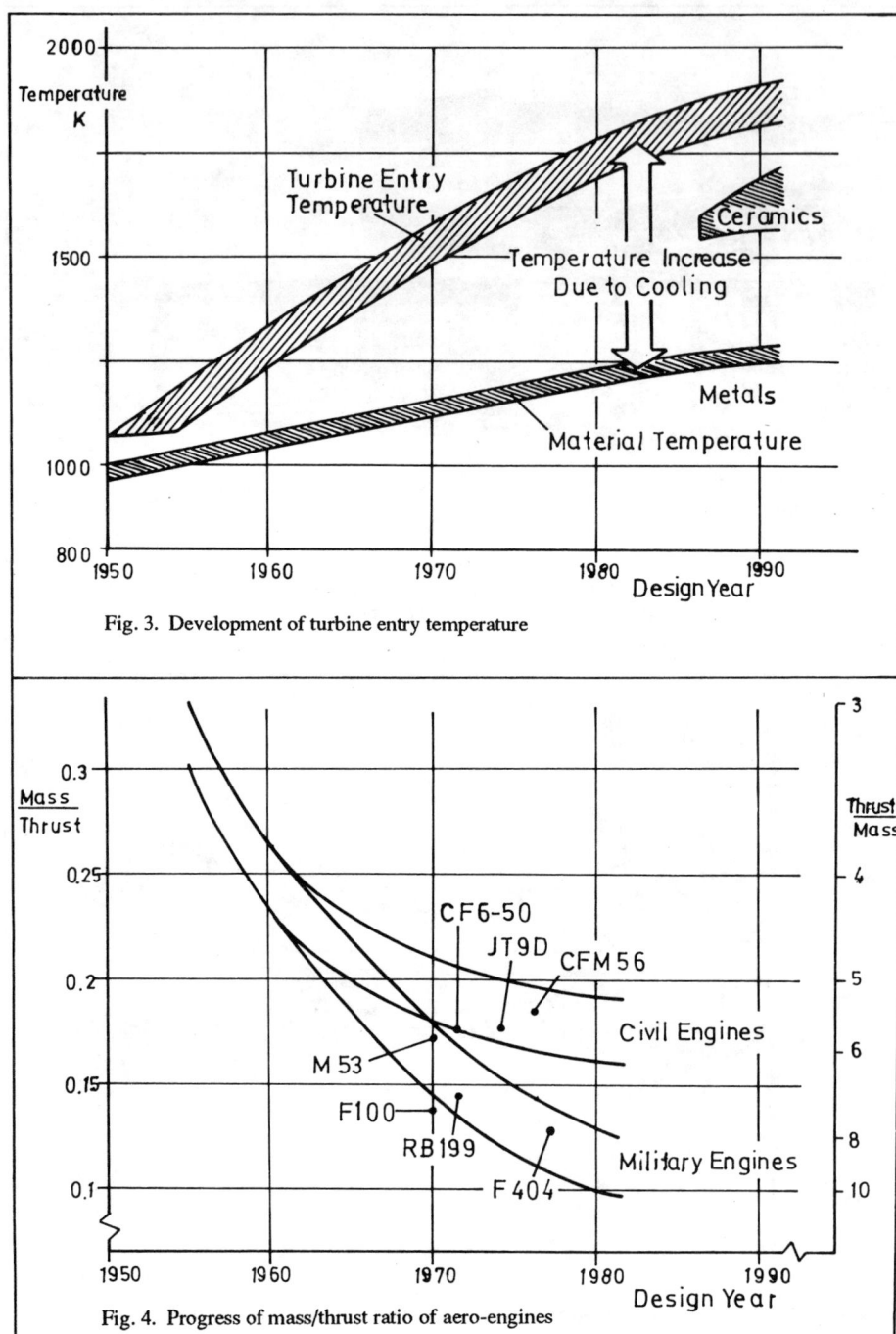

Fig. 3. Development of turbine entry temperature

Fig. 4. Progress of mass/thrust ratio of aero-engines

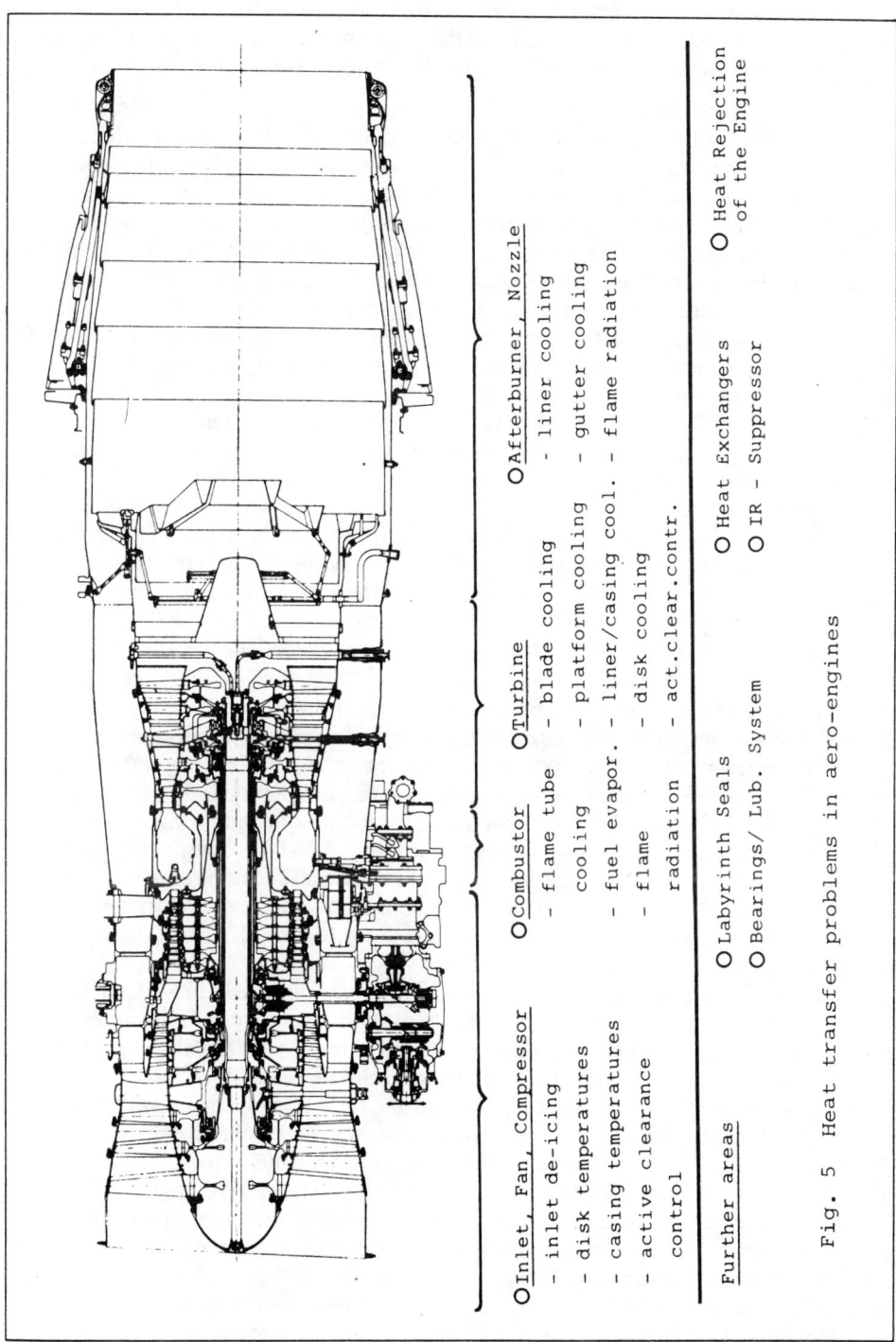

Fig. 5  Heat transfer problems in aero-engines

○ Inlet, Fan, Compressor
- inlet de-icing
- disk temperatures
- casing temperatures
- active clearance control

○ Combustor
- flame tube cooling
- fuel evapor.
- flame radiation

○ Turbine
- blade cooling
- platform cooling
- liner/casing cool.
- disk cooling
- act.clear.contr.

○ Afterburner, Nozzle
- liner cooling
- gutter cooling
- flame radiation

○ Heat Rejection of the Engine

○ Heat Exchangers
○ IR - Suppressor

Further areas
○ Labyrinth Seals
○ Bearings/ Lub. System

- To predict the transient rotor and casing clearances and again, to actively influence the design such that the transient thermal behavior of the parts is optimal, i.e. resulting in tight clearances throughout the flight cycle.

The life calculaion is of particular importance for the rotor. In addition to the stresses resulting from the centrifugal transient conditions. The disk rim, which is close to the main stream, follows a change in power setting much faster than the bulkier hub section which is removed from the main stream. An example is shown in Fig. 7 for a typical compressor disk. The large intermittent temperature difference, causing high thermal stress and possibly severe consumptiuon of disk life, should be noted. The main stream temperature is also plotted. It changes as rapidly as the rotational speed, i.e. typically within about 5 - 10 seconds form idle to full-power.

The resulting radial deflection of the rotor during transient conditions is also relatively slow. In contrast, the radial movement of the casing is usually much faster, because the temperature of the thin-walled casing, which is exposed to the main copal variation in the radial deflection of a typical compressor, (shown in Fig. 6), is sketched in Fig. 8 for and acceleration from steady-state idle to full power and a deceleration back to idle.The deflection of casing and rotor is shown at the top of Fig. 8, with the resulting clearance as a function of time shown at the bottom. Initially, there is a short closing of the gap when the rotor speeds up and expands because of the increased centrifugal during warm-up, leading to a loss of effeciency and possibly even to surge. These gaps close to the desired small size under steady-state full-power. During deceleration, however, they increase briefly owing it the reduced centrifugal load on the rotor and then decrease to negative values, i.e. the rotor blades rub into the abradable coating on the casing. Then the clearances are enlarged by the amount of rub-in, even in the steady-state full-power condition, leading to a further deterioration in the efficiency of the compressor and in the surge margin.

To emphasize the importance of the transient variation of the gaps, the high-pressure spoolspeed and thrust of an actual aero-engine as a function of time are shown in Fig. 9. It should be noted that the thrust increases from its value at idle to full power rather rapidly, then decreases to a minimum and increases slowly, finally to reach its full-power value aganin after 10 - 15 minutes. The minimum of about 15% below the full-power value occurs after about 20-40 seconds, when the aircraft may have reached the end of the runway and is about to take off. This is precisely when maximum thrust is required.

Transient engine behavior, as shown in Fig. 9, is obviously unacceptable. Since it is to a large degree the result of the transient variation of the clearances, one tries to achive in advanced engines a more favorable transient thermal behavior of the components by careful design of the compressor and turbine.With compressors the aim is to speed up the temperature response of the rotor and to slow down that of the casing. For the rotor this can be achieved, for instance, by venting the cavities between the disks by a small amount of air taken from the main stream (a review of the heat transfer in vented rotating cavities simulating those in aero-engine compressors is given in Ref. 4). To slow down the casing one may simply add mass which, of course, is not excatly desirable in aero-engines. A more effective method is to constrict the casing of and outer supporting structure and to hook separate segments to its inner side to form the outer wall of the compressor annulus. In this way the radial thermal movement of the casing is controlled solely by the outer structure, which is not exposed to the main stream and, therefore, racts slowly to changed in the power setting.

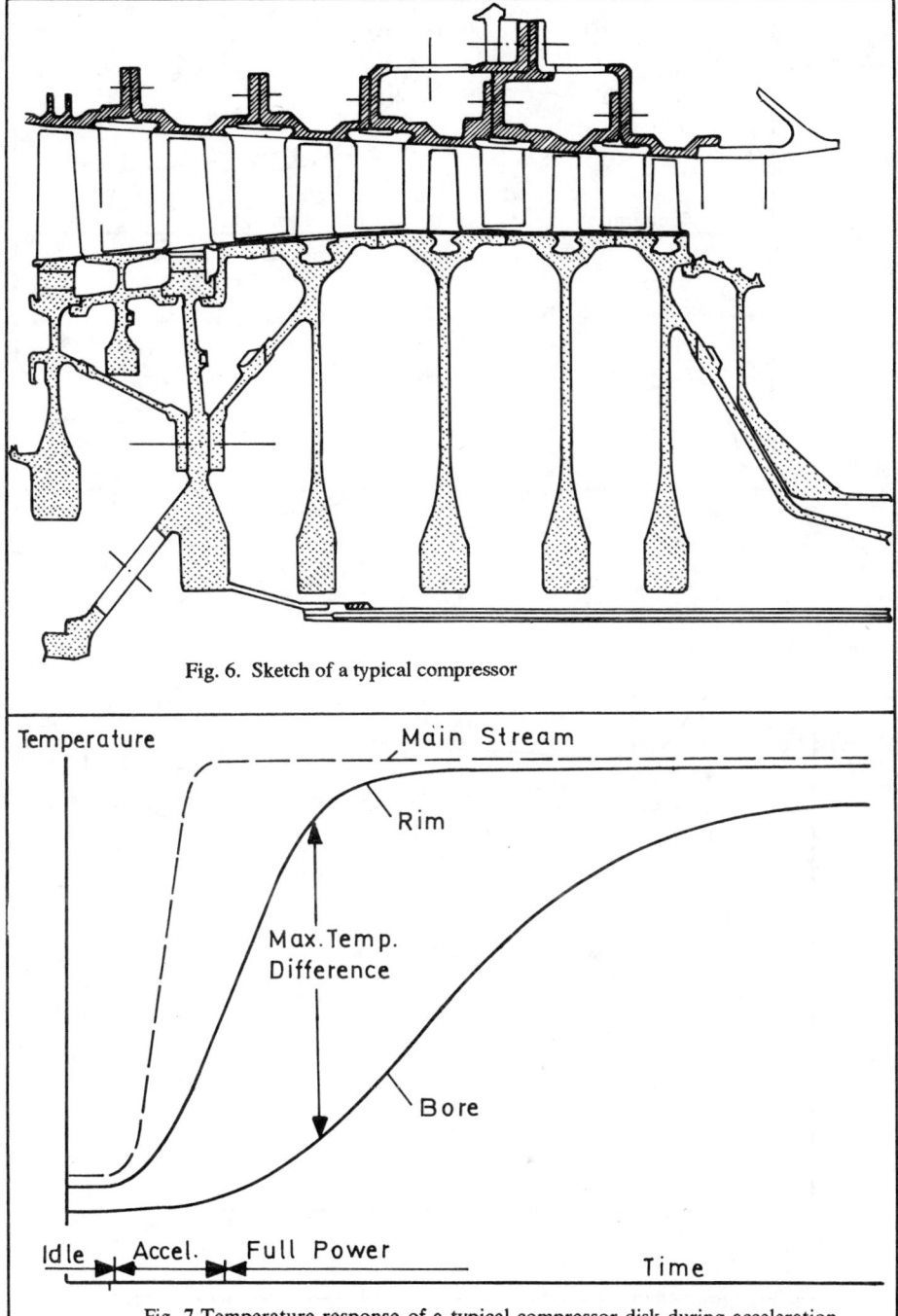

Fig. 6. Sketch of a typical compressor

Fig. 7 Temperature response of a typical compressor disk during acceleration

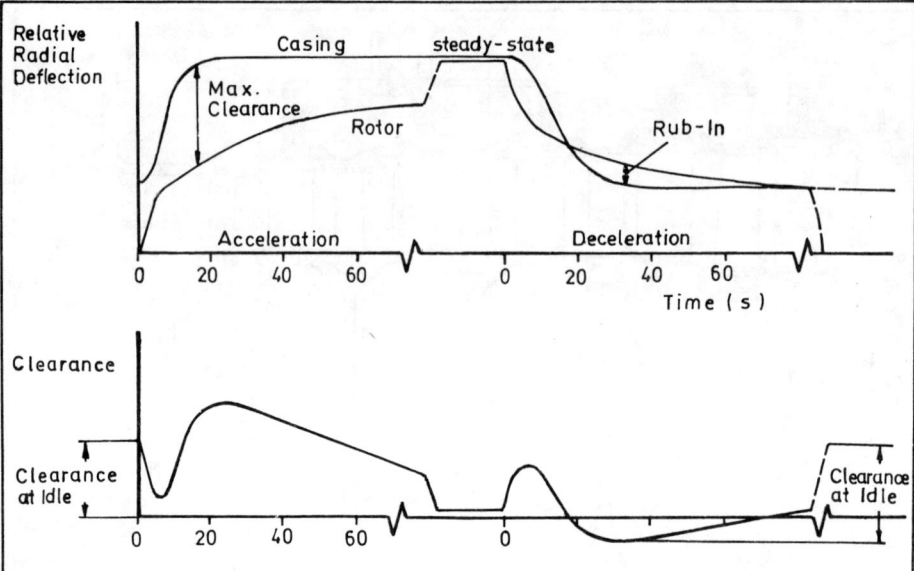

Fig. 8. Transient radial deflection of a compressor casing and rotor and the resulting tip clearance

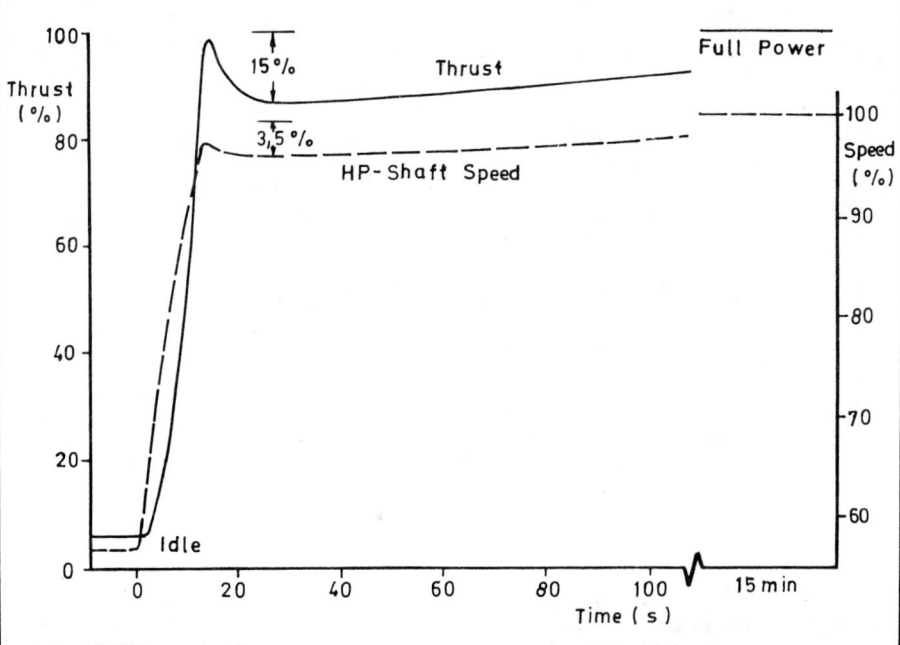

Fig. 9. Transient thrust and high pressure rotor speed of an aero-engine (measured)

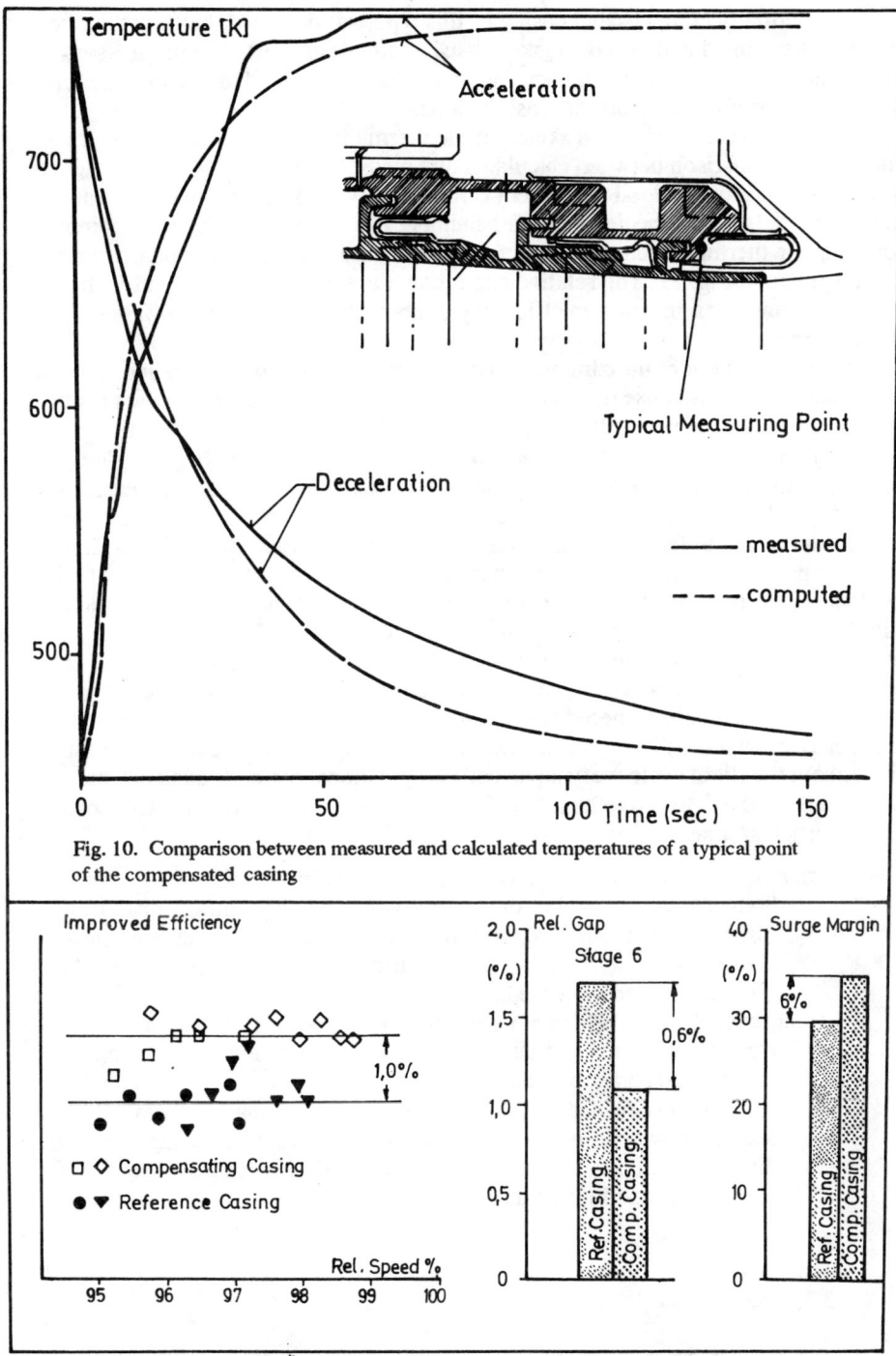

Fig. 10. Comparison between measured and calculated temperatures of a typical point of the compensated casing

Fig. 11. Improvements achieved with thermal compensation of a compressor casing (measured)

A thermally compensated casing of this type was designed, built, comprehensively instrumented and tested. Back-to-back engine tests, with the compressor having a standard casing and the new casing, were carried out to demonstrate the improvment to be obtained from the casing alone.

Figure 10 (inset) shows a sketch of the thermally compensated casing. An example of a comparison between calculated and measured temperatures as a function of time is also shown. Good agreement can be observed indicating that the heat transfer model used in the design is satisfactory. The improvements of the compressor using the thermally compensated casing compared to using the standard casing is demonstrated in Fig. 11. The relative gap (shown for stage 6) is reduced by about 0.6%, resulting in an increase in efficiency of about 1% and a gain of 6% in the surge margin.

This method of optimizing the gaps in turbomachinery may be called "passive clearance control", because the clearances react passively to changes in the power setting. Another approach is referred to as "active clearance control", where the engine control system is used to change the clearances actively with the engine running. This can be achieved, for instance, by placing tubes around the casing, for blowing cold air taken from a lower compressor stage through a row of small holes against the casing to facilitate impingement cooling and shrinkage of casing. In this way it is possible to have larger clearances during certain flight conditions with the coolant flow turned off, avoiding excessive abrasion that would occur as a result of high maneuver loads, thermal deflections, etc. During other flight conditions, such as cruise, when most of the fuel is consumed, the cooling system may be turned on to actively close the clearances and increase the efficiency of the compressor.

Finally, it should be noted that before one can design a compressor with thermal behavior which will achieve the desired life and optimum clearance variation throughout the flight cycle, very advanced computational methods and an accurate knowledge of the convective heat transfer are required. The following problem areas are of particular itnterest:

- Disk heat transfer: rotating disks near stationary walls and rotating cavities, both cases with axial throuhgflow and/or radial net inflow or outflow; disks with different temperatures and, thus, buoyancy-induced flow in a centrifugal force field, especially during the transient phase when the disks are heated or cooled;
- Heat transfer in the bladed annulus: heat transfer to the blades, platforms and blade attachments and into the disks; heat generation when the blades rub into the abradable coating;
- Casing heat transfer: heat transfer within the casing usually controlled by leakage flows through narrow channels and irregular passages with three dimensional flows, often affected by natural convection; influence of bleed flow; conduction through contact surfaces.

## 3.2. COMBUSTOR

The highest temperatures in aero-engines occur in the combustor, where the combustion gases may reach temperatures of about 2400 K in the primary zone. Thus, it is obvious that extensive cooling of the flame tube walls is required.

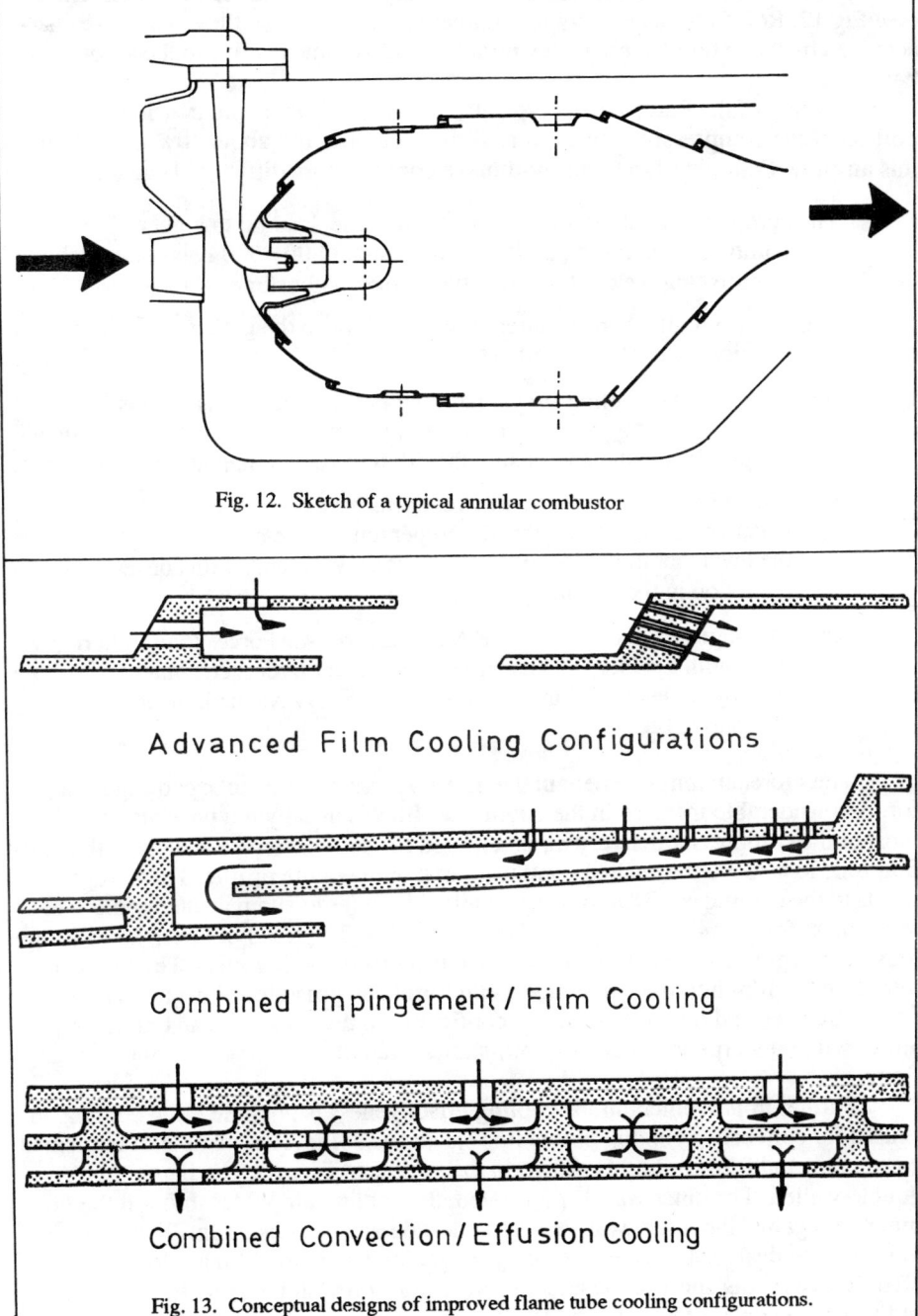

Fig. 12. Sketch of a typical annular combustor

Advanced Film Cooling Configurations

Combined Impingement / Film Cooling

Combined Convection / Effusion Cooling

Fig. 13. Conceptual designs of improved flame tube cooling configurations.

Modern aero-engine combustors are usually annular and either straight axial (see Fig 12, Ref. 5) or, especially for smaller engines, reverse flow. They are characterized by their short length and extremely small volume relative to the amount of heat released.

Cooling of the flame tube is typically effected by film cooling, using a large portion of the compressor delivery air. Today's engines use about 40% to 50% of this air flow. Future design trends will make cooling more difficult, because

- The compression ratio will continue to increase, leading to higher compressor exit temperatures, (now about 800 - 900 K, as stated above), diminishing the cooling potential of the combustor cooling air,

- Higher turbin inlet temperatures (Fig. 3) will raise the gas temperatures in the dilution zone of the combustor,

- Increased turbine cooling flows, required because of the higher gas temperatures, bypass the combustor, reducing the amount of cooling air for the flame tube and raising the dilution zone temperature even further,

- Applicaton of heat exchangers in a regenerative cycle (discussed in Chapter 3.5) will bring the air temperature at the combustor inlet close to the level that the flame tube material can withstand, with consequent reduction in the cooling potential, and

- Fuel shortages are likely to oblige Air Lines and Air Forces to use alternative fuels with a wider, specification and a lower hydrogen content, resulting in a higher level of flame radiation(Ref. 6 surveys the impact of alternative fuels).

Therefore, an improvement in the effectiveness of flame tube cooling is called for. A considerable increase in the cooling air flow is usually not possible, because most of the compressor delivery air is required for combustion as well as in the dilution zone to achieve the desired combustor exit temperature profile. Thus, improvements in the cooling configuration are required in order to attain greater cooling effectiveness for given or even reduced cooling flows. The trend has been from simple louvers, wiggle strips, etc. to higly intricate machined cooling rings. The aim is to introduce the film into the flame tube as uniformly as possible, with a low level of film turbulence and retaining a rugged configuration that can withstand high temperatures without warping. Figure 13 (top) shows examples of these machined cooling rings.

Further improvement in the cooling effectiveness is possible if the cooling air is used for convective cooling before it is ejected as a film. An example of combined convection and film cooling is shown in the middle of Fig. 13. The flame tube is double-walled. The outer wall is perforated, the cooling air passes through the holes, impinges against the inner wall, flows to one side and exits as a film. The perforations may be designed to cool especially the region where the film has lost its high effectiveness. This approach results in a very uniform wall temperature, which is one of the design goals. Effective impingement cooling, however, requires a relatively large pressure drop. Therefore, in some application, different designs of convective cooling may be preferable. These may comprise ribs, pimples or other means to enhance heat transfer in the double-wall structure (Ref. 7).

Still higher levels of cooling effectiveness can be achieved by combining convection and effusion cooling. An example is indicated at the bottom of Fig. 13. The cooling air is passed through several layers of perforated walls separated by ribs and/or pimples to promote turbulence and enlarge surface areas the inner passages. The holes of the outer layer may be designed to meter the air flow, and may be the desired variation of the flow rate of the effused air at the inner surface of the flame tube used to obtain (Ref. 8).

Obviously, the design of complex configurations of this type calls for very sophisticated computational methods with good models for flame radiation, heat transfer coefficients at the inner and outer flame tube surfaces and the hot gas temperature effective for the gas side convection. The latter is particularly complex, because the flow inside the flame tube is usually highly three-dimensional and recirculating and very turbulent with large temperature variations. And the flow strongly by the chemical reactions taking place. In the future, careful designs will require reliable computer codes for solving the full Navier-Stokes equations.

Furthermore, and optimum design necessitates close cooperation between the heat transfer engineer and the materials and manufacturing specialists to ensure that the configuration remains cost-effective.

The use of thermal barrier coatings on ceramic materials, such as zirconium oxide, has become feasible since new spraying and bonding techniques have become available (Ref. 9). The effect of these coatings is twofold: they insulate, thanks to their low heat conductivy, which is particalry effective when additional convective cooling is employed, and they absorb less radiative heat because of their high reflectance, thus reducing the heat flow into the flame tube wall.

Even if future combustors are all ceramic, the heat transfer engineer will still be greatly involved in th design because detailed temperature calculations will be required if the desired life of the flame tube is to be achieved (Ref. 10).

Another heat transfer problem in combustors involves vaporization of the fuel droplets. The speed of this process is one of the factors controlling the subsequent combustion. Therefore, detailed knowledge of the heat and mass transfer during the vaporization of clouds of droplets in complex swirling flow fields is required.

Summing up, the design of durable and reliable combustors calls for detaled methods for predicting:

- The effectiveness of film and convection cooling in complex configurations t achieve a uniform, acceptable wall temperature with a minimum of coolant flow and pressure drop,
- The flame radiation at all points of the flame tube,
- The gas-side heat transfer coefficients, and
- The droplet vaporization process.

In the light of today's highly-loaded combustors and increasing cycle pressures, the current methods are not sufficiently accurate, and a considerable amount of research effort is still required.

### 3.3. TURBINE

Similar to the combustor, the first turbine stages of modern aero-engines operate in an environment where the gas temperatures are well above the allowable metal temperature (Fig. 3). Therefore, efficient cooling is one of the key requirements.

High temperature parts are prone to failure. Therefore, in aero-engines with their extremely severe requirement for reliability, turbine cooling is a particularly important subject. It is not surprising that the largest portion of the maintenance cost of an aero-engine originates form its hot end (Ref. 11). Hence, the design of the turbine cooling system requires particular care. It calls for highly accurate computational methods, sophisticated experimental facilities for realistic testing, as well as advanced materials (Ref. 12) and manufacturing techniques.

Figure 14 shows a typical axial turbine of aero-engine, with indicated parts that require cooling. These are the vanes and blades, platforms, casing liner or shroud, and the disks. Typical values of the cooling air mass flow rate as a percentage of the main stream mass flow are also given.

In general, cooling of a part means the removal of heat and for this a cooling medium is needed. A very efficient coolant would be a liquid, such as water or a liquid metal, because of their high heat transfer capability, large thermal capacity and the possibility to utilize their latent heat of vaporization. In aero-engines, however, air has been used exclusively so far. Although its thermal properties make it a good insulator and a very poor coolant, air still has decisive advantages for aero-engines:

- It is immediate available.
- No special storage or pumping devices are required, if compressor delivery air is used.
- Leakage is not disastrous.
- Air can be fed back into the main gas stream conveniently, so no collection and recirculation devices are needed.
- Air is light, rotating blades don't have to carry the centrifugal load of a heavy liquid.
- Start-up, off-design power and shut-down (all occuring frequently requently with aeroengines) do not pose great problems, in contrast to liquid coolants.

A survey is given in Ref. 13.

For stationary gas turbines the advantages of air over liquid as coolant are not so decisive. Therefore, liquid coolants are considered for actual applications (e.g. Ref. 14). In aero-engines, liquids may be useful for emergency cases. If, for instance, in a twin-engined helicopter, one engine fails the other can deliver much more power for a short time if water injection into the turbine cooling air is used for increasing the turbine entry temperature.

Typical paths of the turbine cooling air are shown in Fig. 14. As can be seen, the air is taken from the compressor and ducted into the vanes from either side, into the blades via their roots, to the platforms, casing and linears, as well as to the disks. Afterwards, the coolant is fed back into the main gas stream to perform aerodynamic work in later turbine stages. It should be noted that the air to cool the blades passes through a preswirler on leaving the stationary parts. This minimizes the degree of heat-up resulting from dissipation. If it is an intermediate- or low-pressure turbine, the cooling air is taken from and intermediate compressor stage, where it is "thermodynamically cheaper" (i.e. not so much compression work has been put into it) and is cooler.

In some cases, the pressure of the coolant ant the blade root is not sufficient to pass the required amount of cooling air through the blade. Special pressure boost systems then come into consideration.

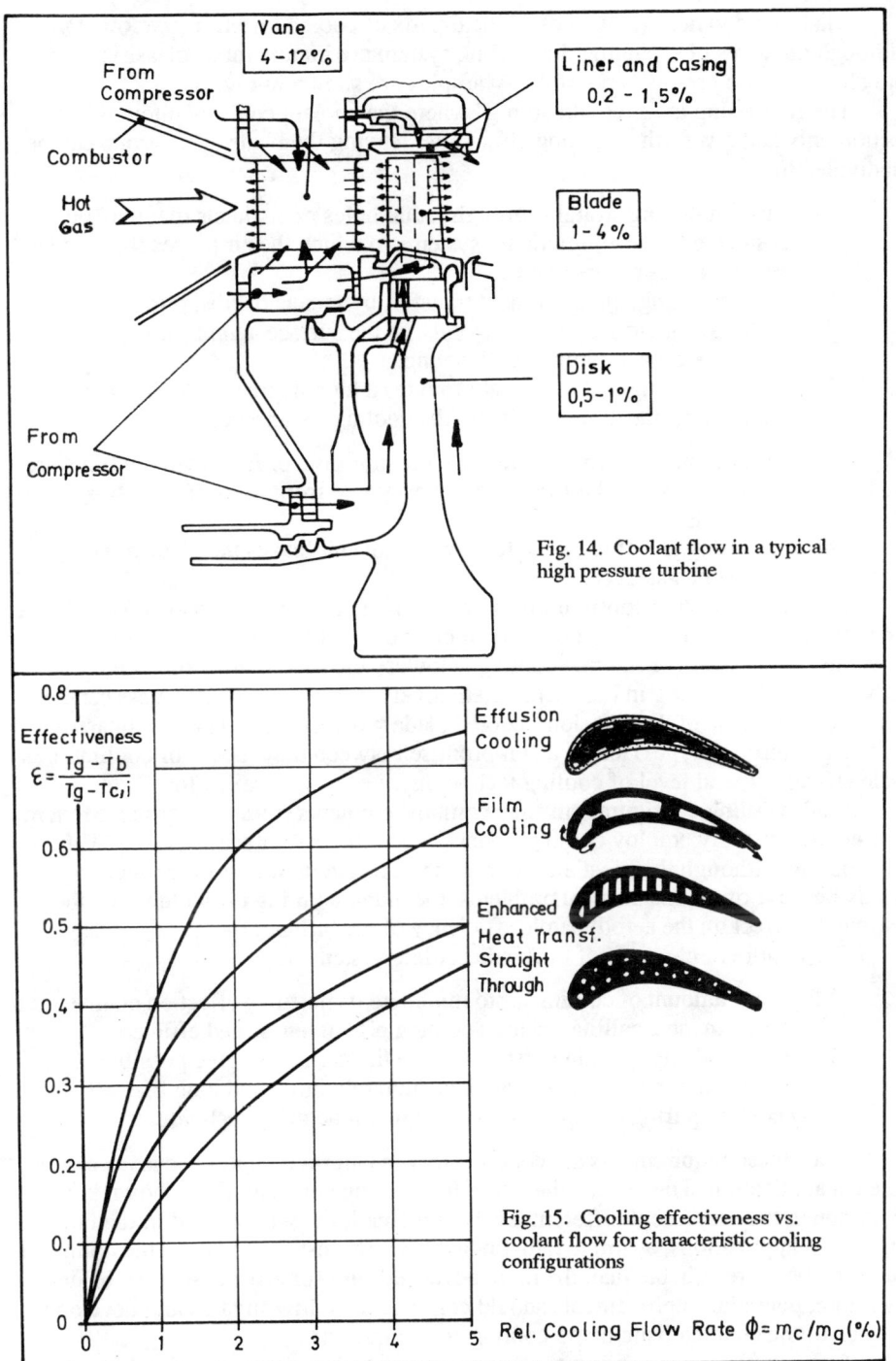

Fig. 14. Coolant flow in a typical high pressure turbine

Fig. 15. Cooling effectiveness vs. coolant flow for characteristic cooling configurations

Blades and vanes are the most critical parts of cooled turbines (see Ref. 1). Although the ways of arranging the cooling system are innumerable, classification into characteristic groups is possible. Examples are shown in Fig. 15.

The first group is "internal cooling" where the coolant cools by internal convection only and has further cooling effect after leaving the blade. This group can be subdivided into:

- Systems employing straight-through radial holes or channels that may be connected to form multipass systems in which the air passes the blade several times before exiting.
- Systems involving enhanced heat transfer, by means of ribs, pimples, pedestals or similar devices to enlarge the surface and promote turbulence, or impimgement cooling.
- Systems employing and inset that provides a high degree of flexibility for distributing the air according to the cooling requirements.

The second group is "film cooling". Single, multiple or full coverage films are used. For structural reasons, the films usually originate from rows of closely-spaced holes rather than slots.

The third group is "effusion cooling" where the air leaves the blade at every point through a porous surface.

For typical cooling configurations, the cooling effectiveness is also plotted as a function of the relative cooling flow rate. It can be seen that, for a given cooling flow, the effectiveness of internal cooling is lowest and that of effusion cooling highest with film cooling in between. Alternatively, a required effectiveness can be achieved in the case of the effusion - cooled blade with less cooling air compared to the straight-through type. Hence, a compromise between the amount of cooling, i.e. cycle efficiency, and level of cooling technology, i.e. cost, is called for.

Blade cooling configurations may combine elements from each group. Modern aero-engines usually employ cooling systems from the first or second group. Effusion cooling, although the most effective, has not found its way into production yet, mainly because of the mechanical problems, the manufacturing difficulties and the detrimental effect on the aerodynamic efficiency of the turbine.

The requirements made of the blade cooling system are:

- Minimum amount of cooling air to minimize its negative effect on engine performance, calling for high cooling effectiveness and efficiency.
- Uniform blade temperature distribution to the greatest degree possible, to reduce thermal stresses to achieve high reliability and long life.
- Low cost, requiring cheap materials and manufacturing techniques.

Since these requirements are contradictory, a careful compromise is necessary for each application. This means that all of the cooling configurations shown in Fig. 15, although they represent different levels of technology, will be used in the future. For certain applications, a simple radial hole blade, for instance, may be the optimum blade and be more suitable than the more advanced film- or effusion-cooled blades. Therefore, potential improvements should continue to be investigated and developed for each of the characteristic blade-cooling configurations.

An example of the cooling configuration of a turbine vane of a high bypass ratio engine is shown in Fig. 16. The inserts facilitating impingement cooling and the leading edge with full coverage film cooling should be noted. As a further example, a

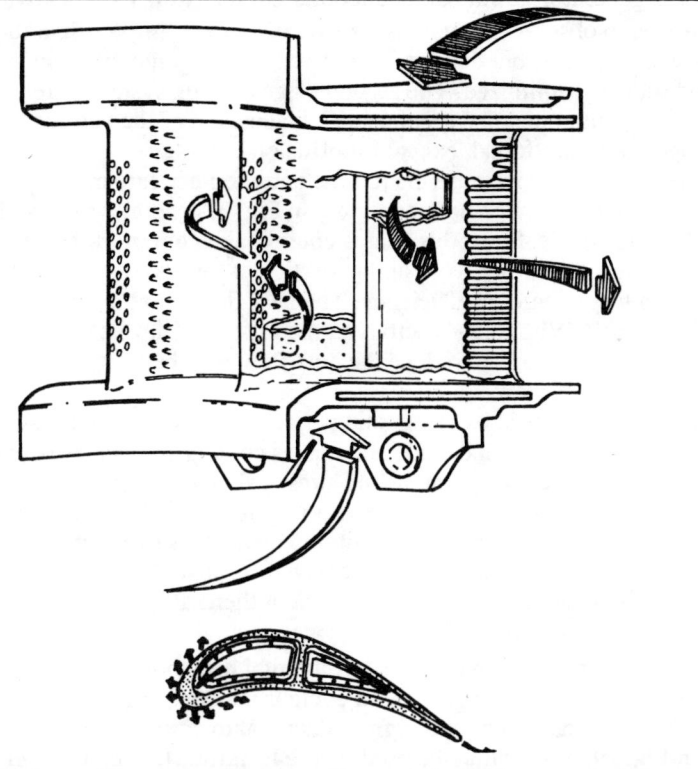

Fig. 16. Modern cooling configuration of a turbine vane of a high bypass ratio engine

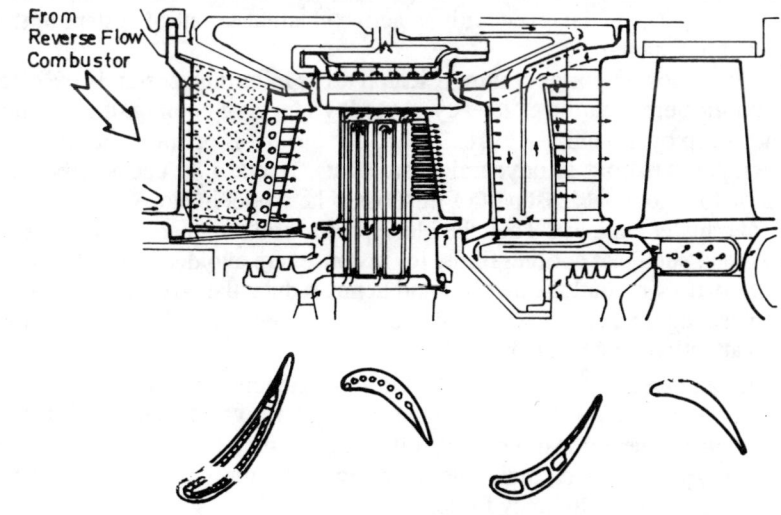

Fig. 17. Cooling configuration in a turbine of a helicopter engine

cooled gas generator turbine of a helicopter engine (Ref. 15) is shown in Fig. 17. It is interesting to observe that not just one particular cooling configuration was chosen for the whole turbine, but that blades and vanes have quite different systems, which were individually optimized. The second stage cooling system differs from the first stage system, with uncooled second stage blades. All of the various features mentioned above are found, except for effusion cooling.

The tendency to increase the turbine entry temperature more rapidly than the allowable material temperature, as shown in Fig. 3, generally means higher heat flow into the turbine blade. Since this heat is energy taken out of the main gas stream, it is lost for the performance of work at that particular turbine stage, which may be expressed as a loss of turbine efficiency. The heat flow into the blades of 1% to 2% of the turbine work. With today's aerodynamic efficiency levels of around 90%, this is not negligible and efforts to reduce the heat flow may be worthwhile, not just from the cooling point of view.

The use of thermal barrier coatings (Ref. 9) presents one possibility for reducing the heat flow into the blade and makes cooling simpler. All-ceramic turbines are still in the research temperature and stress calculations.

Besides the task of cooling the various turbine parts, such as blades, vanes, disks and liners, careful design for optimum clearances is required if the high efficiencies of modern turbines are to be achieved. A procedure similar to that with the compressor is necessary, i.e. detailed transient thermal analysis throughout the flight cycle for the disk and the casing.

As and illustration of the effect of thermal behavior, the movement of the tip fin of a shrouded turbine through a flight cycle is shown in Fig. 18. Large deflections, both in the graph has been plotted from data obtained from x-rays. An example is given in Fig. 19. X-rays may be used by the heat transfer engineer, in addition to temperature measurements, to validate his theoretical models.

Since tight clearances are so very important, new commercial engines, such as the PW 2037 (Fig. 2), are designed with an active clearance control system of the type described in Chapter 3.1.

One of the most important future research areas in turbine design is the effect of cooling on the aerodynamic efficiency. At today's level of cooling, the turbine efficiency may drop by around 5% to 10% when the turbine is operated in a hot environment, compared to cold aerodynamic teste (Refs. 16 and 17). The aerodynamicist can only hope to recover at least part of this loss if he works in close cooperation with the heat transfer engineer. Then the blade profile and the internal cooling system may be designed such that cooling films, for instance, are avoided at locations where they disturb the flow around the aerofoil and hence reduce the efficiency.

Summarizing the heat transfer problems in turbines, the following areas that need special attention may be mentioned:
- Blades and vanes: prediction of the local gas-side heat transfer coefficient still unsatisfactory; effect of secondary and tip flows on blade, casing and platform heat transfer cooling effect in complex internal passages with fins, pimples, etc., including the effect of rotation with the resulting Coriolis and buoyancy forces.
- Rotating disks: cooling with the coolant flowing radially inward or outward;
- Casing: heat transfer in complex structures with leakage flows, contact resistances.
- Interaction between cooling and aerodynamics.

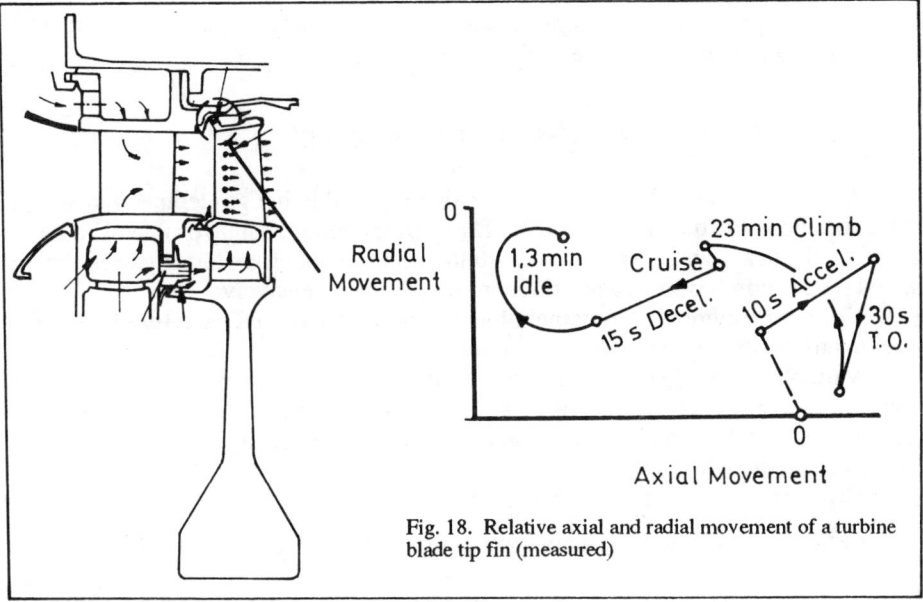

Fig. 18. Relative axial and radial movement of a turbine blade tip fin (measured)

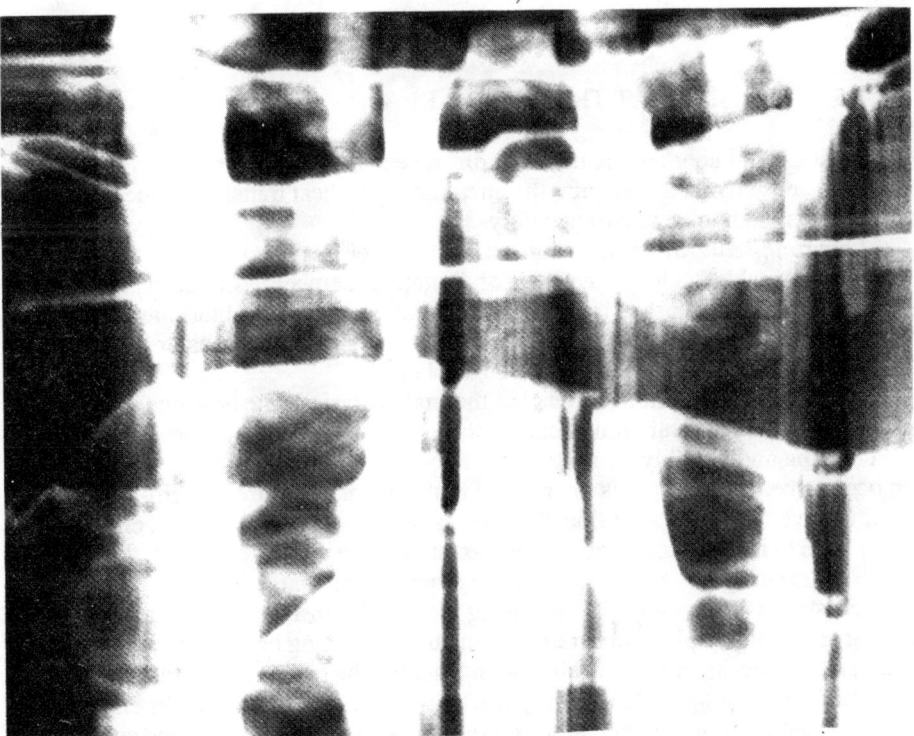

Fig. 19. Example of an X-ray photograph to measure clearances

An idea of the accuracies required can be obtained if one considers that today's turbine blades are operated at temperatures which are so high that a further increase by just 15 K would halve the life of the blade.

## 3.4. AFTERBURNER AND EXHAUST NOZZLE

Afterburner and exhaust nozzles are also exposed to hot gas temperatures well above the allowable metal temperature. The corresponding cooling problems to be solved by the heat transfer engineer are similar to those that have already been discussed for the combustor. Again film cooling is used extensively.

The temperatures of the flameholders, generally, have to be analyzed, too, in order to ensure structural integrity and reliabiltiy.

Variable-geometry exhaust nozzles need special attention, because unfavorable temperature distributions may cause warping of the flaps and failure of the hinges and actuating mechanism. The situation is complicated by leakage flows that are difficult to predict (Ref. 18).

Heat transfer problems of importance are:

- Film cooling with high main stream turbulence and large gas-to-film temperature ratio.
- Film cooling in convergent/divergent nozzles, i.e. reaching from subsonic flow through the throat to supersonic flows.
- Flame radiation.
- Droplet vaporization.

## 3.5. FURTHER HEAT TRANSFER AREAS

Besides the major engine modules discussed above, there are many components in an aero-engine that require the attention of the heat transfer engineer. Lack of space permits them to be mentioned only.

*Labyrinth seals.* Since labyrinth seals have to be designed for very tight clearances, the thermal behavior of the static and rotating member has to be evaluated for the whole flight cycle, just as is the case with the compressor and turbine. Thermal analysis is still hampered by a lack of understanding of the heat transfer inside a modern high-performance step seal (see Ref. 19).

*Bearings and lubrication.* Detailed thermal analysis of the bearings is required, because certain temperature limits must not be exceeded in order to prevent the oil form cracking and possibly igniting. Especially critical situations also to be considered occur when the oil flow is interrupted in the event of failure, in zero-g-condition or during enverted flight. Additionally, the heat balances of the whole oil circulation system have to be evaluated. Papers on this subject are published in Ref. 20.

*Heat exchangers.* Aero-engines have a number of heat exchangers for various purposes, such as for cooling the oil (using the fuel for instance), for cooling the cabin air, etc. Future aplications may also include extracting heat from the exhaust stream and transfering it into the air stream between the compressor and combustor to reduce fuel consumption. This seems particularly attractive for helicopter engines (Ref. 21). Another future application is likely to be the cooling of turbine cooling air. Since the temperature of the cooling air will increase and thus its cooling potential will decrease with increasing compression ratios, cooling of the cooling air using part of the cold by-pass flow, for instance, will be very effective. Of course, the extra complication, weight and safety risk have to be wieghed against the advantages.

The heat exchanger for aero-engines may be characterized by their compactness, light wieght, reliability, and large efficiency values with low pressure drops. A great deal of research is still necessary for designing these heat exchangers, especially for future applications.

*Infrared radiation suppressor.* For many military applications, a reduction in infrared radiation is required to reduce the detectability of the aircraft by sensors. Most of the radiation is emitted by the hot turbine and nozzle surfaces and by the exhaust gas. This is effected by attaching the suppressor to the rear of the engine. The usual principle is to shield the hot parts form direct view and to cool the hot gas by admixing cold air (e.g. Ref. 22). Detailed information is scarce because of the sensitivity of the subject.

*Heat rejection by the engine.* The heat loss from the engine through its outer casing to the environment is also one of the problem areas of the heat transfer engineer. The amount of heat rejected has to be known for the design of the structure and the ventilation of the engine bay or nacelle. In addition, especially for smaller engines, the heat lost has to be accounted for in a proper heat balance when analyzing an engine's performance.

## 4. COMPUTATIONAL METHOD

Prediction of the transient temperature distribution in the various engine parts is made with the aid of an appropriate illustrated in Fig. 20. The left-hand column shows the computation sequence, and the center column the data files and subroutine library. The first step is to take the relevant geometry data and generate a finite element grid. Then the thermal property values are evaluated. After that the boundary conditions (heat transfer coefficients, air temperatures, etc.) for each surface element are computed using cycle, air system and mission data and calling-up the relevant subroutines. These subroutines contain relationships describing the heat transfer process for typical situations, such as flat plates, channels of various geometries, rotating disks, etc. In some cases, especially for performed to evaluate the gas side heat transfer coefficient. The next block, the heart of the program, solves the heat conduction equation for one time step. Then the procedure is repeated for all time steps throughout the flight mission. The temperatures are stored and, as indicated in the right-hand column in Fig. 20, are presented together with the geometry in isotherm plots, in graphs as function of time, and as mean values. In some cases, heat fluxes and temperature gradients are of interest. These results are used by the heat transfer engineer to make an initial evalution as to whether the desired thermal behavior or cooling performance has been achieved. Then, the results are processed further in programs that determine the stress distribution and life.

To describe a generally highly complex geometry with curved surfaces it is convenient to employ isoparametric quadratic or cubic elements as shown in Fig. 21 (Refs. 23 and 24). They may be used directly for stress calculations without the necessity for interpolation onto another finite element grid. In this way, computation onto another finite element grid. In this way, computation is simplified and there is no unnecessary loss of accuracy.

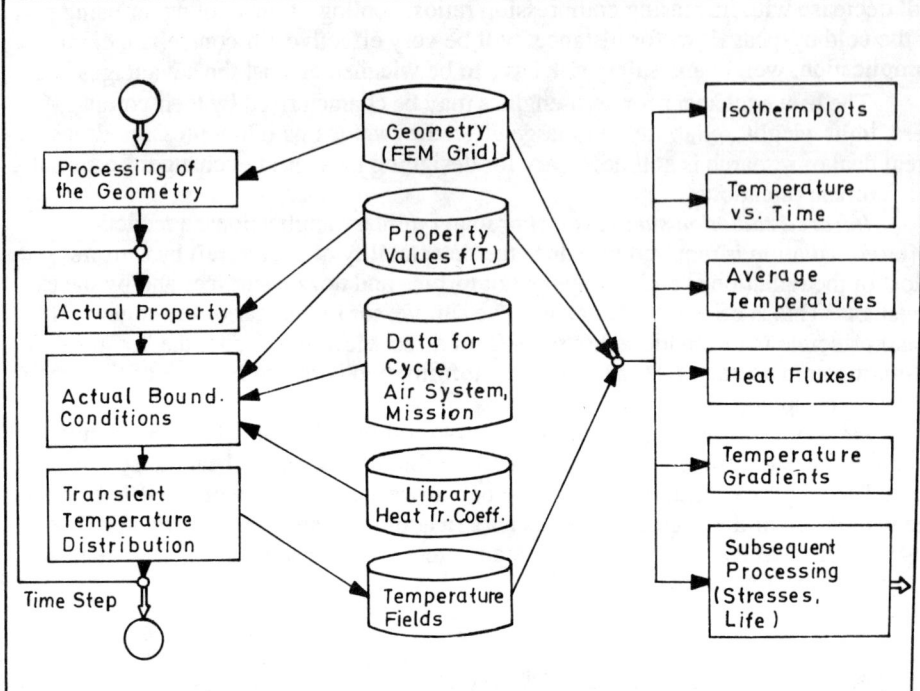

Fig. 20. Structure of computer program system for thermal analysis

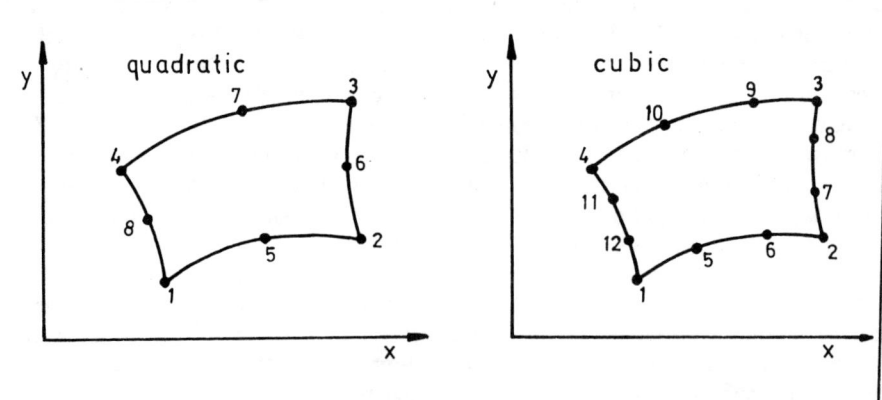

Fig. 21. Typical finite element shapes

## 5. CONCLUSIONS

It has been shown that the activity of the heat transfer engineer in the field of aero-engine development has greatly expanded and has gained importance in recent years. Traditionally, he concentrated mainly on the cooling of the hot parts, notably the combustor and turbine. Now, his responsibilities include the thermal behaviour of nearly all other parts, including the "cold parts", such as the fan, compressor, labyrinth seals, bearings, etc. This is the result of the trends in aero-engines toward higher pressure ratios and turbine temperatures. Furthermore, costly redesigns can be avoided and engine reliability increased with simultaneous reduction in engine size and weight if thermal analyses are performed for the start of a new project. Finally, future engines will have to maintain tight clearances throughout the flight cycle. Only then will it be possible to achieve still higher aerodynamic efficiencies of the compressors and turbines and, thus, to attain the reduction in fuel consumption which is so much strived after.

A discussion of the heat transfer tasks involved in major engine components has demonstrated that nearly all aspects of heat transfer techology are met in aero-engines. Very detailed models, describing the heat transfer processes, will have to be developed in order that the accuracies necessary in engine design may be achieved. Since many phenomena are still not fully understood, continued, intensive research is called for. Particularly important areas have been highlighted in this report.

## REFERENCES

1. HENNECKE, D.K. 1982. Turbine blade cooling in aero engines. Von Karman Institute for Fluid Dynamics. LS 3 on Film cooling and turbine blade heat transfer.
2. WINTERFELD, G. 1979. Prospects for propulsion and energetics. AGARD Highlights 79/1, pp. 12-25.
3. AGARD, 1978. Icing testing for aircraft engines. AGARD-CP-236.
4. OWEN, J.M. 1982. Rotating disks and enclosures. XIV ICHMT Symp. on Heat and Mass Transfer in Rotating Machinery, Dubrovnik, Yugoslavia.
5. KROCKOW, W., Simon, B. and Parnell, E.C. 1979. A chemical reactor model and its application to a practical combustor. AGARD-CP-275, Paper 23.
6. AGARD 1978. Aircraft engine future fuels and energy conservation. AGARD-LS-96.
7. SINGLETON. R.E. 1982. Aeropropulsion research for the U.S.Army. ASME-Paper 82-GT-203.
8. ESSMAN, D.J., VOGEL, R.E. Tomlinson, J.G. and Novick, A.S. 1981. TF41/Lamilloy accelerated mission test. AIAA-81-1349.
9. STEPKA, F.S., LIEBERT, C.H. and Stecura, S. 1977. Summary of NASA research on thermal-barrier coatings. SAE, Int. Automotive Eng. Congr., Detroit .
10. KAPPLER, G. 1979. Status of combustor development from ceramic materials. 6th Army Materials Techn. Conf. on Ceramics for High Performance Application - III. Reliability, Orcas Island, Wash.
11. SWAN, W.C., BOUWER, D.W. and TOLLE, F.F. 1976. Life cycle cost impact on design considerations for civil transport aircraft propulsion systems, in Proceedings 3rd Int. Air Breathing Engines, D.K. Hennecke and G. Winterfeld, Eds. DGLR-Fachbuch Nr. 6.
12. FRECHE, J.C. and AULT, G.M. 1977. Progress in advanced high temperature turbine materials, coatings, and technology AGARD-CP-229, Paper 3.

13. JAPIKSE, D. 1976. Alternative turbine cooling technology, von Karman Inst., LS 83
14. HORNER, M.W., DAY, W.H., SMITH, D.P. and COHN, A. 1978. Development of a water-cooled gas turbine, ASME Paper 78-GT-72.
15. HOURMOUZIADIS, J. and KREINER, H.B. 1981. Advanced component design basis for next generation medium power helicopter engines, AGARD-CP-302.
16. BARRY, G. 1976. Effect of cooling on aerodynamic performance, von Karman Inst., LS 83.
17. MCDONEL, J.D. and EISWERTH, J.E. 1977, Effects of film injection on performance of a cooled turbine, AGARD-CP-229, Paper 29.
18. GRIEB, H., VEDOVA, T., ENDERLE, H. and NAGEL, H. 1981. Comparison of different nozzle concepts for a reheated turbofan. AGARD-CP-301.
19. AGARD 1979. Seal technology in gas turbine engines. AGARD-CP-237.
20. AGARD 1982. Problems in bearings and lubrication. To be published as AGARD Conference Proceedings.
21. GRIEB, H. and KLUSSMANN, W. 1981. Regenerative helicopter engines - Advances in performance and expected development problems. AGARD-CP-302.
22. BARLOW, B. and PETACH, A. 1977. Advanced design infrared suppressor for turboshaft engines. 33rd Annual Nat. Forum of the Am Helicopter Soc., Wash., D.C.
23. ZIENKIEWICZ, O.C. and PARAKH, C.H. 1970. Transient field problems: two-dimensional and three-dimensional analysis by isoparametric finite elements, Int. J. Num. Meth. Engng. 2, p. 67-71.
24. KÖHLER, W. and PITTR, J. 1974. Calculation of transient temperature fields with finite elements in space and time dimensions, Int. J. Num. Meth. Engng. 8, p. 625-631.

# HEAT AND MASS TRANSFER MEASUREMENT TECHNIQUES

O.G. MARTINENKO

*Institute of Heat and Mass Transfer, BSSR Academy of Sciences, Minsk, USSR*

The 15th Symposium, which was organized by the International Center for Heat and Mass Transfer in 1983 and which was concerned with the methods and techniques for measuring heat and mass transfer processes in plazmas, summarized the advances in the development of various measuring techniques. Many interesting papers were presented at the Symposium. Worthy mention among these is an interesting generalized lecture on "Recent Achievements in the Technique of Heat and Mass Transfer Measurements" read by R.I.Soloukhin, Full Member of the BSSR Academy of Sciences, and his paper "The Methods of Measurement in the Physics and Techology of Plasmas". In these works R.I.Soloukhin conducted a detailed analysis of, and called the researchers, attention to the following methods:

## 1. THE LENGMUIR PROBE METHOD

This method affords the possibility to determine the electron concentration, electron temperature and some ion characteristics from the voltampere characteristics of the current to the probe, during a change in its voltage relative to the plasma potential. Most characteristic ranges are $T_I \leq 10^5 K$ (secondary emission-based limitation) and $N_I \leq 10^{15} K$ cm$^{-3}$ (are limited). One of essential drawbacks of probe measurements is the distorting effect of magnetic fields. In his report, R.I.Soloukhin suggests a possible version of probe measurement with a minimum of magnetic field distorsion. In his scheme an experimentally obtained shape of a HF-discharge curve for plasma energy resonance absorption in a magnetic field, makes it possible to obtain both the concentration of the charged plasma component and the frequency of collisions of electrons and ions with neutral particles. The technuque of measuring plasma and magnetic field parameters using magnetic (inductive) probes has been successfully applied to high-temperature and unsteady-state plasma processes.

## 2. INTERFERENCE METHODS.

Given examples are concerned with the IR-interferometry application and study into development of ionization avalanches behind a shock wave front in a rarefied gas.

It is shown that at the 10.6 µm wavelength of a $CO_2$-laser, the interferometer allows one to explore ionization processes in a veakly-ionized gas ($N_l/N \sim 10^{-6} + 10^{-7}$) with a time resolution of ~0.1 µs. Experimental data were obtained within the range $N_l \sim 2 \cdot 10^{12} - 7 \cdot 10^{13}$ cm$^{-3}$.

Though amenable to measurements in the IR–range the range of $N_l$ values is practically inaccessible to common methods of visible and super-high frequency interferometries. The development of laser-interferometric measurements in the far IR-spectrum is also promising. Among techniques of temperature measurement the most developed methods of relative intensities and spectrum line reversal, measurement of spectral-line broadening, methods of laser scattering and absorption measurements should be mentioned. A number of the specific features of spectral measurements in a higher-density plasma were noted. In addition to the aforegoing, among the most common recent methods, the technique of intraresonance laser spectroscopy which makes it possible to measure the concentration of atoms within the range $(1-60) \cdot 10^{10}$ cm$^{-3}$ should be mentioned. The sensitivity of intreference holographic plasma diagnosis in the IR-spectrum made it possible to measure the values of $N_l l$ of about $2 \cdot 10^{14}$ cm$^{-2}$. Among the plasma objects, which are the most complex for investigation, but which have found wide application in technology, is the weaklyionized plasma moving with supersonic speeds. Currently the concetration of electrons within the range $N_l \sim 10^9 \ldots 10^{11}$ cm$^{-3}$ is measured by the SFH-interferometry method and by the method based on plasma conductance measured during the application of voltage pulse supplied with a steep leading front.

The dynamics with which the optical methods of diagnosis were developed and found a successful use in thermophysical experiments should also be mentioned. They play a special part in the study of heat and mass transfer processes, particulary, in studing gasdynamic phenomena in chemically reacting singlephase and multiphase flows.

The development of modern laser techniques spurred the appearance of new optical methods and improvement of traditional ones. By the end of the 1970's three trends were identified in the development of optical diagnostic methods. First are based on records of object self – radiation; second, based on medium exposure to external source radiation and the change in radiation parameters after passage through the studied medium is recorded and, third, those based on records of radiation scattered by a medium.

At the beginning of the 1980's, each approach resutted in the accumulattion of works dealing with the creation and improvement diagnostic devices; however, most intensive recearches were noted with the third groups based on the records of radiation scattered by a medium. Qualitatively, the greates progress during the last decade, has been in methods of high-speed visualization, laser-Doppler anemometry and those based on the records of Raman light scattering.

First of all, high-speed visualization methods are, associated with the diagnostics of fast processes, viz. multi-frame visualization at the nano-second interval with a pico-second exposure.

Laser-Doppler velocity measurements developed in two directions: due to the improvement of optical systems and due to the refinement of techniques for signal processing.

Among the optical methods a special place is occupied by laser spetroscopy of chemically reacting systems, gases, gas flows and low-temperature plasmas by recording Raman scattering [both spontaneous (SRS) and active spectroscopy of Raman scattering (ARS)]. The development of these methods, with respect to heat and mass transfer processes, began at the middle of 1970's, and turned out to be extremely important since they made it possible undisturbingly control composition, temperature, concentrarion (number of molecules) of each separate component of a gas mixture probing it without disruption while aerodynamic combustion and explosions, and making a diagnosis of a plasma, etc.

In the SRS and ARS methods information is obtained by measuring the parameters of spectrum lines, such as the intensity and wave number $\Delta v$ characterizing the displacement of RS lines relative to exciting radiation $v_0$.

Works in the field of SRS were related to an increase of scattering radiation intensity, since these problems are of paramount importance from the view-point of the analytical application of SRS. This, in turn, required further development of the methods of SR-spectrum control, viz. the techniques of input, transformation and the output of readings. One-and multi-channel methods of data processing were developed thus making it possible to simulataneously analyze binary, ternary and multiple gas mixtures.

During the last decade, the studies, development and application of the ARS methods took and important place when the scattered radiation intensity is by two orders of magnitude greater than in the SRS and to a greater extent doesn't depend on the intensities of pumping and probing, but rather on the degree of combination resonance attainment, on the accuracy of coincidence of the difference between laser pumping frequencies and the frequencies of molecules studied. The successes of the ARS method are attributed to the analysis of non-organic gases in fast processes, and the development of the methods is due to the improvement of tunable lasers and accuracy of their tuning.

# THE CALCULATION OF IR RADIAITON TRANSFER IN COMBUSTION GASES CONTAINING $CO_2$ AND CO FOR TEMPERATURE AND CONCENTRATION MEASUREMENTS

S. I. Kryuchkov, N.N. Kudryavtsev, and S.S. Novikov

*Institute of Chemical Physics, Moscow, USSR*

R. I. Soloukhin

*Heat and Mass Transfer Institute, BSSR Academy of Sciences, Minsk, USSR*

## 1. INTRODUCTION

Determination of the radiation characteristics in the IR spectrum of high-temperature media, such as combustion products containing $CO_2$ and CO molecules, is of importance in the investigation of heat and mass transfer processes in energy production facilities, and the use of optical diagnostic techniques in studies of combustion processes. Absorptivity calculations of $CO_2$ and CO bands of thermodynamic equilibrium have been made in a number of works [1-8]. In [1-3], for this purpose "weak" and "strong" line approximations and vibrational-rotational transition parameters obtained using the approximation of a harmonic oscillator-rigid rotator have been used. In [4-8], analytical relations were derived to describe $CO_2$ and CO integral absorptivities with an accuracy to 20-30%.

In this paper, a numerical calculation method is developed that provides an accuracy of 3-5% for the spectral absorptivity, $\overline{A}(\omega)$, at wavelengths of 2.7, 4.3 and 15 μm for $CO_2$ and 4.7 um for CO bands arbitrary gas optical paths. The method is based on the determination of rotational structure-averaged $\overline{A}(\omega)$ values employs a narrow bandwidth model of vibrational-rotational bands in the P-and R-branches and "line-into-line" integration in the Q-branches of individual vibrational transitions. The calculations are based on the experimental and theoretical results for the intensities of individual vibrational transitions forming these bands, the spectrum position and individual vibrational transitions forming these bands, the spectrum position and broadening of the rotational lines obtained from the detailed studies of $CO_2$ and CO molecule structures. The spectral absorption coefficients of the $CO_2$ 2.7, 4.3 and 15 μm and CO 4.7 μm bands and their relevant bandwidths (integral absorptivities) are calculated in a wide range of the governing parameters, namely, a pressure of $P = 0.01$-100 atm, a temperature of $T = 300$-3000K, an optical path of $X = 10^{16}$ –

$10^{20}$ cm$^{-2}$, in particular, of combustion chambers of various energy production facililties. The features of radiation transfer in the bands considered are of importance for diagnostics and the modelling of heat and mass transfer processes.

## 2. $CO_2$ AND CO IR MOLECULAR SPECTRA

Let us consider in brief the radiation spectrum structure of the CO 4.7 μm and $CO_2$ 2.7, 4.3 and 15 μm bands. The rotational levels of the vibrational states with $l>0$ ($l$ being the vibrational momentum component along the molecule axis) are doubly-degenerated, which correspond to two possible directions of vibrational momenta along the molecule axis. The vibrational-rotational interactions result in a splitting of each rotational level of the vibrational states at $l > 0$ into two levels with wave functions [9]:

$$\psi_c = [\psi(v_1,v_2,v_3,|l|,J) + \psi(v_1,v_2,v_3,-|l|,J)]/\sqrt{2},$$
$$\psi_d = [\psi(v_1,v_2,v_3,|l|,J) - \psi(v_1,v_2,v_3,-|l|,J)]/\sqrt{2}, \qquad (1)$$

where $\psi(v_1,v_2,v_3,|l|,J)$ are the unperturbed vibrational-rotational functions, J the rotational quantum number, $v$ the vibrational quantum number. Indices 1,2 and 3 denote symmetric, deformation and asymmetric $CO_2$ vibrations. It is easy to show that a splitting may be represented as $\Delta E \sim [\eta^2 J(J+l)]^l$ [9] where $n$ is a small parameter characterizing the interaction. Thererfore, for π states, all the rotational levels with wave functions having $c$ and $d$ indices have different rotation constants $B^{(c)}$ and $B^{(d)}$. The splitting is much less pronounced for the states with $l>1$, and is accounted for by the higher order terms in the expression for rotational energy.

In a mostly spread isotope $O^{16}$-$C^{12}$-$O^{16}$, the oxygen nuclei are equivalent and have zero spins. Therefore, only the wave functions that are symmetric relative to the replacement of these nuclei are resolved, i.e. a half of the rotational states are forbidden. Depending on whether level $c$ or $d$ turns out to be forbidden in the lower state, transitions $c \to c$ and $d \to d$ for the P- and R-branches and $c \to d$ and $d \to c$ for the Q-branch may take place. In conformity with rotational wave function symmetry, the type of transition alternates depending on the variation of the rotational quantum number, $J$, [10]. The vibrational-rotational band of an individual vibrational transition ($v' \leftarrow v''$) may be considered a superposition of two subbands, i.e. $c \to c$ and $d \to d$ (P-, R-branches), $c \to d$ and $d \to c$, (Q-branches), with transitions either from levels with even or odd quantum numbers occuring in each of them. In the P- and R-branches far from the edge, in the superposition of two subbands such as $c \to c$ and $d \to d$ the mean distance between the rotational lines is $\sim 2B$. Two edges in a band of an individual vibrational transition ($v' \leftarrow v''$) are caused by a difference in the magnitudes $\Delta B^{(c \leftarrow c)}_{v' \leftarrow v''}$ and $\Delta B^{(d \leftarrow d)}_{v' \leftarrow v'''}$.

For the determination of the averaged absorptivity in the P-and R-branches of individual vibrational transitions, a statistical model assuming a random line distribution within the spectrum [11] was employed:

$$\overline{A}(\omega) = 1-\exp\left[-\sum_{v'v''} W_{v'J' \leftarrow v''J''}(\omega)/\delta_{v' \leftarrow v''}(\omega)\right] \quad (2)$$

where

$$W_{v'J' \leftarrow v''J''}(\omega) = \int_{-\infty}^{\infty} \{1-\exp[-K_{v'J' \leftarrow v''J''}(\omega)X]\} d\omega,$$

and $\delta_{v' \leftarrow v''}(\omega)$ is the equivalent width of rotational lines and the distance between them in separate vibrational transitions ($v' \leftarrow v''$), $K(\omega)$ is the absorption coefficient, $X = N_n D'$ the optical path of an emitting volume, $D$ the geometric beam path length in a gas, $N_n$ the number of molecules of an emitting mixture component per unit volume. An equivalent rotational lines width with A Voigt contour was found according to [12]. The absorption in the Q-branches of individual vibrational transitions has been determined by direct "line-into-line" integration over the spectrum:

$$\overline{A}(\omega) = \frac{1}{2\Delta\omega} \int_{\omega-\Delta\omega}^{\omega+\Delta\omega} [1-\exp(-K(\omega')X)] d\omega' \quad (3)$$

where $\Delta\omega = 0.1$ cm$^{-1}$ is the averaging integral, $K(\omega) = \sum_{J''} K(J)$ the total absorption coefficient. An accuracy of the "line–into–line" integration was 0.1%. For the integral absorptivity of a vibrational-rotational line according to the harmonic oscillator-rigid rotator approximation the following relation has been derived [13]:

$$W_{v'J' \leftarrow v''J''} = \frac{8\pi^3 \omega_{v'J' \leftarrow v''J''}}{3 hc} \frac{N_{v''J''}}{P g_{v''J''}} \beta^2_{v' \leftarrow v''} g_{v'J'} \times \quad (4)$$

$$\times \{\Gamma^{J''I'}_{J''I''} 2[1-N_{v'J'} g_{v''J''} / N_{v''J''} g_{v'J'}]$$

where $\omega_{v'J' \leftarrow v''J''}$ is the wave number at the line centre, $\beta^2_{v' \leftarrow v''}$ the squared vibrational matrix element of a dipole transition ($v' \leftarrow v''$) moment, $\Gamma^{J''I'}_{J''I''}$ the molecule rotation amplitude factor, and the last cofactor in (4) allows for induced corrections. For calculations it is convenient to use integral absorption coefficients of the individual vibrational transitions $a_{v' \leftarrow v''}$ which are measured experimentally or computed numerically. A relationship between the integral absorption coefficient of a rotational line and that of a corresponding vibrational transition reads [12] as:

$$S_{v'J' \leftarrow v''J''} = \alpha_{v' \leftarrow v''} \frac{P\alpha}{P} \frac{N_{v''J''}}{N_v^{(\alpha)}} \left(\Gamma^{J''I'}_{J''I''}\right)^2 \times \quad (5)$$

$$\times \frac{1-N_{v'J'}\,gv''J''/N_{v''J''}\,gv'J'}{1-N_{v'}^{(\alpha)}\,gv''/N_{v''}^{(\alpha)}\,gv'}$$

where index $\alpha$ designates gas parameter values corresponing to $\alpha_{v'\leftarrow v''}$ measurement conditions.

The determination integral absorption coefficient for transitions between excited states, $\alpha_{v\leftarrow v''}$, using theoretical vibrational quantum number-dependent squared matrix elements, $\beta^2_{v''v''}$ may be recommended for calculations for diatomic molecule such as CO bands [14] and the polyatomic molecule bands, similar to them in structure, for instance, the $CO_2$ 4.3 µm band [15]. The relevant calculations have been carried out in the present paper with reference to $\alpha$ values reported in [13,16].

The interaction of combined $CO_2$ mode levels due to the Fermi resonance exerts a decisive influence on the squared matrix elements of the vibrational transitions at 2.7 µm [17,18] and governs the appropriate 15 µm bands. The $CO_2$ radiation characteristics of these bands should be calculated on the basis of experimental and theoretical data on integral absorption coefficients of the individual vibrational transitions $\alpha_{v'\leftarrow v''}$. The most reliable measured integral absorption coefficients of individual vibrational $CO_2$ transitions [19] and those calculated with allowance made for the detailed structure of $CO_2$ bands [20] are listed in Tables 1 and 2 for the 2.7 and 15 µm bands, respectively.

The basic mechanisms determining the width of rotational lines in a gas at pressures P>0.1 atm are the Doppler effect and molecular collisions. The Doppler halfwidth of a spectrum line is determined by $\gamma_D = \frac{\omega_0}{c}\sqrt{\left(\frac{kT\log 2}{m}\right)}$ where $m$ is the emitting volume molecule and $\omega_0$ the wave number at line centre. To calculate the collisional halfwidth in a gas mixture, the theoretical dependence widely used [13] is:

$$\gamma_L = P\sqrt{\left(\frac{T_0}{T}\right)\sum_n \xi_n \gamma_{LO}^{(n)}} \qquad (6)$$

where $\xi_n$ are the component molar concentrations and $\gamma_{LO}^{(n)}$ is based on normal condition partial halfwidths for emitting molecule collisions with a mixture component designated with the index $n$. The values of $\gamma^{(n)}$ for a CO molecule are: 0.072, 0.060, 0.060, 0.040, 0.050, 0.050, 0.060, atm$^{-1}$ cm$^{-1}$ [3,21-23]; while for a $CO_2$ molecule: 0.087, 0.067, 0.067, 0.040, 0.055, 0.080, 0.050, $atm$-1 $cm$-1 [3,24-27] for broadening by $CO_2$, $N_2$, CO, $N_2$, $O_2$, $H_2$, $H_2O$, respectively.

The distance between neighbouring rotational lines of the individual vibrational transition subbands can be obtained taking account of the difference in rotational constants for the combining levels:

$$\delta = 4B'' + (B'-B'')\,(4m+2) \quad \text{in the P- and R-branches} \qquad (7)$$

$$\delta = (B'-B'')\,(4m+2) \quad \text{in the Q-branches} \qquad (8)$$

with $m=-J$ for the P-branches, $m=J+1$ for the R-branches and $m=J$ for the Q-branches. The CO 4.7 µm and $CO_2$ 4.3 µm band data were calculated using the squated matrix elements for the individual vibrational transitions as a function of the quantum numbers and the CO 2.7 μm and 15 *um* bands using the data on integral absorption coefficients and spectroscopic parameters of individual vibrational transitions between $CO_2$ low-lying ($E_v \leq 350$ cm$^{-1}$) levels presented in Tables 1 and 2. These transitions are responsible for more than 70-95% of the band radiation under the conditions considered. The integral absorption coefficients of rotational lines were calculated using Eq. (5).

For transitions between $CO_2$ vibrational states with an angular momentum projection $l>0$, the Q-branch is absent, in the bands of the individual vibrational transitons. In [1], it is found that for the re-absorbed lines, the integral radiation intensity of the Q-branch transition of the $CO_2$ 2.7 μm and 4.3 μm bands is comparable with the respective value of individual rotational lines of the P- and R-branches. Therefore, the radiation in the Q-branches of "parallel" $CO_2$ (4.3 and 2.7 μm) bands was not taken into account. In the $CO_2$ (15 μm) "perpendicular" band, the P-, R-branches of individual vibrational transitions have comparable intensity. The Q branch density of the rotational lines, $1/\sigma$ and hence, the absorptivity is as much as 10÷100 higher than the P- and R-branches. Therefore, the spectrum radiation intensity or absorptivity of the 15 μm band was determined neglecting Q-branche and P- and R-branche overlapping: $\overline{A}(\omega) = \overline{A}(Q,(\omega)) + \overline{A}(P,R,(\omega))$. For strong absorption : $A(\omega) \approx 1$ and the contribution of the Q-branches was neglected [22]. The comparison of the present absorptivity calculation results in the CO 4.7μm and $CO_2$ 4.3, 2.7 and 15 μm bands with the experimental and theoretical data of [16, 28-31] shows good agreement within 5 per cent.

The absorptivity of a molecular gas at thermodynamic equlibrium depends on the optical path, $X$, gas pressure, P and gas temperature, T. The absorptivity calculation results are given for an optical path of $X = 10^{19}$ cm$^{-2}$ typical of a number of power plants. This is consistent with the maximum spectral absorptivity, $\overline{A}_{max}(\omega) < 0.1$–$0.5$ for the $CO_2$ 2.7 and 15 μm bands and the CO 4.7 μm band, and with $\overline{A}_{max}(\omega) = 1$ for the $CO_2$ 4.3 μm band. The values of $X$ have to be varied over a wide range of $10^{16}$÷$10^{20}$ cm$^{-2}$ if we are to calculate the integral absorptivity or the equivalent width of the bands.

The rotational-vibrational level populations are a function of the gas temperature which, in turn, affects the Doppler broadening of rotational lines at low gas pressures (P< 0.2 – 0.5 atm). Variations in the vibrational-rotational level molecule distribution substantially influence the spectral absorptivity distribution in the vibrational-rotational bands.

## 3. SPECTRAL CHARACTERISITCS

The $\overline{A}(\omega)$ computation results for the considered bands at different gas temperatures are presented in Fig.1. For CO at 4.7 μm an increase in gas temperature substantially elevates the absorptivity in the short-wave range and, especially, in the long-wave wing of a band. These specific features were indicated by a simultaneous increase in the high-lying vibrational $N_V$ and rotational $N_{VJ}$ level populations at high gas temperatures. In the central part of the band, $\omega = 2000$-$2300$ cm$^{-1}$, the absorptivity decreases as much as 2-3 times when T increases from 300 to 3000 $K$ due to decreasing populations of the low-lying vibrational $N_V$ and rotational $N_{VJ}$

Table 1. Values of the Parameters of the Individual Vibrational Transitions of the 2.7 μm band [19]

| State lower: upper | Transition | $E_{v'}$ cm$^{-1}$ | $\omega_{v' \leftarrow v'''}$ cm$^{-1}$ | $\Delta\omega_{v' \leftarrow v'''}$ cm$^{-1}$ | $\Delta B^{(c \leftarrow c)}_{v' \leftarrow v'''}$ 10$^5$cm$^{-1}$ | $\Delta B^{(d \leftarrow d)}_{v' \leftarrow v'''}$ 10$^5$cm$^{-1}$ | $\alpha_{v' \leftarrow v'''}$ atm$^{-1}$cm$^{-2}$ |
|---|---|---|---|---|---|---|---|
| 00°0→10°1 | $\Sigma_g^+ - \Sigma_u^+$ | 0 | 3714.78 | 47.24 | 315.65 | 315.65 | 36.0 |
| 02°1 | | | 3612.84 | 46.78 | 217.52 | 217.52 | 24.84 |
| 01$^1$0→11$^1$1 | $\Pi_u - \Pi_g$ | 667.38 | 3723.25 | 47.56 | 329.23 | 302.62 | 2.724 |
| 03$^1$1 | | | 3580.33 | 46.39 | 286.93 | 256.63 | 1.93 |
| 02$^2$0→12$^2$1 | $\Delta_g - \Delta_u$ | 1335.13 | 3726.65 | 47.18 | 317.40 | 317.40 | 0.116 |
| 04$^2$1 | | | 3552.84 | 46.07 | 272.40 | 272.40 | 7.50(−2) |
| 10°0→20°1 | $\Sigma_g^+ - \Sigma_u^+$ | 1388.19 | 3711.47 | 48.58 | 270.09 | 270.09 | 8.40(−2) |
| 12°1 | | | 3589.64 | − | 365.99 | 365.09 | 4.28(−2) |
| 02°2→20°1 | $\Sigma_g^+ - \Sigma_u^+$ | 1285.41 | 3814.25 | 49.41 | 299.43 | 299.43 | 1.85(−3) |
| 12°1 | | 3692.42 | 3692.42 | − | 395.33 | 395.33 | 0.102 |
| 04°1 | | | 3566.21 | 47.23 | 229.43 | 229.43 | 8.11(−2) |
| 03$^3$0→13$^3$1 | $\Phi_u - \Phi_g$ | 2003.24 | 3727.38 | − | 315.10 | 315.10 | 5.02(−3) |
| 05$^3$1 | | | 3528.04 | − | 268.80 | 268.80 | 2.92(−3) |
| 11$^1$0→21$^1$1 | $\Pi_u - \Pi_g$ | 2076.87 | 3713.71 | 48.32 | 300.30 | 273.10 | 6.76(−3) |
| 13$^1$1 | | | 3555.90 | 45.04 | 339.80 | 321.60 | 2.64(−3) |
| 05$^1$1 | | | 3398.21 | 45.13 | 226.30 | 192.10 | 9.89(−6) |
| 03$^1$0→21$^1$1 | $\Pi_u - \Pi_g$ | 1932.47 | 3858.11 | 49.80 | 332.30 | 306.70 | 6.02(−5) |
| 13$^1$1 | | | 3700.29 | 46.52 | 371.80 | 355.20 | 8.52(−3) |
| 05$^1$1 | | | 3542.60 | 46.61 | 258.30 | 225.20 | 7.55(−3) |
| 04$^4$0→14$^4$1 | $\Gamma_u - \Gamma_g$ | 2671.69 | 3726.35 | − | 314.50 | 314.50 | 2.16(−4) |
| 06$^4$1 | | | 3504.93 | − | 267.70 | 267.70 | 1.14(−4) |
| 12$^2$0→22$^2$1 | $\Delta_u - \Delta_g$ | 2760.74 | 3713.76 | − | 287.30 | 287.30 | 2.63(−4) |
| 14$^2$1 | | | 3527.76 | − | 306.30 | 306.30 | 8.98(−5) |
| 06$^2$1 | | | − | − | 210.70 | 210.70 | − |
| 04$^2$0→22$^2$1 | $\Delta_u - \Delta_g$ | 2585.03 | − | − | 327.60 | 327.60 | − |
| 14$^2$1 | | | 3703.46 | − | 346.60 | 346.60 | 3.61(−4) |
| 06$^2$1 | | | 3518.64 | − | 251.00 | 251.00 | 3.31(−4) |
| 20°0→30°1 | $\Sigma_g^+ - \Sigma_u^+$ | 2797.15 | 3705.93 | − | 258.90 | 258.90 | 1.32(−4) |
| 22°1 | | | 3550.70 | − | 413.20 | 413.20 | 4.68(−5) |
| 14°1 | | | − | − | 386.60 | 386.60 | − |
| 06°1 | | | − | − | 166.70 | 166.70 | − |
| 12°0→30°1 | $\Sigma_g^+ - \Sigma_u^+$ | 2671.11 | 3813.97 | − | 158.20 | 158.20 | 2.21(−6) |
| 22°1 | | | 3676.74 | − | 312.50 | 312.50 | 2.20(−4) |
| 14°1 | | | 3556.75 | − | 285.90 | 285.90 | 1.51(−4) |
| 06°1 | | | − | − | 066.00 | 066.00 | − |
| 04°0→30°1 | $\Sigma_g^+ - \Sigma_u^+$ | 2548.37 | − | − | 313.27 | 313.27 | − |
| 22°1 | | | 3799.51 | − | 467.57 | 467.57 | 6.72(−6) |
| 14°1 | | | 3679.55 | − | 440.97 | 440.97 | 2.37(−4) |
| 06°1 | | | 3527.61 | − | 221.07 | 221.07 | 2.48(−4) |
| 00°1→10°2 | $\Sigma_g^+ - \Sigma_u^+$ | 2349.15 | 3667.54 | − | 322.24 | 322.24 | 9.19(−4) |
| 02°2 | | | 3566.06 | − | 261.34 | 261.34 | 4.99(−4) |
| 01$^1$1→11$^1$2 | $\Pi_u - \Pi_g$ | 3004.02 | 3675.94 | − | 328.30 | 306.30 | 7.94(−5) |
| 03$^1$2 | | | 353.94 | − | 278.90 | 246.80 | 4.22(−5) |

Table 2. Values of Parameters of Individual Vibrational Transitions of 15 μm Band [17]

| State lower:upper | Transition | $E_{v''}$ cm⁻¹ | $\omega_{v'\leftarrow v''}$ cm⁻¹ | $\Delta\omega_{v'\leftarrow v''}$ cm⁻¹ | $\Delta B^{(cd)}_{v'\leftarrow v''}$ 10⁻⁵cm⁻¹ | $\Delta B^{(dc)}_{v'\leftarrow v''}$ 10⁻⁵cm⁻¹ | $\Delta B^{(cc)}_{v'\leftarrow v''}$ 10⁻⁵cm⁻¹ | $\Delta B^{(dd)}_{v'v'}$ 10⁻⁵cm⁻¹ | $\alpha_{v'\leftarrow v''}$ atm⁻¹cm⁻² |
|---|---|---|---|---|---|---|---|---|---|
| $00^00 \to 01^10$ | $\Sigma_g^+ - \Pi_u$ | 0.0 | 667.38 | 12.51 | 103.70 | — | 42.30 | — | 198.20 |
| $00^00 \to 02^20$ | $\Pi_u - \Delta_g$ | 667.38 | 667.75 | 12.49 | 104.10 | 42.60 | 104.10 | 42.60 | 15.57 |
| $10^00$ | $\Sigma_g^+$ | | 720.81 | 10.04 | — | -106.70 | -45.30 | — | 4.448 |
| $02^00$ | $\Sigma_g^+$ | | 618.81 | 9.21 | — | -77.40 | -16.00 | — | 3.447 |
| $02^20 \to 03^30$ | $\Delta_g - \Phi_u$ | 1335.13 | 668.11 | 12.47 | 73.80 | 73.80 | 73.80 | 73.80 | 0.918 |
| $11^10$ | $\Pi_u$ | | 741.74 | 10.39 | -33.80 | -126.60 | -126.60 | -33.80 | 1.90(−1) |
| $03^10$ | $\Pi_u$ | | 579.34 | 8.91 | -0.20 | -94.60 | -94.60 | -0.20 | 0.125 |
| $10^00 \to 11^10$ | $\Sigma_g^+ - \Pi_u$ | 1338.19 | 688.68 | 12.83 | 115.50 | — | 22.70 | — | 3.57(−1) |
| $03^10$ | $\Pi_u$ | | 544.28 | 11.35 | 149.10 | — | 54.70 | — | 6.54(−3) |
| $02^00 \to 11^10$ | $\Sigma_g^+ - \Pi_u$ | 1285.41 | 791.45 | 13.67 | 86.20 | -6.63 | -6.63 | — | 2.70(−2) |
| $03^10$ | $\Pi_u$ | | 647.06 | 12.19 | 119.80 | 25.40 | 25.40 | — | 0.53 |
| $03^30 \to 04^40$ | $\Phi_u - \Gamma_u$ | 2003.24 | 668.45 | 12.46 | 74.20 | 74.20 | 74.20 | 74.20 | 4.79(−2) |
| $12^20$ | $\Delta_g$ | | 757.50 | 10.63 | -88.50 | 88.50 | -88.50 | -88.50 | 7.89(−3) |
| $04^20$ | $\Delta_g$ | | 581.79 | 8.74 | -48.20 | -48.20 | -48.20 | -48.20 | 4.64(−3) |
| $11^10 \to 12^20$ | $\Pi_u - \Delta_g$ | 2076.87 | 683.87 | 12.72 | 11.90 | 19.10 | 111.90 | 19.10 | 2.17(−2) |
| $04^20$ | $\Delta_g$ | | 508.17 | 10.83 | 152.20 | 59.40 | 152.20 | 59.40 | 1.24(−4) |
| $20^00$ | $\Sigma_g^+$ | | 720.29 | 11.26 | 14.70 | — | -78.10 | — | 1.15(−2) |
| $12^00$ | $\Sigma_g^+$ | | 594.25 | 7.33 | -86.00 | — | -178.80 | — | 2.18(−3) |
| $04^00$ | $\Sigma_g^+$ | | 471.51 | 8.52 | 69.10 | — | -23.70 | — | 2.09(−5) |
| $03^10 \to 12^20$ | $\Pi_u - \Delta_g$ | 1932.47 | 828.27 | 14.19 | 79.90 | -14.50 | 79.90 | -14.50 | 4.83(−4) |
| $04^20$ | $\Delta_g$ | | 652.54 | 12.28 | 120.20 | 25.80 | 120.20 | 25.80 | 3.97(−2) |
| $20^00$ | $\Sigma_g^+$ | | 864.68 | 12.74 | — | -111.70 | -17.30 | — | 1.04(−4) |
| $12^00$ | $\Sigma_g^+$ | | 738.64 | 8.53 | — | -212.40 | -118.00 | — | 7.25(−3) |
| $04^00$ | $\Sigma_g^+$ | | 615.90 | 10.00 | — | -57.30 | 37.10 | — | 1.65(−2) |

Table 3. Values of Parameters of Individual Vibrational Transitions of 15 Band [17] (Continued)

Table 2. Values of Parameters of Individual Vibrational Transitions of 15 μm Band [17]

| State lower→upper | Transition | $E_{v'}$ cm$^{-1}$ | $\omega_{v'\leftarrow v''}$ cm$^{-1}$ | $\Delta\omega_{v'\leftarrow v''}$ cm$^{-1}$ | $\Delta B^{(cd)}_{v'\leftarrow v''}$ 10$^{-5}$cm$^{-1}$ | $\Delta B^{(dc)}_{v'\leftarrow v''}$ 10$^{-5}$cm$^{-1}$ | $\Delta B^{(cc)}_{v'\leftarrow v''}$ 10$^{-5}$cm$^{-1}$ | $\Delta B^{(dd)}_{v'\leftarrow v''}$ 10$^{-5}$cm$^{-1}$ | $\alpha_{v'\leftarrow v''}$ atm$^{-1}$cm$^{-2}$ |
|---|---|---|---|---|---|---|---|---|---|
| $04^40\rightarrow 05^50$ | $\Gamma_g$–$\mu_u$ | 2671.69 | 668.79 | 12.44 | 74.60 | 74.60 | 74.60 | 74.60 | 2.34(-4) |
| $13^30$ | $\Gamma_g$–$\phi_u$ | | 770.36 | 10.65 | 15.00 | 15.00 | 15.00 | 15.00 | 3.24(-4) |
| $05^30$ | $\Gamma_g$–$\phi_u$ | | 568.87 | 8.50 | -46.60 | -46.60 | -46.60 | -46.60 | 1.72(-4) |
| $12^20\rightarrow 13^30$ | $\Delta_g$–$\phi_u$ | 2760.74 | 681.52 | 12.68 | 77.70 | 77.70 | 77.70 | 77.70 | 1.10(-3) |
| $05^30$ | $\phi_u$ | | 479.83 | 10.33 | 116.10 | 116.10 | 116.10 | 116.10 | 2.88(-6) |
| $21^10$ | $\pi_u$ | | 739.86 | 11.05 | 16.50 | -107.40 | -107.40 | 16.50 | 4.23(-4) |
| $13^10$ | $\pi_u$ | | 578.61 | 7.62 | -39.00 | -150.00 | -150.00 | -39.00 | 9.03(-5) |
| $05^10$ | $\pi_u$ | | 420.72 | 7.42 | 78.10 | -50.70 | -50.70 | 78.10 | — |
| $04^20\rightarrow 13^30$ | $\Delta_g$–$\phi_u$ | 2585.03 | 857.22 | 14.58 | 37.40 | 37.40 | 37.40 | 37.40 | 2.64(-5) |
| $05^30$ | $\phi_u$ | | 655.53 | 12.22 | 75.80 | 75.80 | 75.80 | 75.80 | 2.38(-3) |
| $21^10$ | $\pi_u$ | | 915.56 | 12.95 | -23.80 | -147.70 | -147.70 | -23.80 | 4.68(-6) |
| $13^10$ | $\pi_u$ | | 754.31 | 9.52 | -79.30 | -190.30 | -190.30 | -79.30 | 3.86(-4) |
| $05^10$ | $\pi_u$ | | 596.42 | 9.32 | 37.90 | -91.00 | -91.00 | 37.80 | 6.17(-4) |
| $20^00\rightarrow 21^10$ | $\Sigma_g^+$–$\pi_u$ | 2797.15 | 703.44 | 12.52 | 113.70 | — | -10.20 | — | 5.91(-4) |
| $13^10$ | $\pi_u$ | | 542.19 | 9.09 | 58.20 | — | -52.80 | — | 1.71(-5) |
| $05^10$ | $\pi_u$ | | 384.30 | 8.89 | 175.30 | — | 46.50 | — | — |
| $12^20\rightarrow 21^10$ | $\Sigma_g^+$–$\pi_u$ | 2671.11 | 829.48 | 16.73 | 214.40 | — | 90.50 | — | 2.76(-5) |
| $13^10$ | $\pi_u$ | | 668.23 | 13.30 | 158.90 | — | 47.90 | — | 7.48(-4) |
| $05^10$ | $\pi_u$ | | 510.34 | 13.10 | 276.00 | — | 147.20 | — | 9.60(-6) |
| $04^00\rightarrow 21^10$ | $\Sigma_g^+$–$\pi_u$ | 2548.37 | 952.22 | 15.26 | 59.30 | — | 64.60 | — | 1.54(-5) |
| $13^10$ | $\pi_u$ | | 790.97 | 11.83 | 3.83 | — | -107.20 | — | 1.32(-4) |
| $05^10$ | $\pi_u$ | | 633.08 | 11.63 | 120.90 | — | -7.87 | — | 1.56(-3) |
| $00^01\rightarrow 01^11$ | $\Sigma_g^+$–$\pi_g$ | 2349.15 | 654.87 | — | 104.90 | — | 45.20 | — | 2.12(-3) |
| $01^11\rightarrow 02^21$ | $\pi_g$–$\Delta_u$ | 3004.02 | 655.26 | — | 105.40 | 45.70 | 105.40 | 45.70 | 1.82(-4) |
| $01^11\rightarrow 10^01$ | $\pi_g$–$\Sigma_u^+$ | 3004.02 | 710.77 | — | — | -112.70 | -5.30 | — | 4.85(-5) |
| $02^01$ | $\pi_g$–$\Sigma_g^+$ | | 608.83 | — | — | -68.60 | -8.90 | — | 4.20(-5) |

states. The same gas temperature effect is observed for the $CO_2$ 15 μm band (Fig. 1d). Note that the absorptivity in the strong Q-branches of individual vibrational transitions, $\overline{A}$ (ω)≈1 (see, Fig. 1d), is substantially higher than in the P-and R-branches in the same spectral ranges. Due to the width of the Q-branches, $\Delta\omega_0 = 10 \frac{|B'-B''|}{B''} \frac{kT}{hc}$, [13] of about 1-5 cm$^{-1}$ at $T$ = 1000 K their structure cannot be reproduced on the given scale of the wave numbers (Figs. 1d and 2d), and the appropriate results are not given in these figures.

In the 4.3 μm band (Fig. 1b), the absorptivity in the long-wave wing (ω<2350 cm$^{-1}$) increases with increasing temperature. In this case, in the central part of the band, the absorptivity is $\overline{A}$ (ω) = 1 i.e. the gas emits as a black body while in the short-wave part of the band a sharp edge is observed at ω = 2390 cm$^{-1}$. In the 2.7 μm band, the temperature increase results in a strong growth of the absorptivity in the long-wave wing (ω<3600 cm$^{-1}$). At the same time, there smoothing of the frequency-dependent absorptivity occurs. The above noted features are specified, similarly as in the case of $CO_2$ 15 μm and CO 4.7 μm bands, by varying the dependences of the vibrational-rotational $N_{VJ}$ states assuming that the 2.7 μm band consists of two types of vibrational transitions: long-wave $\Delta v_2 = 2$, $\Delta v_3 = 1$ and short-wave $\Delta v_1 = \Delta v_2 = 1$. At low temperature ($T$<300K), the above transitions form two sub-bands with the boundary at ω≈3650 cm$^{-1}$ [32].

The results shown in Fig. 1 are prerequise to the development of optical diagnostic techniques using the IR radiation of $CO_2$ in combustion processes. These data allow a choice of the bands and spectral ranges to measure the dependences of the recorded values of $\overline{A}$ (ω) on high-temperature gas parameters. For example, at a pressure $P$ = 1.0 atm and $CO_2$ molar concentration $\xi_{CO_2}$ = 0.3 according to the data in Fig. 1b $A(\omega) \approx 1$ for an emitting volume thickness of $D$<1 cm. To exclude the effect of $CO_2$ from colder wall layers, it is therefore necessary to record radiation in the long-wave wing of this band. According to the data presented in Fig. 1, the effect of the wall layers with a characteristic thicknesses of $D \approx 1.0$ cm is not substantial for the other investigated bands.

Figure 2a-d shows the spectral radiation intensity and absorptivity of the considered bands at gas pressures varying within $10^{-2}$–100 atm. Note that the gas pressure determines the collisional broadening of the vibrational lines and the degree of their overlap. There exist three limiting cases in which the value of $\overline{A}$ (ω) does not depend on the gas pressure: the re-absorbed radiation is absent when for all transitions the relation $K_{v'J' \leftarrow v''J''} \chi \ll 1$ is valid where $K$ is the absorption coefficient, the radiation of an optically dense gas layer ($\overline{A}$ (ω)→1) occurs and the Doppler mechanism of broadening considerably prevails over the collisional one: $\gamma_D \gg \gamma_L$ (under the conditions considered, at $p \gtrless 0.1$ atm, $\gamma_L > \gamma_D$). The first two cases were encountered under the conditions of the present work. For the CO 4.7 μm (Fig. 2a) are the $CO_2$ 15 μm band (Fig. 2d), the pressure increase from 0.01 to 10 atm greatly affects the absorptivity in the central part of the bands: ω = 1900-2300 cm$^{-1}$ (4.7 μm) and 700-850 cm$^{-1}$ (15 μm). A further pressure increase up to 100 atm does not result in a considerable change of $\overline{A}$ (ω) in these bands. This is due to the broadening of the spectral lines causing a decrease of the radiation absorption in their central parts:

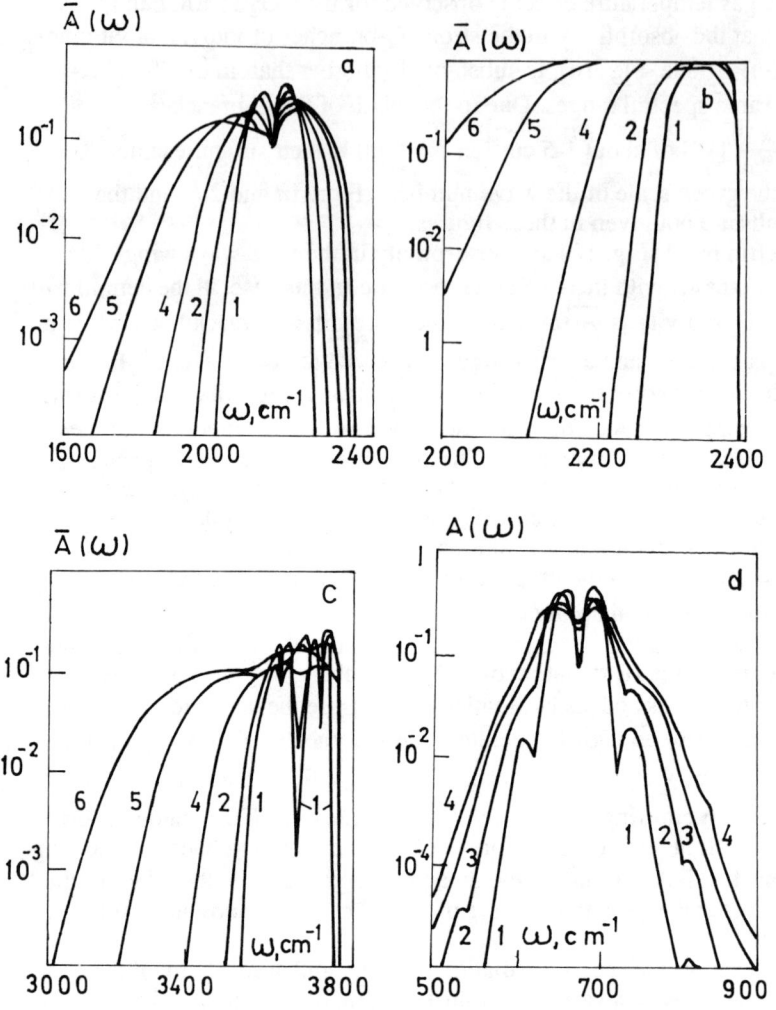

Fig. 1. Spectral absorptivity of the CO 4.7 μm (*a*), 4.3 μm (*b*), 2.7 μm (*c*) and $CO_2$ 15 μm bands as a function of the gas temperature. Curves 1 correspond to $T = 300$ K; 2 – 500; 3 – 750; 4 – 1000; 5 – 2000; 6 – 3000, P = 1.0 atm; $X = 10^{19}$ cm$^{-2}$

$K(\omega) X \ll 1$. For the 2.7 μm band, the absorptivity does not depend on the gas pressure, with a varying from $10^{-2}$ to 100 atm. The estimates show that under the investigated conditions the radiation reabsorption is practically absent in the 2.7 μm band.

The second possibility is realized for the central part of the 4.3 μm band ($\omega = 2230\text{-}2300$ cm$^{-1}$), which corresponds to the black body radiation, $\overline{A}(\omega) = 1$ (Fig. 2*b*) and, hence, does not depend on pressure. A weak dependence of $\overline{A}(\omega)$ on the gas pressure is observed in the 4.3 μm band over two peripheral spectrum parts

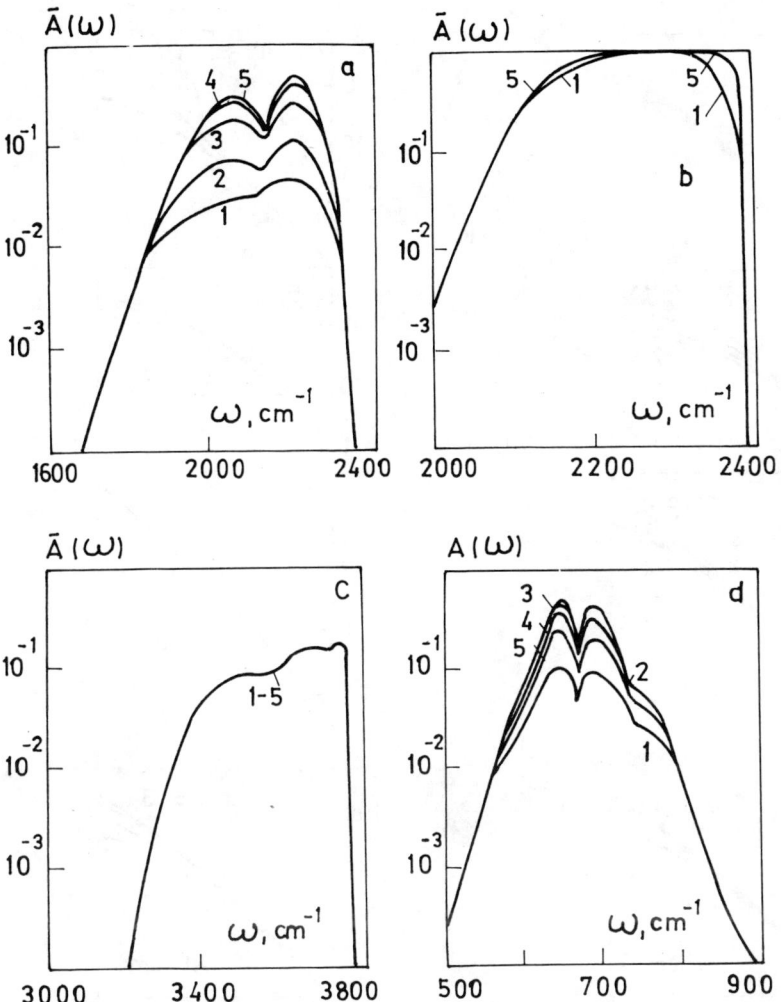

Fig. 2. Spectral absorptivity of the CO 4.7 μm (*a*), 4.3 μm (*b*), 2.7 μm (*c*) and $CO_2$ 15 μm (*d*) bands as a function of the gas pressure. Curves 1 correspond to $10^{-2}$ atm; 2, 0.1; 3, 1.0; 4, 10.0; and 5, 100.0; $X = 10^{19}$ cm$^{-2}$, T = 1000 K (*a-c*) and 750 K (*d*)

($\omega$=2100–2200 cm$^{-1}$ and 2330–2380 cm$^{-1}$), for which there occurs a rapid transition from the black body radiation in the central part of the band to the non-reabsorbed one in its wings

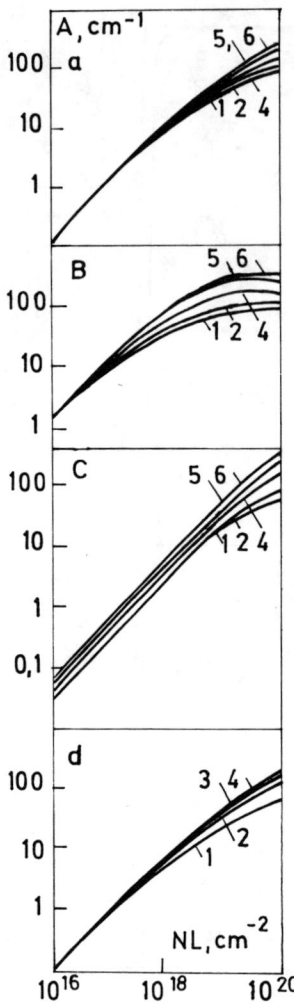

Fig. 3. Integral absorptivity of CO 4.7 μm (a), 4.3 μm (b), 2.7 μm (c) and $CO_2$ 15 μm (d) bands at different gas temperatures. The notation of the curves is the same as in Fig. 1. $P = 1.0$ atm $X = 10^{19}$ cm$^{-2}$

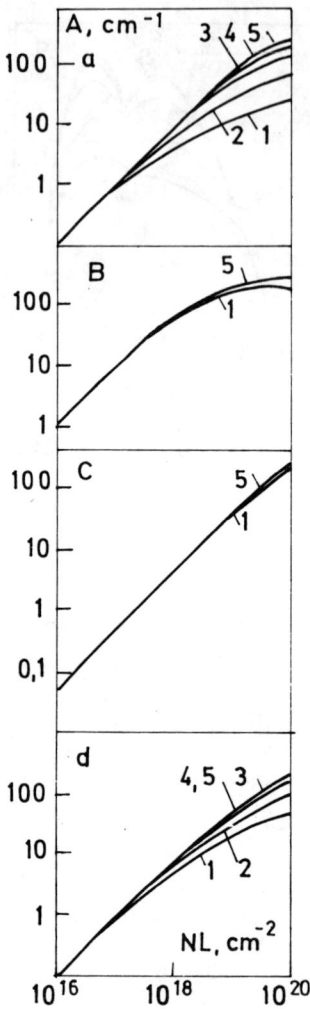

Fig. 4. Integral absorptivity of CO 4.7 μm (a), 4.3 μm (b), 2.7 μm (c) and $CO_2$ 15 μm (d) bands at different gas pressures. The notation of the curves is the same as in Fig. 2. $X = 10^{19}$ cm$^{-2}$, $T = 1000$ K(a-c) and 250 K(d)

## 4. INTEGRAL CHARACTERISTICS

The interest in the study of the integrated band absorption characteristics is attributed to the fact that in a number of cases wide-band optical systems are used for combustion processe diagnostics. Figure 3 shows integral absorption coefficients, atm$^{-1}$ cm$^{-2}$, in the considered bands at different gas temperatures as a function of the optical path length (so-called "growth curves"). For small optical thicknesses when radiation reabsorption is absent, the integral radiation of the main $CO_2$ 4.3 and 15 μm and CO 4.7 μm bands formed by the vibrational transitions inside one mode does not depend on the gas temperature $A \sim S_0$ and $S_0 \neq S_0(T)$ at $N_n$ = const [13]. For the 4.7 μm band, this occurs at $X < 10^{18}$ cm$^{-2}$, for the 4.3 μm band, at $X \leq 10^{17}$ cm$^{-2}$ and for the 15 μm band, at $X \leq 10^{18}$ cm$^{-2}$. The combined 2.7 μm band is formed by the vibrational transitions with a simultaneous change of the quantum numbers in several modes. Under conditions of no radiation reabsorption ($X \leq 10^{17}$ cm$^{-2}$) its integral absorptivity is defined by the fact that for such bands, $S_0 = S_0(T)$ at $N_n$ = $const$ [13]. At higher optical path lengths ($X > 10^{18}$ cm$^{-2}$), the integral absorption of bands greatly depends on the gas temperature, changing approximately as much as 5 times when $T$ is increased form 300 to 3000 K.

Figure 4 illustrates the integral absorption coefficients as a function of the optical path length at different gas pressures. It is seen that the effect of pressure on the integral absorptivity is pronounced for the CO 4.7 μm and $CO_2$ 15 μm bands within a wide range of optical thicknesses, $X > 10^{17}$ cm$^2$ due to the radiation reabsorption. For the $CO_2$ 2.7 and 4.3 μm bands (Fig. 4b,c), the gas pressure practically does not affect the spectral absorptivity, however, owing to different reasons. In the 2.7 um band, this is attributed to no radiation reabsorption while in the 4.3 μm band the opposite situation is observed, i.e. practically complete reabsorption of radiation, $A(\omega) \approx 1$, takes place. Note that the use of the above calculation results on the absorptivities $A$ and $\overline{A}(\omega)$ allows determinaiton of gas temperature through the measured radiaiton intensity, $I = A.B(\omega,T)$ where $B(\omega,T)$ is the Planck function. According to the calculations made for diagnostic purposes the $CO_2$ and CO bands or parts of their spectra may be chosen, in which the absorptivity under specific experimental conditions weakly depends on the gas pressure and temperature. This ensures a high accuracy of the gas temperature "recovery" through measuring the radiation intensity.

## REFERENCES

1. MALKMUS, W. 1963. Infrared emissivity of carbon dioxide (4.3 μm band). J. Opt. Soc. Amer., Vol. 53, pp. 951-961; 1964, Vol. 54, pp. 751-758.
2. KAMENSHCHIKOV, V.A., PLASTININ, YU.A., NIKOLAEV, V.M. and NOVITSKY, L.A., 1971. Radiation properties of gases at high temperatures, Moscow, Mashinostroyeniye.
3. LUDWIG, C.B. 1973. Handbook of infrared radiation from combustion gases. NASA Report SP-3080, Washington.
4. SMIRNOV, B.M. and SHLYAPNIKOV, G.V. 1980. IR radiation transfer in molecular gases. Uspekhi Fiz. Nauk, Vol. 130, pp. 377-414.
5. TIEN, C.L. 1968. Radiation gas properties. Advances in Heat Transfer, Vol. 5. Academic Press, New York, London, Moscow.
6. EDWARDS, D.K. 1969. Radiation characteristics of materials: Teploperedacha, Vol. 91, No. 2.
7. EDWARDS, D.K., GLASSEN, L. K., HAUSER, U.K. and TASHER, J.S. 1967. Radiant heat transfer in non-isothermal non-grey gases. Teploperedacha, Vol. 89, No. 3.

8. PLASS, C.N. 1960. Useful representation for measurements of spectral band absorption. J. Opt. Soc. Amer., Vol. 50, pp. 868-872.
9. ANDRADE E SILVA, M.H. and AMAT, G. 1969 Influence of Fermi resonance on the rotational constants of limar triatomic molecules. J. Molecular Spectr., Vol. 29, pp. 384-396.
10. HERTZBERG, G. 1950. Molecular spectra and molecular structure. I. Spectra of diatomic molecules. Van Nostrand, New York.
11. GOODY, R.M. 1964. Atmospheric radiation. I. Theoretical basis. Oxford University Press, London.
12. ROGERS, C.D. and WILLIAMS, W. 1974. Integrated absorption for Voigt profiles. J. Quant. Spectr. Radiat. Transf., Vol. 14, pp. 319-326.
13. PENNER, S.S. 1959. Quantitative molecular spectroscopy and gas emissivities. Addison-Wesley, London.
14. YOUNG, L.A. 1969. $CO$ infrared spectra. J. Quant. Spectr. Radiat. Transf., Vol. 8, pp. 693-717.
15. HODGSON, J.P. 1968. A survey of the infrared radiation properties of carbon dioxide. Aeronaut. Res. Coun. Current Papers, No. 981.
16. VARANASI, P. and SARANGI, S. 1975. Measurements of intensities and nitrogen-broadened linewidths in the CO Fundamental at low temperatures. J. Quant. Spectr. Radiat. Transf., Vol. 15, pp. 473-483.
17. OSIPOV, V.M. 1979. Effect of the Fermi resonance on relative intensities of vibrational bands of carbon dioxide. Opt. i Spektr., Vol. 46, pp. 48-55.
18. DOWLING, H.D., BROWN, L.R. and HUNT, R.H. 1975. Line intenisties of $CO_2$ in the 2.7 µm region. J. Quant. Spectr. Radiat. Transf., 15, pp. 205-211.
19. ROTHMAN, L. S. and YOUNG, L.D.G. 1981. J. Quant. Spectr. Radiat. Transf., 25, pp. 505-524.
20. GOLOVNEV, I.F., Sevast'yanenko, V.G. and SOLOUKHIN, R.I. 1979. Mathematical modelling of optical characteristics of carbon dioxide. J. Engng Phys., Vol. 36, pp. 197-203.
21. BOUANICH, J.P. 1973. Largeurs des raies spectrales de CO autopertube at pertube par $N_2$, $O_2$, $H_2$, HCl, NO et $CO_2$. J. Quant. Spectr. Radiat. Transf., Vol. 13, pp. 953-961.
22. OSTROUKHOVA, I.I. and SHLYAPNIKOV, G.V. 1978. IP radiation of molecular gases in the Q-branch. Teplofiz. Vysok. Temp., Vol. 16, pp. 279-281.
23. DRAEGERT, D.A. and WILLIAMS, S.D. 1968. Collisional broadening of $CO$ absorption lines by foreign gases. J. Opt. Soc. Amer., 58, pp. 1399-1411.
24. DRAYSON, S.R. and YOUNG, C. 1967. Band strength and line halfwidth of the 10.4 µm $CO_2$ band. J. Quant. Spectr. Radiat. Transf., Vol. 7, pp. 993-997.
25. Young, C. and Charman, R.E. 1974. Line-widths and band strengths for the 9.4 and 10.4 µm $CO_2$ bands. J. Quant. Spectr. Radiat. Transf., Vol. 14, pp. 679-686.
26. PETROV, S.B. and PODKLADENKO, M.V. 1975. Study of the broadening effects in $CO_2$ absorption 4.3 µm bands in a wide temperature range. ZhPS, Vol. 22, pp. 473-476.
27. YAMAMOTO, G., TANAKA, M. and AOKI, T. 1969. Estimation of the rotational line widths of carbon dioxide bands. J. Quant. Spectr. Radiat. Transf., Vol. 9, pp. 371-382.
28. ABU-RAMIA, M.M. and TIEN, C.L. 1966. Measurements and correlations of infrared radiation of carbon monoxide at elevated temperatures. J. Quant. Spectr. Radiat. Transf., Vol. 6, pp. 143-169.
29. BURCH, D.E. and GRYVNAK, D.A. 1966. Laboratory investigation of the absorption and emission of the infrared radiation. J. Quant. Spectr. Radiat. Transf., Vol., 6, pp. 229-241.
30. BURCH, D.E., GRYVNAK, D.A. and Williams, D. 1962. Infrared absorption of carbon dioxide. Applied Optics, Vol. pp. 759-770.
31. DRAYSON, S.R. 1966. Atmospheric transition in the $CO_2$ bands between 12 µ and 18 µ. Applied. Optics, Vol. 5, pp. 385-404.
32. KUDRYAVTSEV, N.N. and NOVIKOV, S.S. 1982. Theoretical and experimental investigations of IR radiation transfer in vibrationally non-equilibrium molecular gas containing $CO_2$ and CO. Int. J. Heat and Mass Transfer, Vol. 25, pp. 1541-1558.

# HEAT AND MASS TRANSFER IN FIXED AND FLUIDIZED BEDS

W. P. M. van Swaaij

University of Twente, Enchede, The Netherlands

The International Centre for Heat and Mass Transfer (ICHMT) selected for its 16th ICHMT Symposium the subject "Heat and Mass Transfer in Fixed and Fluidized Beds". It was held in Dubrovnik, Yugoslavia, September 3-7, 1984. Fluidized beds and packed beds are extensively used as fluid-solid contactors and/or chemical reactors in the process industry, energy conversion and in metallurgical processes. A few examples are: drying, absorption processes, heterogeneously catalyzed chemical reactions, solid fuel combustion, gasification, black furnaces and other are processing equipment, and regenerative heat exchangers both for continuous and discontinuous operation etc.

Considerable research has been devoted to basic insight into the mechanisms and rates of transport of heat and mass in these systems. New concepts, refined measurements, and new application areas have stimulated the development of ideas and knowledge on transport phenomena and produced sophisticated theories and design procedures for fluidized beds and packed beds. Nevertheless the problems are far from solved. This specially holds for fluidized beds due to the extremely complex phenomena associated with bubble formation and bubble growth but also due to the wide variety of conditions and regimes that are covered by the term fluidization.

Although the situation for packed beds is much less complex, for many critical conditions such as packed bed reactors with steep radial heat gradients or operating close to runaway conditions, transport rates connot yet be predicted with sufficient accuracy.

At the same time one can see the fields of applicaiton expanding rapidly. Especially the application of fluidized beds in coal and other solid fuel combustion stimulated the fundamental work on transport phenomena in fluidized beds. The main emphasis here is on heat transfer, reactor modeling, and transport processes around a single burning coal particle.

The heat transfer to submerged surfaces is an important parameter for the design of fluidized bed combustors because it determines the heat exchanger area required and it is often the limiting factor in the operation flexibility and turndown ratio. Apart from collecting more empirical data and improving the understanding by fundamental studies, in practice, solutions are also sought on different fluidization regimes fast or circulating beds with fluidized external heat transfer, etc. For the investigator involved in fundamental studies this increases the number of important regimes but it also indicates the need for a more general fundamental approach not too closely related to a particular application.

The general situation in the field of heat and mass transfer in packed and fluidized beds was reflected in the selected contributions to the 1984 ICHMT symposium. Apart from many papers on the fundamentals of heat and mass transfer a large fraction dealt with applications, special processes in packed and fluidized beds, and special operation regimes. They reflect a defferent angle of attack to the problems renging from industrial practice to fundamental university laboratory studies. Also indicated is the need for a continuous discussion between the investigators, inventors, designers, and engineers from industry, etc. to mobilize all resources for progress in both fundamentals and the application of heat and mass transfer in packed beds, fluidized beds, and related operations.

Even now, nearly four years after the symposium it is difficult to selected a particular single paper as representative of the symposium or as a fine example of current problems. Recent emphasis and progress on computer modelling cannot replace the careful selection of basic mechanisms and physical concepts and the paper of Holger Martin at the symposium is a fine example of what can be realized with simple concepts even in such a complex problem as heat transfer to gas fluidized beds.

# THE EFFECTS OF PRESSURE AND TEMPERATURE ON HEAT TRANSFER TO GAS – FLUIDIZED BEDS OF SOLID PARTICLES

H. Martin

*Institut fur Thermische Verfahrenstechnik
der Universitat (TH) Fridericiana Karlsruhe, FRG*

## 1. INTRODUCTION

A theoretical model for wall–to–bed heat transfer in gas fluidized beds of solid particles has been developed some years ago [1]. The model makes use of some of the fundamental concepts of molecular kinetic theory as applied to solid particles in a fluidized bed. Recently this model has been slightly modified to improve its applicability in a wide range of variables [2,3]. It has been shown, that the predictions of this model are in fair agreement with most experimentally observed phenomena, especially with the characteristic nonmonotonous variations of heat transfer coefficients with particle size and with bed voidage (or gas flowrate) [3–7]. Many applications of fluid bed technology involve high temperatures and sometimes also high pressures. It is the aim of this paper to discuss the effects of pressure and temperature both from the viewpoint of model predictions and the available experimental evidence.

In the first part of the paper (MODEL) the basic ideas and concepts of the theoretical approach as well as the assumptions and hypotheses used to derive the model equations shall be briefly reviewed.

The second part (PARAMETERS) discusses the main dependecies of wall–to–bed heat transfer coefficients on particle size, bed voidage and particle as well as gas properties, as predicted from the model equations in comparison with a few typical sets of experimental data.

The third part (VARIABLES) deals with the influence of pressure (lower as well as higher than atmospheric) and temperature on heat transfer to gas fluidized beds of solid particles.

In the fourth part (APPLICATION) experimental data obtained by Bergbauforschung, Essen, in a pilot scale allothermal coal gasificaiton plant with both high pressure (up to 40 bar) and high temperature (~800°C) will be presented and compared to the model predictions.

## 2. MODEL

The theoretical approach is based on the following assumptions:
The three mechanisms of thermal energy transfer in fluidized beds, i.e. particle convection, gas convection, and radiation may be regarded essentially independent of each other, so that the total heat transfer coefficient $\alpha = \dot{q}/(T_0 - T_B)$ can be written as the sum of three terms [8]:

$$\alpha = \alpha_P + \alpha_G + \alpha_R \tag{1}$$

The gas convective contribution can be calculated from Baskakov's equation [9]:

$$Nu_G = 0.009 Pr^{1/3} Ar^{1/2} \tag{2}$$

and the radiative contribution from the linearized Stefan–Boltzmann equation

$$\alpha_R = 4\varepsilon\sigma \overline{T}^3 \tag{3}$$

The model equation for the particle convective contribution may be written [1,3]:

$$\alpha_P = \frac{1}{6}(1-\psi)(\rho c)_P \overline{w}_P [1 - e^{-N}], \tag{4}$$

where $\overline{w}_P$ is an average velocity of the random displacement of particles between bulk and surface while the term in square brackets denotes the degree of thermal accommodation of a particle during its contact with the heated or cooled wall, and $N$ is a dimensionless contact time (or "number of transfer units")

$$N = \frac{\alpha_{WP} A_{PW}}{(\rho c)_P V_P} t_c = \frac{\alpha_{WP}}{\frac{1}{6}(\rho c)_P \overline{w}_P} \cdot \frac{\overline{w}_P t_c}{4d} \tag{5}$$

The heat transfer coefficient $\alpha_{WP}$ between a plane wall and a single particle in point contact may be thought of as made up of two resistances in series:

$$\frac{1}{\alpha_{WP}} = \frac{1}{\alpha_{WP(max)}} + \frac{1}{\alpha_{i,p} t_c} \tag{6}$$

The first and in most cases the dominant resistance $1/\alpha_{WP(max)}$, due to conduction in the gas–filled wedge between wall and particle is obtained from Schlunders formula [10]:

$$Nu_{WP(max)} = \frac{\alpha_{WP(max)} d}{\lambda_g} = 4\left\{ \left(1 + \frac{2l}{d}\right) \ln\left(1 + \frac{d}{2l}\right) - 1 \right\} \tag{7}$$

with the modified mean free path of the gas molecules $l$ calculated according to [2,11]:

$$l = \left(\frac{2}{\gamma} - 1\right) \frac{\lambda_g}{p} \frac{\sqrt{2\pi \tilde{R} T / \tilde{M}}}{\left(c_{p,g} - \frac{1}{2}\tilde{R}/\tilde{M}\right)} \qquad (8)$$

For air at 1 bar and 25°C, $l$ has a value of 274 nm (with an accommodation coefficient of $\gamma = 0.9$) leading to a value of $Nu_{WP(max)}$ of $24 + 20\%$ in the range of particle diameters from 200 μm to 2 mm, based on the projection area $A_{PW}$ of a particle on the wall. It has been shown [3] that this agrees very well with experimental results recently obtained by Gloski, Decker and Glicksman [12].

The transient internal conduction resistance in the solid particle $1/\alpha_{i,P}(t_c)$ is calculated approximatively by assuming the isothermals in the solid to be concentric spheres with the center in the contact point [3].

The model equations (4 and 5 through 8) still contain two so far unknown quantities, the contact time $t_c$ of a particle with the wall and the average velocity of random displacement $\overline{w}_P$. To eliminate these two unknowns two additional hypotheses had to be introduced into the model:

(I) The contract time was assumed to be proportional to the time necessary to (randomly) displace on e particle over a distance proportional to its diameter, i.e. the group $\overline{w}_P \cdot t_c / d$ (see r.h. side of Eq. (5)) was taken to be a constant.

$$\frac{4d}{\overline{w}_p t_c} = C \qquad (9)$$

(II) The average velocity of particle displacement $\overline{w}_P$ is related to a "mean free path" or a mean free falling path of the particles as

$$\frac{1}{2} \overline{w_p^2} = g l_{\text{free}} \qquad (10)$$

with $l_{\text{free}}$ calculated just as in the kinetic theory from the diameter and the number density of the particles in the bed. The number density, wiht respect to the free, i.e. mobile volume, thus relates the unknown $\overline{w}_P$ to particle size ($d$) and to bed voidage $\psi$. The characteristic dimensionless quantity for the particle convective (or rather "conductive") mechanism obtained form these assumptions becomes:

$$Z \equiv \frac{1}{6}\frac{(\rho c)_P}{\lambda_g} \overline{w}_P d = \frac{1}{6}\frac{(\rho c)_P}{\lambda_g}\sqrt{\frac{gd^3(\psi-\psi_L)}{5(1-\psi_L)(1-\psi)}} \qquad (11)$$

Now, model equations (4), (5) and (6) may written as [2,3]:

$$\alpha_p = \frac{\lambda_g}{d}((1-\psi)Z[1-e^{-N}] \qquad (4a)$$

$$N = Nu_{WP}/(C \cdot Z) \qquad (5a)$$

$$\frac{1}{Nu_{WP}} = \frac{1}{Nu_{WP(max)}} + \frac{\lambda_g/\lambda_P}{4\left[1+\sqrt{\frac{3C}{2\pi}\frac{\lambda_g}{\lambda_P}Z}\right]}. \qquad (6a)$$

These equations yield $\alpha p$ as a function of particle properties ($d$, $(\rho c)_P$, $\lambda_P$, $\psi_L$), gas properties ($\lambda_g$, $c_{p,g}$, $\tilde{M}$, $\gamma$), bed voidage $\psi$ (which in turn depends on gas velocity $u$) and the operational variables, pressure ($p$) and temperature ($T$). They contain one single adjustable parameter $C$ (Eq. (9)) that has to be determide by comparison with experimental data.

## 3. PARAMETERS

The predictions of the model equations (1–11) have been compared with a large volume of experimental data [1–7]. The model parameter $C$ (see Eq. (9)) has been determinde by comparison with Wunders data [13] for spherical glass particles [2,3]:

$$C = 2.6 \qquad (12)$$

This value has been kept constant for all subsequent comparisons with other data.

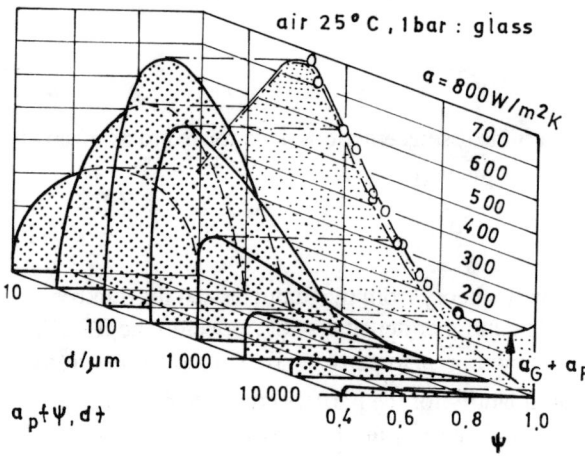

Figure 1 Particle convective heat transfer coefficient $\alpha_P$ as a function of bed voidage $\psi$ and particle diameter $d$. Curves calc'd open symbols: Wunders data, glass, spherical [13].

Figure 1 shows the particle convective heat transfer coefficient $\alpha_P$ calculated from the model equations (4a, 5a, 6a) as a function of bed voidage $\psi$ in the range from 0.4 to 1.0) and particle diameter d (from 3 μm to 10 mm) [3]. The curves were calculated with the physical properties of air at 25 °C, 1 bar and glass.

The maximum heat transfer coefficients obtained under these conditions in Wunders experiments [13] are shown as circles in the rear $\alpha_{max}$, $d$– plane. The gas convective and radiative contributions (the latter is unimportant at that temperature) from Eqs. (2) and (3) have been added to $\alpha_{pmax}$ (see Eq.(1)). It is interesting to note that the model predicts an absolute maximum for particle diameters in the range of 40 μm (for glass, air 25°C, 1 bar) and a minimum at particle diameters of about 3 mm. This is in agreement with experimental observations (see e.g. [9],[13],[14]).

Figure 2 shows a direct comparison of experimental data [13] in a plot of heat transfer coefficients α vs. bed voidage ψ for glass beads with diameters form 47 μm to 1400 μm (circles). It may be seen, that the data agree very well in magnitude as well as in characteristic shape with the curves predicted by the model, though usually the choice of a slightly higher viodage at minimum fluidization in the calculations

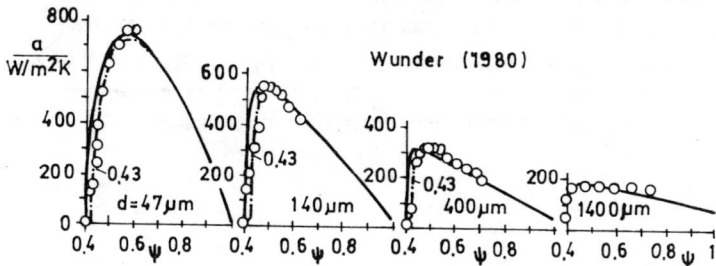

Figure 2. Heat transfer coefficients vs. bed voidage.

would have led to even better agreement with the data in the ascending branch of the curves (the dotted lines have been calc'd with $\psi_L = 0.43$ instead of 0.4). A similar comparison with data obtained more than 30 years earlier by Mickley and Trilling [15] is presented in [3,4].

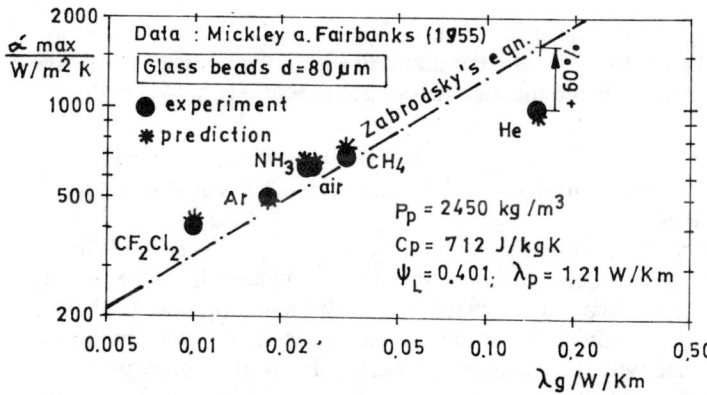

Figure 3. Influence of gas properties on heat transfer.

Figure 3 gives a plot of maximum heat transfer coefficients $\alpha_{max}$ vs. the thermal conductivity of the fluidizing gas $\lambda_g$. The large dots represent experimental data obtained with six different gases and 80 μm glass beads by Mickley and Fairbands in 1955 [16]. The corresponding model predictions are shown as asterisks while the Zabrodsky equation ($\alpha_{max} \sim \lambda^{0.6}$) prediction is shown as a dotted line (for more details of that comparison, see [3]).

## 4. VARIABLES

### 4.1. PRESSURE

Gas pressure has an effect on both the gas convective and the particle convective contributions to the total heat transfer coefficient. In Baskakov's equation (2) for $\alpha_G$, the quantity depending on pressure is the gas density $\rho_g$ (in the Archimedes number). Therefore, as long as $\rho_p$ remains much larger than $\rho_g$, and $\eta_g$ as well as $\lambda_g$ do not significantly vary with pressure, one would expect the gas convective heat transfer coefficient to vary approximately as the square root of pressure:

$$\frac{\alpha_G(p)}{\alpha_G(p_o)} = \sqrt{\frac{p}{p_o}} \qquad (13)$$

The particle convective contribution, from the model equations, would be affected by a change in pressure via the inverse proportionality of the mean free path (Eq. (8)), as long as $1/\alpha_{WP(max)}$ (Egs.(6,7)) is the dominant resistance. In that range, the dependence on pressure may be roughly estimated form Egs.(4a) and (7) (with $d/(2l) \gg 1$) to be:

$$\frac{\alpha_P(p)}{\alpha_P(p_o)} \cong 1 + \frac{\ln(p/p_o)}{\ln(d/2l_o)} \qquad (14)$$

with $l_0 = l(p_0)$ form Eq. (8). For air at 25 °C, 1 bar $2l_0 \cong 0.55$ µm. For extremely small particles, in the range below the absolute maximum in Fig. 1, the particle convective heat transfer coefficient is predicted independent of gas pressure (Eq. (4a) with $N \to \infty$).

Experiments carried out in 1959 by R. Ernst [16] with fine catalyst particles (20–100 µm) in a pressurized fluidized bed at 2 bar and at 100 bar showed no influence of pressure at all.

Figure 4 shows the maximum heat transfer coefficients (of about 560 W/m²K) obtained in these experiments in comparison with curves calculated from the model equations for 1 bar and for 100 bar. Unfortunately the physical properties of the particles were not given by the author, so that the comparison is more or less qualitative in nature. From the model, with the estimated properties as given in the figure one would still expect a significant change with pressure from 2 to 100 bar in that range of particle sizes.

If one assumes, however, a certain roughness of the particles, i.e. a lower limit for the "effective" mean free path $l_{min}$, the dependence on pressure in the range of particle sizes below 100 µm would be considerably reduced (broken line for 100 bar with $l_{min} = 2.5 \cdot 10^{-7}$m).

Experiments at subatmospheric pressures in fluidized beds have been reported by Shlapkova [17] in 1966 with 200 µm sand in a range of pressures form atmospheric down to 266 Pa. The maximum heat transfer coefficients from these experiments are plotted vs. the mean pressure in the bed (and vs. the Knudsen number $2l/d$) in Figure 5 and compared with a curve calculated from the model equations.

Though the disagreement between data and prediction is considerable, expecially at the lowest pressures, the general trend, nevertheless, seems to be correctly represented by the model.

More recently Xavier, King, Davidson and Harrison [18] published the results of experiments in a pressurized fluidized bed with $N_2$ and $CO_2$ and different particles in a range of pressures from 1 to 25 bar.

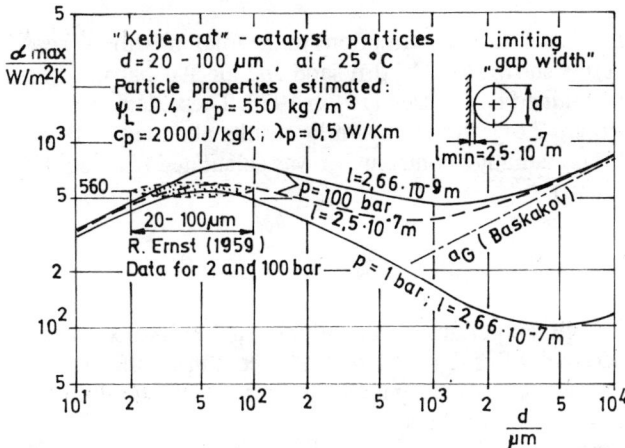

Figure 4. Effect of pressure on heat transfer: Curves calc'd from model equations.

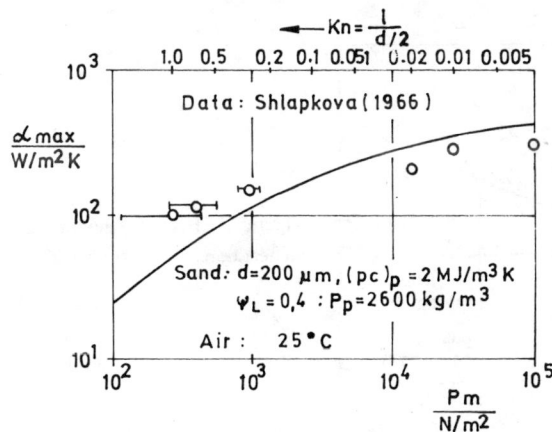

Figure 5. Effect of subatmospheric pressures on heat transfer.

The ratio of $\alpha_{max}$ (25 bar) /$\alpha_{max}$ (1 bar) from these data is found to be about 1.57 for the smallest particles (61 µm glass, $CO_2$) and about 1.8 for the largest (688 µm polymer). The authors [18] compared their results with their own model based on a modified packet theory for particle convective heat transfer containing a "fictitious gap" proportional to particle size. From such a model $\alpha_p$ is predicted to be (almost) independent of pressure. Therefore the authors could not explain the measured increase in $\alpha_{max}$ from 510 (at 1 bar) to about 800 W/M²K (at 25 bar) for their smallest particles. Form equation (14) with $l_0 = 0.155$ µm for $CO_2$ at 1 bar and $d=61$ µm one obtains a ratio of 1.6, very close to the observed value (see [3] for more details of that comparison).

## 4.2. TEMPERATURE

The effect of temperature on the three contributions to the total maximum heat transfer coefficient (see Eq. (1)) is shown for 717 µm sand and air at 1 bar in Figure 6 and compared with data obtained by Janssen [19] in 1973. These data are shown again (as curve a ) in Figure 7 together with similar results obtained in 1980 by Botterill and Teoman [20]. The radiative contribution was calculated from eq. (3) with an effective emissivity of $\varepsilon = 0.5$.

## 5. APPLICATION

High pressure as well as high temperatures were applied in the fluid bed steam gasification experiments carried out at Bergbauforschung, Essen, during the last several years. Heat required for the endothermic gasification reaction is supplied to the fluidized bed via an immersed heat exchanger.

In the earlier lab–scale experiments[21] this was an electrically heated vertical cylinder in the center of the bed. In the pilot–plant scale experiments, conducted

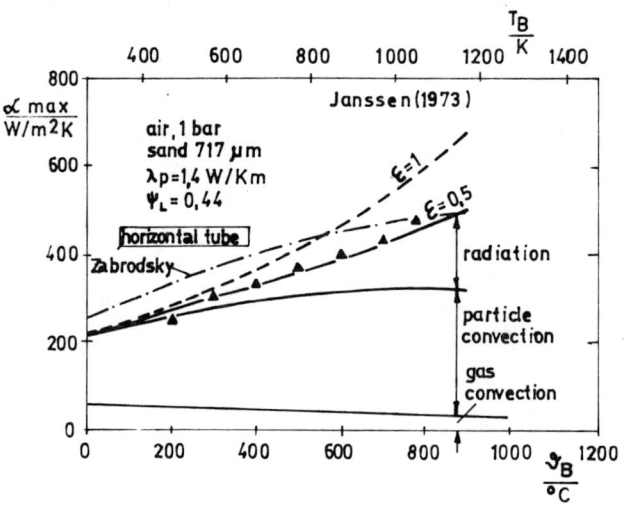

Figure 6. Effect of temperature on heat transfer, curves calc'd.

since 1976, the 0.8×0.9 m² bed with a height of about 4 m contains a bundle of vertical tubes of 33 m² surface area with hot helium flowing inside the tubes [22]. A comparison of earlier data for 800 °C and 1 bar or 40 bar with the model predictions has been given in [3,4].

Data, more recently obtained by Bergbauforschung with bituminous, caking coal ("Gasflammkohle") and with anthracite are presented in Tables 1 and 2 below.

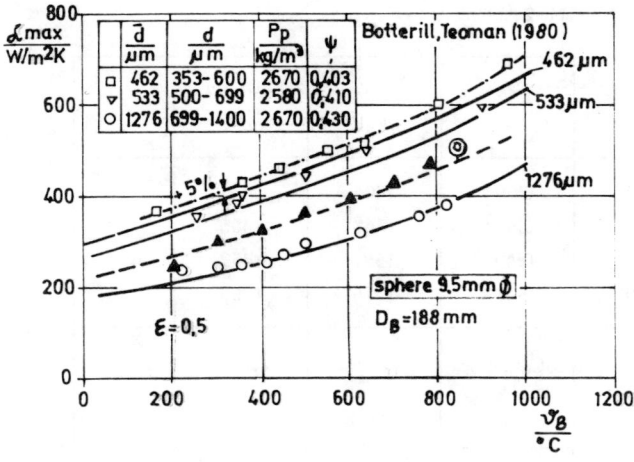

Figure 7. Effect of temperature on heat transfer, curves calc'd from model equaitons. Curve with Janssen's data. a form Fig. 6

The average diameters $d_{-0.5,3}$ (i.e. the mass average over the reciprocal of the square root of the diameters) and $d_{1,3}$ (the mass average over the diameters) have been calculated from a sieve analyses of samples taken from the bed during operation. The maximum heat transfer coefficients were measured in or close to the center of the bed. Bundle average values have been found to be considerably lower. The maximum heat transfer coefficients calculated from the model equations are given for $d_{-0.5,3}$ as a characteristic average diameter. The gas properties were taken as those of pure steam because the product gas contained at least 60% steam and the properties of the mixture of other gases ($H_2$, CO, $CO_2$, $N_2$, $CH_4$) are not too far from those of $H_2O$. The comparison between calculated and measured maximum heat transfer coefficients from Tables 1 and 2 is shown graphically in Figure 8.

Table 1 Bitounous, caking coal (GFK)

| No. | $\bar{d}_{-0.5,3}$ μm | $\bar{d}_{1,3}$ μm | $\psi_L$ — | $C_p$ J/kgK | $\rho_p$ kg/m³ | |
|---|---|---|---|---|---|---|
| 1 | 750 | 802 | 0.897 | 1390 | 1377 | |
| 2 | 172 | 301 | 0.847 | 1216 | 1542 | |
| 3 | 339 | 580 | 0.807 | 1294 | 1597 | |
| 4 | 386 | 610 | 0.788 | 1296 | 1491 | |

| No. — | p bar | T K | u m/s | $\alpha_{max}^{(exp)}$ W/m²K | $\alpha_{max}^{(calc)}$ W/m²K | Δ % |
|---|---|---|---|---|---|---|
| 1 | 37 | 1061 | 0.11 | 426 | 470 | 10 |
| 2 | 38 | 1009 | 0.10 | 631 | 654 | 4 |
| 3 | 38 | 996 | 0.10 | 649 | 647 | −0.3 |
| 4 | 38 | 1001 | 0.10 | 709 | 652 | −8 |

Table 2 Anthracite

| No. | $\overline{d}_{-0.5,3}$ | $\overline{d}_{1,3}$ | $\psi_L$ | $C_p$ | $\rho_p$ |
| --- | --- | --- | --- | --- | --- |
| – | μm | μm | – | J/kgK | kg/m³ |
| 5 | 260 | 350 | 0.591 | 1404 | 1437 |
| 6 | 226 | 310 | 0.491 | 1368 | 1406 |
| 7 | 242 | 330 | 0.493 | 1367 | 1452 |
| 8 | 239 | 360 | 0.489 | 1216 | 1700 |

| No. | $p$ | $T$ | $u$ | $\alpha_{max}^{(exp)}$ | $\alpha_{max}^{(calc)}$ | $\Delta$ |
| --- | --- | --- | --- | --- | --- | --- |
| – | bar | K | m/s | W/m²K | W/m²K | % |
| 5 | 37 | 1030 | 0.099 | 823 | 1108 | 35 |
| 6 | 39 | 1089 | 0.087 | 1075 | 1374 | 28 |
| 7 | 38 | 1089 | 0.112 | 1505 | 1368 | −9 |
| 8 | 39 | 1109 | 0.116 | 1760 | 1421 | −19 |

$$\overline{d}_{-0.5,3} = \left[\frac{1}{\sum_i \frac{x_i}{\sqrt{d_i}}}\right]^2 \; ; \; \overline{d}_{1,3} = \sum_i x_i d_i \; ; \; x_i = \text{mass fraction of size } d_i$$

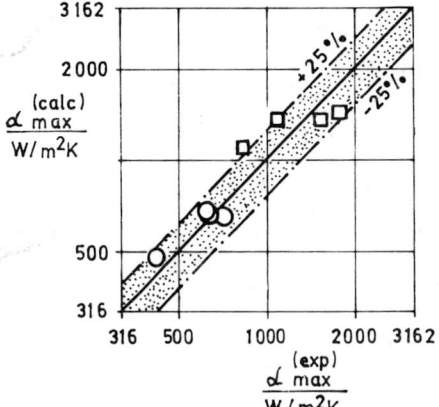

Maximum Heat Transfer Coefficients
Vertical Tube Bundle
Fluid Bed Gasification of Coal with
Steam at 1000K, 40 bar, ($u \cong 0.1$ m/s)
Data: Bergbauforschung, Essen (1983)
Bitumenous, caking coal ($\psi_L = 0.79...0.89$)
Anthracite ($\psi_L = 0.49...0.59$)

Figure 8

## 6. CONCLUSIONS

Though the maximum heat transfer coefficients in large scale fluidized beds may be predicted with reasonable accuracy, there remains a major uncertainty because of the severe maldistribution of local bed voidage. Future research in the field of fluid bed heat transfer sholuld be devoted, therefore to the question of how to achieve a more uniform distribution of solids throughout the whole volume so that the maximum heat transfer coefficients might be obtained not only locally at a certain position but also as an average value over the whole heat transfer surface.

## 7. ACKNOWLEDGEMENT

The author wants to express his thanks to H. Juntgen, K.H. van Heek and R. Krichhoff, Bergbauforschung, Essen, for generously providing, and granting permission to use, their data on heat transfer in fluid bed gasification of coal (see Tables 1 and 2 and Fig. 8).

## 8. NOTE ADDED IN PROOF

Some experimental results on the effect of bed temperature on the maximum heat transfer coefficients between a gas fluidized bed of alumina particles and a

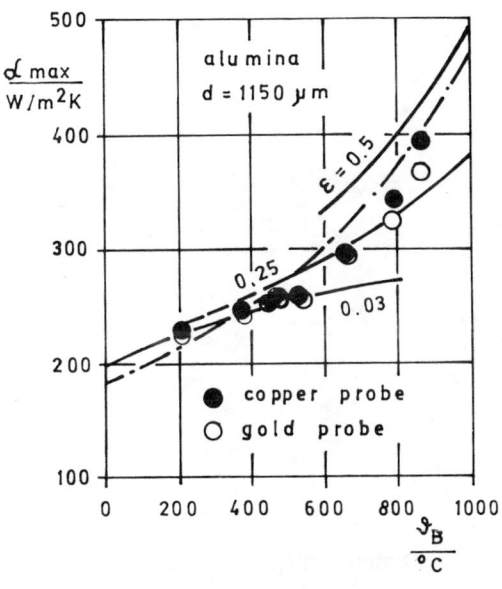

Figure 9. (Replotted from Fig. 4 of paper 2.4 by Botterill et. al.) Measured bed–to–probe heat transfer coefficients showing discountinuity in behaviour at 550°C. Curves calc'd from model equations with emissivity $\varepsilon$ as a parameter. Dotted curve from Fig. 7 for 1276 μm sand.

spherical probe have been presented in paper 2.4 by Botterill, Teoman and Yüregir. These data might be compared with model equations $(1-12)$ as has been done with similar results for beds of quartz sand in Fig. 7.

Such a comparison is shown in Fig. 9 (replotted from Fig. 4 in paper 2.4) with the curves calculated from the model equations $(1-12)$ and the properties of the alumina particles estimated to be $\rho_p = 3850$ kg/m³, $c_p = 800$ J/kgK, $\lambda_p = 10$ W/mK and $\psi_L = 0.45$. The emissivity of gold is $\varepsilon = 0.03$ or less for temperatures below 500°C while it increases sharply to values of up to 0.62 at 1093°C (2000°F)*. Other pure metals show a similar behaviour at temperatures between 500°C and 1000°C. The "discontinuity in behaviour at 550°C" observed by Botterill et al might be due to that increase in emissivity of the probe surface. There seems to be no need to explain that "discountinuity" by a "change in fluidization behaviour" at that temperature.

## NOMENCLATURE

| | | |
|---|---|---|
| $A$ | | surface area |
| $C$ | – | constant in eq. (9) ($C = 2.6$) |
| $c$ | J/kgK | heat capacity per mass |
| $d$ | m | (average) particle diameter |
| $g$ | m/s$^2$ | gravitional acceleration |
| $L$ | m | height of fluidezed bed |
| $l$ | m | modified mean free parth (eq.8) |
| $M$ | g/mol | molar mass |
| $N$ | – | "number of transfer units" in eq. (4,5) |
| $p$ | Pa | pressure |
| $\dot{q}$ | W/m$^2$ | heat flux |
| $R$ | J/kmolK | gas constant (universal) |
| $T$ | K | temperature (absolute) |
| $t$ | s | time |
| $u$ | m/s | gas velocity |
| $V$ | m$^3$ | volume |
| $w$ | m/s | particle velocity |
| $\alpha$ | W/m$^2$K | heat transfer coefficient |
| $\gamma$ | – | accommodation coefficient |
| $\varepsilon$ | – | effective emissivity |
| $\eta$ | Pas | viscosity |
| $\upsilon$ | °C | temperature |
| $\lambda$ | W/mK | heat conductivity |
| $\rho$ | kg/m$^3$ | density |
| $\sigma$ | W/m$^2$K$^4$ | Stefan Boltzmann constant |
| $\psi$ | – | void fraction |

Subscripts                                   Dimensionless numbers

$B$    bed, bulk                       $$Ar = \frac{gd^3}{\eta_g^2} \cdot \rho_g(\rho_p - \rho)$$

$c$    contact
$g$    gas
$G$    gas convective

---
* Tables of Emissivity of Surfaces (J. R. Singham ed.) Int. J. Heat Mass Transfer, vol. 5, pp. 67-76, 1962

| | | |
|---|---|---|
| $i$ | internal | $Nu = \dfrac{\alpha d}{\lambda_g}$ |
| $L$ | lower limit of fluidization (minimum fluidization) | |
| $p$ | particle convective, particle | $Pr = \eta_g c_{pg}/\lambda_g$ |
| $R$ | radiative | |
| $Wp$ | wall–to–particle | |

## REFERENCES

1. MARTIN, H., Wärme–und Stoffübertragung in der Wirbelschicht, *Chemie–Ing. –Tech.*, vol. 52, pp. 199–209, 1980 and Über die Mechanismen der Wärmeübertragung zwischen Gas–Fest–stoff–Wirbelschichten und festen Wänden, vol. 54, pp. 156–157, 1982.
2. MARTIN, H., Wärmeübertragung and Gas/Festsoff–Wirbelschichten *Chemie–Ing. –Tech.*, vol. 56, pp. 225–227, 1984.
3. MARTIN, H., Heat Transfer Between Gas–Fluidized Beds of Solid Particles and the Surfaces of Immersed Heat Exchanger Elements, Part, I and II, *Chemical Engineering and Processing* (Lausanne), vol. 18, no. 3 and 4.
4. MARTIN, H., Fluid Bed Heat Exchangers, in *Heat Exchangers, Theory and Practice*, eds. J. TABOREK, G.F. HEWITT and N. AFGAN, pp. Hemisphere, Washington, D.C., 1983.
5. KUBIAK, H., Modellierung einer allothermen Wasserdampfvergasung in einer Wirbelbettrinne, Dissertation, Univ. Essen, 1981.
6. GREWAL, N.S. and SAXENA, S.C., Experimental Studies of Heat Transfer between a Bundle of Horizontal Tubes and a Gas–Solid Fluidized Bed of Small Particles, *Ind.Eng.Chem.Process Des.Dev.*, vol. 22, pp. 367–376, 1983.
7. MARTIN, H., Flud–Bed Heat Exchangers–A New Model for Particle Convective Heat Transfer, *Chem. Eng.Commun.*, vol. 13, pp. 1–16 and 21–22, 1981.
8. BOTTERILL, J.S.M., *Fluid Bed Heat Transfer.*, Academic Press, New York, 1975.
9. BASKAKOV, A.P., BERG, B.V., VITT, O.K., FI;O[[PVSLO. N.F., KIRAKOSYAN, V.A., GOLDOBIN, J.M., and MASKAEV, V.K., Heat Transfer to Objects Immersed in Fluidized BEDs, *Powder Technology*, vol. 8, pp. 273–282, 1973.
10. SCHLÜNDER, E.U., Wärmeübergang an bewegte Kugelschüttungen bei kurzfristigem Kontakt, *Chemie–Ing., –Tech.*, vol. 43, pp. 651–654, 1971.
11. MOLLEKOPF, N., and MARTIN, H., Zur Theorie des Wärmeübergangs an bewegte Kugelschüttungen bei kurzfristigem Kontakt, *Verfahrenstechnik*, vol. 16. pp. 701–706, 1982.
12. GLOSKI, D., GLICKSMAN, L., and DECKER, N., Thermal Resistance at a Surface Immersed in a Fluidized Bed, Manuscript obtained by personal communicaiton, 1982.
13. WUNDER, R., Wärmeübergang an vertikalen Wärmetauscherflächen in Gaswirbelschichten, Dissertation, TU Munchen, 1980.
14. BAERNS, M., Fluidization of Fine Particles, in Proceedings of the Int. Symp. on Fluidization, June 1967, Eindhoven, ed. Drinkenburg, A.A.H., pp. 403–415, Netherlands Univ. Press, Amsterdam, 1967.
15. MICKLEY, H.S., and TRILLING, Ch. A., Heat Transfer Characteristics of Fluidized Beds, *Ind.Eng.Chem.*, vol. 41, pp. 1135–1147, 1949.
16. ERNST, R. Der Mechanismus des Wärmeübergangs an Wärmetauschern in Fließbetten, Dissertation, TH Aachen, 1959.
17. SHLAPKOVA, YA.P., Heat Transfer to a Cylindrical Skurface Immersed in a Fluidized Bed at Low Pressure, *J. Eng. Phys.*, vol. 10, pp. 318–320, 1966.

18. XAVIER, A.M., KING, D.F., DAVIDSON, J.F., and HARRISON, D., Surface Bed Heat Transfer in a Fluidized Bed at High Pressure, in Proc. 1980 Int. Fluidiz. Conf., eds. Grace, J.R. and Matsen, J.M., Plenum, New York, 1980.
19. JANSSEN, K., Beitrag zur Berechnung von Wärmeuubergangszahlen zwischen Fluidatbetten und darin eintauchenden Wärmetauscherflächen in Abhängigkeit von den Strömungsbedingungen inhomogener Fluidisierungszustände, Dissertation, TH Aachen, 1973.
20. BOTTERILL, J.S.M. and TEOMAN, Y., Fluid Bed Behaviour at Elevated Temperatures, in Proc. 1980 Fluidz. Conf., eds. Grace, J.R. and Matsen, J.M., Plenum, New York, 1980.
21. PETROVIC, V., Messung und Berechnung des Wärmeübergangs von einem Heizrogr an eine Kohlewirbelschicht in Wasserdampg un Inertgas, Dissertation, TH Aachen, 1978.
22. VAN HEEK, K.H., JÜNTGEN, H., and PETERS, W., Wasserdampfvergasung von Kohle mit Hilfe von Prozeßwärme aus Hochtemperatur–Kernreaktoren, *Atomkernenergie*/Kemtechnik, vol. 40, pp. 225–246, 1982.

# HIGH TEMPERATURE HEAT EXCHANGERS

Y. Mori

*University of Electro-Communications, Tokyo, Japan*

At the 17th Symposium on High Temperature Heat Exchangers, 45 papers including 8 invited papers were reported in four groups: High Temperature Heat Exchanger Development, Basic Problems in High Temperature Heat Exchangers, Heat Exchangers for Future Power Plants and Industrial Applications.

The Symposium was organized to focus attention on heat and mass transfer processes associated with high temperature heat exchangers. An understanding of the processes plays an essential and important role in the rapid development of high temperature heat exchangers towards improved energy utilization and a higher efficiency of energy conservation at various power and industrial plants. In high temperature heat exchangers, gas-gas heat transfer dominates. Research and development is under way on many kinds of high temperature heat exchengers : recuperative, regenerative and direct-contact. In addition to heat transfer enhancement problems, there are several very important, basic and serious problems peculiar to high temeprature heat exchangers. These are reliability, thermal stress, thermal expansion, difficulty in scale-up and physical property variation of fluids. When discussing high temperature heat exchangers, the difference between problems encountered in heat exchangers at moderate temperatures and those at high temperatures should be clearly recognized. Heat transfer enhancement for high temperature heat exchangers, is disscussed mainly in connection with radiation. In gaf–gas heat exchangers the viscosity of a gas increases with temperature and other physical properties also vary. Viscosity increase is associated with relaminarization of gas flow at high temperatures due to the decrease of the Reynolds number. This phenomenon peculiar to high temperature heat transfer is discussed. The gas radiative emissivity has been assumed to be grey so far, but radiation heat transfer problems are discussed for several channel flow cases by proposing an accurate analytical model taking into account the multi-band feature of a given gas with some numerical results. In discussion of constrictions of heat exchangers, high temperature heat exchangers have to be provided with a structure of symmetry and to fulfill the condition of a uniform outlet temperature of heated gas pipes or ducts as to avoid hot spots in the exchanger structure.

The satisfy these conditions, several heat exchanger desings and new heat enhancement methods were proposed. The reported maximum inlet heated gas temperature in various heat exchangers is summarized as follows. In a regenerative heat exchanger provided with wire net layers so as to fully utilize radiation for heat transfer enhancement, the maximum heated gas temperature is about $1000°C$ when the main construction is of the shell-tube type. In order to meet the requirement of the increase of blast furnace temperature the use of the melting heat of packed particles or plates is proposed for a regenerative heat exchanger and a maximum gas temperature of $1700°C$ is reported. In a ceramic heat exchanger of the compact channel type of waste heat recovery, the waste gas temperature is $1350°C$ and the heated air temperature is $820°C$. In a ceramic shell-tube type heat exchanger consisiting of $37.5mm$ outer diameter tubes whose wall thickness, is $35mm$ when the outlet exhaust gas temperature is $1380°C$, the heated air temparature which is preheated up to $720°C$ is $1150°C$. For a closed cycle *MHD* combined power plant using combustion gas heat, a pebble bed regenerative heat exchanger is used and the maximum measured temperature of the heated argon is $1770°C$. Shell-tube type heat exchangers whose tube outer diameter is $375mm$ and walls are $3.5mm$ thick an outlet exhaust gas temperature of $1380°C$ is obtained by preheating air to $720°C$ and heating it to $1150°C$. For high temperature heat exchangers of the shell-tube type for high temperature nuclear reactors, a temperature of $950°C$ is reported when the inlet helium temperature of the primary loop is $1000°C$, and in another paper a $905°C$ secondary helium temperature is obtained for $940°C$ primary helium. In fuel cell plants, the compact plate-fin heat exchanger is required due to the fact that plants are located in populated areas., the hot gas inlet temperature of $650°C$ is tested.

In the field of basic research, sutdies of heat transfer enhancement are numerous and a comparison of many enhancement methods that place importance on radiative heat transfer is reported and a description of several different heat exchanger designs towards radiative heat transfer enhancement is given. Applicable and utilixed fields of high temperature heat exchangers are reported together with proposals for new heat transfer enhancement methods or exchanger performance analyses. Nuclear process heat application, high temperature heat exchangers of several constructions and suitable heat transfer enhancement methods are discussed that transfer heat from the hot helium of the reactor loop through the intermediat loop to the steam-reforming or coal gasification loop. Cycle analyses are reported for efficiency improvement of energy conversion and waste heat recovery with various modifications and expected future developments and for thermal energy conversion applied to chemical industrial furnaces. New ideas on high temperature heat exchangers for blast furnace were fowarded. As a study in direct contact high temperature heat transfer, the development of an oxygen furnace slag granulation heat recovery system is reported with two features which effectively recover about $80°C$ of the heat of the slag and the slag is granulated into useful particles.

For the development of high temperature metal heat exchangers, data on the rupture strength, creep life and corrosion performance of metals and of welded parts, and fastening methods to relax thermal expansion are discussed.

For ceramic heat exchangers, several sealing mechanisms of tube ends or plate tips are reported. In the development of ceramic compact heat exchangers, an accurate stress analysis is required. Brazing data for plate-fin compact heat exchangers are reported. Finally it is noted that hydrogen permeation experimental data throutgh hot metals is reported for the safety purposes of HTR.

# HIGH-TEMPERATURE HEAT EXCHANGERS FOR ADVANCED POWER-GENERATING FACILITIES

A.E. Sheindlin

*High Temperature Institute, USSR Academy of Sciences*
*Moscow, USSR*

Present-day power engineering is characterized by the permanent tendency towards increasing the operating parameters of plant cycles. The increase of these parameters, along with the utilization of waste heat and its recovery, is called upon to provide for a substantial increase in the economy of energy resources. The accomplishment of these objectives is inseparably linked with the development of high-temperature heat exchangers.

From the technical and economic point of view, the parameters of conventional fossil fuel-fired steam power plants appear to have reached their maximum possible values or to be at least close to such values. The long history of the development of high-temperature components of such plants (first of all, components such as radiation and convection steam superheaters in boilers) has been well described in the extensive literature available on the subject while the specific features of the design and operation of such components need not be additionally described.

The problems of the development of high-temperature heat transfer apparati are acquiring an ever growing importance for nuclear power engineering. Interest is growing in gas-cooled reactors, including those cooled with helium, characterized by its chemical inertness and relatively high heat conductivity. A high-temperature gas-cooled reactor combined with a steam-turbine plant permits an efficiency increase to 40% as compared with 34% typical of water-cooled power reactors. Advances made in the field of materials studies, the theory of heat transfer and the progress of the methods for intensifying heat transfer make, the large-scale commercialization of gas-cooled power reactors rated for temperatures on the order of 900-950°C [1] realistic, even now. We shall not dwell in more detail on the problem of the high-temperature components of gas-cooled reactors because they will be covered in a special lecture. However, it should be noted that the steam generating surfaces of high-temperature exchangers in such reactors operate under conditions that are considerably less demanding than those encountered for the corresponding surfaces of fossil fuel-fired boilers.

A temperature of about 800°C in nuclear reactors may also be reached using salt coolants. However, it should be stressed that a number of technological problems associated with the use of such coolants are still awaiting their solution.

The development of fast breeder reactors cooled with a liquid-metal coolant (sodium) is very rapid. Inasmuch as such reactors are combined with the conventional steam power cycle, no increase of the coolant working temperature above 550-650°C is currently required. The technological and thermophysical problems involved in the development of high-temperature heat transfer apparati operating at these temperatures have been largely solved, as is evident in the available literature.

In the sixties, considerable attention was paid in U.S. publications to power-generating facilities utilizing alkali metal (potassium) vapors. By the early seventies, the basic heat cycle equipment of such facilities (potassium turbine, steam generator, radiator-condenser) were developed and had passed 5 to 15 thousand hours continuous testing [2]. Concurrently with the perfection of the cycles of spaceborne facilities, a number of authors, in particular, the author of Ref. [3], proposed the use of Rankine cycles utilizing alkali metal vapors on the ground as a topping for the conventional steam/water cycle. Both nuclear and fossil fuel were regarded as possible sources of energy. The scheme of one of the proposed potassium-steam-gas turbine cycles is shown in Fig. 1. The development of high-temperature alloys for continuous operation with potassium vapors (although much is still to be done in this direction), as well as the successful use of liquid-metal coolants in nuclear power plants, justifies, in our opinion, extensive research into alkali metal-steam binary cycles. Accordingly, high-temperature heat transfer apparati such as the potassium vapor generator with its specific problems of boiling incipience and alkali metal boiling stability and of the intensification of heat transfer in the high quality region, the potassium vapor condenser-steam generator with the attendant the problems of postdryout heat transfer on the water side, and a number of other components in the system. Adequate scientific expertise exists in this field.

Speaking of gas-turbine power-generating plant heat exchangers, the efficiency of these plants can be increased substantially through the use of exhaust gas heat for oxidizer and, if required, fuel preheating. However, gas-air heat exchangers have not yet found wide application in gas-turbine plants. This is primarily due to the fact that gas-turbine plants are mainly used nowadays as flexible peak load units. Under these conditions, it is, as a rule, inexpedient to increase design complexity and the cost of the plant by introducing a heat exchanger-waste heat regenerator. In binary gas-turbine-steam power plant cycles, waste gases from the gas turbine are delivered directly to the boiler and there is no need for a high-temperature heat exchanger.

In recent years, interest has grown in gas-turbine plants in combination with gas-cooled reactors with reactor core outlet gas temperatures of up to 950°C. Existing designs situate the gas turbine inside the reactor vessel of prestressed concrete. Provision is made for the use of a high-temperature recuperative heat exchangers for turbine exhaust gas heat regeneration and for the preheat of the working medium after the compressor. Technical difficulties involved in the development of such a system are formidable and, therefore, the current efforts of both researchers and developers are still concentrated on gas-cooled reactors with steam-turbine plants.

As is known, one of the ways to improve the efficiency of electric power generation is the use of a MHD power-generating unit in the high-temperature part of the cycle. High-temperature oxidizer preheaters are among the basic components of both open- and closed-cycle MHD power plants. The principal scheme of an open-cycle MHD power-generating unit is shown in Fig. 2. This design provides for the both a first-stage oxidizer preheater utilizing the heat of MHD generator exhaust gases (most often, a recuperative heat exchanger) and a second-stage autonomous high–temperature heater provided with its own combustor.

For second–stage preheat, regenerative preheaters with refractory matrises are usually used. Possible versions of oxidizer preheat schemes are shown in more detail in Fig. 3.

If recuperative heaters are used for oxidizer preheat, the heating surfaces must be sufficiently refractory to withstand internal pressures and thermal stresses. In addition, the metal used must be easily welded. Modern metal materials which satisfy these requirements may only be used at oxidizer preheat temperatures of up to 850°C. Tubular heaters rated for such temperatures can be used in plants in which the oxidizer is air oxygen enriched up to 50% or more* , as well as in the case of combination oxidizer preheat (cf. Fig. 1.3, c). In this latter case, a tubular preheater is the first stage heater and uses exhaust gas heat (regeneration) and a blast furnace-type autonomous heater for subsequent heating to a higher temperatures. Thanks to partial regenerative oxidizer heating, the efficiency of such a plant is higher than that of a plant utilizing a non-regenerative scheme (cf. Fig. 1.3, a).

Considerable attention has recently been paid to the development of ceramic recuperative preheaters, particularly, silicon carbide preheaters. In such preheaters, the oxidizer preheat temperature is about 1200°C. The principal problems include

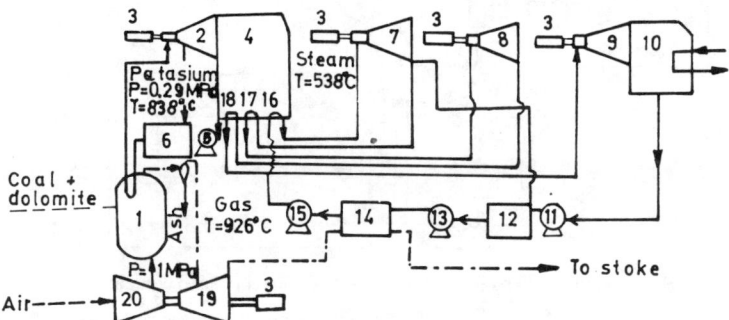

Figure 1. Scheme of combined potassium-steam-gas turbine cycle. 1 - potassium vapour generator; 2 - potassium turbine; 3 - generator; 4 - potassium condenser-steam generator; 5 - potassium feed pump; 6 - potassium preheater; 7 - high-pressure steam turbine; 8 - medium-pressure turbine; 9 - low-pressure turbine; 10 - steam condenser; 11, 13, 15 - water feed pump of the Ist, IInd and IIIrd stage, respectively; 12 - regenerative feedwater preheater; 14 - feedwater preheater; 16 - superheated steam generator; 17, 18 - 1st and 2nd intermediate steam superheaters; 19 - gas turbine; 20 - compressors.

component mechanical strength related problems applies, especially to the housing at substantial oxidizer pressures, high working medium flow rates at moderate pressure losses, adequate heat transfer (desirably, overall heat transfer coefficients of 100-200 W/$m^2$ K), prevention of the accumulation of solid particles or seed in the preheater and its slagging. The first of these calls for the limiting of the preheater dimensions and a modular apparatus, i.e., a combination of relatively small ceramic elements. The specifications of such a preheater are listed in Ref. [4]. A consits of 20 modules,

---

* Investigations have shown [4] that the preheat of 50% oxygen enriched air to a temperature of 750°C in MHD plants is as efficient from the technical and economic point of view as the preheat of pure air to 930°C.

each having a diameter of about 0.3 m and a length of about 2.5 m with a wall thickness of 6.35 mm. Ceramic preheaters may also be of the plate type, with the proper organization of gas flow (plate perforation, turbulizers) ensuring intensified heat transfer. It should be noted, however, that the air tightness of ceramic preheater

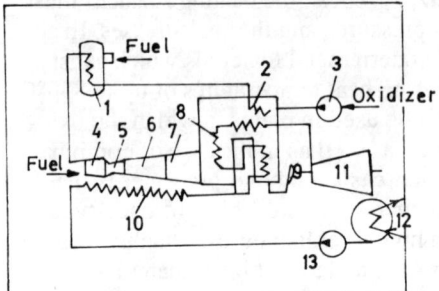

Figure 2. Simplified thermal layout of an open-cycle MHD power-generating unit.
1 - autonomous oxidizer heater; 2 - oxidizer preheater utilizing MHD exhaust gases; 3 - compressor; 4 - combustor; 5 - nozzle; 6 - MHD generator; 7 - diffusor; 8 - steam-generating surfaces; 9 - exhaust; 10 - cooling system of the combustor, nozzle of the MHD channel and diffusor; 11 - turbine; 12 - condenser; 13 - pump.

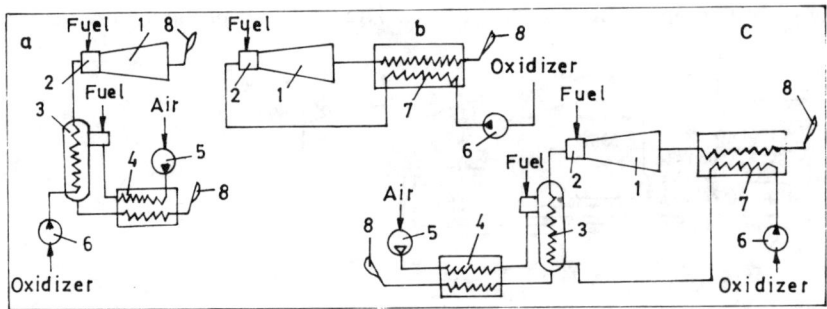

Figure 3. Schematic of a straight forward MHD facility with indirect (a), direct (b) and combined (c) oxidizer preheat.
1 - MHD generator; 2 - combustor; 3 - autonomous oxidizer heater; 4 - preheater of autonomous heater combustion air; 5 - fan; 6 - compressor; 7 - oxidizer preheater utilizing MHD exhaust gases; 8 - exhaust.

housings and the moderate leakage of air to the gas flow train still pose a serious technical problem.

Regenerative (direct) exhaust gas oxidizer heating (cf. Fig. 1.3, b) eliminates the need for burning fuel in an autonomous preheater. In facilities with direct oxidizer heating to 1700–2000°C, 37–40% fuel is saved as compared to a non-regenerative scheme. However, regenerative oxidizer heating involves a number of technical difficulties. The seed, as well as slag in the case of coal and oil combustion, solidifies at combustion product cooling, which may lead to the clogging (slagging) of gas passages in the heat-accumulating matrix. In order to preclude slagging, the combustion products must leave the heater at the free-flowing temperature of the suspension, i.e., at 1100-1300°C, which considerably complicates heater design. Additional difficulties are caused by the fact that the decrease of gas temperature to 1200°C leads to seed condensation, with the resulting liquid phase penetrating the pores of ceramic refractories and damaging due to the attendant physicochemical processes. It is believed that this effect may be avoided through the use of extremely dense ceramics made of high-purity magnesium. Relevant studies are underway.

Considerable experience in the operation of experimental and pilot air heaters for MHD facilities has accumulated in the U.S.S.R., U.S.A. and some other countries. However, in the case of commercial plants, one should always bear in mind that increasingly stringent service life and reliability requirements are placed upon the air heaters proper and upon the other components of the air preheat system (high-temperature valves, combustors and so on).

In a commercial 1000 MW MHD plant about 350-400 kg/s of air at a pressure of about 1 MPa have to be preheated to a temperature of 1700-2000°C. As already mentioned, the preheat temperature may be lower if the oxidizer is oxygen enriched. This corresponds to a doubled blast flow rate in the world's largest blast furnaces, approximately 5,000 $m^3$ in volume, an air pressure two times higher than the blast pressure in these furnaces and an afore-cited temperature level 300-400°C higher than ever reached anywhere in the world in commercial units of this type.

At present, only gas-fired units should be considered for first-generation MHD regenerative high-temperature heaters operating at the above-mentioned temperatures since, as already noted, the problems associated with the design and operation of high-temperature heaters utilizing MHD exhaust gases or solid fuel combustion products are still far from solved.

Regenerative high-temperature heaters may have a moving or stationary heat-accumulating matrix, rotary matrix chambers or rotary gas-distributing chambers; and regenerators with matrices undergoing phase transition (melting and solidifying) exist as well. A detailed description of potential high-temperature heaters can be found in Refs. [5, 6] thus there is no need to list all of these techniques.

At the present stage of science and technology development, with due regard for the need to ensure a high heat availability factor and maintain the chemical composition of the oxidizer constant (absence of heating combustion product leaks to the oxidizer and vice–versa), high-temperature heaters with a stationary matrix operating in the alternating heating and cooling mode with autonomous heating appear to hold the most promise.

The principal problems associated with the development of these high-temperature heaters that await solution are:

1. Development of refractory materials for air heaters matrix and lining, for continuous operation at heating gas temperatures of 1850-2150°C and at variable (varying within the 0.1-1 MPa range) gas media pressure.

2. Development of heat-insulating materials for compensation beds of low heat conductivity, and high elastic deformation properties at adequately high (1000°C) operating temperatures.

3. Development of optimum air heater matrix designs ensuring a high heat transfer coefficient at reasonable air heater construction costs and minimum possible pressure losses in the matrix, as well as a high air heater efficiency.

4. Development of air heater steel jackets about 10 m in dia. for operation at:

a) static loads at oxidizer pressures up to 1 MPa and temperature expansion of the lining,

b) dynamic loads due to cyclic (with a period of 0.5-1.0 h.) pressure variations in the vessel,

c) in aggressive media containing nitric acid formed from nitrogen oxides,

d) local heating of the jacket to temperatures of 300-400°C.

5. Development of burner design with a heat capacity of up to $7 \times 10^8$ kJ/h and stable fuel combustion.

6. Development of stop valve designs up to 2 m in dia. for oxidizer delivery lines to the MHD combustor operating at 1700-2000°C and a pressure difference of up to 1 MPa.

7. Development of temperature measuring equipment for monitoring the gas media, matrix and lining temperatures of 1850-2150°C.

8. Development of methods for nitrogen oxide suppression and removal from autonomous burner combustion products of the high-temperature heater.

At present, the requirements placed upon high-temperature heaters for commercial MHD power plants appear to be best satisfied by regenerative air heaters of the most modern blast furnaces with a volume of about 5000 m$^3$.

These air heaters have :

- metal jacket diameters of 9-11 m
- metal jacket thickneses of about 40 mm
- a height of up to 50 m
- up to 12 thousand tons of refractories in a single unit, of which about 30% are the most expensive and scarcest high-temperature materials
- oxidizer preheat temperatures up to 1400°C
- oxidizer pressures up to 0.5 MPa
- 200 kg/s of oxidizer being heated

Every blast furnace has at least three such units, two heating and one- blasting. Quite a few furnaces also have a fourth (stand-by) unit. Therefore, in the great majority of cases the heating time of the matrix of each unit (1.5-2 hours) is actually twice longer than oxidizer heating (45-60 minutes). True, it should be stressed that such practices are to some extent due to modern blast furnace heater low burner capacity as a vesult of the low available blast-furnace gas pressure in metallurgical plants.

The general operational – worthiness of blast-furnace high-temperature heaters for operation in a pilot MHD power plant has been tested at the U-25 Facility of the Institute of High Temperatures of the U.S.S.R. Academy of Sciences equipped with four regenerative air heaters of the type shown in Fig. 4.

Over a period of 15 years, these air heaters have supported the reliable operation of the U-25 Facility with the following oxidizer parameters: flow rate up to 50 kg/s; temperature 1200°C; pressure up to 0.35 MPa; oxygen enrichment degree up to 45%.Natural gas was used for heating.

The experience gained from the operation of blast-furnace high-temperature air heaters at the U-25 Facility led to a number of operationally important conclusions. High–temperature heater switch–over and oxygen introduction may be accompanied by a drop in the power of the MHD facility, which calls for design solutions and effective measures aimed at precluding a decrease in the rate of oxidizer flow to the MHD combustor during the switch over of the high-temperature heaters. Similarly, in the case of oxygen-enriched blast, one should preclude the possibility of supplying the MHD comustor with air with reduced oxygen content upon switching over from one heater to another. In other words, the procedure of the switching on of a heater should be properly organized.

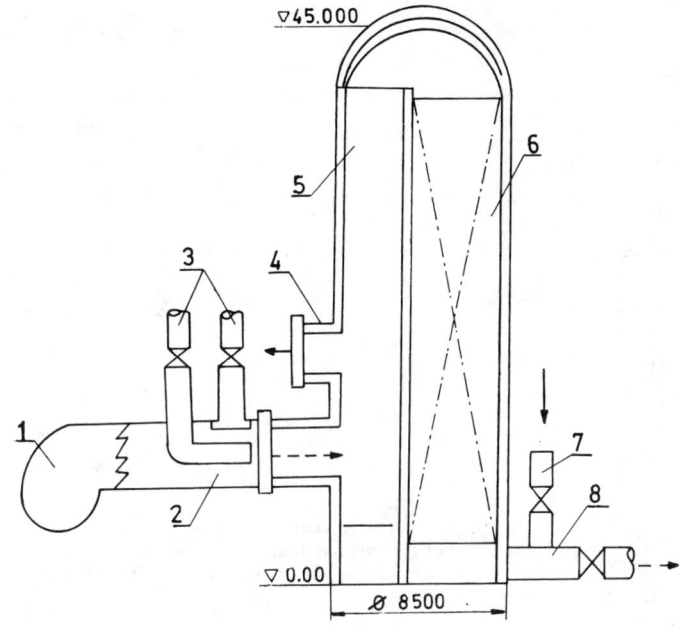

Figure 4. Air heater of the U-25 Facility at the Institute of High Temperatures of the U.S.S.R. Academy of Sciences.
1 - burner fan; 2 - burner; 3 - natural gas delivery pipelines; 4 - removal of hot oxidizer; 5 - combustor; 6 - matrix chamber; 7 - delivery of oxidizer being heated; 8 - removal of combustion products.

When operating at a reduced oxidizer flow rate, one should take into account the inevitable decrease of the hot blast temperature. This factor is of importance in designing peak-load MHD plants.

Let us consider briefly some structural peculiarities of high-temperature heaters for use in commercial MHD power plants. While air heaters rated for 1300°C often have combustors built in a common metal housing with the matrix chamber, combustors for air heaters with higher oxidizer preheat must either have a separate metal housing (cf. Fig. 5) or be placed in a common metal housing above the matrix chamber. By removing the combustor from the matrix chamber housing, two objectives are pursued.

First, the operation at higher temperatures and, apparently, somewhat greater losses of pressure in the matrix as compared to the current air heaters results in sharp deterioriation of the operating conditions of the brick partition between the combustor and matrix chamber, having an adverse effect on the strength and tightness of this partition and the overall performance of the high-temperature heater.

Second, the removal of the comustor reduces the diameter of the matrix chamber metal housing which, in air heaters operating at high oxidizer pressures (over 1 MPa), reduces the thickness of the high-temperature heater metal housing from 80-100 mm to 40-60 mm. This considerably facilitates manufacture and assembly (primarily, the welding) of the housing.

The air heater matrices should have a more developed heating surface than the presently used matrix (checker). It is possible to manufacture and use a matrix with a specific heating surface of 45-55 $m^2/m^3$ instead of the currently used matrices with a specific heating surface of 25-35 $m^2/m^3$. This results in air heaters of smaller overall dimensions (among other things, a smaller diameter of the metal housing) while maintaining the same heat capacity.

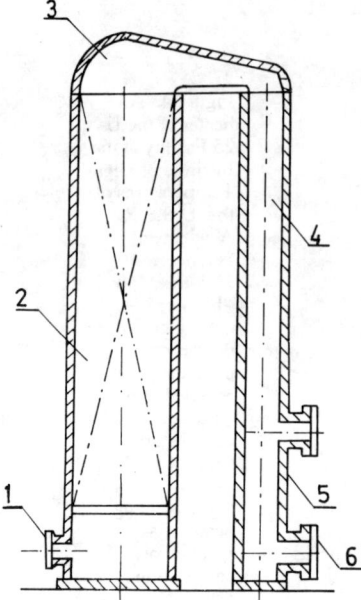

Figure 5. Air heater with external combustor.
1 - removal of combustion products; 2 - matrix chamber; 3 - dome; 4 - combustor; 5 - metal jacket; 6 - delivery of heating combustion products.

Oxidizer preheat to 1700°C calls combustion product temperatures of 1800-1850°C to heat the matrix of the high-temperature heater. At normal fuel combustion at $\alpha < 1.05$ this results in the formation of considerable quantities of nitrogen oxides. Special measures must be taken to preclude the discharge of these oxides to the atmosphere, as well as to protect the metal housing of the air heater against acid corrosion.

High oxidizer flow rates require large diameter (2 m and more) air lines and, consequently, large flow area stop valves. When we consider that these valves have to operate at pressures on the order of 1 MPa (with the hot-blast valve subjected to a temperature of about 1700°C.), we must admit that the development and manufacture of valves present a serious technical problem. The dimensions and weght of such valves is considerable, and their operation time - limited. All this has direct bearing on the choice of optium working cycle duration of air heaters, as will be discussed below.

The cost of corundum particles to be used in the high-temperature zone of high-temperature heaters is several times higher than that of currently used materials. Therefore, one must take all possible measures to reduce their mass. In addition to the afore-mentioned increase of the specific heating surface of the matrix, such measures include the reduction air heater operating cycle duration . While most air heaters currently used in metallurgy are rated for one-hour blasting operation, it is possible, in principle, to reduce the blast time to 30 minutes.

The temperature of combustion products heating the air heater matrix, required for an oxidizer temperature of 1700°C, is close to the maximum temperature of cold fuel combustion in nonpreheated air. Therefore, combustion air and, in some cases, fuel as well, have to be preheated, using systems of utilizing high-temperature heater waste heat.

In view of the high cost of air heaters for a commercial MHD power plant, serious consideration should be given to the possibility of supporting plant operation with a minimum number of high-temperature heaters. For instance, it appears possible to support the operation of a 1000 MW MHD power plant with two units operating in parallel in the blast mode. Suitable high-capacity burners that operato of reliably are required for this purpose. There is reason to believe that the problem of developing such burners has now been practically solved. The cost of high-temperature heaters for MHD power plants per heated air unit mass will be two to three times higher than that of currently available commercial high-temperature heaters. It should be stressed, however, that the heat transfered to a unit mass of air will be about 40% greater.

The large overall dimensions of the described high-temperature heaters for MHD power plants lead to the excessive increase of wall and matrix refractory specific loads. This necessitates the use in lower beds and, in spite of relatively moderate working temperatures, more expensive and less available materials than in the case of conventional modern heaters.

As already noted above, a serious disadvantage of modern air heaters is the low intensity of heat transfer in these heaters. Large diameter straight channels with a thick wall between the channels and a small specific heat transfer surface cannot provide for intensive heat transfer and effective accumulation of heat. This leads to an increase in air heaters dimensions. For over a hundred years the performance of high-temperature heaters considerably exceeded the heating temperature requirements of blast furnaces and, therefore, developments and new design techniques were rather slow to appear. It was only in the late fifties of the XXth century, with the advent of natural gas and oil fired blast furnaces, that the development of high-temperature heaters accelerated somewhat.

One type of matrix with intensified of heat transfer is the pebble-bed matrix. Heaters with a pebble-bed matrix are being tested in many countries. In the U.S.S.R., heaters with a pebble-bed matrix are available at the Institute of High Temperatures and some other organizations. Tests have been carried out on heaters with the pebble diameter ranging from 6 to 40 mm; 20 mm dia. pebbles seem to be best. The specific surface of such a matrix is 180 $m^2/m^3$. Thanks to a well developed heat transfer surface and high values of the heat transfer coefficient, such a matrix has a volumetric heat transfer coefficient 30-50 times higher than a brick matrix. The change – over from brick to pebble-bed matrix ensures rapid matrix heating and heat recycling to the air being heated, i.e., the duration of oxidizer heating by a single unit may, in principle, last 5-8 minutes while the heat exchanger volume is reduced by well over an order of magnitude. However, this is accompanied by an increase in the number of air heater switchings. The arrangement of such a heater is illustrated in Fig. 6. Another advantage of the pebble-bed matrix is the simplicity of assembly whereas brick matrix has to be laid manually.

It should be noted that pebble-bed matrix has a high heat transfer coefficient has an increased hydraulic resistance. While the matrix of a typical high-temperature heater with 25 mm dia. straight channels a height of the matrix bed of 30 m has a resistance of 440 mm $H_2O$ at the characteristic specific flow rate of the gas, the pebble-bed matrix of a heater of the same capacity with a bed height of 4 m has a hydraulic resistance of 770 mm $H_2O$.

The mentioned increased frequency of switchings of pebble-bed matrix high-temperature heaters calls for the development of new switching valves. At the same time, it should be emphasized that the reduction of the overall dimensions of heaters proper upon transition to a pebble-bed matrix does not entail a reduction of the dimensions of the high temperature valves. These latter dimensions depend on the total oxidizer flow rate which remains unchanged.

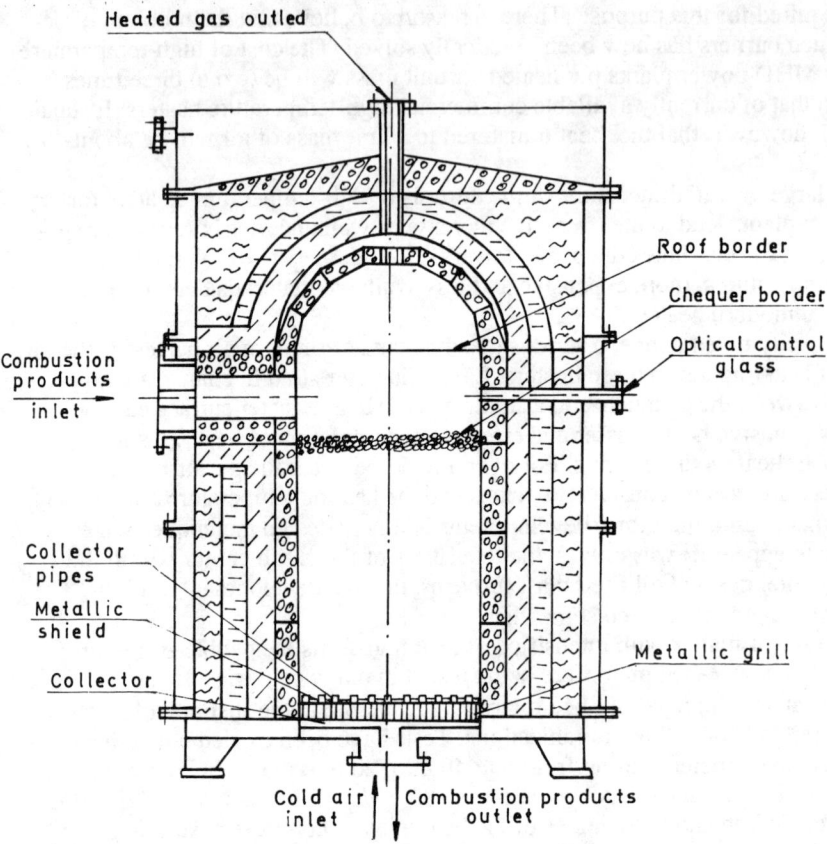

Figure. 6. High-temperature regenerative gas heater with globular chequer.

The operating characteristics (stability) of a pebble-bed matrix in large, continuously operating high-temperature heaters have not yet been sufficiently studied. This and the lack of suitable high temperature valves rated for a large number of switchings represent the main reasons why pebble-bed matrix high-temperature heaters have not been included in the designs of the first commercial MHD power plants. These are the basic immediate problems, the solution of which will represent the next step towards reducing the dimensions and cost of high-temperature heaters. To solve these problems, new and more advanced apparati are being developed and built at the Institute of High Temperatures in addition to those existing for many years. Their operation is expected to yield answers to the numerous questions raised above.

Studies into heat transfer processes in the matrix of regenerative heaters have shown that possibilities exist for improving the utilization of the accumulating capacity of the matrix, which have not yet been taken advantage of. Heaters usually operate with a linear or close to linear distribution of temperatures over the matrix height. However, by properly selecting the heating gas and oxidizer flow rates, one can obtain various temperature distribution curves and, consequently, considerably affect air heater performance. Figure 7 illustrates the operation of a pebble-bed matrix air heater with a linear (dashed line) and "bulging" distribution of temperature over the matrix height. The heater characteristies are a matrix bed height of 5.5 m, a matrix chamber diameter of 9 m, a flow rate of combustion products of 190 kg/s and the duration of the periods of heating and cooling of the matrix (i.e., heating the oxidizer) of 30 min each. As seen from the figure, for a linear distribution of temperatures over the height of the matrix bed the hot oxidizer temperature drop is 570°C in 30 min, and in the case of a "bulging" distribution - only 70°C. In order to maintain the temperature drop over a blow cycle constant (heating of oxidizer in a single apparatus), the height of a bed of pebble-bed matrix in a heater operating with a "bulging" temperature distribution may be one and a half times smaller than in a heater operating with a linear temperature distribution over the height of the matrix bed. This provides a reserve for the further reduction of the dimensions and cost of air heaters.

However, operation with a "bulging" temperature distribution involves a considerable increase of the matrix temperature variation amplitude during the cycle, reaching at times 1000–1200°C in the middle beds of the matrix (such a case is also shown in Fig. 7). The possibility of matrix operation at such high amplitudes needs to be experimentally verified. The transition to matrices with large specific heating surfaces leads to an increase of the temperature gradient over the height of the matrix bed and over the thickness of the matrix and insulating elements. This causes a rise of the thermal stresses in refractories. As a result, there arises the need to obtain data on permissible stresses in the matrix body and walls of made of various materials at different levels of the matrix chamber. Reliable and adequately fast answers to these questions can only be obtained as a result of simultaneous calculation studies, integrated test-bed experiments and the testing of the commercial-scale units. A considerable reduction of the dimensions of air heaters will facilitate the solution of the problems of the further increase of oxidizer temperature because, with the total costs remaining unchanged, it will be possible to use more expensive high-temperature material such as stabilized zirconia etc.

The use of bulk matrices may further contribute to the solution of the problems due to the use of nongaseous fuel and MHD exhaust gases for oxidizer preheat, the possible rapid replacement of the air heater matrix and appropriate subsequent treatment of the matrix to restore its condition and remove foreign inclusions.

Let us now consider some other problems related to the use of high-temperature heaters in MHD waste heat regeneration.

At a temperature above 1750°C dissociation makes a considerable contribution to the heat capacity of combustion products. This means that the combustion products from the MHD generator channel contain more heat than is necessary for the preheating of the oxidizer delivered to the combustor. Therefore, in the cycle under consideration, the thermodynamically possible degree of regeneration is considerably less than unity. It is possible to increase the degree of regeneration of the heat of combustion products by effecting the so-called chemical regeneration, i.e., when the

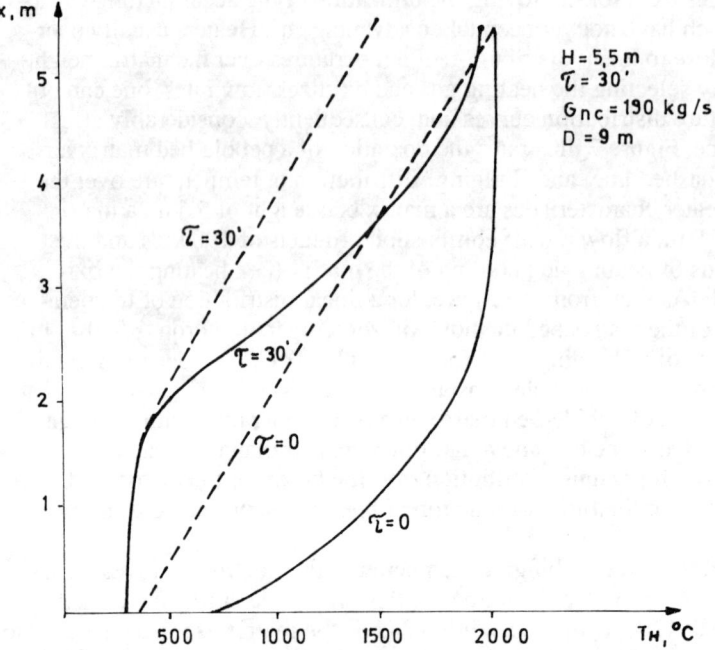

Figure. 7. Variation of the matrix temperature with height in the heating-cooling cycle.
- - - - - linear profile;
_____ "bulging" profile.

chemical energy of the heat transfer agent increases with its internal energy. A possible way of effecting chemical regeneration is through natural gas pyrolysis and conversion endothermal reactions:

$$CH_4 \xrightarrow{+Q} C + 2H_2 \qquad \text{—methane pyrolysis}$$

$$CH_4 + H_2O \xrightarrow{+Q} CO + 3H_2 \qquad \text{—steam conversion}$$

$$CH_4 + CO_2 \xrightarrow{+Q} 2CO + 2H_2 \qquad \text{—}CO_2 \text{ conversion}$$

The resulting carbonblack – and – hydrogen mixture or CO and $H_2$ mixture may be used (in which case we shall deal with an energy-technology facility) or they may be delivered for direct combustion to the MHD combustor.

The combustion in the combustor of the carbon black-and-hydrogen mixture resulting from the pyrolysis of methane, or of carbon monoxide and hydrogen resulting from its conversion, leads to higher temperatures in the channel and, consequently, a plasma higher electrical conductivity.

Shown in Table 1 is the dependence of the theoretical combustion temperature of $CH_4$ (with air as oxidizer) and of the $CO + 3H_2$ mixture on the oxidizer preheat temperature.

Table 1

| Preheat temperature, K | 500 | 1,000 | 1,500 | 2,000 | 2,500 |
|---|---|---|---|---|---|
| Combustion temperature of $CH_4$, K | 2,400 | 2,570 | 2,720 | 2,860 | 2,970 |
| Combustion temperature of $CO+3H_2$, K | 2,700 | 2,820 | 2,920 | 3,000 | 3,080 |

High-temperature heaters utilizing MHD combustion products may help to solve another problem closely associated with the ecological aspects of the MHD energy conversion, namely, the prevention of $NO_x$ emissions to the atmosphere. To this end, the rate of combustion product cooling should be such that of their residence time in the high temperature zone be sufficient for complete decomposition of the formed nitrogen oxides. For the decomposition of nitrogen oxides in schemes involving autonomous heating of high-temperature heaters, a special chamber for holding the combustion products must be constructed.

The development of regenerative high-temperature heaters utilizing MHD exhaust gases is associated, with the solution of the following two problems: ensuring the heating of the matrix with combustion products containing ionizing seed and the optimization of the regimes of high-temperature noncatalytic conversion and high-temperature pyrolysis of natural gas. The main difficulty lies in the need to develop refractory materials resistant to reduction- oxidation media in the presence of alkali seed used to attain high electrical conductivity of the plasma in the MHD channel. Today, this work is at the stage of investigation.

The problem may be regarded as solved to a considerable degree in respect to its other aspect, of the development of regenerative gas heaters for the high-temperature conversion and pyrolysis of natural gas. For this purpose, it is expedient to use regenerative preheaters with a pebble-bed matrix. Experimental facitilites of this type are being successfully used at the Institute of High Temperatures of the U.S.S.R. Academy of Sciences.

Presented in Fig. 6 is the design of the high-temperature heater of the "Regenerator" installation - pilot facility for methane converison. The high-temperature heater is a lined cylindrical vessel 1.8 m in dia. and 3 m high, with forced water cooling of the outer housing. In its bottom portion, the housing accommodates a gas-distribution grid suporting a pebble bed matrix of corundum pebbles 20 mm in dia. The matrix bed is 1.3 m high and 0.8 m in dia. The bottom bed of the matrix is made up of metal pebbles capable of withstanding sharp thermal shocks. The combustor is built in the spherical dome of the heater, with the heating gas moving downwards.

The results a cycle of studies of the high-temperature noncatalytic conversion of methane, are the temperature conditions at which the conversion reaction passes to completion and, the temperature dependence of converted gas and a high-temperature heater design.

Figure 8 shows experimental data of the distribution of the volume concentration of converted gas components the height of the matrix at various initial distributions of temperature. As seen, this concentration strongly affects the composition of the converted gas.

One can form an opinion on the completeness of a reaction from the residual content of methane in the conversion products. In the case of curve 3 in Fig. 8a with

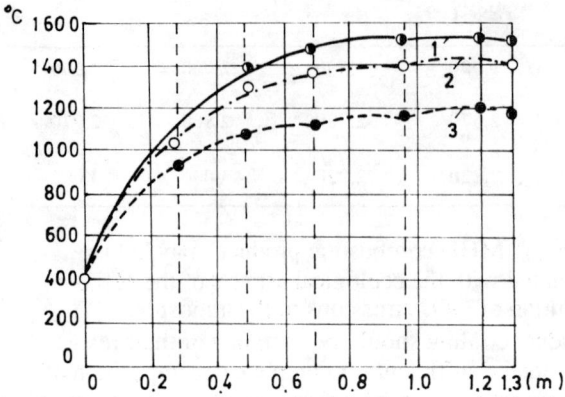

Figure. 8a. Temperature distribution over the matrix height

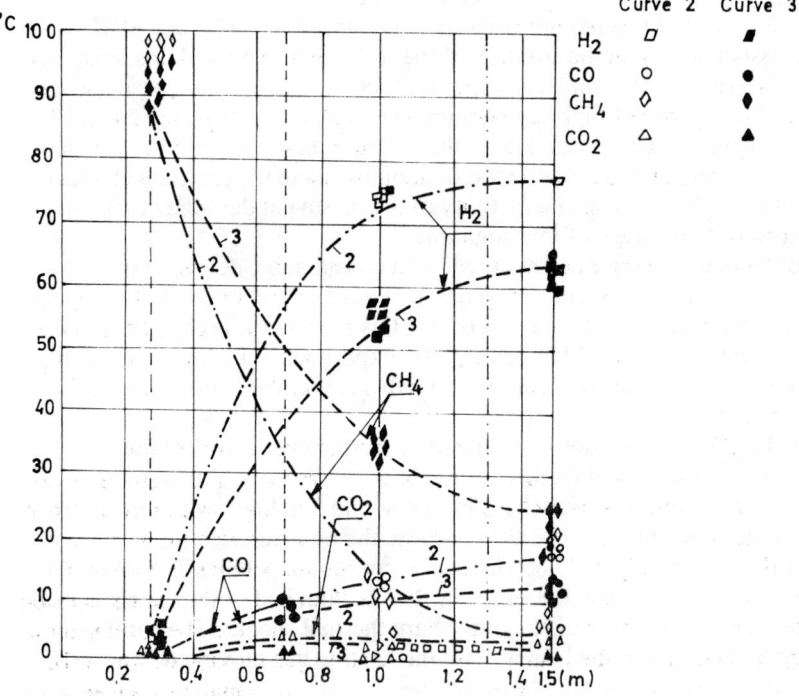

Figure. 8b. Distribution of component concentrations over the height of heat exchanger.

a maximum matrix heating temperature of 1,200°C the residual content of methane amounted to about 23%, with the hydrogen content in the mixture not exceeding 65%. For curve 2 with a top matrix bed temperature of about 1,400°C the residual methane content did not exceed 5%, with a carbon monoxide content of about 18% and hydrogen content of 76%. Under all operating conditions of the stand, the $CO_2$ content in the converted gas did not exceed several per cent. This low concentration of carbon dioxide in the conversion products is due to the fact that the equilibrium of the $H_2O + CO \rightarrow CO_2 + H_2$ reaction shifts to the left with an increase in temperature.

At the 1000-1100°C temperature range the conversion rate is such that several hours is required for the reaction to complete its course. At a temperature of the matrix top bed of about 1450°C, the time of gas passage through the matrix of the high-temperature heater is sufficient for the completion of reaction and equal to several seconds.

The technology of natural gas pyrolysis for the purpose of obtaining carbon and hydrogen in a regenerative preheater has been known for quite some time. Facilities of this type were built as early as the thirties and forties. However, since the maximum preheat temperature did not exceed, 1200-1300°C, the degree of natural gas decomposition was rather low. In the pyrolysis of natural gas in a high-temperature heater, carbon may form in its two modifications, namely, carbon black and pyrocarbon. Carbon black consists of fine dispersed particles formed by the homogeneous disintegration of methane. In the heterogeneous decomposition of natural gas, i.e., in the case of reactions occurring on the surface, pyrocarbon is formed and a carbon film is coating reaction surfaces. Upon the thermal decomposition of natural gas in regenerative high-temperature heaters, most of the resulting carbon black is removed from the reaction volume together with gaseous of products pyrolysis. Pyrocarbon in the form of solid deposits on the matrix surface cannot be removed from the matrix together with the pyrolysis products. In principle, the efficiency of regeneration will depend on the amount of carbon in the form of black carbon which gets to the MHD combustor together with the pyrolysis products and on the amount of carbon remaining in the matrix in the form of pyrocarbon. The carbon remaining in the matrix is removed by reaction with oxidizers in the course of heating the matrix, in accordance with the reactions:

$$C + CO_2 \rightleftarrows 2CO$$

$$C + H_2O \rightleftarrows CO + H_2$$

For complete removal of carbon deposited in the matrix, it is necessary to maintain an adequately high temperature of the combustion products providing the complete gasification of carbon separated in the matrix. Otherwise, carbon will accumulate from cycle to cycle in the matrix volume of high-temperature heater, finally leading to the loss of gas permeability. Wit a view to attaining a more complete re generation of heat in the cycle of MHD power-generating unit, it is expedient that the process of thermal decomposition of methane be conducted such that the greater part of separated carbon be formed as an oxide and removed from the matrix volume together with the gaseous combustion products.

These are main directions for the commercialization of high-temperature heat exchangers in power engineering. As is evident, certain technical difficulties are to be overcome for attaining these objectives, however, the economic benefits involved fully justify the efforts to be made.

## REFERENCES

1. А. П. АЛЕКСАНДРОВ. Атомная энергетика т. 56, вып. 6 с. 339
2. А. И. ЦыПКИН. Ракетная техника и космонавтика 1967, 5, №8, с. 140-147
3. FRAAS A. P. Chem. Eng. Progr., 1973, 69, N 3, p. 58
4. D. SAARI. 8-ая Международная конференция по МГД-преобразованию эхергии, Москва 1983, т. 3, стр. 112
5. Магнитогидродинамический генератор открытого цикла, пер. с англ., изд-во Мир, 1972., 836 с.
6. Магнитогидродинамическое преобразование энергии. Откгтый цикл. Б. Я. ШУМяЦКИй и М. ПЕТРИК. М. Наука, 1979

International Symposium 1986

# HEAT AND MASS TRANSFER IN CRYOENGINEERING AND REFRIGERATION

## J. Bougard

*Faculte Polytechnique de Mons, Mons, Belgium*

The desing of efficient cryogenic and refrigerating equipment relies on the knowledge of heat and mass transfer coefficients as well as of the thermophysical properties of materials in various working conditions.

The Symposium addressed the problems of heat and mass transfer in a very wide range of temperatures, from 400 K to the fascinating region of the mK, at the meeting point of thermodynamics, transport processes and technology.

The following topics were discussed: thermal insulation, heat and mass transfer in pute refrigerants and mixtures, thermodynamics and thermophysical properties, freezing and melting , heat and mass transfer at very low temperature, and cooling of supraconducting devices.

In the field of super insulation, Prof. Tien discussed anomalous conduction and thermal radiation phenomena occuring when the mean free path of phonons or electrons is of the same order of magnitude as the characteristic length of the material, for instance in cryogenic multilayer insulation or microsphere packing.

Some papers dealt with interesting results concerning classical insulants: the connection between thermal properties and structure, the calculation of heat transfer in complex systems and methods for measuring transport properties.

The performances of refrigeration and liquefaction units are strongly dependent on the quality of heat exchangers and their related energy dissipation. A good balance between thermodynamic performance and heat and mass transfer processes must be found.

The up-grading of heat transfer coefficients of refrigerants and cryogenic fluids is thus important. Significant enhancement of heat transfer during boiling has been effected by using porous surfaces or the application of an electric field.

It is known that the use of mixtures improves the thermodynamic qualities of cycles, leading for instance to a decrease of energy consumption, to the use of low grade sources and to an extension of the working temperature range. However as heat transfer coefficients are generally reduced the study of coupled heat and mass transfer raises complex problems.

A dozen high level papers presented new experimental and theoretical results on transport properties and the heat and mass transfer coefficients in mixtures, and the desing of absorber as well as liquid vapour equilibrium data up to the critical line.

Heat transfer and the diffusion of water in porous systems such as soils near the freezing point are very difficult to predict and to measure. However good results are now obtained.

And last three sessions were devoted to the problems of cryogenic cooling and expecially, superconducting equipment.

Heat transfer coefficients in boiling cryogenic liquids have been improved by an order of magnitude using porous coating techniques in different laboratories in the USSR, USA and France.

In superconducting systems, the challenge is to overcome the complexity of several coupled problems such as the thermal stability of superconductors, convection and the influence of Coriolis forces.

Interesting results were presented.

In summary, 49 papers were discussed by 71 participants from 15 different countries.

It is worth noting that this Symposium was also a good opportunity to strengthen scientific cooperation and the bonds of friendship between the ICHMT and the International Institute of Refrigeration.

# HEAT TRANSFER IN LOW-TEMPERATURE INSULATION

C.L. Tien and A.J. Stretton

*University of California,
Berkeley, USA*

## 1. INTRODUCTION

Thermal insulation has long been a subject of great importance to engineers and was indeed one of the major concerns in the early development of the heat transfer technology. During the period between the 1930's and 1960's, however, a multitude of new heat transfer research and developments moved the subject of insulation heat transfer from its earlier prominence to a level of stagnant obscurity. Thermal insulation had become a classical subject that was considered by many as already well developed and of concern only to the manufacturing and design engineers. In the meantime, developments in aerospace, nuclear, computing, cryogenic and other modern technologies have extended a great number of formidable engineering problems at the extreme temperature limits. Consequently, the past twenty years have registered an intensive surge of interest in thermal insulation for both high and low temperature applications.

Insulation for low temperature conditions differs in several ways from insulations for other temperature ranges. The different temperature ranges dictate the use of different insulation materials and methods that in turn result in fundamentally different thermophysical characteristics as well as transport phenomena. For example, cryogenic insulation is normally operated under vacuum conditions while high-temperature insulation often encounters oxidizing or reducing atmospheres, where material sublimation and reactivity are a major problem. Insulation technology for energy conservation at room temperatures is usually inadequate for the stringent control of heat flow necessary in cryogenics.

The importance of insulation in very low temperature applications is easily realized by noting that the latent heat of vaporization of cryogenic liquids as well as the specific heat of solids at cryogenic temperatures are much smaller than the corresponding properties at room temperatures. Therefore, it takes very little inflow of heat to boil off cryogenic liquids or to raise the system temperature. The development of better insulations for low temperatures has created tremendous growth in the field of cryogenic science, and has supported the largescale use of liquefied gases in industry and many new low-temperature devices for special applications such as cryogenic infrared sensors and Josephson junctions.

The purpose of this paper is to give some insight into the physical processes occurring in low temperature insulations and to discuss several insulation types which represent state-of-the-art in this field. The relatively young field of cryogenics provides ample opportunity for further fundamental research to bridge the gap between the wide acceptance of newly developed cryogenic insulations and the lack of understanding about the basic underlying physical mechanisms involved. In view of the increasing interest in low temperature insulation, several monographs and review articles have been written over the past two decades [1-5]. Many important topics which could affect the thermal performance of cryogenic insulation will not be treated here: manufacturing processes, mechanical support and penetration, evaluation and test standards, systems design, etc. Discussion on these topics can be found in the literature [6, 7].

## 2. BASIC INSULATION CONCEPTS AND REQUIREMENTS

Thermal insulation refers to either a single homogeneous material or a composite of materials, often making use of air or evacuated spaces, that is designed to reduce heat flow between surfaces with different temperatures. The basic heat transfer mechanisms are conduction, convection, radiation, and in some unevacuated situations, phase change such as condensation. Evacuation is the most effective way to eliminate conduction, convection, and phase change, while radiation may be re-

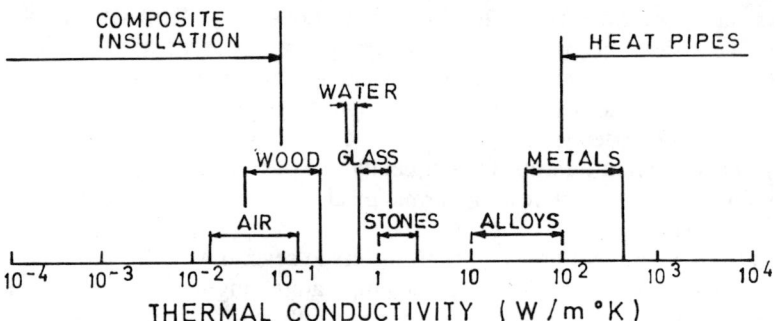

Figure 1. Spectrum of material thermal conductivities.

duced through the use of highly reflective boundary surfaces or by the addition of materials which are opaque to radiation in the insulation medium. In many refrigeration applications, an insulation thermal conductivity equivalent to that of air is sufficient, since air has quite a low thermal conductivity (see Fig. 1). This can be achieved with the use of porous or composite materials, fibers, powders, or foams, which provide effective resistance to convection and radiation heat transfer with little contribution to heat flow by conduction in the solid material.

It is common practice to define an "effective" (or "apparent") conductivity to describe heat flow through an insulating medium, thus avoiding the complex interactions of the various mechanisms and the details of the geometry. The effective conductivity is based on the simple one-dimensional model of conduction through a slab, and is defined by

$$k_e = \frac{QL}{A(T_1 - T_2)} \tag{1}$$

where $Q$ is the total heat flow through the insulation, $L$ is the thickness of the insulation, $A$ is the heat-flow cross-sectional area, and $T_1$, $T_2$ are the bondary surface temperatures. For an anisotropic insulation (i.e., directionally dependent properties), more than one value of effective conductivity may be necessary. Typical values for effective conductivity range from $2 \times 10^{-5}$ W/m K for evacuated multilayer insulation to $4 \times 10^{-2}$ W/m K for foams and fiberglass insulation.

Despite its primary function of reducing heat flow, thermal insulation must also satisfy many other requirements in practical service. Thermal requirements normally consist of temperature limits, thermal diffusivity, specific heat capacity, thermal shock resistance, and thermal expansion characteristics. Economic requirements are also a major factor in engineering design selection, and can be dominated by indirect costs such as energy savings, investment savings and maintenance expenses, as well as traditional costs such as fabrication, shipping, storage, and installation. The mechanical requirements are based on properties, usually temperature dependent, such as compressive strength, flexural strength, shear strength, shock and wear resistance, and crack propagation characteristics. Other factors which can be important are permeability to moisture and gases, corrosion resistance to various chemicals, flammability and toxicity.

Thermal insulation can generally be categorized by its composition into five basic types. They are: low-conductance solids, powder (flake or granular) insulation, fibrous insulation, cellular (foam or honeycomb) insulation, and reflective insulation. Each category is schematically illustrated in Fig. 2, and the effective thermal property range of each is shown in Fig. 3. Low-conductance solid insulation refers to rigid or semi-rigid boards, blocks, sheets, and cement that have low thermal conductivity and low porosity, such as firebricks, asbestos, and cork. Powder insulations are generally composed of small particles in the form of spheres, flakes, or irregular grains, with porosities ranging from 30% to above 90%. Fibrous insulation is made of interwoven small diameter fibers, forming a flexible layer with a very high porosity of about 90%. Cellular insulation contains many small individual cells sealed from each other, either in the form of a honeycomb or foam or both, which has the advantages of low density, low heat capacity, and good compressive strength. Reflective insulation is composed of parallel thin sheets of high reflectance, and will be discussed in more detail later in this review.

## 3. ANOMALOUS HEAT TRANSFER EFFECTS AT CRYOGENIC TEMPERATURES

It is important to be aware of the various heat transfer processes responsible for heat flow in insulations so that accurate predictions of thermal performance can be made. Cryogenic insulations typically involve evacuation and very small length scales where heat transfer is subject to effects which are normally neglected in higher temperature applications.

The contribution of gas conduction can be significant in highly evacuated insulation due to outgassing and gas entrainment in internal cavities. Free convection within the voids of the insulation is always negligible even in unevacuated insulations because the characteristic length of the voids is so small ($\approx 10^{-2}$ cm or less) that the Rayleigh number is much less than the critical value ($Ra_c \approx 10^3$) for onset of convection. Gas conduction in cryogenic insulations is often in the free molecular regime where kinetic theory is required for analysis [8]. One notable feature of free-molecule flow is that the conductive heat flux is independent of the gas-layer thickness, which

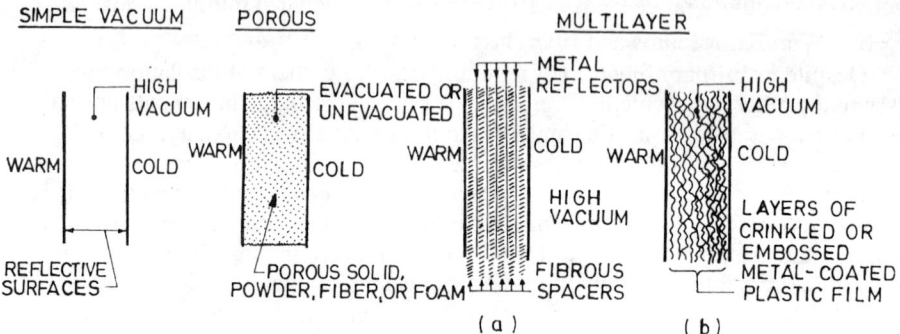

Figure 2. Various types of thermal insulation.

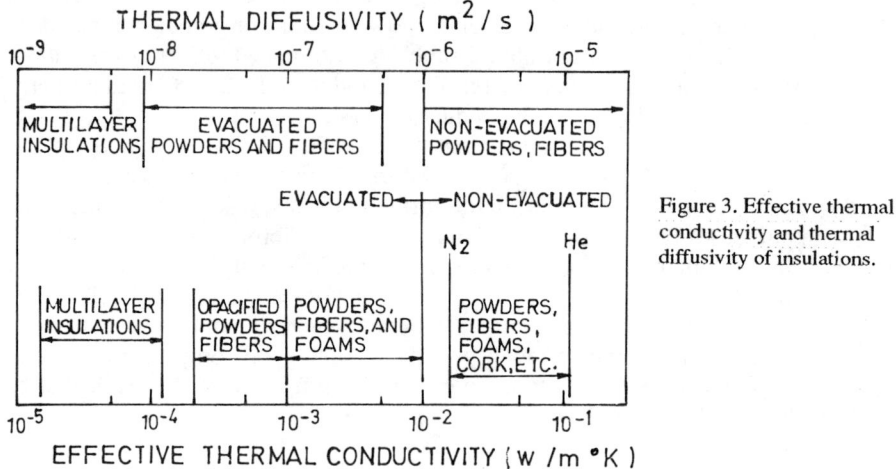

Figure 3. Effective thermal conductivity and thermal diffusivity of insulations.

is analogous to radiation transport in a nonparticipating medium. The multiple shielding concept in radiation is also applicable to residual gas conduction. This makes it possible to achieve equivalent insulation effectiveness with a lesser vacuum requirement when more shields are used.

The contribution of heat conduction in the solid is due primarily to two major mechanisms according to theory: mechanical interaction between molecules (i.e., lattice vibrations) and translation of free conduction electrons. Because lattice vibrations can be treated as phonons, thermal transport in solids can be regarded as energy transport in phonon and electron gases, and the mean-free-path concept from the kinetic theory of gases is directly applicable [9]. The free-electron contribution dominates in energy transport in metals and the phonon contribution is predominant in dielectric solids, whereas in very impure metals or in disordered metals, the phonon contribution may be comparable with the free-electron contribution. A significant feature of thermal conduction in cryogenic solids is that the mean free path of the phonons and electrons increases as the temperature decreases. The relationship can be expressed as $l_p \propto T^{-1}$ and $l_e \propto T^{-1}$ to $T^{-3}$ where at room temperatures $l_p \approx 10\text{-}100$ Å for crystalline dielectrics and $l_e \approx 100$ Å for pure metals [9]. At cryogenic temperatures, these mean free paths are comparable to the characteristic dimensions of the dielectric powders, fibers, and metallic coatings commonly used in cryogenic insulations. Under these conditions, many of the phonon or free-electron paths will be shortened as a result of termination at the boundary surface of the solid elements, and the thermal conductivity is expected to be less than that of the bulk solid. The significance of this size effect has been quantitatively demonstrated for thin metallic films at cryogenic temperatures [10], and the effect is illustrated schematically in Fig. 4 (a).

One of the most significant features of heat transfer at cryogenic temperatures is that the majority of the radiant energy lies at very long wavelengths. Consider, for instance, blackbody radiation at 10 K. Most of the electromagnetic energy is contained between 100 and 1000 μm, which is the same magnitude or higher than the characteristic dimensions of the solid elements and voids in high-performance cryogenic insulations. The long-wavelength radiation results in a reconsideration of many radiation phenomena, which can be legitimately neglected at room or high temperatures but become increasingly important at low temperatures. These effects include "non-blackbody" radiation in small cavities [11], the enhancement of radiative transfer due to electromagnetic wave interference at close spacings [12], radiation dissipation in metals by the anomalous skin effect [13] and the lowering of emissivity from bulk values by the size effect [4]. There also exist better-known effects of long-wavelength radiation, such as the increase in specularity of reflected radiation from rough surfaces [15], and the electromagnetic scattering effect from curved bodies [16]. Figures 4 (b-e) illustrate some of the various thermal radiation effects that have been mentioned. A careful assessment of these phenomena and their associated effects is often necessary in the consideration of heat transfer through cryogenic insulations.

## 4. MULTILAYER INSULATION

Evacuated multilayer insulation, on account of its high thermal effectiveness, is often called "super-insulation". It normally consists of a laminated assembly of numerous thin (~0.4-8 μm) plastic films coated on one or both sides by a thin vapor-deposited layer (~0.04 μm) of high reflectance metal, usually aluminim or gold. Plastic films are employed instead of solid metal films because of their high mechanical strength, low density, and low thermal conductivity. These radiation shields (~10-50 per cm) are separated from each other either by spacers of low thermal conductivity such as low density paper or netting (1.5-15 μm), by crinkling/embossing the shields so that when they are placed against each other they touch at only a few discrete points, or by tufts of short fibers bonded to one of the surfaces. Under well-controlled conditions, the effective thermal conductivity in the direction normal to the layers is on the order of $1 \times 10^{-5}$ W/m K, which is one or two orders less than that of evacuated powders or fibers. However, multilayer insulation often suffers considerable deterioration and poor predictability in thermal performance after installation in a practical cryogenic system. These undesirable features are caused by a number of factors, namely the difficulty of installing the insulation in certain geometries, the strong dependence of thermal performance on the compressive load, and the highly anisotropic thermal conductivity. In particular,

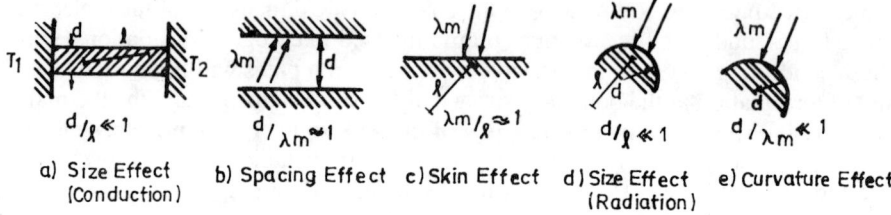

a) Size Effect (Conduction)   b) Spacing Effect   c) Skin Effect   d) Size Effect (Radiation)   e) Curvature Effect

Figure 4. Anomalous conduction and thermal radiation phenomena in solids at cryogenic temperatures.

the large effective lateral conductivity (i.e., parallel to the layers), which is generally about three to six orders of magnitude greater than the effective normal conductivity, causes large lateral heat leaks to the junctions where structural members or feed-lines penetrate the insulation.

The normal direction heat transfer characteristics in multilayer insulation often constitute the major criterion in the thermal design and performance evaluation. The primary system parameters that affect normal heat transfer consist of the layer density (including the thickness of shields and spacers, and the imposed pressure) and the thermal properties (conduction and radiation) of the shields and spacers. In typical multilayer insulation systems, the radiation and conduction transfer across two adjacent shields can be calculated separately and then added up [2, 17]. The conduction contribution is not influenced much by the spacer conductivity but primarily by the interface contact conductance which in turn depends on the contact pressure and deformation characteristics [18, 19]. The two-shield case can be extended to the multilayer system by way of the continuum model [20], which assumes the discrete multilayer as a continuous homogeneous medium and defines a local equivalent thermal conductivity characterizing the combined conduction and radiation transport. It gives a simple convenient method of calculation and the results agree well with measurements [20] as well as complex numerical calculations [21, 22]. For convenience of evacuation, perforated shields are often used, and a careful study on the perforation effect is available [23].

Lateral heat transfer in multilayer insulation is governed by conduction along the shield and lateral radiation tunneling, that is, multiple reflection along the shields. The use of highly scattering fibrous spacer material can reduce the lateral radiation contribution considerably [24], while the lateral conduction can be reduced by selective slitting of the shields to increase resistance in the direction of heat flow [25]. The slitting results in a two-dimensional conduction pattern, but the resulting two-dimensional conduction-radiation interaction becomes extremely complicated and has not yet been analyzed. In actual installations, the system geometry and mechanical penetrations often necessitate the use of stitches, patches, and overlaps, all of which may exert a significant effect on lateral heat transport and must be evaluated carefully [26].

## 5. MICROSPHERE INSULATION

Packed, hollow glass spheres have provided an alternative "super-insulation" without some of the shortcomings of multilayer insulation. Spheres for cryogenic insulation range in diameter from 15 to 150 µm and in nominal wall thickness from 0.5-10 µm with an optional exterior film of low-emittance metal such as aluminum. While the effective thermal conductivity is several times that of multilayer insulation, the isotropic nature, ease of installation, and repeatable, predictable performance make microspheres the superior system in many applications. The high performance of microsphere insulation comes from constriction resistance to conduction in the thin walls, radiation attenuation by the metal coating, and the low heat capacity of the hollow spheres. Typical performance characteristics are presented in Table 1 for comparison with those of multilayers.

From a fundamental point of view, conduction through packed spheres consists of two physically distinctive stagers: conduction through a spherical contact, and conduction through a packed bed. For a spherical contact, the conduction process is characterized by three series-connected thermal resistances, namely the macroscopic constriction resistance due to the contraction of conduction passages, the microscopic contact resistance due to surface roughness at the contact area, and

Table 1. Comparison of microspheres and multilayer insulation.

| Characteristic | Microsphere | Multilayer |
|---|---|---|
| Predictability of performance | ±10% | ±100% (installed) |
| Ratio of normal to lateral conductivity | 1 | $10^{-1}$ to $10^{-4}$ |
| Bulk density, kg/m$^3$ | 70 to 80 | 50 to 60 |
| Effective thermal conductivity, (installed) W/m K | 2.2 to 4.0×$10^{-4}$ | 5.0 to 10.0×$10^{-5}$ |
| Compressive strength, N/m$^2$ | > 3×$10^6$ | nil |

the film resistance due to surface contamination. The macroscopic resistance is predominant for uncoated glass microspheres, while for aluminum-coated microspheres the microscopic resistance is probably the most dominant due to the relatively rough surface texture [27]. The conductance of packed spheres has been calculated analytically for several packing arrangements [28], and the results exhibit a monotonic dependence on the solid (or void) fraction. Another study, using a general random stochastic treatment to approximate real agglomerates with a particle size distribution, showed the overall conduction of the packed spheres was proportional to bed compressive loading raised to a power n, where n ranges from 0.33 for solid or thick-walled hollow spheres to 0.7-0.9 for thin-walled rough-coated spheres [29].

The radiation contribution to the effective thermal conductivity can be analyzed with a diffusion approximation to the equation of radiative transfer, since the bed of microspheres is opaque to the predominantly long-wavelength radiation at cryogenic temperatures [3]. For uncoated microspheres, the radiation contribution is dominant over the conduction contribution, and this is manifested in the strong dependence of the effective thermal conductivity on temperature in the experimental data [27, 30]. For aluminum coated microspheres, conduction accounts for much more than radiation, and the effective thermal conductivity is less dependent on temperature. The temperature influence on the effective thermal conductivity should be accounted for, particularly in the case of uncoated microspheres, when there are large differences in the boundary temperatures. The effect of the boundary emissivities on the radiative transfer can also be important [31].

## 6. RECENT DEVELOPMENTS

In the past fifteen years, tremendous progress has been made in the study of fundamental heat transfer processes in thermal insulation. The current thrust of the research at the University of California at Berkeley is primarily concerned with un-evacuated porous insulations, due to future energy concerns. The following section presents a brief description of Berkeley research results.

Fibrous insulation is commonly used in medium temperature applications (200-500 K) such as building insulation and refrigeration systems. For convective heat transfer through porous insulation, several problems have been addressed, including two-dimensional convection in a vertical insulation layer [32], in a horizontal layer [33], and in concentric cylinders and spheres [34], and infiltration effects through insulation [35]. Moreover, for calculations of convection heat transfer in insulation, the limits of Darcy's law have been investigated in flows near solid boundaries and in high-porosity insulation subject to significant inertial effects [36, 37]. Water-moisture transport and condensation at room temperatures is a major concern which often causes a significant deterioration in the thermal performance of porous insulations. Fundamental research has resulted in a mathematical formulation for moisture transport and condensation in a wall insulation configuration [38], with a solution for the one-dimensional case which clearly exhibits the effects of various physical and system parameters.

The role of radiation in the thermal performance of light-weight fibrous insulation (LWFI) is gaining recognition as an important contributor to the overall heat flow. A combined theoretical and experimental research program at Berkeley is being carried out to study this problem. The radiative properties of LWFI have been calculated based on the extinction coefficients for commercial LWFI have been measured via an infrared transmission apparatus and the total heat flux has been measured by a guarded hot plate apparatus [40]. The results indicate that careful manufacture of fibers with certain optimal sizes can minimize the thermal radiative transfer through LWFI and reduce the effective thermal conductivity by a significant fraction. Thermal radiation contributes as much as 40-50% to the total energy transfer, even for moderate temperature applications. The other major heat transfer mode is gas conduction through the spaces between the fibers.

The packed-spheres in microsphere insulation do not scatter radiation independently in low temperature applications, contrary to the assumptions behind most theoretical calculations to this date. Conventional results from electromagnetic wave theory assume that the particles can be treated as being far apart and that each particle scatters radiation in the same manner as a single isolated particle. A rigorous experimental work on packed/fluidized beds was conducted and compared to the theoretical calculations, which resulted in the conclusion that the porosity (or distance between the particles) alone is irrelevant [41]. The critical parameter is the ratio of the average clearance between particles to the wavelength, where independent scattering occurs for ratios above the cut-off value of approximately 0.3-0.5. Under such a criterion, packed spheres at high temperatures (such as in a fluidized bed combustor) can be treated as independent ascatterers, while a bed of packed microspheres at cryogenic temperatures is in the dependent scattering regime. The investigations at Berkeley in this area are continuing to further the understanding of the physics of such fundamental phenomena.

One of the most exciting new concepts for insulation in recent years concerns the attainment of vaccum-like performance without evacuating the system and without complex technology. Since gas conduction is characterized by the molecular mean free path, it can be reduced greatly if the pore size between the insulating particles is less than, and thus limits, the mean free path of the gas molecules. Preliminary investigations of silicon dioxide ultra-fine powders (diameter ~70 A) have given this concept very encouraging support. The loosely packed structure has a porosity of 97-98%, but the particles can be compressed to yield a denser structure with a porosity of about 90% to further reduce both gas conduction and radiative energy transmission. Similar studies show that additional improvements in effective thermal conductivity are possible if highly reflecting sheets are inserted within the powder.

Another promising concept deals with thermal insulation for flow systems. By installing highly porous insulation materials across a fluid flow, a portion of the fluid energy and incoming radiative energy is absorbed and converted to change the temperature of the solid matrix, which in turn emits radiation in both the upstream and downstream directions. The insulation layer across the flow effectively reduces the flow of heat in the fluid by reradiating some of the thermal energy back to the source. A one-dimensional model on the combined radiative and convective heat transfer in a porous medium has been developed [42]. Applications of this concept can be instituted in low temperature flow systems by installing porous radiation shields in appropriate locations.

## REFERENCES

1. KAGANER, M.G., **Thermal Insulation in Cryogenic Engineering** (English translation from Russian by A. Moscona), Israel Program for Scientific Translations, Jerusalem, 1969.
2. TIEN, C.L. and CUNNINGTON, G.R., Cryogenic Insulation Heat Transfer, **Advances in Heat Transfer**, Vol. 9, pp. 349-417, 1973.
3. TIEN, C.L. and CUNNINGTON, G.R., Recent Advances in High-Performance Cryogenic Thermal Insulation, **Cryogenics**, Vol. 12, pp. 419-421, 1972.
4. TIEN, C.L. and CUNNINGTON, G.R., Glass Microsphere Cryogenic Insulation, **Cryogenics**, Vol. 16, pp. 583-586, 1976.

5. TIEN, C.L. and WANG, K.Y., Thermal Insulation Heat Transfer, in **Special Issue for the U.S.-China Binational Heat Transfer Workshop** (in English), pp. 1-11, Chinese Society of Engineering Thermophysics, Beijing, China, 1984.
6. KROPSCHOT, R.H., Advances in Thermal Insulation, **Advances in Cryogenic Engineering**, Vol. 16, pp. 104-108, 1971.
7. CUNNINGTON, G.R. and TIEN, C.L., Heat Transfer in Microsphere Cryogenic Insulation, **Advances in Cryogenic Engineering**, Vol. 18, pp. 103-111, 1973.
8. SPRINGER, G.S., Heat Transfer in Rarefied Gases, **Advances in Heat Transfer**, Vol. 7, pp. 163-218, 1971.
9. TIEN, C.L. and LIENHARD, J.H., **Statistical Thermodynamics** (Rev. Ed.), Chapter 11, pp. 303-332, Hemisphere, Washington, D.C., 1979.
10. TIEN, C.L., ARMALY, B.F., and JAGANNATHAN, P.S., Thermal Conductivity of Thin Metallic Films and Wires at Cryogenic Temperatures, **Proceedings of the 8th Thermal Conductivity Conference**, pp. 13-19, 1969.
11. CASE, K.M. and CHIU, S.C., Electromagnetic Fluctuations in a Cavity, **Physical Rewiev A**, Vol. 1, pp. 1170-1174, 1970.
12. CRAVALHO, E.G., TIEN, C.L., and CAREN, R.P., Effect of Small Spacings on Radiative Transfer Between Two Dielectrics, **Journal of Heat Transfer**, Vol. 89, pp. 351-358, 1967.
13. TIEN, C.L. and CRAVALHO, E.G., Thermal Radiation of Solids at Cryogenic Temperatures, **Chemical Engineering Progress Symposium Series**, Vol. 64, No. 87, pp. 56-66, 1968.
14. ARMALY, B.F. and TIEN, C.L., Emissivities of Thin Metallic Films at Cryogenic Temperatures, **Proceedings of the 4th International Heat Transfer Conference**, Vol. 3, pp. R1.1 1-9, 1970.
15. BECKMAN, P. and SPIZZICHINO, A., **The Scattering of Electromagnetic Waves from Rough Surfaces**, MacMillan, New York, 1963.
16. HULST, H.C. VAN DE, **Light Scattering by Small Particles**, Wiley, New York, 1957.
17. WANG, L.S. and TIEN, C.L., A Study of Various Limits in Radiation Heat Transfer Problems, **International Journal of Heat and Mass Transfer**, Vol. 10, pp. 1327-1338, 1967.
18. TIEN, C.L., A Correlation for Thermal Contact Conductance of Nominally-Flat Surfaces in a Vacuum, **Proceedings of the 7th Thermal Conductivity Conference**, NBS Special Publication 302, pp. 755-759, 1968.
19. FLETCHER, L.S., SMUDA, P.A., and GYOROG, D.A., Thermal Contact Resistance of Selected Low-Conductance Interstitial Materials, **AIAA Journal**, Vol. 7, pp. 1302-1309, 1969.
20. CUNNINGTON, G.R. and TIEN, C.L., A Study of Heat Transfer Processes in Multilayer Insulations, in **Thermophysics: Applications to Thermal Design of Spacecraft**, ed. J.T. Bevans, pp. 111-126, Academic Press, New York, 1970.
21. MACGREGOR, R.K., POGSON, J.T., and RUSSELL, D.J., Numerical Evaluation of Multilayer Insulation System Performance, in **Heat Transfer and Spacecraft Thermal Control**, ed. J.W. Lucas, pp. 502-518, MIT Press, Cambridge, 1971.
22. FORSBERG, C.H. and DOMOTO, G.A., Thermal-Radiation Properties of Thin Metallic Films on Dielectrics, **Journal of Heat Transfer**, Vol. 94, pp. 467-472, 1972.
23. TIEN, C.L. and CUNNINGTON, G.R., Radiation Heat Transfer in Multilayer Insulation Having Perforated Shields, in **Thermophysics and Spacecraft Thermal Control**, ed. R.G. Hering, pp. 65-74, MIT Press, Cambridge, 1974.
24. TIEN, C.L., JAGANNATHAN, P.S., and CHAN, C.K., Lateral Heat Transfer in Cryogenic Multilayer Insulation, **Advances in Cryogenic Engineering**, Vol. 18, pp. 118-123, 1973.

25. POGSON, J.T. and MACGREGOR, R.K., Effective Conductance Along Parallel Radiation Shields, in **Heat Transfer and Spacecraft Thermal Control,** ed. J.W. Lucas, pp. 473-486, MIT Press, Cambridge, 1971.
26. STIMPSON, L.D. and JAWORSKI, W., Effects of Overlaps, Stitches and Patches on Multilayer Insulation, in **Thermal Control and Radiation,** ed. C.L. Tien, pp. 247-266, MIT Press, Cambridge, 1973.
27. NAYAK, A.L. and TIEN, C.L., Thermal Conductivity of Microsphere Cryogenic Insulation, **Advances in Cryogenic Engineering,** Vol. 22, pp. 251-262, 1977.
28. CHAN, C.K. and TIEN, C.L., Conductance of Packed Spheres in Vacuum, **Journal of Heat Transfer,** Vol. 95, pp. 302-308, 1973.
29. NAYAK, A.L. and TIEN, C.L., A Statistical Thermodynamic Theory for Coordination-Number Distribution and Effective Thermal Conductivity of Random Packed Beds, **International Journal of Heat and Mass Transfer,** Vol. 21, pp. 669-676, 1978.
30. REINKER, R.P., TIMMERHAUS, K.D., and KROPSCHOT, R.H., Thermal Conductivity and Diffusivity of Selected Porous Insulations between 4 and 300 K, **Advances in Cryogenic Engineering,** Vol. 20, pp. 343-354, 1975.
31. CUNNINGTON, G.R. and TIEN, C.L., Apparent Thermal Conductivity of Uncoated Microsphere Cryogenic Insulation, **Advances in Cryogenic Engineering,** Vol. 22, pp. 263-271, 1977.
32. BURNS, P.J., CHOW, L.C., and TIEN, C.L., Convection in a Vertical Slot Filled with Porous Insulation, **International Journal of Heat and Mass Transfer,** Vol. 20, pp. 919-926, 1977.
33. BEJAN, A. and TIEN, C.L., Natural Convection in a Horizontal Porous Medium Subjected to an End-to-End Temperature Difference, **Journal of Heat Transfer,** Vol. 100, pp. 191-198, 1978.
34. BURNS, P.J. and TIEN, C.L., Natural Convection in Porous Media Bounded by Concentric Spheres and Horizontal Cylinders, **International Journal of Heat and Mass Transfer,** Vol. 22, pp. 929-939, 1979.
35. BURNS, P.J. and TIEN, C.L., Effects of Infiltration on Heat Transfer through Vertical Slot Porous Insulation, in **Energy Conservation in Heating, Cooling, and Ventilating Buildings,** Vol. 1, ed. C.J. Hoogendoorn and N.H. Afgan, pp. 93-105, Hemisphere, Washington, D.C., 1978.
36. VAFAI, K. and TIEN, C.L., Bondary and Inertia Effects on Flow and Heat Transfer in Porous Media, **International Journal of Heat and Mass Transfer,** Vol. 24, pp. 195-203, 1981.
37. VAFAI, K. and TIEN, C.L., Boundary and Inertia Effects on Convective Mass Transfer in Porous Media, **International Journal of Heat and Mass Transfer,** Vol. 25, pp. 1183-1190, 1982.
38. OGNIEWICZ, Y. and TIEN, C.L., Analysis of Condensation in Porous Insulation, **International Journal of Heat and Mass Transfer,** Vol. 24, pp. 421-429, 1981.
39. TONG, T.W. and TIEN, C.L., Radiative Heat Transfer in Fibrous Insulations - Part I: Analytical Study, **Journal of Heat Transfer,** Vol. 105, pp. 70-75, 1983.
40. TONG, T.W., YANG, Q.S., and TIEN, C.L., Radiative Heat Transfer in Fibrous Insulations - Part II: Experimental Study, **Journal of Heat Transfer**, Vol. 105, pp. 76-81, 1983.
41. BREWSTER, M.Q. and TIEN, C.L., Radiative Heat Transfer in Packed Fluidized Beds: Dependent Versus Independent Scattering, **Journal of Heat Transfer,** Vol. 104, pp. 573-579, 1982.
42. WANG, K.Y. and TIEN, C.L., Thermal Insulation in Flow Systems: Combined Convection and Radiation through a Porous Segment, **Journal of Heat Transfer,** Vol. 106, pp. 453-459, 1984.

International Symposium 1987

# HEAT AND MASS TRANSFER IN GASOLINE AND DIESEL ENGINES

**D.B. Spalding**

*Imperial College, London, U. K.*

Whenever a layman asks me what the science of heat and mass transfer is all about, I direct his attention to his automobile engine. All drivers know that cars become hot under normal circumstances, and that sometimes they over-heat, with dire effects. Some are acquainted with starting difficulties in cold weather; and those who have forgotten to fill the cooling system with anti-freeze will have sometimes encountered ice-formation, which interrupts the flow of water and may even lead to rupture of the passages which should contain it. Reflecting on these experiences, most people discover that they are better acquainted with heat- and mass-transfer knowledge than they supposed.

The reciprocating engine is indeed a device which exhibits every major phenomenon in the HMT repertoire, from "simple" heat conduction in the engine block, through single-phase convection in the (normally operating) cooling-water system and in the air passing over the extended surfaces of the (inappropriately named) "radiator", to the phase-change processes in the inlet manifold and the engine cylinder itself. The fluid-flow aspects are of great importance; and exothermic chemical reaction is vital; but it is the heat- and mass-transfer processes which enable the hydrodynamics and the chemistry to interact.

The 1987 Symposium of ICHMT showed how well the autmotive industry, and those who work for it in universities and research institutes, recognise the importance of heat- and mass-transfer processes as determinants of the effectiveness of engine design and performance. Inspection of the list of papers supplied in the Appendix to this review reveals this.

The Symposium is of too recent date to allow the passage of time to distinguish the more-ephemeral papers from those of enduring value. However, there were so many good ones that it is not hard to select one which will assuredly endure - provided that it is not necessary to categorize it as the *most* valuable. Even if that proviso is disallowed, I am sure that I cannot be far wrong in selecting for prominence, and reproduction here, the review paper with which the Symposium began, namely that by Professor R. Pischinger of the University of Graz in Austria, entitled: "The importance of heat transfer in engine design and operation". Its author was invited to deliver the opening lecture because of his autoritative previous publications and because he enjoyed the reputation of being a meticulous preparer and lucid deliverer of lectureres. All who were present at the Dubrovnik meeting will agree that Professor Pischinger's authority and reputation stood even higher at its end than its beginning.

# THE IMPORTANCE OF HEAT TRANSFER TO IC ENGINE DESIGN AND OPERATION

R. Pischinger

*Technical University of Graz, Graz, Austria*

## 1. INTRODUCTION

The ideal IC engine cycle comprises isentropic compression, combustion, isentropic expansion and gas exchange and does not include any form of heat transfer (Fig.1). In contrast to other types of heat engines, heat transfer in the IC engine is neither necessary nor desirable.

Heat transfer cannot be avoided in practical IC engine processes and, in fact, plays an important role in the conceptual and detail design of the engine. It leads to considerable thermal loading of the engine components and requires that some form of cooling be employed, which in turn leads to increased heat transfer.

The development of new high temperature materials, mainly ceramics, raises the possibility of disposing with cooling altogether and building an adiabatic engine with zero heat loss. The cyclic nature of the IC engine process hinders efforts in this direction considerably, so that the ideal situation of an adiabatic engine is virtually unachievable. At best, a thermally insulated engine results where indeed no heat is lost through the walls, but where heat is stored within the insulating material during the combustion process only to be transfered back to the gas during the gas exchange phase.

Heat transfer thus plays a significant role in thermally insulated as well as conventional engines. It must therefore be considered in the conceptual and detail design of the engine and has considerable influence on operational performance and durability.

The calculation of heat transfer is difficult even under steady-state conditions. Under the rapidly changing conditions found within engine cylinders, functions for the gas-side heat transfer can only be defined empirically or semi-empirically. The problem is exacerbated by the difficulty of modelling the combustion process.

It is therefore not surprising that the cooling of IC engines has always been designed on the basis of experience. It is now possible however, as a result of the understanding of the heat transfer process and the availability of more powerful measurement and calculation techniques, to make tolerably realistic heat transfer predic-

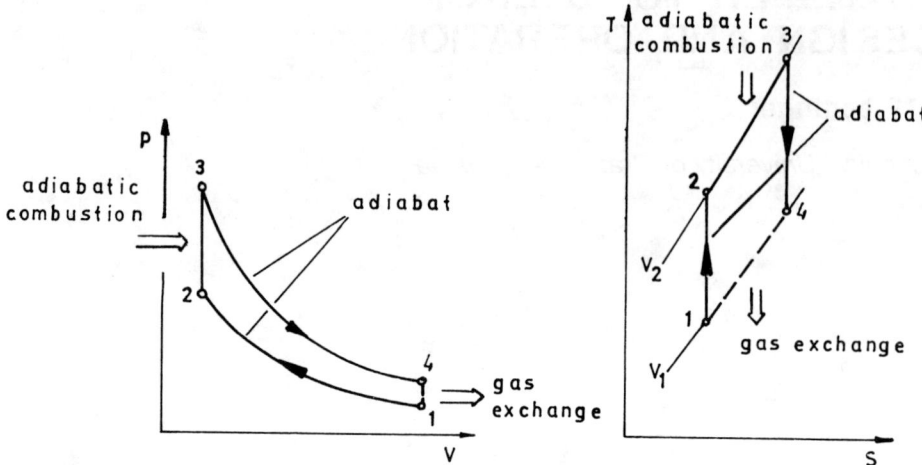

Fig.1. p-v and T-s diagrams for the ideal engine with constant volume combustion.

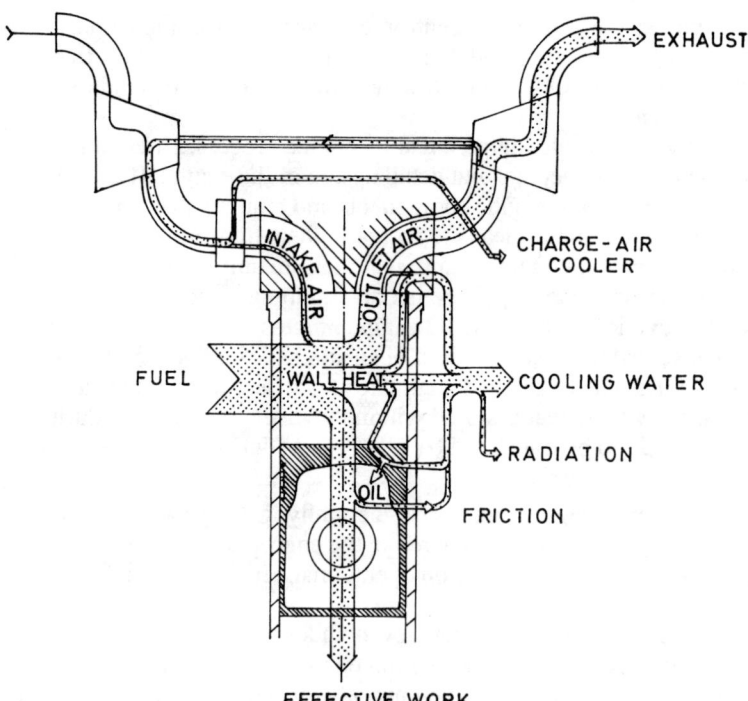

Fig.2. Energy flow in a turbocharged engine with air-charge cooling.

tions for not only conventional but also insulated and adiabatic engines. It is therefore possible to carry out development with confidence when, as before, attention is given to understanding the processes involved.

## 2. ENERGY FLOW IN THE IC ENGINE

Figure 2 shows schematically the energy flow within an IC engine. It is clear from the figure that the system boundaries must be defined precisely when applying the energy balance.

The internal energy balance is used in analysing the thermodynamic processes, whereby the space enclosed by the piston, cylinder and cylinder head is considered. In the case of turbocharged engines, the turbocharger and charge air cooler (if existent) are also included.

For the total energy balance the whole engine including auxiliaries is considered. The energy supplied in the form of fuel appears as useful work, heat to exhaust, heat to coolant (cooling water, air, or oil), and radiation to the surroundings.

It can be seen from the figure that the heat lost to the walls, deduced from the internal energy balance, is not the same as that removed by the coolant. As will be shown, these heat flows originate to a large extent from the exhaust heat and to a lesser extent from the heat which would otherwise be available for work.

## 3. BASIC HEAT TRANSFER CONSIDERATIONS

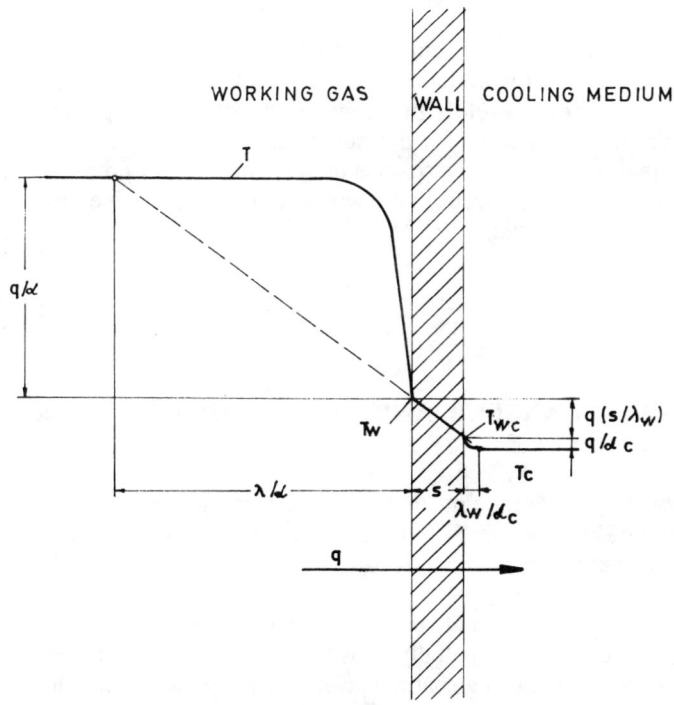

Fig.3. Temperature profile resulting form heat flow through a wall.

Figure 3 shows the temperature profile for steady–state heat transfer through a wall. Heat conduction through the wall is given by the expression

$$q = \lambda \frac{\Delta T}{s} \tag{1}$$

Heat transfer from the working gas to the wall and from the wall to the cooling medium obeys the following relationship

$$q = \alpha . \Delta T \tag{2}$$

For the heat transfer coefficient the following relationships may be employed

$$Nu = f(Re, Pr) \tag{3}$$

where:

$Nu = \dfrac{\alpha . d}{\lambda}$  the dimensionless heat transfer coefficient  (4)

$Re = \dfrac{w . d}{\nu}$  the dimensionless flow velocity  (5)

$Pr = \dfrac{\nu}{\alpha}$  a material constant  (6)

The following relationship has been determined for flow through pipes:

$$Nu = 0.03622 \left(\frac{d}{l}\right)^{0.054} Re^{0.786} Pr^{0.786} \tag{7}$$

During the combustion phase the temperature difference between the working gas and the wall is considerably greater than that between the wall and the cooling medium. The gas–side heat transfer is therefore the dominating factor, but is particularly difficult to determine as it varies strongly with time and location.

As a result of the large time dependent variations in the temperature of the working gas, temperature variations also occur in the walls of the cylinder given by the instantaneous heat conduction equation

$$\frac{\partial T}{\partial t} = a \frac{\partial^2 T}{\partial x^2} \tag{8}$$

## 4. GAS–SIDE HEAT TRANSFER AND ITS INFLUENCE UPON THE COMBUSTION CYCLE

Analysis of the combustion cycle requires that the internal energy balance be applied and that the variation of heat transfer with time be known. Alternatively, cycle calculations can be carried out to determine the influence of heat transfer upon the process.

In most cases it is satisfactory to calculate using the single–zone model which assumes that pressure and temperature are spatially constant and vary only with time. Accordingly, heat transfer is calculated using spatially averaged temperatures and local variations due to flame propagation are not considered. Although the heat transfer equations must be set up using, en masse, temperatures which in reality do not exist, their application in the determination of the functions to equations (3) and (4) has proven satisfactory. A list of the various heat transfer equations can be found in /1/.

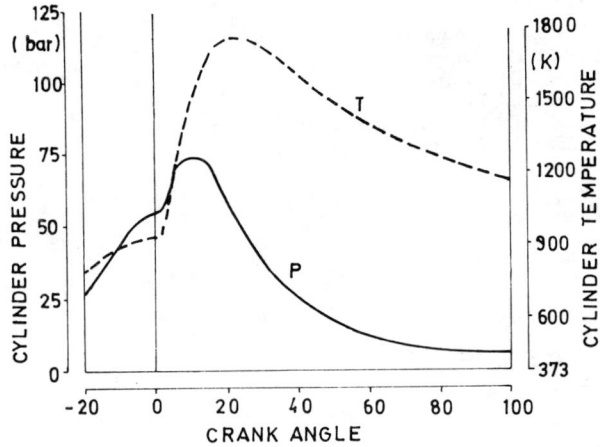

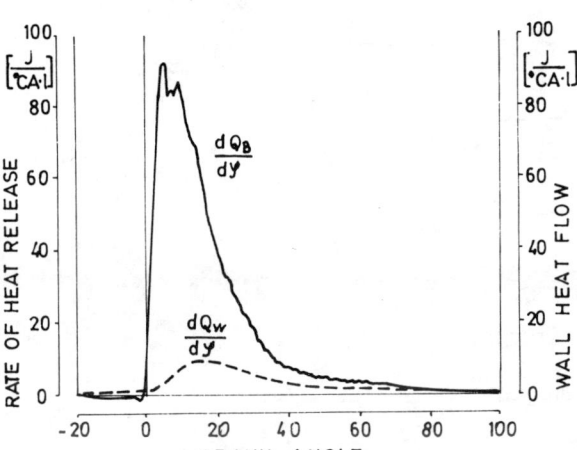

Fig.4. Pressure, temperature, heat release and wall heat flow characteristics of a direct injection diesel engine.

Figure 4 shows the results of such a cycle calculation. Pressure and temperature are shown in the upper part of the diagram while the rate of heat release ($dQ_B/d\varphi$) and the heat flow rate to the walls ($dQ_w/d\varphi$) are shown below. It can be seen that the average cylinder temperature during combustion reaches a very high value in excess of 2000 K. The flame temperature, which is not calculated in the single – zone model, lies considerably higher. The heat flow to the walls ($dQ_w/d\varphi$) is influenced mainly by the cylinder temperature. Although the maximum heat flux lies considerably below the maximum rate of heat release during combustion, heat loss persists over a long period so that the total heat lost to the walls represents a significant proportion of the combustion heat.

Figure 5 (/2/, /3/) shows the magnitude of the wall heat loss at peak torque as a function of mean piston speed for various engines. The characteristics are similar for all engines: the heat lost falls initially with increasing piston speed, and the remains essentially constant at speeds above 8 m/s. The proportion of the heat lost to the

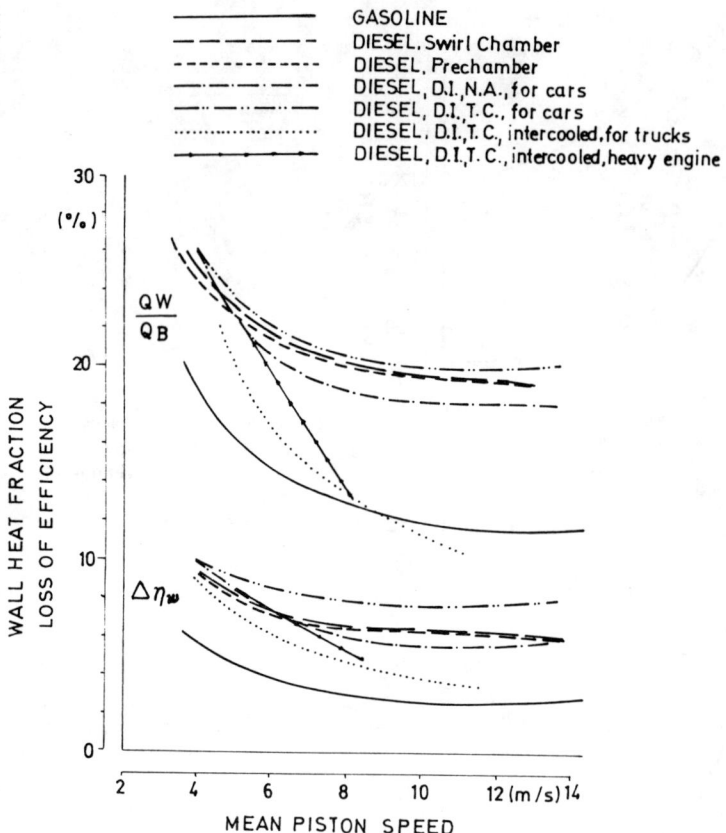

Fig.5. Wall heat fraction $\dfrac{Q_W}{Q_B}$ and the resulting efficiency loss $\Delta\eta_\omega$ as a function of mean piston speed for various engines.

walls lies between 10 and 25%. It is smaller for naturally aspirated engines with low compression ratios, e.g. gasolene engines. High compression diesel engines have high wall heat losses. Pressure charging without intercooling increases the losses moderately, while considerable reductions occur when intercooling is applied. Under part load conditions (not shown in Fig.5) the wall heat loss increases to reach values in excess of 30% at zero load.

The loss in useful work $\Delta\eta_\omega$ resulting from the wall heat loss is shown in the lower part of Fig. 5. At full load it represents between 2.5 and 10% of the heat supplied by the fuel. At zero load (not shown in Fig.5) the losses lie between 3.5 and 20%. Between approximately 23 and 40% of the heat lost to the walls is therefore obtained from the energy which would otherwise be converted into work, and the rest is accounted for by a reduction in exhaust gas enthalpy. It is unfortunately the case that the conditions of high wall heat loss coincide with those where the larger proportion is obtained from the work energy.

The ratio of the surface area available for heat transfer to cylinder volume reduces with an increase in cylinder size and it may be assumed that the proportion of heat lost to the walls also falls. This influence is shown in Fig. 6 for a naturally aspirated direct injection diesel engine. It can be seen that the wall losses do not decrease by nearly as much as would be expected from the reduced surface to volume ratio. The same is true of the loss in work resulting from the wall heat loss.

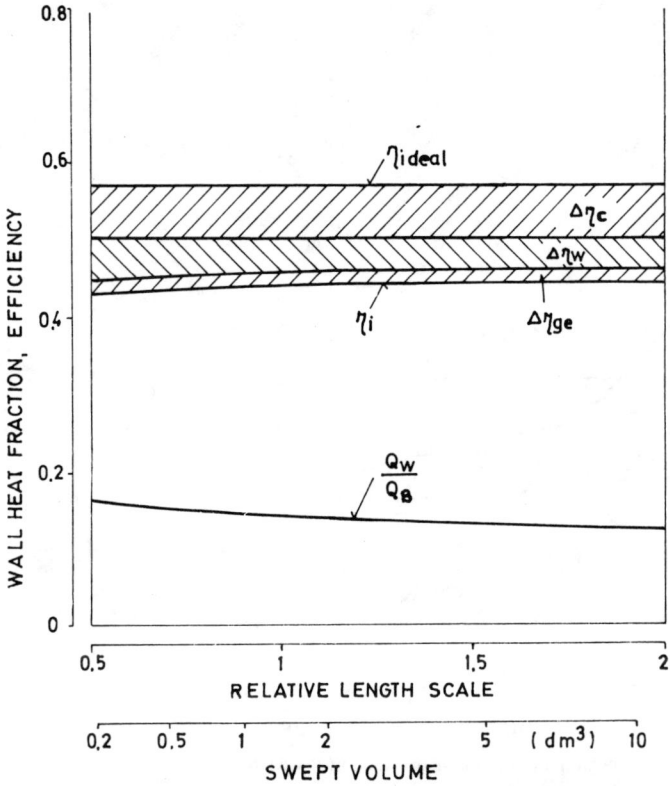

Fig.6. Influence of cylinder size on the heat loss to the walls of direct injection diesel engines.

Figure 7 shows the effect of shifting the beginning of combustion for otherwise identical heat release characteristics. The baseline curve is for a perfect engine with constant volume combustion. Retarded combustion causes an increase in combustion losses but a reduction in thermal losses. The result is a very flat efficiency curve with an optimum combustion beginning at 9° btdc. In practice a somewhat later combustion begin would be favoured in order to reduce the peak firing pressure, the peak temperature, and thus the nitrogen oxide emissions. Similar effects can be achieved through changes to the heat release characteristics. It has been shown many times that fast combustion near to top dead centre, which is thermodynamically advantageous, causes the heat transfer losses to increase. Constant volume combustion at top dead centre is, in view of the heat transfer losses, not the most efficient.

It would be desirable to be able to break this relationship by using modern ceramic materials to insulate the combustion space. The adiabatic engine, in which no heat transfer occurs, would provide the ideal solution. In this way, the heat transfer losses (2.5 to 10% at full load) could be avoided completely. Even with perfect thermal insulation however, combustion temperatures are considerably higher than those of the walls and heat transfer is unavoidable. In addition, this heat transfer occurs during a phase of the cycle which is critical with respect to the conversion of energy into work.

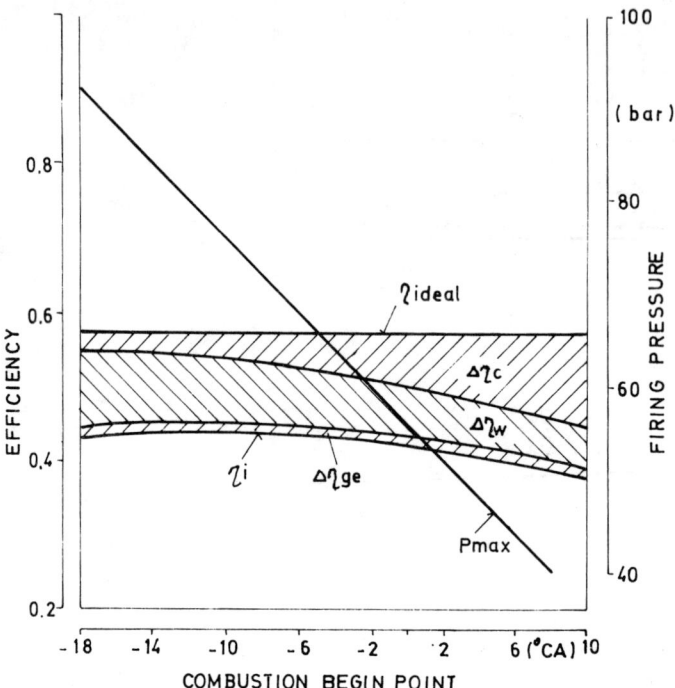

Fig.7. Influence of the combustion begin point on the distribution of losses in a direct injection diesel engine.

In reality, the thermally insulated engine must therefore be considered. It is unfortunate that the terms adiabatic (no heat transfer to the walls) and insulated (no heat removed by the coolant) are often used loosely without concern for their exact definitions. Figure 8 shows the influence of elevated wall temperatures resulting from thermal insulation on the heat flow to the walls, the ensuing losses and the volumetric efficiency. When the insulation is perfect ($Q_W/Q_B = 0$), the wall temperature reaches 1400 K. The efficiency increases only slightly and the volumetric efficiency drops significantly. It was assumed in these calculations that the heat release characteristics remain unchanged. In reality, the reduced volumetric efficiency results in a lower air/fuel ratio which causes combustion to proceed more slowly. In addition, combustion itself would be influenced negatively by the elevated gas and wall temperatures. For a naturally aspirated engine, therefore, rather than an improvement, a loss in efficiency is to be expected, and the nitrogen oxide and smoke emissions will rise as a result of the higher gas temperatures and reduced air/fuel ratio (see also /4/, /5/).

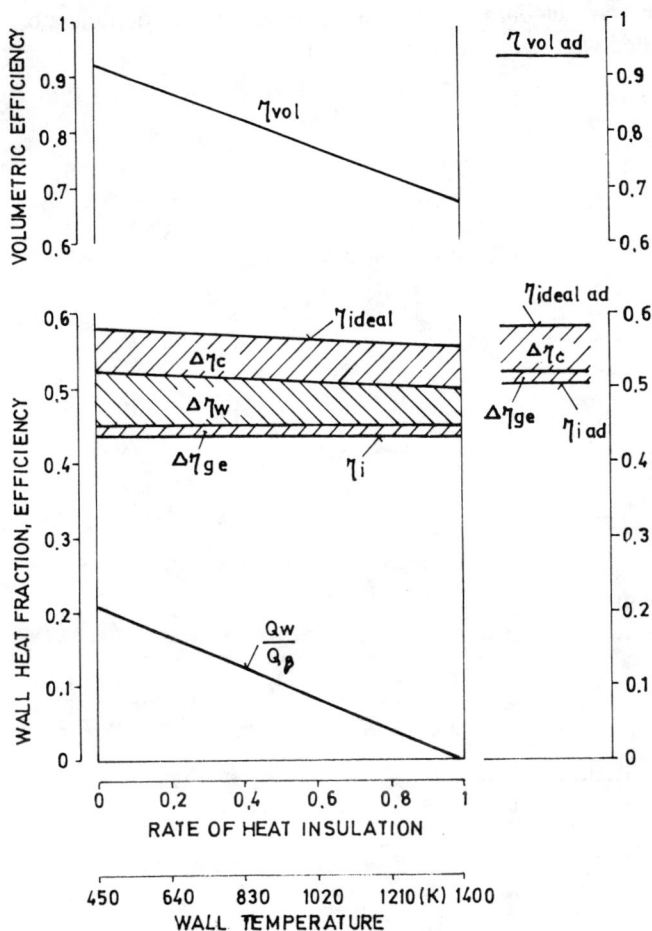

Fig.8. Influence of the wall temperature on the wall heat losses of a direct injection diesel engine.

Through thermal insulation the heat that would otherwise pass to the coolant is diverted to the exhaust gas. This increase in exhaust gas energy can be used to good effect in an exhaust turbine. Figure 9 shows various methods of harnessing this exhaust energy. Although these systems can be used for uninsulated engines, they become more effective as insulation is employed. The poor volumetric efficiency of the thermally insulated engine can be compensated for in turbocharged engines (Fig. 9b) by increasing the boost pressure. In the case of the turbo–compounded machine (Fig. 9c), the additional exhaust energy is converted to work by means of a separate power turbine. This is connected by means of gears to the engine crankshaft. There are other methods of coupling the machines however, which are equivalent thermodynamically, but enable a better matching of the torque characteristics.

Figure 10 shows the results of calculations carried out by Way and Wallace /6/ to determine the efficiency of a turbocharged and a turbo–compounded engine under the assumption of zero insulation (standard engine), halved heat transfer coefficient $\alpha$ (adiabatic engine), and elevated wall temperature (insulated engine).

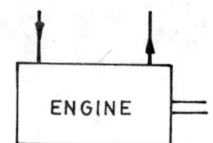

naturally aspirated engine

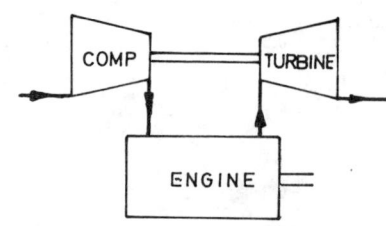

turbocharged engine

Fig.9. Exhaust energy utilization concepts.

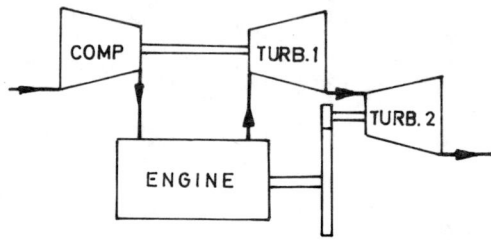

turbocharged/compound engine

In comparison to the standard engine, the adiabatic turbocharged engine is more efficient by 8% at low engine speeds and 2% at high speeds. The insulated engine is, however, only 1% better over the whole speed range. In all three versions, the turbo– compounded machine is 2 to 3% better over the whole speed range than the turbo-charged engine. These calculations show that significant improvements are only achieved with the adiabatic engine, the gains in efficiency through thermal insulation are very small, and that turbo–compounding produces larger improvements than insulation.

It is interesting to compare these calculations with experimental results /7/, /8/, /9/, /10/, /11/, /12/. Depending upon the baseline, the degree of the design and

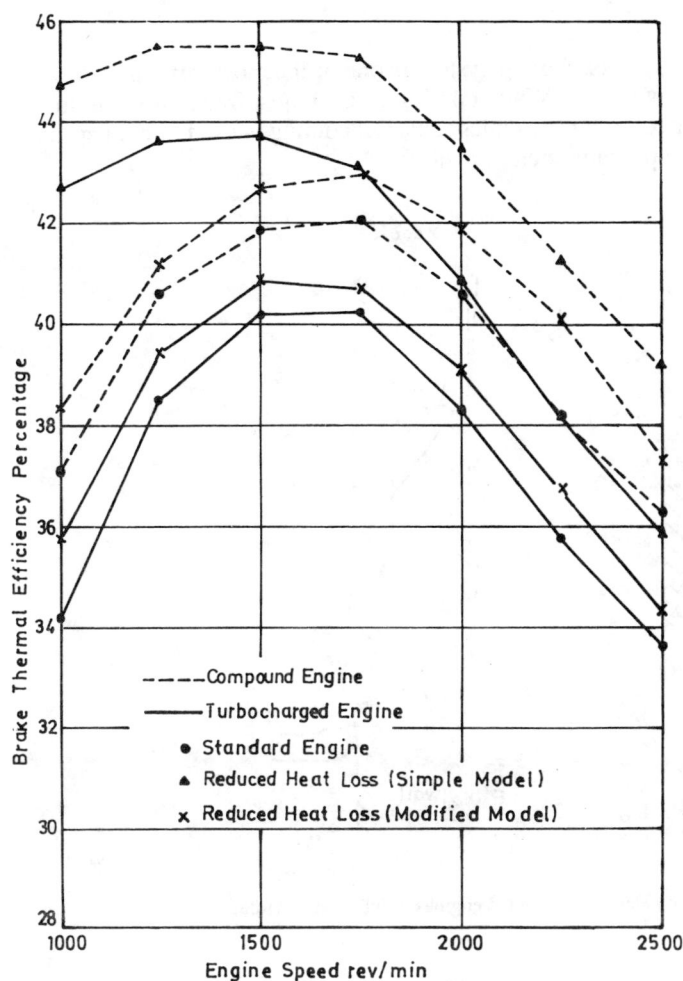

Fig.10. Overall efficiency of standard, adiabatic, and thermally insulated engines in turbocharged and turbo–compound forms (/6/).
simple model = adiabatic
modified model = heat insulated

development measures taken, the results achieved vary quite considerably. It should be noted that the combustion system must be re-optimised and that recent investigations /13/ have shown that the heat transfer coefficient $\alpha$ increases with increasing wall temperature beyond values previously assumed. When these considerations are taken into account, the thermodynamic calculations agree to a large extent with the results
obtained by experiment. Attention must be drawn to the fact that the development target is not only a reduction in fuel consumption, but also simplified engine design and other changes such as the elimination of the cooling system.

## 5. COOLING

The selection of the type of cooling system is one of the most important decisions in the design of a new engine. Water cooling is used most frequently, but in particular applications air cooling is advantageous. Oil cooling is used only in special cases or for particular components such as pistons.

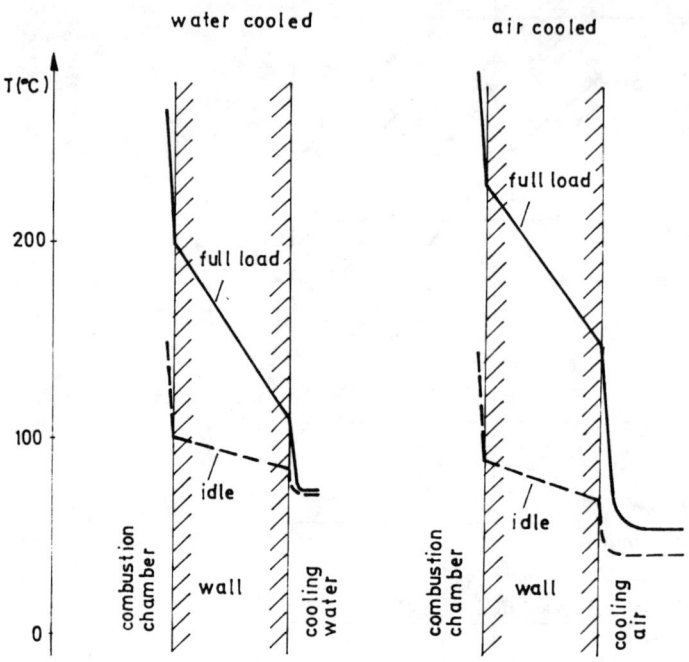

Fig.11. Temperature profiles for water and air cooled engines at full load and idle.

It may be deduced from the relationship, equation (7) that, for normal water velocities (up to 3 m/s) and air velocities (up to 50 m/s), substitution of the material properties of water and air gives a ratio of the heat transfer coefficients (/1/, /14/):

$$\frac{\alpha_{water}}{\alpha_{air}} \approx 100$$

With water cooling, the temperature difference between the cylinder wall and the coolant is approximately 25° to 50°C. With air cooling the difference is 100°C. The temperature ratio is therefore

$$\frac{\Delta T_{air}}{\Delta T_{water}} \approx 2 \text{ to } 4$$

The ratio of the heat transfer surface areas should therefore be

$$\frac{A_{air}}{A_{water}} \approx 25 \text{ to } 50$$

Existing cylinder heads have area ratios between 30 and 50, cylinder liners approximately 10 /15/.

Figure 11 shows the wall temperature profile for water and air cooling under full and idle conditions.

At full load the air cooling produces higher gas–side temperatures which result in somewhat lower volumetric efficiency. At part load however, air cooling results in lower gas–side temperatures. The temperature differences between full and part load are greater with air cooling and this must be allowed for in specifying piston clearances. By regulating the cooling fan, the temperature variations can be reduced.

Ambient air temperature variations also have a strong influence on the wall temperatures of air–cooled engines. The use of a thermostat eliminates this effect in water–cooled engines. However, the design of the radiator must take the highest ambient temperatures into account.

The ribs of air–cooled engines should extend from the engine with a conical cross–section to ensure an adequate flow of heat to the rib ends. For the same reason it is not effective to use ribs which are too long. Because of their better thermal conductivity, aluminium ribs can be longer (up to 70 mm) than cast iron ribs (up to 40mm).

Figure 12 shows as an example the cylinder of a VW engine which was mass produced. The ribs are long at the top of the cylinder to match the local heat flux and to achieve a uniform temperature distribution.

The power requirement of the cooling fan of air–cooled engines lies between 4 and 6% of the brake power and is therefore of the same order as that of water–cooled engines.

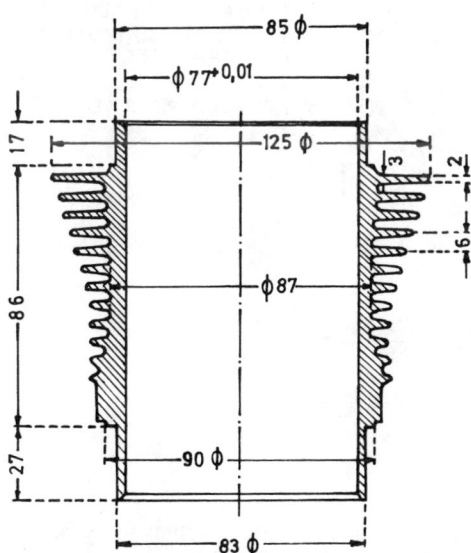

Fig. 12. Grey iron air–cooled gasoline engine cylinder (Volkswagen).

The flow conditions of the water in water–cooled engines are usually poorly defined. Often free convection occurs, whereby the heat transfer coefficient increases only slowly with an increase in the temperature difference. Where the temperature rises locally above the boiling point, very intense heat transfer occurs. This self–regulating effect functions so that wall temperatures vary only a little with changing load, (see Fig. 11). When high heat fluxes occur under poor flow conditions however, it is possible that the stream bubbles are not removed, but join together to form an insulating layer which hinders heat transfer and leads to overheating of the wall.

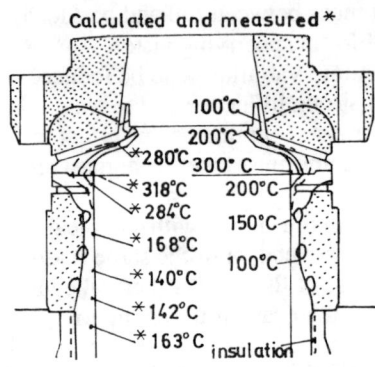

Fig. 13. Large 2z–stroke engine cylinder liner and cover – Sulzer RTA 58. Bore 580 mm, stroke 1700 mm, 15,3 bar bmep at 123 rpm, coolant outlet temperature 82 °C (/10/).

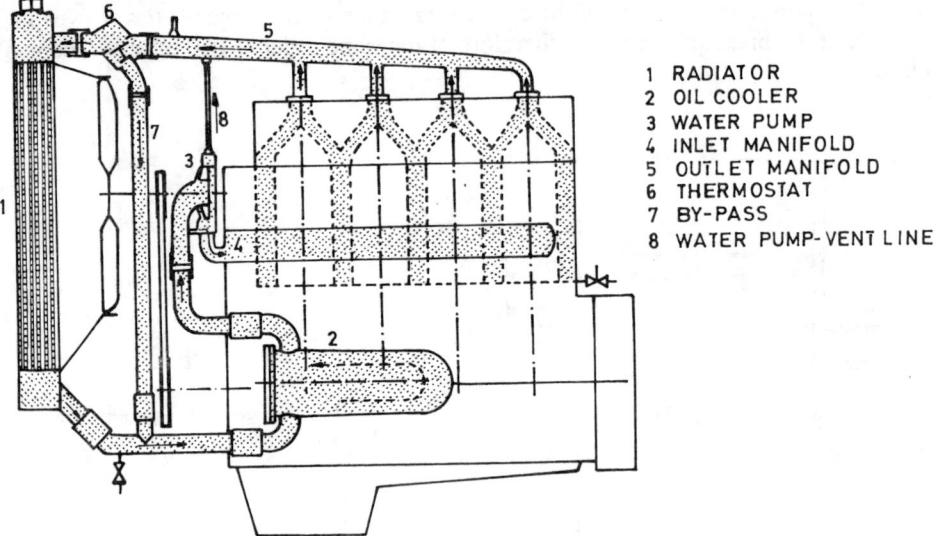

1 RADIATOR
2 OIL COOLER
3 WATER PUMP
4 INLET MANIFOLD
5 OUTLET MANIFOLD
6 THERMOSTAT
7 BY-PASS
8 WATER PUMP-VENT LINE

Fig.14. Scheme of vehicle engine cooling system (/11/).

Defined flow conditions and thereby defined removal can be achieved by bore cooling (Fig. 13). By that an even temperature field can be realized and overheating can be avoided.

The consequences of choosing water cooling are considerable. It can be seen in Fig. 14 that the volume requirements of the system are significant. The number of cooling system related components involved is large, as are the costs. Water cooling has nevertheless been widely accepted because of its advantages with respect to thermostatic control and the described self–regulating effect, both of which improve engine operational performance. In addition, water cooling offers ideal conditions for waste heat utilization. The heat removed by the coolant of vehicle engines can be used to provide passenger space heating, that from large stationary engines to supply the heat requirements of co–generation plants.

## 6. THERMAL LOADING OF ENGINE COMPONENTS

The heat flow in an engine affects the engine components in two ways:
- Component temperatures increase and lead to a reduction in strength depending upon the thermal properties of the material
- As a result of both the steady and unsteady heat fluxes, deformation and thermal stresses occur which cause additional loading of the components.

The table at the end of the paper is a summary of the properties of several metallic and non–metallic materials.

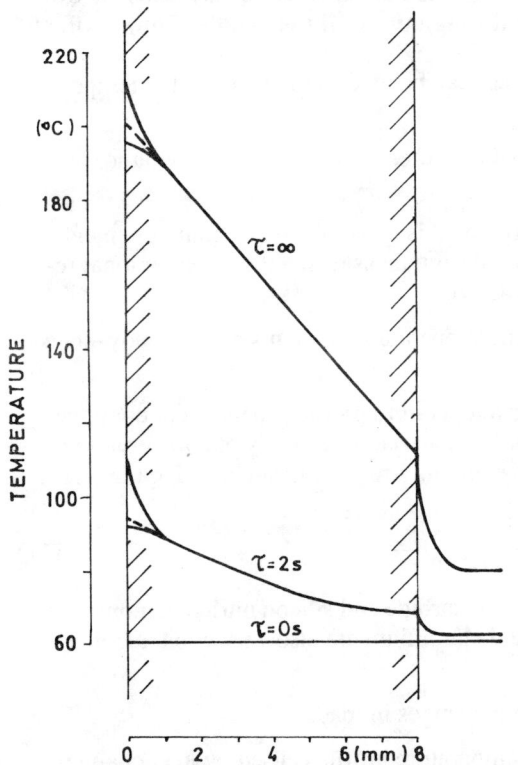

Fig. 15. Wall temperature characteristics under warm–up and steady–state conditions.

It can be seen that the most commonly used materials, aluminium and cast iron, have a temperature limit of about $500°C$. For higher temperatures, high alloy steels with a limit of about $850°C$ must be used. Such steels are used, for example, for highly loaded exhaust valves. Ceramic materials can be used up to a temperature of 1650°C. Since, as a result of their lower thermal conductivity, the wall temperatures are higher, these materials do not necessarily provide a larger margin of safety against thermal overloading. Silicon carbide is relatively good in this respect, aluminium titanate poor.

The following simplified comments are made on the complicated subject of thermal stresses.

Expansion is proportional to the temperature change. In a rigidly located wall however, expansion is not possible in the direction in which the wall is located and thermal stresses result, given by:

$$\sigma_{th} = E.\beta.\Delta T \qquad (10)$$

Figure 15 shows the temperature profile within a wall under warm–up and steady–state conditions. The average temperature rises during warm–up and at the same time the temperature profile changes. Under steady–state conditions the temperature profile is linear, its gradient being dependent upon mean heat flow through the wall and the thermal conductivity of the material. Cyclic variation of the gas–side temperature profile occurs under both warm–up and steady–state conditions. For metallic combustion chamber surfaces the temperature variations have an amplitude of about 15°C and penetrate through the wall to a depth of approximately 2mm.

Four different types of thermal stress can be identified related to the temperature profiles shown in Fig. 15:

– Thermal stresses resulting from an increase in the mean temperature of a component.

These stresses only occur when the component is constrained and cannot expand. They can therefore be influenced through design measures, although functional requirements (e.g. sealing) must be considered.

– Thermal stresses resulting from temperature gradients under steady–state conditions.

The gradient of the approximately linear temperature profile is dependent upon the heat flux and the thermal conductivity of the material. The ability of a material to conduct a certain heat flow without exceeding its stress limit can be judged from the thermal shock parameter

$$R_2 = \frac{\lambda \cdot 0.7 \cdot \sigma_{max}}{E \cdot \beta} \qquad (11)$$

It can be seen from Table 1 that both silicon carbide and silicon nitride demonstrate good characteristics, but that cast iron and aluminium are also very good, albeit at lower temperatures.

– Thermal stresses resulting from changes in load.

The stresses resulting from changes in temperature profile caused by load changes (also warm–up) are mostly smaller than those acting during steady–state operation. It is possible however, that high stress levels occur when rapid changes in temperature occur locally, either from heating or cooling, and the mean component temperature changes only slowly thus hindering expansion (or contraction).

– Thermal stresses resulting from cyclic temperature changes in the combustion chamber wall.

The ability of a material to withstand this type of loading is judged using the thermal shock parameter

$$R_1 = \frac{0.7 \cdot \sigma_{max}}{E.\beta} \qquad (12)$$

Silicon nitride and aluminium titanate are good in this respect.

The stresses caused by cyclic temperature variations are usually small in comparison to the steady-state full load thermal stresses. They can however be dangerous in critical areas of the combustion chamber, such as the bowl edge, where the temperatures are at their highest, the load changes occur at high frequency and the resulting stresses add to a relatively high base stress level.

The total load to which a component is subjected results from the superposition of thermal and mechanical stresses, whereby various steady and unsteady operating conditions can be critical.

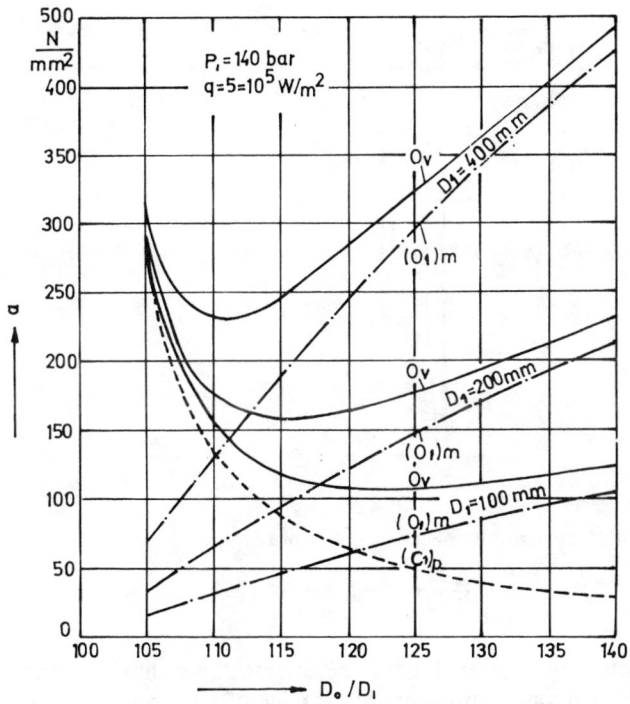

Fig. 16. Resultant stresses produced through mechanical and thermal loading of a circular pipe.

A simple example is the superposition of compressive and thermal stresses shown in Fig. 16 (/1/). It can be seen that as the wall thickness increases, the compressive stresses reduce while the thermal stresses increase. The resultant stress, determined according to the maximum shear strain criterion, reaches a maximum at a diameter ration between 1.1 and 1.2. This agrees well with design experience.

The majority of engine components and the loadings to which they are subjected are so complicated that satisfactory solutions cannot be found using simple analytical methods. Such problems are solved today using the Finite Element Method. Figure 17 shows a simple example of a wet cylinder liner located in the cylinder block. The heat input to the gas side of the liner must be known. This changes along the length of the liner since piston position is time dependent and relates to different phases of combustion, which in turn determines the gas temperature. The calculated temperatures, which agree well with the measured values, are shown in the left part of the diagram The FE model and the thermal distortion, which plays an important

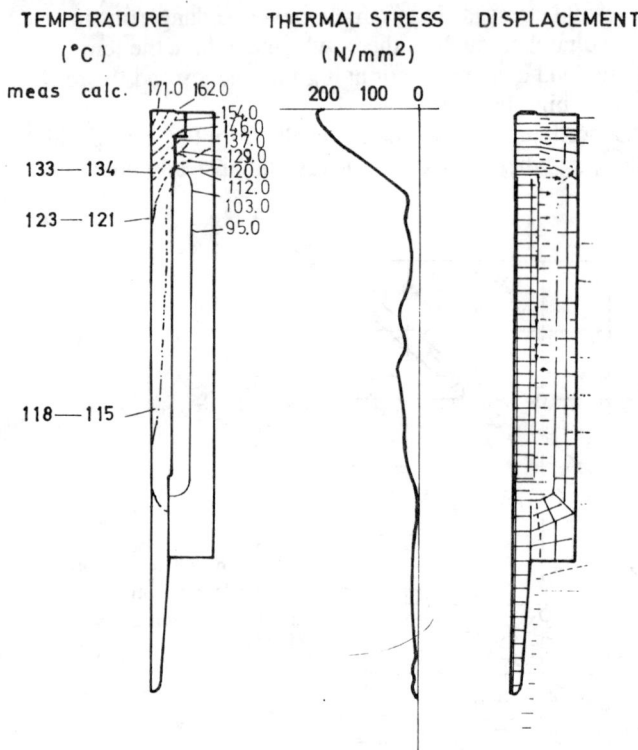

Fig.17. Finite element calculation of the temperature distribution and deformation of a rigidly located wet cylinder liner.

role in determining whether satisfactory operation will be achieved, are shown on the right. The resulting stresses reach a maximum near the top edge of the liner. For the sake of clarity, only the thermal loading is shown in the diagram. Superposition of the mechanical loads, however, does not represent a problem. For the rotationally symmetrical liner shown, a 2–D calculation can be used. The effort involved in creating a 3–D model and executing the calculations is considerably greater.

Sophisticated analytical methods can, however, only be used to evaluate and optimize given designs. The creation of the basic concept will always rely on the solid experience and imagination of the design engineer.

## 7. CONCLUSIONS

Although heat transfer does not occur in the ideal IC engine cycle, heat transfer plays an important role in engine design and operation.

The following points should be considered in designing an engine:

– Fuel consumption and performance

Consumption and performance are mainly determined by the thermodynamic working process. Under full load conditions approximately 10 to 25% of the heat supplied by the fuel is lost to the walls, but only 2.5 to 10% is lost from the portion available for work. This fraction could also be utilised in a truly adiabatic engine if such a machine could be realised. Thermal insulation does not provide any direct advantage since the heat energy barred from the coolant passes out with the exhaust gas. The thermally insulated engine only becomes sensible if the additional exhaust energy is utilised through turbocharging or turbo–compounding

– Cooling

Although the use of water cooling adds to engine complexity, it enables the maintenance of relatively constant operating temperatures and offers the possibility of waste heat recovery. Water cooling has established itself therefore as the preferred system for most applications.

Air cooling offers advantages when the use of water would cause difficulties, when great value is placed upon simplicity, or when motion induced air flow makes the use of a cooling fan unnecessary. All of these points are relevant in the case of motorcycles. Attention must be paid to the correct design of the cooling ribs, to guiding the cooling air stream, and to the large variations in operating temperatures.

– Design

Consideration is to be given to the following:
- Observance of the maximum allowed operating temperatures for the materials used, the cylinder charge and the lubricants.
- Observance of the maximum allowed mechanical and thermal stress levels.
- Observance of the maximum permissible deformation.

Since none of these points can be considered independently from the others, all factors must be treated together as part of one complete system. Despite the availability of sophisticated measurement and analysis techniques, the experienced engineer armed with a good basic understanding of the technology and with imagination is still an essential requirement.

PHYSICAL PROPERTIES OF METALS AND CERAMICS

| material | melting point $T_m$ K | max. appl. temperature $T_{max}$ °C | density $\rho$ g/cm³ | thermal conductivity $\lambda$ W/mK | thermal expansion $\beta$ $10^{-6}$(m/mK) | specific heat $c$ J/gK | thermal diffusivity $a$ $10^{-6}$(m²/s) | thermal inertia $\rho \cdot \lambda \cdot c$ $10^{-6}(\frac{W}{m^2K})^2 \cdot s$ | Young's Modulus $E$ N/mm² | strength at temperature $\sigma_b$ N/mm² | thermal shock parameter $R_1 \frac{0.7 \cdot \sigma_{max}}{E \cdot \beta}$ K | $R_2 \frac{\lambda \cdot 0.7 \cdot \sigma_{max}}{E \cdot \beta}$ W/m |
|---|---|---|---|---|---|---|---|---|---|---|---|---|
| aluminium-silicon-alloy AlSi12 cast | 870 | 300 | 2,65 | 146,5 | 20,5...21,5 (20...200°C) | 0,921 | 60 | 358 | 80.000 (200°C) | 100...150 (250°C) | ~60 | ~8.600 |
| grey cast iron GGG | 1420...1570 | 500 | 7,2 | 54 (0...1000°C) | 9 | 0,54 | 13,9 | 210 | 80.000...130.000 | 240...480 (200°C) | 250 | 13.600 |
| chromium steel X22CrMoV21 wrought | 1700 | 650 | 7,7 | 30 | 12,3 (500°C) | 0,54 (0...800°C) | 7,2 | 125 | 200.000 | 344 (500°C) | 140 | 4.200 |
| nickel-base-alloy Nimonic 80 A wrought | 1700 | 850 | 8,2 | 12,5...30 (100...900°C) | 18,1 (20...1000°C) | 0,545 (20...900°C) | 2,8...6,7 | 70 | 200.000 | 642 (500°C) | 235 | 4.700 |
| nickel-base-alloy INCO 713 C cast | 1580 | 850 | 7,9 | 25 (...1000°C) | 15 (...1000°C) | | | | 170.000 (1000°C) | 900 (620°C) | 350 (1000°C) | 8.750 (1000°C) |
| titanium Ti6AlV4 | 1800 | 500 | 4,54 | 17,2 | 10,8 | 0,53 | 7,15 | 41 | 103.000 | 520 | 243 | 16.700 |
| silicon-carbide SiC sintered | 2520 | 1650 | 3,15 | 70 (100°C) | 4,2 | 1 | 22,2 | 220 | 410.000 | 450 | 243 | 16.700 |
| silicon-nitride Si₃N₄ reaction-bounded | 2170 | 1400 | 2,55 | 20 (100°C) | 3,3 | 0,7 | 11,2 | 36 | 180.000 | 300 | 560 | 11.110 |
| aluminium-titanat Al₂O₃·TiO₂ | 2320 | 1000 | 3,2 | 2 (100°C) | 3 (20...700°C) | 0,7 | 0,89 | 4,5 | 23.000 | 40 (530°C) | 580 | 1.160 |
| zirconium-oxide ZrO₂ MgO stabilized | 3000 | 900 | 5,73 | 2,5 (100°C) | 10,2 | 0,4 | 1,09 | 6 | 200.000 | 600 (530°C) | 255 | 640 |

# NOMENCLATURE

| | | |
|---|---|---|
| $a = \dfrac{\lambda}{c_p \cdot \rho}$ | $\dfrac{m^2}{s}$ | thermal diffusivity |
| $E$ | $\dfrac{N}{m^2}$ | Young's Modulus |
| $d$ | m | diameter, characteristic length |
| $Nu = \dfrac{a \cdot d}{\lambda}$ | – | Nusselt number |
| $Re = \dfrac{w \cdot d}{\nu}$ | – | Reynolds number |
| $R_1 = \dfrac{0.7 \cdot \sigma_{max}}{E \cdot \beta}$ | K | thermal shock parameter |
| $R_2 = \dfrac{\lambda \cdot 0.7 \cdot \sigma_{max}}{E \cdot \beta}$ | $\dfrac{W}{m}$ | thermal shock parameter |
| $Q_B$ | J | fuel energy per cycle |
| $Q_W$ | J | wall heat per cycle |
| $q$ | $\dfrac{W}{m^2}$ | heat flow |
| $T$ | K, °C | temperature |
| $\alpha$ | $\dfrac{W}{m^2 K}$ | heat transfer coefficient |
| $\beta$ | $\dfrac{1}{K}$ | coefficient of thermal expansion |
| $\varphi$ | ° | crank angle |
| $\lambda$ | $\dfrac{W}{mK}$ | coefficient of thermal conductivity |
| $\nu$ | $\dfrac{m^2}{s}$ | kinematic viscosity |
| $\sigma$ | $\dfrac{N}{m^2}$ | stress |
| $\eta_e$ | – | brake thermal efficiency |
| $\eta_i$ | – | indicated thermal efficiency |
| $\eta_{ideal}$ | – | efficiency of the ideal engine |
| $\Delta\eta_c$ | – | loss in efficiency due to retarded combustion |
| $\Delta\eta_w$ | – | loss in efficiency due to wall heat loss |
| $\Delta\eta_{ge}$ | – | loss in efficiency due to gas exchange |

# REFERENCES

1. PFLAUM, W.; MOLLENHAUER, K.: Wärmeübergang in der Verbrennungskraftmaschine (Heat Transfer in IC Engine). Springer Verlag, Wien–New York, 1977
2. SAMS, T.: Thermodynamische Analyse des Motorprozesses (Thermodynamic Analysis of the Working Process of IC Engines). Mitteilungen des Institutes fur Verbrennungskraftmaschinen und Thermodynamik, Heft 46, TU–Graz, 1986

3. PISCHINGER, R.; KRAßNIG, G.; SAMS., T.: Vergleich des thermodynamischen Arbeitsprozesses von Dieselmotoren (Comparison of the Thermodynamic Working Process of Diesel Engines). Technische Akademie Esslingen, Symposium Dieselmotorentechnik, 1986
4. ZAPF, H.: Grenzen und Möglichkeiten eines wärmedichten Brennraumes bei Dieselmotoren (Limits and Chances of a Heat Insulated Combustion Chamber of Diesel Engines). VDI–Bericht Nr. 238 (1975)
5. WALLACE, F.J.; WAY, R.J.B.; VOLLMERT, H.: Effect of Partial Suppression of Heat Loss to Coolant on the High Output Diesel Engine Cycle. SAE–Paper 790823, 1973
6. WAY, R.J.B.; WALLACE, F.J.: Results of Matching Calculations for Turbocharged and Compound Engines with Reduced Head Loss. SAE Paper 790824, 1979
7. – The Adiabatic Engine: Past, Present and Future Developments. SAE, PT–28, 1984.
8. – Advances in Adiabatic Engines. SAE, SP–610, 1985
9. – Adiabatic Engines and Systems. SAE, SP–650, 1986
10.– Adiabatic Engines and Systems. SAE, SP–700, 1987
11. STINGL, P. et al.: Herstellung und Erprobung von keramischen Bauteilen für Otto– und Dieselmotoren (Production and Test of Ceramic Components of Gasoline and Diesel Engines). Entwicklungslinien in Kraftfahrzeugtechnik und Straßenverkehr, Fahrzeugbilanz 1986, Verlag TÜV Rheinburg, 1987
12. ANISITS, F.: Anwendung der Brennverlaufsanalyse für die Aufgabe der Verbrennungsoptimierung. (Application of Heat Release Analysis for the Optimization of Combustion). Mitteilungen des Institutes für Verbrennungskraftmaschinen und Thermodynamik, Heft 49, TU Graz, 1987
13. WOSCHNI, G.; KOLESA, K.; SPINDLER, W.: Isolierung der Brennraumwände – Ein lohnendes Entwicklungsziel bei Verbrennungsmotoren? (Insulation of Combustion Chamber Walls – An Advantageous Target for the Development at IC Engines?) MTZ 1986 /12
14. WEIDENMÜLLER, M.: Zum Entwicklungsstand luftgekühlter Dieselmotoren (Stage of development of Air Cooled Diesel Engines). 1973 /4
15. SCHEITERLEIN, A.: Der Aufbau der raschlaufenden Verbrennungskraftmaschine (Construction of the High Speed IC Engine). Springer Verlag, Wien, 1964
16. WOLF, G.; MARTI, A.: Entwicklungsaspekte bei Großdieselmotoren (Developments Aspects for Large–Bore Diesel Engines). MTZ 1985/4

TJ 260 .A68 1988 v.1

Archives of heat transfer

MAY 1 2 1989